US CUSTOMARY AND METRIC EQUIVALENTS
In the following tables, the equivalents are rounded off.

US to Metric

Length Equivalents

1 in.	= 2.54 cm (exact)		1 cm	= .394 in.
1 ft	= 0.305 m		1 m	= 3.28 ft
1 yd	= 0.914 m		1 m	= 1.09 yd
1 mi	= 1.61 km		1 km	= 0.62 mi

Area Equivalents

1 in.^{2}	= 6.45 cm^2		1 cm^2	= 0.155 in.^{2}
1 ft^2	= 0.093 m^2		1 m^2	= 10.764 ft^2
1 yd^2	= 0.836 m^2		1 m^2	= 1.196 yd^2
1 acre	= 0.405 ha		1 ha	= 2.47 acres

Volume Equivalents

1 in.^{3}	= 16.387 cm^3		1 cm^3	= 0.06 in.^{3}
1 ft^3	= 0.028 m^3		1 m^3	= 35.315 ft^3
1 qt	= 0.946 L		1 L	= 1.06 qt
1 gal	= 3.785 L		1 L	= 0.264 gal

Mass Equivalents

1 oz	= 28.35 g		1 g	= 0.035 oz
1 lb	= 0.454 kg		1 kg	= 2.205 lb

CENTIGRADE AND FAHRENHEIT EQUIVALENTS

$$F = \frac{9}{5}C + 32 \qquad C = \frac{5}{9}(F - 32)$$

Centigrade	Fahrenheit
100°	212° ← water boils at sea level
95°	203°
90°	194°
85°	185°
80°	176°
75°	167°
70°	158°
65°	149°
60°	140°
55°	131°
50°	122°
45°	113°
40°	104°
35°	95°
comfort range { 30°	comfort range { 86°
25°	77°
20°	68°
15°	59°
10°	50°
5°	41°
0°	32° ← water freezes at sea level

Basic Mathematics for

College Students

SEVENTH EDITION

D. Franklin Wright
Cerritos College

HAWKES
PUBLISHING

Acquisitions Editor: Charles W. Hartford
Developmental Editor: Philip Charles Lanza
Production Editor: Melissa Ray
Designer: Judith Miller
Photo Researcher: Lauretta Surprenant
Art Editor: Diane B. Grossman
Production Coordinator: Lisa Merrill

Copyright © 2003 by Hawkes Publishing.

Previous editions copyright © 1995, 1991, 1987, 1983, 1979, 1975, 1969.

Printed in the United States of America.

International Standard Book Number: 0–669–35286–1

Library of Congress Catalog Number: 94–75752

123456789-DOC-06 05 04 03 02

PREFACE

Purpose and Style

The purpose of *Basic Mathematics for College Students*, Seventh Edition, is to provide students with a learning tool that will help them

1. develop basic arithmetic skills,

2. develop reasoning and problem-solving skills, and

3. achieve satisfaction in learning so that they will be encouraged to continue their education in mathematics.

The writing style gives carefully worded, thorough explanations that are direct, easy to understand, and mathematically accurate. The use of color, subheadings, and shaded boxes helps students understand and reference important topics.

The emphasis is on *why* basic operations and procedures work as they do, as well as on *how* to perform these operations and procedures. Point-by-point explanations are incorporated within the examples for better understanding, and directions are given in an easy-to-follow format. Classroom Practice problems (with answers) appear in almost every section to reinforce students' understanding of the concepts in that section and to provide the instructor with immediate classroom feedback.

The NCTM curriculum standards have been taken into consideration in the development of the topics throughout the text.

New Special Features

In each chapter:

- *Mathematics at Work!* presents a brief discussion related to an upcoming concept from the chapter ahead and an example of mathematics used in daily life situations.

- *What to Expect in this Chapter* opens each chapter and offers an overview of the topics that will be covered.

- *Learning Objectives* are now listed at the beginning of each section.

In the Exercise sets:

- *Writing and Thinking about Mathematics* exercises encourage students to express their ideas, interpretations, and understanding in writing.

- *Check Your Number Sense* exercises are designed to help students develop confidence in their judgment, estimation, and metal calculation skills.

- *Collaborative Learning Exercises* are designed to be done in interactive groups.

- *The Recycle Bin* is a skills refresher that provides maintenance exercises on important topics from previous chapters.

Customizing Options

Any of four chapters, **Measurement, Geometry, Solving Equations,** and **Additional Topics from Algebra,** can be included as part of the text. The option of combining any or all of the chapters with the beginning nine chapters allows instructors to better meet the diverse requirements of their courses without using a book that contains material that may not be assigned.

Pedagogical Features

In each section, the presentation and development are based on the following format for learning and teaching:

1. Each section begins with a list of learning objectives for that section.

2. Subsection topics are introduced by color subheadings for easy reference.

3. Thorough and mathematically accurate discussions feature several detailed examples completely worked out with step-by-step explanations.

4. Classroom Practice problems serve as a warm-up for exercise sets and provide students and teachers with immediate classroom feedback.

5. Graded exercises offer variety in their style and level of difficulty including drill, multiple choice, matching, applications, written responses, estimating, and group discussion exercises.

6. Real-life applications are presented in most sections, with several sections devoted solely to applications.

7. Color is used to provide visual support to the text's pedagogy.

Each chapter contains

1. Chapter-opening Mathematics at Work!

2. What to Expect in this Chapter

3. A Summary of Key Terms and Ideas, Rules and Properties, and Procedures

4. A set of Review Questions

5. A Cumulative Review of topics discussed in that chapter and previous chapters (beginning with Chapter 2)

Important Topics and Ideas

- Emphasis has been placed on the development of reading and writing skills as they relate to mathematics.

- Effort has been made to make the exercises motivating and interesting. Applications are varied and practical and contain many facts of interest.

- The use of calculators is encouraged starting with Chapter 5. A new section on scientific notation has been added to help students understand this type of display on their calculators.

- Estimating is an integral part of many discussions throughout the text.

- Geometric concepts such as finding perimeter and area and recognizing geometric figures are integrated throughout the text.

Content

There is sufficient material for a three- or four-semester-hour course. The topics in Chapters 1–7 form the core material for a three-hour course. The topics in Chapters 8 and 9 provide additional flexibility in the course depending on students' background and the goals of the course.

Chapter 1, Whole Numbers, reviews the fundamental operations of addition, subtraction, multiplication, and division with whole numbers. Estimation is used to develop better understanding of whole number concepts, and word problems help to reinforce the need for these ideas and skills in common situations such as finding averages and making purchases.

Chapter 2, Prime Numbers, introduces exponents, shows how to use the rules for order of operations, and defines prime numbers. The concepts of divisibility and factors are emphasized and related to finding prime factorizations, which are, in turn, used to develop skills needed for finding the least common multiple (LCM) of a set of numbers. All of these ideas form the foundation of the development of fractions in Chapter 3.

Chapter 3, Fractions, discusses the operations of multiplication, division, addition, and subtraction with fractions. A special effort is made to demonstrate the validity of the use of improper fractions, and exponents and the rules for order of operations are applied to fractions. Knowledge of prime numbers (and prime factorizations) underlies all of the discussions about fractions.

Chapter 4, Mixed Numbers, shows how mixed numbers are related to whole numbers and fractions. Topics include relating improper fractions and mixed numbers, the basic operations, complex fractions, and the rules for order of operations.

Chapter 5, Decimal Numbers, covers the basic operations with decimal numbers, estimating, and the use of calculators (including scientific notation). To emphasize number concepts, the last section presents operating with decimal numbers, fractions, and mixed numbers in a single expression.

Chapter 6, Ratios and Proportions, develops an understanding of ratios and introduces the idea of a variable. Techniques for solving equations are developed through finding the unknown term in a proportion.

Chapter 7, Percent (Calculators Recommended), approaches percent as hundredths and uses this idea to discuss percent of profit and to find equivalent numbers in the form of percents, decimals, and fractions. The applications with percent are developed around proportions, the formula $R \times B = A$, and the skills of solving equations. A special section on estimating with percent is included to reinforce basic understanding.

Chapter 8, Consumer Applications, addresses the topics of simple interest, compound interest, balancing a checking account, buying and owning a car, reading graphs, and statistics (mean, median, mode, and range). The use of calculators is encouraged and even necessary, such as in the case of using the formula for compound interest: $A = P(1 + r/n)^{nt}$. Calculating with this formula serves to emphasize the importance of following the rules for order of operations.

Chapter 9, Introduction to Algebra: Signed Numbers, provides a head start for those students planning to continue in their mathematics studies, either in a prealgebra course or a beginning course in algebra. Signed numbers are introduced with the aid of number lines and are graphed on number lines. Topics included are absolute value, operations with signed numbers, order of operations, translating phrases, solving equations, and problem solving.

Customized Chapters

The following chapters are designed to complement the basic text material (Chapters 1–9) to fit any additional curriculum features at your school or any particular syllabus topics desired by instructors. Any of these chapters may be requested by the instructor from the publisher to be included as part of the text, in the order shown here.

Measurement: This chapter discusses the metric system of measurement, the U.S. customary system of measurement, and how to change units of measure within each system and between the systems. The last section discusses the difference between abstract numbers and denominate numbers and operations with denominate numbers.

Geometry: This chapter provides more thorough coverage of geometry beyond what is integrated in the previous chapters. Included are length, perimeter, area, volume, and formulas related to specific figures. Also included is a discussion of types of angles and types of triangles (along with similar triangles). The chapter concludes with a discussion of square roots and the Pythagorean Theorem (with emphasis on how to use a calculator to find roots).

Solving Equations: This chapter forms the basis for continued studies in algebra. Included are simplifying and evaluating expressions (combining like terms), solving equations (with several steps), solving word problems (including consecutive integers), and working with formulas (evaluating formulas and solving for other variables).

Additional Topics from Algebra: This chapter introduces some topics that will be discussed in more detail in a beginning algebra course. Included are solving inequalities, graphing in two dimensions (ordered pairs and straight lines), and operating with polynomials.

Some Possible Course Offerings

Short Course (Chapters 1–7)	Longer Course (Chapter 1–9)	Optional Course (Chapters 2–9 with selected topics from Customized Chapters.)
Whole Numbers	Whole Numbers	Prime Numbers
Prime Numbers	Prime Numbers	Fractions
Fractions	Fractions	Mixed Numbers
Mixed Numbers	Mixed Numbers	Decimal Numbers
Decimal Numbers	Decimal Numbers	Ratios and Proportions
Ratios and Proportions	Ratios and Proportions	Percent (Calculators Recommended)
Percent (Calculators Recommended)	Percent (Calculators Recommended)	Consumer Applications
	Consumer Applications	Introduction to Algebra: Signed Numbers
	Introduction to Algebra: Signed Numbers	Selected topics from Measurement, Geometry, Solving Equations, and Additional Topics from Algebra

Practice and Review

There are more than 5500 practice exercises, lesson exercises, and review items overall. Lesson exercises are carefully chosen and graded, proceeding from easy exercises to more difficult ones. Many sections contain a feature entitled The Recycle Bin which generally has six to ten review exercises from previous chapters. Each chapter includes a Chapter Review, a Chapter Test, and a Cumulative Review (beginning with Chapter 2). The tests are similar in length and content to the tests provided in the *Instructor's Resource Manual*.

Many sections have exercises entitled Writing and Thinking about Mathematics, Check Your Number Sense, and Collaborative Learning Exercises. These exercises are an important part of the text and provide a chance for each student to improve communication skills, develop an understanding of general concepts, and communicate his or her ideas to the instructor. Written responses can be a great help to the instructor in identifying just what students do and do not understand. Many of these questions are designed for the student to investigate ideas other than those presented in the text, with responses that are to be based on each student's own experiences and perceptions. In most cases there is no right answer.

Answers to most exercises, all Chapter Review questions, all Chapter Test questions, and all Cumulative Review questions are provided in the back of the book. Answers to Classroom Practice exercises are given just below the problems themselves.

Supplements

Basic Mathematics for College Students, Seventh Edition, is accompanied by an expanded supplement support package, with each item designed and created to provide maximum benefit to the students and instructors who use them.

Instructor's Resource Manual: This manual includes several prepared forms of chapter quizzes and tests, prepared forms of cumulative tests, and answers for all quizzes and tests. It also includes an answer key to all even text exercises.

Student Study and Solutions Manual: This supplement has been prepared to provide students with additional practice and review questions. Answers to all items are included in the manual, along with solutions to selected items that illustrate the working of all key problem types. Solutions are also included for selected exercises from the text, and for all Chapter Review, Chapter Test, and Cumulative Review exercises.

Videotapes: A series of videotapes, covering all major topics, provides concept review and additional examples to reinforce the text presentation.

Adventure Learning Systems: Basic Mathematics: This multimedia courseware provides step-by-step learning, generates an unlimited number of practice problems, and tests knowledge of the skills of each section.

Acknowledgments

I would like to thank Philip Lanza, developmental editor, and Melissa Ray, production editor, for their hard work and invaluable assistance in the development of this text. Again, Phil has contributed so much and taken such personal interest in the project that he has become my hidden "coauthor." Melissa really does understand that mathematics textbooks have special requirements in style and format and did not hesitate to follow up on my many author's "suggestions."

Many thanks go to the following manuscript reviewers who offered their constructive and critical comments: James Barr of Laramie County Community College; Diana Donaldson of Rock Valley College; Karen Driskell of John C. Calhoun State Community College; Robert Egge of North Dakota State College of Science; Nita Graham of Forest Park Community College; Millicent B. Paisley of Pace University; and Barbara Wilbourn of Jefferson State Community College.

Finally, special thanks go to Charles Hartford, mathematics editor, for his faith in this project and willingness to commit so many resources throughout the development and production process to guarantee a top-quality product for students and teachers.

D. Franklin Wright

CONTENTS

3

Fractions   111

4

Mixed Numbers   165

5   Decimal Numbers   213

8 Consumer Applications   373

Whole Numbers

Mathematics at Work!

View of the main quad at the College of Arts and Sciences, Cornell University.

Whole numbers are everywhere. Businessmen and women, carpenters, auto mechanics, teachers, and bus drivers all deal with whole numbers as part of their daily lives.

As a student you may be interested in information about the fees and other expenses related to college life. The following list contains this type of information about 15 well-known colleges and universities. Annual (yearly) tuition, room, and board for 1992 are given, to the nearest ten dollars. More recent information can be obtained from the school's registrar.

Institution	Enrollment	Tuition		Room/Board
		Resident	Nonresident	
Univ. of Alabama	15,940	$ 2,010	$ 5,020	$3,300
Bryn Mawr College	1,180	16,170	16,170	6,150
Univ. of Colorado	10,410	1,970	9,900	3,540
Cornell University	7,600	16,190	16,190	5,410
Duke University	6,020	16,120	16,120	5,240
Harvard and Radcliffe Colleges	6,620	17,670	17,670	5,840
Univ. of Illinois	26,370	3,060	6,800	3,900
Indiana University	25,310	2,370	6,900	3,370
Univ. of Miami	8,640	15,050	15,050	5,910
Univ. of Notre Dame	7,520	13,500	13,500	3,600
Stanford University	6,510	16,540	16,540	6,310
Univ. of Texas	37,000	1,100	4,220	3,400
UCLA	24,370	2,900	10,600	5,410
Vassar College	2,310	17,210	17,210	5,500
Yale University	5,150	17,500	17,500	6,200

For each of these universities and colleges, find (a) the annual (yearly) expenses that a state resident might expect to pay for tuition, room, and board, and (b) the annual expenses that an out-of-state (nonresident) student might expect to pay. (c) Determine which of the universities and colleges on this list is the least expensive and which is the most expensive for a resident's tuition, room, and board. (d) Determine which is the least expensive and which is the most expensive for a nonresident's tuition and room and board. (See also Section 1.7, Exercise # 29.)

What to Expect in Chapter 1

Chapter 1 provides a review of the basic operations (addition, subtraction, multiplication, and division) with whole numbers. Section 1.1 discusses number systems in general, with emphasis on reading and writing whole numbers in the decimal system. The skills related to rounding off and estimating with whole numbers are developed in Section 1.4. These and other general number concepts are an integral part of the text and the exercises and are emphasized in a feature entitled **Check Your Number Sense.**

Section 1.7 (Problem Solving with Whole Numbers) closes this chapter. The four-step process for solving word problems developed by George Pólya, a famous educator and mathematician from Stanford University, is outlined here and reinforced throughout the text as basic to problem solving of all types.

1.1   Reading and Writing Whole Numbers

OBJECTIVES

1. Understand the concept of a place value number system.
2. Know the powers of ten.
3. Know the values of the places for whole numbers in the decimal system.
4. Read and write whole numbers in the decimal system.

The Decimal System

The Hindu-Arabic system that we use was invented about A.D. 800. It is called the **decimal system** (**deci** means ten in Latin). This system allows us to add, subtract, multiply, and divide more easily and faster than any of the ancient number systems. For example, multiplying 22 by 3 in the Roman system would appear as

$$
\begin{array}{lll}
\text{XXII} & & 22 \\
\underline{\text{III}} & \text{while we might write} & \underline{\times\ 3} \\
\text{LXVI} & & 66
\end{array}
$$

A discussion of several ancient number systems is included in Appendix I. As we develop operations with the decimal system throughout the text, advantages of the decimal system over the other systems will become more and more evident.

> The decimal system (or base ten system) is a **place value system** that depends on three things:
>
> **1.** the **ten digits:** 0, 1, 2, 3, 4, 5, 6, 7, 8, 9;
>
> **2.** the **placement** of each digit; and
>
> **3.** the **value** of each place.

The **decimal point** is the beginning point for writing digits. The value of each place to the left of the decimal point has a **power of ten** (1, 10, 100, 1000, and so on) as its place value. (See Figure 1.1.)

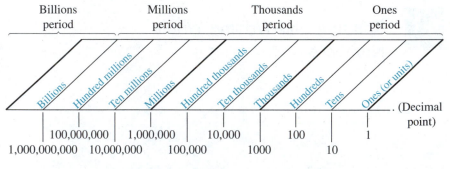

Figure 1.1

When a digit is written in a place, the value of the digit is to be multiplied by the value of the place. The value of any number is found by adding the results of multiplying the digits and place values. For example,

6   9   3   ← digits (standard notation)

100   10   1   ← place values

$693 = 6(100) + 9(10) + 3(1) = 600 + 90 + 3$ (expanded notation)

> **Note:** Writing a number next to a number in parentheses means to multiply.

Reading and Writing Whole Numbers

In **expanded notation** the values represented by each digit in standard notation are added. The English word equivalents can then be easily read from these sums. If a number has more than four digits, commas are placed to separate every three digits beginning from the right. Commas are used in the same manner in the word equivalents.

> **Note:** In this text, we have chosen not to use commas in numbers with four digits in most cases. This choice is a matter of style. You may use commas in four-digit numbers if you prefer.

In Examples 1–4, each number is written in standard notation and in expanded notation and in its English word equivalent. If no decimal point is written, it is understood to be to the right of the rightmost digit.

EXAMPLE 1

573   (standard notation)

$573 = 5(100) + 7(10) + 3(1)$
$= 500 + 70 + 3$ (expanded notation)

five hundred seventy-three

EXAMPLE 2

4862

$4862 = 4(1000) + 8(100) + 6(10) + 2(1)$
$= 4000 + 800 + 60 + 2$ (expanded notation)

four thousand eight hundred sixty-two

EXAMPLE 3

8007

$8007 = 8(1000) + 0(100) + 0(10) + 7(1)$
$= 8000 + 0 + 0 + 7$

eight thousand seven

EXAMPLE 4

1,590,768

$1,590,768 = 1(1,000,000) + 5(100,000) + 9(10,000) + 0(1000)$
$+ 7(100) + 6(10) + 8(1)$
$= 1,000,000 + 500,000 + 90,000 + 0 + 700 + 60 + 8$

one million, five hundred ninety thousand, seven hundred sixty-eight

Whole numbers are those numbers used for counting and the number 0. They are the decimal numbers with digits written to the left of the decimal point and with only 0's (or no digits at all) written to the right of the decimal point.

Decimal numbers with digits to the right of the decimal point will be discussed in Chapter 5.

We use the letter W to represent all whole numbers:

0, 1, 2, 3, 4, 5, 6, 7, 8, 9, 10, 11, 12, . . .

The three dots indicate that the pattern is to continue without end.

Note four things when reading or writing whole numbers:

1. The word **and** does not appear in English word equivalents. **And** is said only when reading a decimal point. (See Chapter 5.)

2. Digits are read in groups of three.

3. Commas are used to separate groups of three digits **if a number has more than four digits.**

4. Hyphens (-) are used to write English words for the two-digit numbers from 21 to 99 except for those that end in 0.

CLASSROOM PRACTICE

Write the following numbers in expanded notation and in their English word equivalents.

1. 512   **2.** 6394

3. Write one hundred eighty thousand, five hundred forty-three as a decimal numeral.

ANSWERS: **1.** 500 + 10 + 2; five hundred twelve   **2.** 6000 + 3000 + 90 + 4; six thousand three hundred ninety-four   **3.** 180,543

Exercises 1.1

Did you read the explanation and work through the examples before beginning these exercises?

Write the following decimal numbers in expanded notation and in their English word equivalents.

1. 37	**2.** 84	**3.** 98	**4.** 56
5. 122	**6.** 493	**7.** 821	**8.** 1976
9. 1892	**10.** 5496	**11.** 12,517	**12.** 42,100
13. 243,400	**14.** 891,540	**15.** 43,655	**16.** 99,999

17. 8,400,810

18. 5,663,701

19. 16,302,590

20. 71,500,000

21. 83,000,605

22. 152,403,672

23. 679,078,100

24. 4,830,231,010

25. 8,572,003,425

Write the following numbers as decimal numbers.

26. seventy-six

27. one hundred thirty-two

28. five hundred eighty

29. three thousand eight hundred forty-two

30. two thousand five

31. one hundred ninety-two thousand, one hundred fifty-one

32. seventy-eight thousand, nine hundred two

33. twenty-one thousand, four hundred

34. thirty-three thousand, three hundred thirty-three

35. five million, forty-five thousand

36. five million, forty-five

37. ten million, six hundred thirty-nine thousand, five hundred eighty-two

38. two hundred eighty-one million, three hundred thousand, five hundred one

39. five hundred thirty million, seven hundred

40. seven hundred fifty-eight million, three hundred fifty thousand, sixty

41. ninety million, ninety thousand, ninety

42. eighty-two million, seven hundred thousand

43. one hundred seventy-five million, two

44. thirty-six

45. seven hundred fifty-seven

46. Name the position of each nonzero digit in the following number: 109,750

Write the English word equivalent for the numbers in each sentence.

47. The population of Los Angeles is 3,485,398.

48. The distance from Earth to the sun is about 93,000,000 miles (or 149,730,000 kilometers).

49. On average, the country of Chile is 100 miles wide and 2700 miles long.

50. The Republic of Venezuela covers an area of 352,143 square miles (or about 912,050 square kilometers).

Writing and Thinking about Mathematics

51. A *googol* is the name of a very large power of ten. Look in a dictionary or in a book that discusses the history of mathematics for the meaning of this term. (Note: Ten to the power of a googol is called a *googolplex.*)

52. According to the 1990 census, Illinois, with 11,430,602 people, is the sixth most populous state in the United States. Name the value of each nonzero digit in this number. In an almanac or other source, find the populations of the first five most populous states.

1.2 Addition

OBJECTIVES

1. Know the basic facts of addition with whole numbers.
2. Be able to add whole numbers in both a vertical format and a horizontal format.
3. Know the following properties of addition: commutative, associative, additive identity (0).

Addition with Whole Numbers

Addition with whole numbers is indicated either by writing the numbers horizontally with a plus sign (+) between them or by writing the numbers vertically in columns with instructions to add.

$$6 + 23 + 17 \quad \text{or} \quad \begin{array}{r} \text{Add} \quad 6 \\ 23 \\ \underline{17} \end{array} \quad \text{or} \quad \begin{array}{r} 6 \\ 23 \\ \underline{+\ 17} \end{array}$$

The numbers being added are called **addends,** and the result of the addition is called the **sum.**

$$\begin{array}{r} \text{Add} \quad 6 \quad \text{addend} \\ 23 \quad \text{addend} \\ \underline{17} \quad \text{addend} \\ 46 \quad \text{sum} \end{array}$$

Be sure to keep the digits aligned (in column form) so you will be adding units to units, tens to tens, and so on. Neatness is a necessity in mathematics.

To be able to add with speed and accuracy, you must **memorize** the basic addition facts, which are given in Table 1.1. Practice all the combinations so you can give the answers immediately. Concentrate.

Table 1.1					Basic Addition Facts					
+	**0**	**1**	**2**	**3**	**4**	**5**	**6**	**7**	**8**	**9**
0	**0**	1	2	3	4	5	6	7	8	9
1	1	**2**	3	4	5	6	7	8	9	10
2	2	3	**4**	5	6	7	8	9	10	11
3	3	4	5	**6**	7	8	9	10	11	12
4	4	5	6	7	**8**	9	10	11	12	13
5	5	6	7	8	9	**10**	11	12	13	14
6	6	7	8	9	10	11	**12**	13	14	15
7	7	8	9	10	11	12	13	**14**	15	16
8	8	9	10	11	12	13	14	15	**16**	17
9	9	10	11	12	13	14	15	16	17	**18**

As an exercise to help find out which combinations you need special practice with, write, on a piece of paper in mixed order, all one hundred possible combinations to be added. Perform the operations as quickly as possible. Then, using the table, check to find the ones you missed. Study these frequently until you are confident that you know them as well as all the others.

Your adding speed can be increased if you learn to look for combinations of digits that total 10.

EXAMPLE 1   To add the following numbers, we note the combinations that total 10 and find the sums quickly.

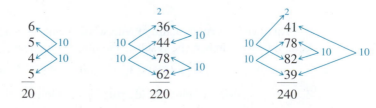

When two numbers are added, the order of the numbers does not matter. That is, $4 + 9 = 13$ and $9 + 4 = 13$. Also,

$$
\begin{array}{r}
5 \\
+7 \\
\hline
12
\end{array}
\quad \text{and} \quad
\begin{array}{r}
7 \\
+5 \\
\hline
12
\end{array}
$$

By looking at Table 1.1 again, we can see that reversing the order of any two addends will not change their sum. This fact is called the **commutative property of addition.** We can state this property using letters to represent whole numbers.

Commutative Property of Addition

If a and b are two whole numbers, then $a + b = b + a$.

To add three numbers, such as $6 + 3 + 5$, we can add only two at a time, then add the third. Which two should we add first? The answer is that the sum is the same either way:

$$6 + 3 + 5 = (6 + 3) + 5 = 9 + 5 = 14$$

and

$$6 + 3 + 5 = 6 + (3 + 5) = 6 + 8 = 14$$

We can write

$$8 + 4 + 7 = (8 + 4) + 7 = 8 + (4 + 7)$$

These examples illustrate the **associative property of addition.**

Associative Property of Addition

If a, b, and c are whole numbers, then

$$a + b + c = (a + b) + c = a + (b + c)$$

Another property of addition with whole numbers is addition with 0. Whenever we add 0 to a number, the result is the original number:

$$8 + 0 = 8, \qquad 0 + 13 = 13, \qquad \text{and} \qquad 38 + 0 = 38$$

Zero (0) is called the **additive identity** or the **identity element for addition.**

> ### Additive Identity
>
> If a is a whole number, then there is a unique whole number 0 with the property that $a + 0 = a$.

EXAMPLE 2

(a) $(15 + 20) + 6 = 15 + (20 + 6)$ associative property

(b) $(15 + 20) + 6 = (20 + 15) + 6$ commutative property

(c) $17 + 0 = 17$ additive identity

Note that Example 2 shows a change in the **grouping** (association) of the numbers, while Example 3 shows a change in the **order** of two numbers.

EXAMPLE 3

Another illustration of the associative property:

$$
\begin{array}{r}
8 \\
2 \\
+\ 3 \\
\hline
\end{array}
\quad
\begin{array}{r}
10 \\
3 \\
\hline
13
\end{array}
\qquad
\begin{array}{r}
8 \\
2 \\
3 \\
\hline
\end{array}
\quad
\begin{array}{r}
8 \\
5 \\
\hline
13
\end{array}
$$

Adding large numbers or several numbers can be done by writing the numbers vertically so that the place values are **lined up** in columns. Then only the digits with the same place value are added.

EXAMPLE 4

If no sum of digits is more than 9:

Add $5623 + 3172$.

$$
\begin{array}{r}
5\ 6\ 2\ 3 \\
+\ 3\ 1\ 7\ 2 \\
\hline
5
\end{array}
$$
$\leftarrow$ addend
$\leftarrow$ addend
Add ones.

$$
\begin{array}{r}
5\ 6\ 2\ 3 \\
+\ 3\ 1\ 7\ 2 \\
\hline
9\ 5
\end{array}
$$
Add tens.

$$
\begin{array}{r}
5\ 6\ 2\ 3 \\
+\ 3\ 1\ 7\ 2 \\
\hline
7\ 9\ 5
\end{array}
$$
Add hundreds.

$$
\begin{array}{r}
5\ 6\ 2\ 3 \\
+\ 3\ 1\ 7\ 2 \\
\hline
8\ 7\ 9\ 5
\end{array}
$$
Add thousands.

$$\begin{array}{r} 5\,6\,2\,3 \\ +\,3\,1\,7\,2 \\ \hline 8\,7\,9\,5 \end{array}$$ ← You do not write all the steps. You
write only the addends and the sum.

← sum

EXAMPLE 5

If the sum of the digits in one column is more than 9:

(a) Write the ones digits in that column.

(b) **Carry** the other digit to the column to the left.

$$\begin{array}{r} 2 \\ 2\,1\,7 \\ 3\,8\,9 \\ 6\,3\,4 \\ 5\,3\,6 \\ \hline 6 \end{array}$$

Add 7 + 9 + ④ + ⑥ = 7 + 9 + 10
= 26

Carry the 2.

10

$$\begin{array}{r} 1\ 2 \\ 2\,1\,7 \\ 3\,8\,9 \\ 6\,3\,4 \\ 5\,3\,6 \\ \hline 7\,6 \end{array}$$

10

Add ② + 1 + ⑧ + 3 + 3 = 10 + 1 + 3 + 3
= 17

$$\begin{array}{r} 1\ 2 \\ 2\,1\,7 \\ 3\,8\,9 \\ 6\,3\,4 \\ 5\,3\,6 \\ \hline 1\,7\,7\,6 \end{array}$$

10

Add ① + 2 + ③ + ⑥ + 5 = 10 + 2 + 5
= 17

Exercises 1.2

Did you read the explanation and work through the examples before beginning these exercises?

Do the following exercises mentally and write only the answers.

1. $6 + (3 + 7)$

2. $(4 + 5) + 6$

3. $(2 + 3) + 8$

4. $(2 + 6) + (4 + 5)$

5. $8 + (3 + 4) + 6$

6. $9 + (8 + 3)$

7. $9 + 2 + 8$

8. $7 + (6 + 3)$

9. $(2 + 3) + 7$

10. $8 + 7 + 2$

11. $9 + 6 + 3$

12. $4 + 3 + 6$

13. $4 + 4 + 4$

14. $9 + 1 + 5 + 6$

15. $5 + 3 + 7 + 2$

16. $3 + 6 + 2 + 8$

17. $5 + 4 + 6 + 4 + 8$

18. $6 + 2 + 9 + 1$

19. $4 + 3 + 6 + 6 + 2$

20. $8 + 8 + 7 + 6 + 3$

Show that the following statements are true by performing the addition. State which property of addition is being illustrated.

21. $9 + 3 = 3 + 9$

22. $8 + 7 = 7 + 8$

23. $4 + (5 + 3) = (4 + 5) + 3$

24. $4 + 8 = 8 + 4$

25. $2 + (1 + 6) = (2 + 1) + 6$

26. $(8 + 7) + 3 = 8 + (7 + 3)$

27. $9 + 0 = 9$

28. $(2 + 3) + 4 = (3 + 2) + 4$

29. $7 + (6 + 0) = 7 + 6$

30. $8 + 20 + 1 = 8 + 21$

Copy the following problems and add.

31.
```
 65
 43
 54
```

32.
```
 24
 78
 95
```

33.
```
 73
 68
 98
```

34.
```
165
276
394
```

35.
```
876
279
143
```

36.
```
268
 93
 74
192
```

37.
```
981
146
 92
 17
```

38.
```
2112
 147
 904
1005
```

39.
```
 114
5402
 710
 643
```

40.
```
1403
7010
 622
  29
```

41.
```
213,116
116,018
722,988
 24,336
526,968
```

42.
```
 21,442
 32,462
564,792
801,801
 43,433
```

43.
```
  438,966
1,572,486
  327,462
  181,753
   90,000
```

44.
```
123,456
456,123
879,282
617,500
740,765
```

45. Mr. Juarez kept the mileage records indicated in the table shown here. How many miles did he drive during the six months?

Month	Mileage
Jan.	546
Feb.	378
Mar.	496
Apr.	357
May	503
June	482

46. The Modern Products Corp. showed profits as indicated in the table for the years 1987–1990. What were the company's total profits for the years 1987–1990?

Year	Profits
1987	$1,078,416
1988	1,270,842
1989	2,000,593
1990	1,963,472

47. During six years of college, including two years of graduate school, Fred's expenses were $2035; $2842; $2786; $3300; $4000; $3500. What were his total expenses for six years of schooling? (NOTE: He had some financial aid.)

48. Apple County has the following items budgeted: highways, $270,455; salaries, $95,479; maintenance, $127,220. What is the county's total budget for these three items?

49. The following numbers of students at South Junior College are enrolled in mathematics courses: 303 in arithmetic; 476 in algebra; 293 in trigonometry; 257 in college algebra; 189 in calculus. Find the total number of students taking mathematics.

50. In one year, the High Price Manufacturing Co. made 2476 refrigerators; 4217 gas stoves; 3947 electric stoves; 9576 toasters; 11,872 electric fans; 1742 air conditioners. What was the total number of appliances High Price produced that year?

Complete each of the following addition tables. You may want to make up your own tables for practicing difficult combinations.

51.

+	5	8	7	9
3				
6				
5				
2				

52.

+	6	7	1	4	9
3					
5					
4					
8					
2					

1.3   Subtraction

OBJECTIVES

1. Be able to subtract whole numbers with no borrowing.
2. Be able to subtract whole numbers with borrowing.

Subtraction with Whole Numbers

A long-distance runner who is 25 years old had run 17 miles of his usual 19-mile workout when a storm forced him to quit for the day. How many miles short was he of his usual daily training?

What *thinking* did you do to answer the question? You may have thought something like this: "Well, I don't need to know how old the runner is to answer the question, so his age is just extra information. Since he had already run 17 miles, I need to know what to add to 17 miles to get 19 miles. Since $17 + 2 = 19$, he was 2 miles short of his usual workout."

In this problem, the sum of two addends was given, and only one of the addends was given. The other addend was the unknown quantity.

$$17 \quad + \quad \square \quad = \quad 19$$
addend   missing addend   sum

As you may know, this kind of addition problem is called **subtraction** and can be written:

$$19 \quad - \quad 17 \quad = \quad \square \quad \text{(Read: "19 minus 17 equals blank.")}$$
sum   addend   missing addend
                   (or difference)

OR

$$\begin{array}{r} 19 \\ -\ 17 \\ \hline \square \end{array}$$

19   sum
– 17   addend
□   difference or missing addend

In other words, subtraction is a reverse addition, and the missing addend is called the **difference** between the sum and one addend.

Perform the following subtraction mentally: $17 - 8 = \square$. You should think, "8 plus what number gives 17? Since $8 + 9 = 17$, we have $17 - 8 = 9$."

Suppose you want to find the difference $496 - 342$. Thinking of a number to add to 342 to get 496 is difficult. A better technique, using the place values, is developed and discussed in the following examples.

EXAMPLE 1 | Subtract 496 − 342.

Solution

STEP 1:   4 9 6   Subtract ones.   STEP 2:   4 9 6   Subtract tens.
         − 3 4 2   6 − 2 = 4   − 3 4 2   9 − 4 = 5
              4   5 4

STEP 3:   4 9 6   Subtract hundreds.
         − 3 4 2   4 − 3 = 1
           1 5 4   ← difference

Or, using expanded notation, we get the same result:

$$\begin{array}{rl} 4\,9\,6 = & 400 + 90 + 6 \\ -\,3\,4\,2 = & -\,[300 + 40 + 2] \\ \hline & 100 + 50 + 4 \;=\; 154 \quad \leftarrow \text{difference} \end{array}$$

EXAMPLE 2 | Subtract 65 − 28 using expanded notation.

Solution

$$\begin{array}{rl} 65 = & 60 + 5 \\ -\,28 = & -\,[20 + 8] \end{array}$$

Starting from the right, 5 is smaller than 8. We cannot subtract 8 from 5, so we **borrow** 10 from 60.

                   Borrow 10   10 borrowed from
                   from 60.   60, plus 5.
                      ↓   ↓

$$\begin{array}{rlllll} 65 = & 60 + 5 \;= & 50 + 10 + 5 \;= & 50 + 15 & \text{Now} \\ -\,28 = & -\,[20 + 8] \;= & -\,[20 + + 8] \;= & -[20 + 8] & \text{subtract.} \\ \hline & & & 30 + 7 \;=\; 37 \end{array}$$

EXAMPLE 3 | Use expanded notation and borrowing to find the difference 536 − 258.

Solution

$$\begin{array}{rl} 536 = & 500 + 30 + 6 \\ -\,258 = & -\,[200 + 50 + 8] \end{array}$$

                                 Since 6 is smaller   Since 20 is smaller
                                 than 8, borrow   than 50, borrow
                                 10 from 30.   100 from 500.

$$\begin{array}{rlll} 536 = & 500 + 30 + 6 \;= & 500 + 20 + 16 \;= & 400 + 120 + 16 \\ -\,258 = & -\,[200 + 50 + 8] \;= & -\,[200 + 50 + 8] \;= & -[200 + 50 + 8] \\ \hline & & & 200 + 70 + 8 \\ & & & = 278 \end{array}$$

A common practice, with which you are probably familiar, is to indicate the borrowing by crossing out digits and writing new digits instead of expanded notation. The expanded notation technique has been shown so that you can have a "picture" in your mind to help you understand the procedure of subtraction with borrowing.

EXAMPLE 4

10 is borrowed from 60.
10 is added to 5.
↓

50 in place of 60.
Borrowed 10
↓

$$
\begin{array}{r}
65 = \quad 60 + 5 = \quad 50 + 15 \\
- 28 = - [20 + 8] = - [20 + \ \ 8]
\end{array}
$$

can be written

$$
\begin{array}{r}
5 \ \ 1 \\
\cancel{6} \ \ 5 \\
- 2 \ \ 8 \\
\hline
3 \ \ 7
\end{array}
$$

EXAMPLE 5

$$
\begin{array}{r}
736 \\
- 258
\end{array}
$$

STEP 1:   Since 6 is smaller than 8, borrow 10 from 30 and add this 10 to 6 to get 16. This leaves 20, so cross out 3 and write 2.

$$
\begin{array}{r}
2 \ \ 1 \\
7 \cancel{3} 6 \\
- 2 5 8
\end{array}
$$

STEP 2:   Since 2 is smaller than 5, borrow 100 from 700. This leaves 600, so cross out 7 and write 6.

$$
\begin{array}{r}
6 \ 12 \ 1 \\
\cancel{7} \cancel{3} 6 \\
- 2 5 8
\end{array}
$$

STEP 3:   Now subtract.

$$
\begin{array}{r}
6 \ 12 \ 1 \\
\cancel{7} \cancel{3} 6 \\
- 2 5 8 \\
\hline
4 7 8
\end{array}
$$

EXAMPLE 6

$$
\begin{array}{r}
8000 \\
- 657
\end{array}
$$

STEP 1:   Trying to borrow from 0 each time, we end up borrowing 1000 from 8000. Cross out 8 and write 7.

$$
\begin{array}{r}
7 \ \ 1 \\
\cancel{8} \ \ 0 \ \ 0 \ \ 0 \\
- \quad 6 \ \ 5 \ \ 7
\end{array}
$$

STEP 2:   Now borrow 100 from 1000. Cross out 10 and write 9.

$$
\begin{array}{r}
9 \\
7 \cancel{1} 1 \\
\cancel{8} \cancel{0} 0 0 \\
- \ 6 5 7
\end{array}
$$

STEP 3:   Borrow 10 from 100. Cross out 10 and write 9.

$$
\begin{array}{r}
9 \ 9 \\
7 \cancel{1} \cancel{1} 1 \\
\cancel{8} \cancel{0} \cancel{0} 0 \\
- \ 6 5 7
\end{array}
$$

STEP 4:   Now subtract.

$$
\begin{array}{r}
\overset{9\ 9}{\underset{}{7\ \cancel{8}\ \cancel{0}\ 1}} \\
\cancel{8}\cancel{0}\cancel{0}\ 0 \\
-\quad 6\ 5\ 7 \\
\hline
7\ 3\ 4\ 3
\end{array}
$$

EXAMPLE 7

What number should be added to 546 to get a sum of 732?

Solution

We know the sum and one addend. To find the missing addend, subtract 546 from 732.

$$
\begin{array}{r}
\overset{1}{\underset{}{6\ 2\ 1}} \\
\cancel{7}\ \cancel{3}\ 2 \\
-\ 5\ 4\ 6 \\
\hline
1\ 8\ 6 \quad \text{difference}
\end{array}
$$

The number to be added is 186.

EXAMPLE 8

The cost of repairing Ed's used TV set was going to be $250 for parts (including tax) plus $85 for labor. To buy a new set, he was going to pay $450 plus $27 in tax and the dealer was going to pay him $85 for his old set. How much more would Ed have to pay for a new set than to have his old set repaired?

Solution

Used Set		New Set		Difference	
$250	parts	$450	cost	$392	
+ 85	labor	+ 27	tax	− 335	
$335	total	$477		$ 57	
		− 85	trade-in		
		$392	total		

Ed would pay $57 more for the new set than for having his old set repaired. What would you do?

CLASSROOM PRACTICE

Find the following differences.

1.   83
    − 54

2.   600
    − 368

3.   7856
    − 6397

ANSWERS: **1.** 29   **2.** 232   **3.** 1459

Exercises 1.3

Subtract. Do as many problems as you can mentally.

1. $8 - 5$　　　　**2.** $19 - 6$　　　　**3.** $14 - 14$　　　　**4.** $17 - 9$

5. $20 - 11$　　　**6.** $17 - 0$　　　　**7.** $17 - 8$　　　　**8.** $16 - 16$

9. $11 - 6$　　　　**10.** $13 - 7$

11. $\begin{array}{r} 17 \\ -\,17 \end{array}$	**12.** $\begin{array}{r} 42 \\ -\,31 \end{array}$	**13.** $\begin{array}{r} 89 \\ -\,76 \end{array}$	**14.** $\begin{array}{r} 53 \\ -\,33 \end{array}$	**15.** $\begin{array}{r} 47 \\ -\,27 \end{array}$
16. $\begin{array}{r} 96 \\ -\,27 \end{array}$	**17.** $\begin{array}{r} 23 \\ -\,18 \end{array}$	**18.** $\begin{array}{r} 74 \\ -\,29 \end{array}$	**19.** $\begin{array}{r} 61 \\ -\,48 \end{array}$	**20.** $\begin{array}{r} 52 \\ -\,27 \end{array}$
21. $\begin{array}{r} 126 \\ -\,32 \end{array}$	**22.** $\begin{array}{r} 174 \\ -\,48 \end{array}$	**23.** $\begin{array}{r} 347 \\ -\,129 \end{array}$	**24.** $\begin{array}{r} 256 \\ -\,118 \end{array}$	
25. $\begin{array}{r} 692 \\ -\,217 \end{array}$	**26.** $\begin{array}{r} 543 \\ -\,167 \end{array}$	**27.** $\begin{array}{r} 900 \\ -\,307 \end{array}$	**28.** $\begin{array}{r} 603 \\ -\,208 \end{array}$	
29. $\begin{array}{r} 474 \\ -\,286 \end{array}$	**30.** $\begin{array}{r} 657 \\ -\,179 \end{array}$	**31.** $\begin{array}{r} 7843 \\ -\,6274 \end{array}$	**32.** $\begin{array}{r} 6793 \\ -\,5827 \end{array}$	
33. $\begin{array}{r} 4376 \\ -\,2808 \end{array}$	**34.** $\begin{array}{r} 3275 \\ -\,1744 \end{array}$	**35.** $\begin{array}{r} 3546 \\ -\,3546 \end{array}$	**36.** $\begin{array}{r} 4900 \\ -\,3476 \end{array}$	
37. $\begin{array}{r} 5070 \\ -\,4376 \end{array}$	**38.** $\begin{array}{r} 8007 \\ -\,2136 \end{array}$	**39.** $\begin{array}{r} 4065 \\ -\,1548 \end{array}$	**40.** $\begin{array}{r} 7602 \\ -\,2985 \end{array}$	
41. $\begin{array}{r} 7,085,076 \\ -\,4,278,432 \end{array}$	**42.** $\begin{array}{r} 6,543,222 \\ -\,2,742,663 \end{array}$	**43.** $\begin{array}{r} 4,000,000 \\ -\,2,993,042 \end{array}$		
44. $\begin{array}{r} 8,000,000 \\ -\,647,561 \end{array}$	**45.** $\begin{array}{r} 6,000,000 \\ -\,328,989 \end{array}$			

46. What number should be added to 978 to get a sum of 1200?

47. What number should be added to 860 to get a sum of 1000?

48. If the sum of two numbers is 537 and one of the numbers is 139, what is the other number?

49. A man is 36 years old, and his wife is 34 years old. Including high school, they have each attended school for 16 years. How many total years of schooling have they had?

50. Basketball Team A has twelve players and won its first three games by the following scores: 84 to 73, 97 to 78, and 101 to 63. Team B has ten players and won its first three games by 76 to 75, 83 to 70, and 94 to 84. What is the difference between the total of the differences of Team A's scores and those of its opponents and the total of the differences of Team B's scores and those of its opponents?

51. The Kingston Construction Co. made a bid of $7,043,272 to build a stretch of freeway, but the Beach City Construction Co. made a lower bid of $6,792,868. How much lower was the Beach City bid?

52. Two landscaping companies made bids on the landscaping of a new apartment complex. Company A bid $550,000 for materials and plants and $225,000 for labor. Company B bid $600,000 for materials and plants and $182,000 for labor. Which company had the lower total bid? How much lower?

53. In June, Ms. White opened a checking account and deposited $1342, $238, $57, and $486. She also wrote checks for $132, $76, $25, $42, $480, $90, $17, and $327. What was her balance at the end of June?

54. A manufacturing company had assets of $5,027,479, which included $1,500,000 in real estate. The liabilities were $4,792,023. By how much was the company "in the black"?

55. A man and woman sold their house for $135,000. They paid the realtor $8100, and other expenses of the sale came to $800. If they owed the bank $87,000 for the mortgage, what were their net proceeds from the sale?

56. A woman bought a condominium for a price of $150,000. She also had to pay other expenses of $750. If the local Savings and Loan agreed to give her a mortgage loan of $105,500, how much cash did she need to make the purchase?

57. Junior bought a red car with a sticker price of $10,000. But the salesman added $1200 for taxes, license, and extras. The bank agreed to give Junior a loan of $7500 on the car. How much cash did Junior need to buy the car?

58. In pricing a four-door car, Pat found she would have to pay a base price of $9500 plus $570 in taxes and $250 for license fees. For a two-door of the same make, she would pay a base price of $8700 plus $522 in taxes and $230 for license fees. Including all expenses, how much cheaper was the two-door model?

Writing and Thinking about Mathematics

59. Nothing was said in the text about a commutative property for subtraction. Do you think that this omission was intentional? Is there a commutative property of subtraction? Give several examples that justify your answer and discuss this idea in class.

60. Nothing was said in the text about an associative property for subtraction. Do you think that this omission was intentional? Is there an associative property of subtraction? Give several examples that justify your answer and discuss this idea in class.

61. You may have heard someone say that "you cannot subtract a larger number from a smaller number." Do you think that this statement is true? Does a subtraction problem such as 6 − 10 make sense to you? Can you think of any situation in which such a difference might seem reasonable?

Separate the class into teams of two to four students. Each team is to read and fill out the partial 1040 Forms in Exercises 62–64. After these are completed, the team leader is to read the team's results and discuss any difficulties they had in reading and following directions while filling out the forms.

62. Even the Internal Revenue Service allows us to calculate on our income tax returns with whole numbers. Calculate the taxable income on the portion of the Form 1040 shown here. Assume that you have 3 exemptions.

32	Amount from line 31 (adjusted gross income)	**32**	56,458
33a	Check if: ☐ **You** were 65 or older, ☐ Blind; ☐ **Spouse** was 65 or older; ☐ Blind. Add the number of boxes checked above and enter the total here . . . . ▶ **33a**		
b	If your parent (or someone else) can claim you as a dependent, check here . ▶ **33b** ☐		
c	If you are married filing separately and your spouse itemizes deductions or you are a dual-status alien, see page 22 and check here. . . . . . . ▶ **33c** ☐		
34	Enter the **larger** of your: **Itemized deductions** from Schedule A, line 26, **OR** **Standard deduction** shown below for your filing status. **But if you checked any box on line 33a or b**, go to page 22 to find your standard deduction. If you checked **box 33c**, your standard deduction is zero. • Single–$3,600 • Head of household–$5250 • Married filing jointly or Qualifying widow(er)–$6,000 • Married filing separately–$3,000	**34**	6,000
35	Subtract line 34 from line 32	**35**	
36	If line 32 is $78,950 or less, multiply $2,300 by the total number of exemptions claimed on line 6e. If line 32 is over $78,950, see the worksheet on page 23 for the amount to enter .	**36**	
37	**Taxable income.** Subtract line 36 from line 35. If line 36 is more than line 35, enter "0" .	**37**	

63. Calculate the taxable income on the portion of the Form 1040 shown here. Assume that you have 4 exemptions.

32	Amount from line 31 (adjusted gross income)	**32**	43,552
33a	Check if: ☐ **You** were 65 or older, ☐ Blind; ☐ **Spouse** was 65 or older; ☐ Blind. Add the number of boxes checked above and enter the total here . . . . ▶ **33a**		
b	If your parent (or someone else) can claim you as a dependent, check here . ▶ **33b** ☐		
c	If you are married filing separately and your spouse itemizes deductions or you are a dual-status alien, see page 22 and check here. . . . . . . ▶ **33c** ☐		
34	Enter the **larger** of your: **Itemized deductions** from Schedule A, line 26, **OR** **Standard deduction** shown below for your filing status. **But if you checked any box on line 33a or b**, go to page 22 to find your standard deduction. If you checked **box 33c**, your standard deduction is zero. • Single–$3,600 • Head of household–$5250 • Married filing jointly or Qualifying widow(er)–$6,000 • Married filing separately–$3,000	**34**	6,000
35	Subtract line 34 from line 32	**35**	
36	If line 32 is $78,950 or less, multiply $2,300 by the total number of exemptions claimed on line 6e. If line 32 is over $78,950, see the worksheet on page 23 for the amount to enter .	**36**	
37	**Taxable income.** Subtract line 36 from line 35. If line 36 is more than line 35, enter "0" .	**37**	

64. Calculate the taxable income on the portion of the Form 1040 shown here. Assume that you have 2 exemptions.

32	Amount from line 31 (adjusted gross income)	**32**	75,950
33a	Check if: ☐ **You** were 65 or older, ☐ Blind; ☐ **Spouse** was 65 or older; ☐ Blind. Add the number of boxes checked above and enter the total here . . . . ▶ **33a**		
b	If your parent (or someone else) can claim you as a dependent, check here . ▶ **33b** ☐		
c	If you are married filing separately and your spouse itemizes deductions or you are a dual-status alien, see page 22 and check here. . . . . . . ▶ **33c** ☐		
34	Enter the larger of your: **Itemized deductions** from Schedule A, line 26, **OR** **Standard deduction** shown below for your filing status. **But if you checked any box on line 33a or b**, go to page 22 to find your standard deduction. If you checked **box 33c**, your standard deduction is zero. • Single–$3,600 • Head of household–$5250 • Married filing jointly or Qualifying widow(er)–$6,000 • Married filing separately–$3,000	**34**	6,000
35	Subtract line 34 from line 32	**35**	
36	If line 32 is $78,950 or less, multiply $2,300 by the total number of exemptions claimed on line 6e. If line 32 is over $78,950, see the worksheet on page 23 for the amount to enter .	**36**	
37	**Taxable income.** Subtract line 36 from line 35. If line 36 is more than line 35, enter "0" .	**37**	

1.4 Rounding Off and Estimating

OBJECTIVES

1. Know the rule for rounding off whole numbers to any given place.
2. Use the rule of rounding off to estimate sums and differences.
3. Develop a sense of the expected size of a sum or difference so that major errors can be easily spotted.

Rounding Off Whole Numbers

To **round off** a given number means to find another number close to the given number. The desired place of accuracy must be stated. For example, if you were asked to round off 872, you would not know what to do unless you were told the position or place of accuracy desired. The number lines in Figure 1.2 help to illustrate the problem.

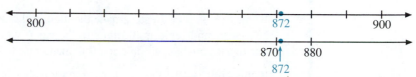

Figure 1.2

We can see that 872 is closer to 900 than to 800. So, **to the nearest hundred,** 872 rounds off to 900. Also, 872 is closer to 870 than to 800. So, **to the nearest ten,** 872 rounds off to 870.

Inaccuracy in numbers is common, and rounded-off answers can be quite acceptable. For example, what do you think is the distance across the United States from the east coast to the west coast? Most people will use the rounded-off answer of 3000 miles. This answer is appropriate and acceptable.

Number lines are used as visual aids in rounding off in the following Examples.

EXAMPLE 1 Round 47 to the nearest ten.

Solution

47 is closer to 50 than to 40. So, 47 rounds off to 50 (to the nearest ten).

EXAMPLE 2 Round 238 to the nearest hundred.

Solution

238 is closer to 200 than to 300. So, 238 rounds off to 200 (to the nearest hundred).

Using figures as an aid to understanding is fine but, for practical purposes, the following rule is more useful.

Rounding-Off Rule for Whole Numbers

1. Look at the single digit just to the right of the digit that is in the place of desired accuracy.

2. If this digit is 5 or greater, make the digit in the desired place of accuracy one larger and replace all digits to the right with zeros. All digits to the left remain unchanged unless a 9 is made one larger.

3. If this digit is less than 5, leave the digit that is in the place of desired accuracy as it is and replace all digits to the right with zeros. All digits to the left remain unchanged.

The Rounding-Off Rule is used in the following examples.

EXAMPLE 3   Round 5749 to the nearest hundred.

Solution

5749	5749	5700
↑	↑	↑
place of desired accuracy	Look at one digit to the right; 4 is less than 5.	Leave 7 and fill in zeros.

So, 5749 rounds off to 5700 (to the nearest hundred).

EXAMPLE 4   Round 6500 to the nearest thousand.

Solution

6500	6500	7000
↑	↑	↑
place of desired accuracy	Look at 5; 5 is 5 or greater.	Increase 6 to 7 (one larger) and fill in zeros.

So, 6500 rounds off to 7000 (to the nearest thousand).

EXAMPLE 5   Round 397 to the nearest ten.

Solution

397	397	400
↑	↑	↑
place of desired accuracy	Look at 7; 7 is 5 or greater.	Increase 9 to 10. (This affects two digits, both 3 and 9.)

So, 397 rounds off to 400 (to the nearest ten).

Estimating Answers

One very good use for rounded-off numbers is to **estimate** an answer before any calculations with the given numbers are made. Thus, answers that are not at all reasonable, possibly due to someone pushing a wrong button on a calculator or due to some other large calculation error, can be spotted. Usually, we simply repeat the calculations and the error is found. There are also times, particularly in working word problems, when the wrong operation has been performed. For example, someone may have added when he should have subtracted.

To estimate an answer means to use rounded-off numbers in a calculation to get some idea of what the size of the actual answer should be. This is a form of checking your work before you do it.

To Estimate an Answer:

1. Round off each number to the place of the **leftmost** digit.

2. Perform the indicated operation with these rounded-off numbers.

EXAMPLE 6

Estimate the sum; then find the sum.

$$
\begin{array}{r}
4\,6\,8 \\
9\,3\,6 \\
+\,6\,8\,7 \\
\end{array}
$$

Solution

(a) Estimate the sum first by rounding off each number to the nearest hundred (in this case), and then by adding. In actual practice, many of these steps can be done mentally.

$$
\begin{array}{rcl}
4\,6\,8 & \rightarrow & 5\,0\,0 \\
9\,3\,6 & \rightarrow & 9\,0\,0 \\
+\,6\,8\,7 & \rightarrow & +\,7\,0\,0 \\
\hline
 & & 2\,1\,0\,0 \quad \text{estimated sum}
\end{array}
$$

(b) Now we find the sum with the knowledge that the answer should be close to 2100.

$$
\begin{array}{r}
{\scriptstyle 1\ 2} \\
4\,6\,8 \\
9\,3\,6 \\
+\ \ 6\,8\,7 \\
\hline
2\,0\,9\,1 \quad \leftarrow \text{This sum is quite close to 2100.}
\end{array}
$$

EXAMPLE 7   Estimate the difference; then find the difference.

$$
\begin{array}{r}
758 \\
-\,289 \\
\end{array}
$$

Solution

(a) Estimate the difference first by rounding off each number to the place of the leftmost digit (in this case, the nearest hundred) and then subtracting.

$$
\begin{array}{r}
758 \\
-\,289 \\
\end{array}
\quad
\begin{array}{r}
\rightarrow \\
\rightarrow \\
\end{array}
\quad
\begin{array}{r}
800 \\
-\,300 \\
\hline
500 \\
\end{array}
\quad \text{estimated difference}
$$

(b) Now we find the difference with the knowledge that the answer should be close to 500.

$$
\begin{array}{r}
{}^{\,1}\;6\;\;4\;1 \\
\cancel{7}\,\cancel{5}\,8 \\
-\,2\;8\;9 \\
\hline
4\;6\;9 \\
\end{array}
\quad \leftarrow \text{This difference is close to 500.}
$$

Exercises 1.4

Round off as indicated.

To the nearest ten:

1. 763	**2.** 31	**3.** 82	**4.** 503
5. 296	**6.** 722	**7.** 987	**8.** 347

To the nearest hundred:

9. 4163	**10.** 4475	**11.** 495	**12.** 572
13. 637	**14.** 3789	**15.** 76,523	**16.** 7007

To the nearest thousand:

17. 6912	**18.** 5500	**19.** 7500	**20.** 7499
21. 13,499	**22.** 13,501	**23.** 62,265	**24.** 47,800

To the nearest ten thousand:

25. 78,419	**26.** 125,000	**27.** 256,000	**28.** 62,200
29. 118,200	**30.** 312,500	**31.** 184,900	**32.** 615,000

33. 87 to the nearest hundred   **34.** 46 to the nearest hundred

35. 532 to the nearest thousand

First estimate the answers by using rounded-off numbers; then find the following sums and differences.

36.	83	**37.**	96	**38.**	146	**39.**	475
	62		46		259		126
	+ 78		+ 25		+ 384		+ 572

40.	600	**41.**	851	**42.**	5742	**43.**	483
	542		736		6271		1681
	+ 483		+ 294		8156		3054
					+ 972		+ 4006

44.	22,506	**45.**	8742	**46.**	6421	**47.**	10,531
	38,700		− 3275		− 1652		− 4,600
	+ 10,465						

48.	275,600	**49.**	63,504	**50.**	74,305
	− 94,300		− 42,700		−33,082

51. If the population of the People's Republic of China is 1,165,800,000 and the population of the United States is 255,600,000, about how many more people live in mainland China than live in the United States?

52. In 1990, the University of California campuses had the following enrollments:

UC Berkeley	22,262	UC Riverside	7,310
UC Davis	17,877	UC San Diego	14,392
UC Irvine	13,811	UC Santa Barbara	15,975
UC Los Angeles	24,368	UC Santa Cruz	8,883

Approximately how many students were enrolled in the University of California system in 1990?

53. The perimeter (distance around) of a triangle (a three-sided plane figure) is found by adding the lengths of its sides. Suppose a triangle has sides of length 12 centimeters, 16 centimeters, and 25 centimeters. Without actually adding the numbers, which of the following numbers do you think is closest to the actual perimeter of the triangle:
20 centimeters, 30 centimeters, 60 centimeters, or 100 centimeters?

54. A quadrilateral is a four-sided plane figure. If a quadrilateral has sides of length 8 meters, 9 meters, 10 meters, and 11 meters, what is your estimate of the perimeter of (distance around) the quadrilateral:
20 meters, 40 meters, or 55 meters?

55. If you look at the book list at school and find that the prices of the textbooks that you need to buy this semester are $45, $56, $32, and $17, about how much cash should you take with you to buy all of them: $50, $100, $150, $200, $250? (Don't forget that you will need to pay some sales tax.)

1.5 Multiplication

OBJECTIVES

1. Know the basic multiplication facts (the products of the numbers from 0 to 9).
2. Develop the skill of mentally multiplying numbers involving powers of ten.
3. Be able to multiply whole numbers by using partial products.
4. Be able to multiply whole numbers by using the standard short method.
5. Estimate products.

Basic Multiplication and Powers of Ten

Multiplication is a process that shortens repeated addition with the same number. For example, we can write

$$8 + 8 + 8 + 8 + 8 = 5 \cdot 8 = 40$$

The repeated addend (8) and the number of times it is used (5) are both called **factors** of 40, and the result (40) is called the **product** of the two factors.

$$8 + 8 + 8 + 8 + 8 = 5 \cdot 8 = 40$$
$$\qquad\qquad\qquad\uparrow\quad\uparrow\quad\uparrow$$
$$\qquad\qquad\text{factor factor product}$$

Terms Related to Multiplication with Whole Numbers

Multiplication: Repeated addition

Product: Result of multiplication

Factors: Numbers that are multiplied (Each whole number multiplied is a **factor** of the product.)

Several notations can be used to indicate multiplication. In this text, we will use the raised dot (as with $5 \cdot 8$) and parentheses much of the time. The cross sign ($\times$) will also be used; however, we must be careful not to confuse it with the letter x used in some of the later chapters and in algebra.

Symbols for Multiplication

Symbol		Example
·	raised dot	5 · 8
( )	numbers inside or next to parentheses	5(8) or (5)8 or (5)(8) 5 × 8 or 8
×	cross sign	× 5

To change a multiplication problem to a repeated addition problem every time we are to multiply two numbers would be ridiculous. For example, 48 · 137 would mean using 137 as an addend 48 times. The first step in learning the multiplication process is to **memorize** the basic multiplication facts in Table 1.2. The factors in the table are only the digits 0 through 9. Using other factors involves the place value concept, as we shall see.

Table 1.2		**Basic Multiplication Facts**								
·	0	1	2	3	4	5	6	7	8	9
0	0	0	0	0	0	0	0	0	0	0
1	0	1	2	3	4	5	6	7	8	9
2	0	2	4	6	8	10	12	14	16	18
3	0	3	6	9	12	15	18	21	24	27
4	0	4	8	12	16	20	24	28	32	36
5	0	5	10	15	20	25	30	35	40	45
6	0	6	12	18	24	30	36	42	48	54
7	0	7	14	21	28	35	42	49	56	63
8	0	8	16	24	32	40	48	56	64	72
9	0	9	18	27	36	45	54	63	72	81

If you have difficulty with **any** of the basic facts in the table, write all the possible combinations in a mixed-up order on a sheet of paper. Write the products down as quickly as you can and then find the ones you missed. Practice these in your spare time until you are sure you know them.

The table indicates three properties of multiplication. Since the table is a mirror image of itself on either side of the main diagonal (the numbers 0, 1, 4, 9, 16, 25,

36, 49, 64, 81), multiplication is **commutative.** Multiplication by 0 gives 0, and this result is called the **zero factor law.** Multiplication by 1 gives the number being multiplied, and thus 1 is called the **multiplicative identity** or the **identity element for multiplication.**

Commutative Property of Multiplication

If a and b are whole numbers, then $a \cdot b = b \cdot a$.

EXAMPLE 1

$5 \cdot 3 = 15$ and $3 \cdot 5 = 15$.

Thus, $5 \cdot 3 = 3 \cdot 5$.

EXAMPLE 2

$$\begin{array}{c} 9 \\ \times\,8 \\ \hline 72 \end{array} \qquad \begin{array}{c} 8 \\ \times\,9 \\ \hline 72 \end{array}$$

Thus, $8 \times 9 = 9 \times 8$.

Zero Factor Law

If a is a whole number, then $a \cdot 0 = 0$.

Multiplicative Identity

If a is a whole number, then there is a unique whole number 1 with the property that $a \cdot 1 = a$.

EXAMPLE 3

Zero Factor Law

(a) $0 \cdot 7 = 0$
(b) $9 \cdot 0 = 0$
(c) $83 \cdot 0 = 0$
(d) $0 \cdot 654 = 0$

Multiplicative Identity

(e) $1 \cdot 7 = 7$
(f) $9 \cdot 1 = 9$
(g) $83 \cdot 1 = 83$
(h) $1 \cdot 654 = 654$

The number 1 is called the **multiplicative identity** or **identity element for multiplication.**

If three or more numbers are to be multiplied, we have to decide which two to **group** or **associate** together first. The **associative property of multiplication** indicates that we can multiply different pairs of numbers and arrive at the same product.

> **Associative Property of Multiplication**
>
> If a, b, and c are whole numbers, then $a \cdot b \cdot c = a(b \cdot c) = (a \cdot b)c$.

EXAMPLE 4

$$2 \cdot 3 \cdot 7 \qquad 2 \cdot 3 \cdot 7$$
$$= 6 \cdot 7 \qquad\quad = 2 \cdot 21$$
$$= 42 \qquad\qquad = 42$$

Thus, $(2 \cdot 3)7 = 2(3 \cdot 7)$

EXAMPLE 5

$$9 \cdot 2 \cdot 5 = (9 \cdot 2) \cdot 5 = 18 \cdot 5 = 90$$
$$9 \cdot 2 \cdot 5 = 9 \cdot (2 \cdot 5) = 9 \cdot 10 = 90$$

CLASSROOM PRACTICE

Which property of multiplication is illustrated?

1. $5 \cdot (3 \cdot 7) = (5 \cdot 3) \cdot 7$ **2.** $32 \cdot 0 = 0$ **3.** $6 \cdot 8 = 8 \cdot 6$

ANSWERS: **1.** Associative property **2.** Zero factor law **3.** Commutative property

Powers of 10

A **power** of any number is 1, that number, or that number multiplied by itself one or more times. We will discuss this idea in more detail in Chapter 2. Some of the powers of 10 are: 1, 10, 100, 1000, 10,000, and 100,000, and so on.

$$1$$
$$10$$
$$10 \cdot 10 = 100$$
$$10 \cdot 10 \cdot 10 = 1000$$
$$10 \cdot 10 \cdot 10 \cdot 10 = 10,000$$
$$10 \cdot 10 \cdot 10 \cdot 10 \cdot 10 = 100,000$$

and so on

Multiplication by powers of 10 is useful in explaining multiplication with whole numbers in general, as we will do in Section 1.7. Such multiplication should be done mentally and quickly. The following examples illustrate an important pattern:

$$6 \cdot 1 = 6$$
$$6 \cdot 10 = 60$$
$$6 \cdot 100 = 600$$
$$6 \cdot 1000 = 6000$$

If one of two whole number factors is 1000, the product will be the other factor with three zeros (000) written to the right of it. Two zeros (00) are written to the right of the other factor when multiplying by 100, and one zero (0) is written when multiplying by 10. Will multiplication by one million (1,000,000) result in writing six zeros to the right of the other factor? The answer is yes.

To multiply a whole number

by 10, write 0 to the right;

by 100, write 00 to the right;

by 1000, write 000 to the right;

by 10,000, write 0000 to the right;

and so on.

Many products can be found mentally by using the properties of multiplication and the techniques of multiplying by powers of 10. The processes are written out in the following examples, but they can easily be done mentally with practice.

To Multiply Two or More Whole Numbers that End with 0's:

1. Count the ending 0's.

2. Multiply the numbers without the ending 0's.

3. Write the product from Step 2 with the counted number of 0's at the end.

EXAMPLE 6

(a) $6 \cdot 90 = 6(9 \cdot 10) = (6 \cdot 9)10 = 54 \cdot 10 = 540$

(b) $3 \cdot 400 = 3(4 \cdot 100) = (3 \cdot 4)100 = 12 \cdot 100 = 1200$

(c) $2 \cdot 300 = 2(3 \cdot 100) = (2 \cdot 3)100 = 6 \cdot 100 = 600$

(d) $6 \cdot 700 = 6(7 \cdot 100) = (6 \cdot 7)100 = 42 \cdot 100 = 4200$

(e) $40 \cdot 30 = (4 \cdot 10)(3 \cdot 10) = (4 \cdot 3)(10 \cdot 10) = 12 \cdot 100 = 1200$

(f) $50 \cdot 700 = (5 \cdot 10)(7 \cdot 100) = (5 \cdot 7)(10 \cdot 100) = 35 \cdot 1000 = 35,000$

EXAMPLE 7

$$200 \cdot 800 = (2 \cdot 100)(8 \cdot 100) = (2 \cdot 8)(100 \cdot 100) = 16 \cdot 10,000$$
$$= 160,000$$

EXAMPLE 8

$$7000 \cdot 9000 = (7 \cdot 1000)(9 \cdot 1000) = (7 \cdot 9)(1000 \cdot 1000) = 63 \cdot 1,000,000$$
$$= 63,000,000$$

A whole number that ends with 0's has a power of 10 as a factor. Thus, we can easily find at least two factors of a whole number that ends with one or more 0's.

EXAMPLE 9

$$70 = \underset{\uparrow}{7} \cdot \underset{\uparrow}{10}$$
$$\text{factor} \quad \text{factor}$$

EXAMPLE 10

$$500 = \underset{\uparrow}{5} \cdot \underset{\uparrow}{100}$$
$$\text{factor} \quad \text{factor}$$

EXAMPLE 11

$$46,000 = \underset{\uparrow}{46} \cdot \underset{\uparrow}{1000}$$
$$\text{factor} \quad \text{factor}$$

Multiplication with Whole Numbers

We can use expanded notation and our skills with multiplication by powers of 10 to help in understanding the technique for multiplying two whole numbers. We also need the **distributive property of multiplication over addition** illustrated in the following discussion.

To multiply $3(70 + 2)$, we can add first, then multiply:

$$3(70 + 2) = 3(72) = 216$$

But we can also multiply first, then add, in the following manner:

$$3(70 + 2) = 3 \cdot 70 + 3 \cdot 2$$
$$= 210 + 6 \qquad \text{(210 and 6 are called \textbf{partial products}.)}$$
$$= 216$$

Or, vertically,

$$70 + 2$$
$$\underline{\phantom{70 + {}} 3}$$
$$210 + 6 = 216$$

$$\text{partial products} \qquad \text{product of 72 and 3}$$

We say that 3 is **distributed** over the sum of 70 and 2.

> **Distributive Property of Multiplication Over Addition**
>
> If a, b, and c are whole numbers, then $a(b + c) = a \cdot b + a \cdot c$.

EXAMPLE 12

Multiply $6 \cdot 39$.

$$
\begin{array}{r}
39 \\
\times\ 6 \\
\hline
\end{array}
\qquad
\begin{array}{c}
30 + 9 \\
\qquad\ \ 6 \\
\hline
180 + 54 = 234
\end{array}
\qquad \text{or} \qquad
\begin{array}{r}
39 \\
6 \\
\hline
54 \quad \leftarrow 6 \cdot 9 \\
180 \quad \leftarrow 6 \cdot 30 \\
\hline
234 \quad \leftarrow \text{product}
\end{array}
$$

partial products   product $\Big\}$ partial products

The expanded notation is shown so that you can see that 6 is multiplied by 30 and not just by 3. However, the vertical notation is generally used because it is easier to add the partial products in this form.

EXAMPLE 13

Multiply $37 \cdot 42$.

$$
\begin{array}{r}
42 \\
\times\ 37 \\
\hline
\end{array}
\qquad
\begin{array}{c}
40 + 2 \\
30 + 7 \\
\hline
280 + 14 \\
1200 + \ \ 60 \\
\hline
1200 + 340 + 14 = 1554
\end{array}
\qquad \text{or} \qquad
\begin{array}{rl}
42 & \text{factor} \\
37 & \text{factor} \\
\hline
14 & (7 \cdot 2 = 14) \\
280 & (7 \cdot 40 = 280) \\
60 & (30 \cdot 2 = 60) \\
1200 & (30 \cdot 40 = 1200) \\
\hline
1554 & \text{product}
\end{array}
$$

EXAMPLE 14

Multiply $26 \cdot 276$.

$$
\begin{array}{r}
276 \\
\times\ 26 \\
\hline
\end{array}
\qquad
\begin{array}{c}
200 + \ 70 + \ 6 \\
20 + \ 6 \\
\hline
1200 + 420 + 36 \\
4000 + 1400 + 120 \\
\hline
4000 + 2600 + 540 + 36 = 7176
\end{array}
\qquad \text{or} \qquad
\begin{array}{rl}
276 & \text{factor} \\
26 & \text{factor} \\
\hline
36 & (6 \cdot 6 = 36) \\
420 & (6 \cdot 70 = 420) \\
1200 & (6 \cdot 200 = 1200) \\
120 & (20 \cdot 6 = 120) \\
1400 & (20 \cdot 70 = 1400) \\
4000 & (20 \cdot 200 = 4000) \\
\hline
7176 & \text{product}
\end{array}
$$

There is a shorter and faster method of multiplication used most of the time. In this method:

(a) digits are carried to be added to the next partial product.

(b) sums of some partial products are found mentally.

EXAMPLE 15

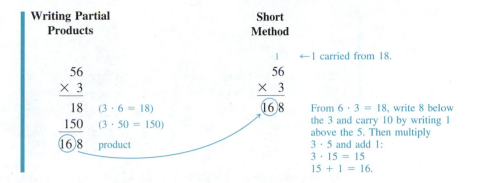

Writing Partial Products

$$
\begin{array}{r}
56 \\
\times\ 3 \\
\hline
18 \\
150 \\
\hline
168
\end{array}
$$

18    $(3 \cdot 6 = 18)$
150    $(3 \cdot 50 = 150)$
16̸8   product

Short Method

1   ←1 carried from 18.

$$
\begin{array}{r}
56 \\
\times\ 3 \\
\hline
168
\end{array}
$$

From $6 \cdot 3 = 18$, write 8 below the 3 and carry 10 by writing 1 above the 5. Then multiply $3 \cdot 5$ and add 1:
$3 \cdot 15 = 15$
$15 + 1 = 16$.

EXAMPLE 16

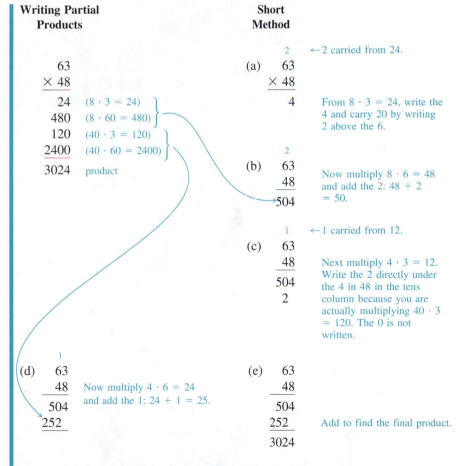

Writing Partial Products

$$
\begin{array}{r}
63 \\
\times\ 48 \\
\hline
24 \\
480 \\
120 \\
2400 \\
\hline
3024
\end{array}
$$

24    $(8 \cdot 3 = 24)$
480    $(8 \cdot 60 = 480)$
120    $(40 \cdot 3 = 120)$
2400 $(40 \cdot 60 = 2400)$
3024   product

Short Method

2   ←2 carried from 24.

(a)
$$
\begin{array}{r}
63 \\
\times\ 48 \\
\hline
4
\end{array}
$$

From $8 \cdot 3 = 24$, write the 4 and carry 20 by writing 2 above the 6.

(b)
2
$$
\begin{array}{r}
63 \\
48 \\
\hline
504
\end{array}
$$

Now multiply $8 \cdot 6 = 48$ and add the 2: $48 + 2 = 50$.

(c)
1   ←1 carried from 12.
$$
\begin{array}{r}
63 \\
48 \\
\hline
504 \\
2
\end{array}
$$

Next multiply $4 \cdot 3 = 12$. Write the 2 directly under the 4 in 48 in the tens column because you are actually multiplying $40 \cdot 3 = 120$. The 0 is not written.

(d)
1
$$
\begin{array}{r}
63 \\
48 \\
\hline
504 \\
252
\end{array}
$$

Now multiply $4 \cdot 6 = 24$ and add the 1: $24 + 1 = 25$.

(e)
$$
\begin{array}{r}
63 \\
48 \\
\hline
504 \\
252 \\
\hline
3024
\end{array}
$$

Add to find the final product.

Steps (a), (b), (c), and (d) are shown for clarification of the Short Method. You will write only step (e) as shown in Example 17.

EXAMPLE 17

4
1
87
26
522   ← 6 · 7 = 42. Write 2, carry 4.   6 · 8 = 48. Add 4: 48 + 4 = 52.
174   ← 2 · 7 = 14. Write 4, carry 1.   2 · 8 = 16. Add 1: 16 + 1 = 17.
2262   ← product

Suppose you want to multiply 2000 by 423. Would you write

```
  2000      or        423
   423              2000
  6000               000
  4000               000
  8000               000
846000               846
                  846000
```

Both answers are correct; neither technique is wrong. However, writing all the 0's is a waste of time, so knowledge about powers of ten is helpful. We can write

```
  423
  2 000
846,000
```

We know 2000 = 2 · 1000, so we are simply multiplying 423 · 2 and then multiplying the result by 1000.

EXAMPLE 18

Multiply 596 · 3000.

```
    596
  3 000
1,788,000
```

EXAMPLE 19

Multiply 265 · 15,000.

```
    265
  15,000
1 325 000
2 65
3,975,000
```

	Find the following products.		
CLASSROOM PRACTICE	**1.** 18	**2.** 300	**3.** 129
	24	500	39

ANSWERS: **1.** 432   **2.** 150,000   **3.** 5031

Estimating Products with Whole Numbers

Estimating the product of two numbers before actually performing the multiplication can be of help in detecting errors. Just as with addition and subtraction, estimations are done with rounded-off numbers.

> **To Estimate a Product:**
>
> **1.** Round off each number to the place of the **leftmost** digit.
>
> **2.** Multiply the rounded-off numbers by using your knowledge of multiplying by powers of 10.

EXAMPLE 20

Find the product $62 \cdot 38$, but first estimate the product by multiplying the numbers in rounded-off form.

Solution

(a)
$$
\begin{array}{rcl}
62 & \rightarrow & 60 \quad \text{rounded-off value of 62} \\
\times\, 38 & \rightarrow & \times\, 40 \quad \text{rounded-off value of 38} \\
\hline
 & & 2400 \quad \text{estimated product}
\end{array}
$$

(b)
$$
\begin{array}{r}
1 \\
62 \\
\times\, 38 \\
\hline
496 \\
186 \\
\hline
2356 \\
\end{array}
$$
← The actual product is close to 2400.

EXAMPLE 21

Use rounded-off values to estimate the product $267 \cdot 93$; then find the product.

Solution

(a) Estimation:
$$
\begin{array}{rcl}
267 & \rightarrow & 300 \\
\times\, 93 & \rightarrow & \times\, 90 \\
\hline
 & & 27{,}000
\end{array}
$$

(b) The product should be near 27,000.

$$\begin{array}{r} 267 \\ \times\ 93 \\ \hline 801 \\ 24\ 03 \\ \hline 24{,}831 \end{array}$$

In this case, the estimated value of 27,000 is considerably larger than 24,831. However, we only expect an estimation to indicate that our answer is reasonable. For example, if we had arrived at an answer of 248,310, then we would know that a mistake had been made because this number is not reasonably close to 27,000.

The Concept of Area

Area is the measure of the interior, or enclosed region, of a plane surface and is measured in **square units.** The concept of area is illustrated in Figure 1.3 in terms of square inches.

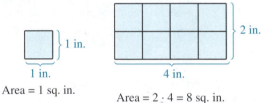

Area = 1 sq. in.

Area = $2 \cdot 4 = 8$ sq. in.

There are 8 squares that are each 1 sq. in. for a total of 8 sq. in.

Figure 1.3

In the metric system, some of the units of area are square meters, square decimeters, square centimeters, and square millimeters. In the U.S. customary system, some of the units of area are square feet, square inches, and square yards.

EXAMPLE 22

The area of a rectangle (measured in square units) is found by multiplying its length by its width. (a) Estimate the area of a rectangular lot for a house if the lot has a length of 123 feet and a width of 86 feet. (b) Find the exact area.

Solution

(a) We round off 123 to 100 and round off 86 to 90 and then multiply $100 \cdot 90$.

$$\begin{array}{r} 100 \\ \times\ \ 90 \\ \hline 9000 \end{array}$$ square feet   estimated area of lot size

86 ft

123 ft

(b) To find the exact area, we multiply 123 · 86.

$$
\begin{array}{r}
123 \\
\times\ \ 86 \\
\hline
738 \\
9\ 84 \\
\hline
\end{array}
$$

10,578 square feet   exact area

Exercises 1.5

Do the following problems mentally and write only the answers.

1. 8 · 9   **2.** 7 · 6   **3.** 6(4)   **4.** 5(0)

5. 1(7)   **6.** 9 · 5   **7.** 0(3)   **8.** (7)(7)

Find each product and tell what property of multiplication is illustrated.

9. 4 · 7 = 7 · 4   **10.** 2(1 · 6) = (2 · 1)6

11. 5 · 1   **12.** 8 · 0

Use the technique of multiplying by powers of ten to find the following products mentally.

13. 25 · 10   **14.** 47 · 1000   **15.** 20 · 200   **16.** 500 · 700

17. 200 · 80   **18.** 40 · 2000   **19.** 30 · 30   **20.** 300 · 600

21. 4000 · 4000   **22.** 900 · 3000   **23.** 10 · 500   **24.** 120 · 300

25.
$$
\begin{array}{r}
6000 \\
\times\ \ \ \ 6 \\
\hline
\end{array}
$$
26.
$$
\begin{array}{r}
90 \\
\times\ \ 90 \\
\hline
\end{array}
$$
27.
$$
\begin{array}{r}
300 \\
\times\ \ 500 \\
\hline
\end{array}
$$
28.
$$
\begin{array}{r}
700 \\
\times\ \ 80 \\
\hline
\end{array}
$$

In the following problems, first (a) estimate each product, and then (b) find the product.

29.
$$
\begin{array}{r}
56 \\
\times\ 4 \\
\hline
\end{array}
$$
30.
$$
\begin{array}{r}
27 \\
\times\ 6 \\
\hline
\end{array}
$$
31.
$$
\begin{array}{r}
48 \\
\times\ 9 \\
\hline
\end{array}
$$
32.
$$
\begin{array}{r}
65 \\
\times\ 5 \\
\hline
\end{array}
$$

33.
$$
\begin{array}{r}
84 \\
\times\ 3 \\
\hline
\end{array}
$$
34.
$$
\begin{array}{r}
95 \\
\times\ 8 \\
\hline
\end{array}
$$
35.
$$
\begin{array}{r}
42 \\
\times\ 56 \\
\hline
\end{array}
$$
36.
$$
\begin{array}{r}
25 \\
\times\ 33 \\
\hline
\end{array}
$$

37.
$$
\begin{array}{r}
48 \\
\times\ 20 \\
\hline
\end{array}
$$
38.
$$
\begin{array}{r}
93 \\
\times\ 30 \\
\hline
\end{array}
$$
39.
$$
\begin{array}{r}
83 \\
\times\ 85 \\
\hline
\end{array}
$$
40.
$$
\begin{array}{r}
96 \\
\times\ 62 \\
\hline
\end{array}
$$

41.
$$
\begin{array}{r}
17 \\
\times\ 32 \\
\hline
\end{array}
$$
42.
$$
\begin{array}{r}
28 \\
\times\ 91 \\
\hline
\end{array}
$$
43.
$$
\begin{array}{r}
20 \\
\times\ 44 \\
\hline
\end{array}
$$
44.
$$
\begin{array}{r}
16 \\
\times\ 26 \\
\hline
\end{array}
$$
45.
$$
\begin{array}{r}
25 \\
\times\ 15 \\
\hline
\end{array}
$$

46.
$$
\begin{array}{r}
93 \\
\times\ 47 \\
\hline
\end{array}
$$
47.
$$
\begin{array}{r}
24 \\
\times\ 86 \\
\hline
\end{array}
$$
48.
$$
\begin{array}{r}
72 \\
\times\ 65 \\
\hline
\end{array}
$$
49.
$$
\begin{array}{r}
12 \\
\times\ 13 \\
\hline
\end{array}
$$
50.
$$
\begin{array}{r}
81 \\
\times\ 36 \\
\hline
\end{array}
$$

51.	126	52.	232	53.	114	54.	72	55.	207
	× 41		× 76		× 25		× 106		× 143

56.	420	57.	200	58.	849	59.	673	60.	192
	× 104		× 49		× 205		× 186		× 467

61. Find the sum of eighty-four and one hundred forty-seven. Find the difference between ninety-six and thirty-eight. Then find the product of the sum and the difference.

62. Find the sum of thirty-four and fifty-five. Find the sum of one hundred seventeen and two hundred twenty. Find the product of the two sums.

63. If your beginning salary is $2300 per month and you are to get a raise once a year of $230 per month, by how much will your pay increase from year to year? How much money will you make over a five-year period?

64. If you rent an apartment with three bedrooms for $720 per month and you know that the rent will increase $40 per month every 12 months, what will you pay in rent during the first three years of living in this apartment? By how much does the total rent increase each year?

65. Your company bought 18 new cars, each with air conditioning and antilock brakes, at a price of $12,800 per car. How much did your company pay for these cars?

In Exercises 66–70, first make an estimate of the answers. Then, after you have performed the actual calculation, check to see that your estimate and your exact answer are reasonably close.

66. A rectangular lot for a house measures 208 feet long by 175 feet wide. Find the area of the lot in square feet. (Area of a rectangle is found by multiplying the length times the width.)

67. Find the number of square meters of area for a rectangular swimming pool and deck if the plans call for a length of 215 meters and a width of 82 meters. (Area of a rectangle is found by multiplying the length times the width.)

68. Network television has 12 minutes of commercial time in each hour. How many minutes of commercial time does a network have in a one-day programming schedule of 20 hours? In one week?

69. According to the U.S. Fish and Wildlife Service, about 800,000 live birds (including about 250,000 parrots) are imported each year at a value of about $19 per bird. What is the total value of these imported birds?

70. According to the Department of Transportation, U.S. citizens drove some sort of vehicle almost 6000 miles per person in 1990. If the 1990 census population is 248,709,873, how many miles did all of the U.S. citizens drive?

Check Your Number Sense

71. You want to pay cash for your textbooks at the bookstore because you know that the cash line is shorter and moves faster than the other lines. If you are going to buy four books and their prices are between $35 and $45 each, about how much cash do you need to take with you: $100, $200, $300, or $400? (Don't forget sales tax will be involved.)

72. Match each indicated product with the closest estimate of that product. Perform any calculations mentally.

Product	Estimate
____ (a) 16 · 18	A. 210
____ (b) 6(78)	B. 300
____ (c) 11 · 32	C. 400
____ (d) (8)(69)	D. 480
____ (e) 25(7)	E. 560

73. Match each indicated product with the closest estimate of that product. Perform any calculations mentally.

Product	Estimate
____ (a) 37(500)	A. 200
____ (b) 37(50)	B. 2000
____ (c) 37(5)	C. 20,000
____ (d) 37(5000)	D. 200,000

Writing and Thinking about Mathematics

74. Write, in your own words, the meaning of the zero factor law.

75. Explain, in your own words, why 1 is called the multiplicative identity.

the # keeps its identity when multiplied by 1.

1.6   Division

OBJECTIVES

1. Know the terms related to division: dividend, divisor, quotient, remainder.
2. Be able to divide with whole numbers.
3. Understand that when the remainder is O, both the divisor and quotient are factors of the dividend.
4. Estimate quotients.
5. Know that division by O is not possible.

Division with Whole Numbers

We know that $6 \cdot 10 = 60$ and that 6 and 10 are **factors** of 60. They are also called **divisors.** The process of division can be thought of as the reverse of multiplication. In division, we want to find how many times one number is contained in another. How many 6's are in 60? There are 10 sixes in 60, and we say that 60 **divided by** 6 is 10 (or $60 \div 6 = 10$).

The relationship between multiplication and division can be seen from the following table format:

				Division					Multiplication	
Dividend		**Divisor**		**Quotient**				**Factors**		**Product**
21	÷	7	=	3	since			$7 \cdot 3$	=	21
24	÷	6	=	4	since			$6 \cdot 4$	=	24
36	÷	4	=	9	since			$4 \cdot 9$	=	36
12	÷	2	=	6	since			$2 \cdot 6$	=	12

This table indicates that the number being divided is called the **dividend;** the number doing the dividing is called the **divisor;** and the result of division is called the **quotient.**

Division does not always involve factors (or exact divisors). Suppose we want to divide 23 by 4. By using repeated subtraction, we can find how many 4's are in 23. We continuously subtract 4 until the number left (called the **remainder**) is less than 4.

$$
\begin{array}{ccccc}
23 & 19 & 15 & 11 & 7 \\
-\ 4 & -\ 4 & -\ 4 & -\ 4 & -\ 4 \\
\hline
19 & 15 & 11 & 7 & 3 \leftarrow \text{remainder}
\end{array}
$$

(subtraction 5 times)

Another form used to indicate division as repeated subtraction is shown in Examples 1 and 2.

EXAMPLE 1 How many 7's are in 185?

$$
\begin{array}{r}
7\,)\overline{\,185} \\
-140 \quad \leftarrow \text{subtract} \quad 20 \text{ sevens} \\
\hline
45 \\
-42 \quad \leftarrow \text{subtract} \quad \underline{6} \text{ sevens} \\
\hline
3 \qquad\qquad 26 \text{ sevens total}
\end{array}
$$

$(20 \cdot 7 = 140)$
$(30 \cdot 7 = 210;$ too much, since 210 is greater than 185$)$
$(6 \cdot 7 = 42)$

↑ remainder   ↑ quotient

Check

Division is checked by multiplying the quotient and the divisor and then adding the remainder. The result should be the dividend.

$$
\begin{array}{r}
26 \quad \text{quotient} \\
\times\ 7 \quad \text{divisor} \\
\hline
182
\end{array}
\qquad
\begin{array}{r}
182 \\
+\quad 3 \quad \text{remainder} \\
\hline
185 \quad \text{dividend}
\end{array}
$$

EXAMPLE 2

Find $275 \div 6$ using repeated subtraction, and check your work.

$$
\begin{array}{r}
6\,)\overline{275} \\
-180 \\
\hline
95 \\
-60 \\
\hline
35 \\
-18 \\
\hline
17 \\
-12 \\
\hline
5
\end{array}
$$

Subtract   30   sixes. $(30 \cdot 6 = 180)$

Subtract   10   sixes. $(10 \cdot 6 = 60)$

Subtract   3   sixes. $(3 \cdot 6 = 18)$

Subtract   2   sixes. $(2 \cdot 6 = 12)$

Subtract   45   sixes total

↑   ↑
remainder   quotient

Note: You can subtract any number of sixes less than the quotient. But this will not lead to a good explanation of the shorter division algorithm. In general, at each step you should subtract the largest number of thousands, hundreds, tens, or units of the divisor that you can.

Repeated subtraction provides a good basis for understanding the **division algorithm,*** a much shorter method of division.

EXAMPLE 3

Find $2076 \div 8$.

STEP 1:
$$
\begin{array}{r}
2 \\
8\,)\overline{2076} \\
-1600 \\
\hline
476
\end{array}
$$

← Write 2 in hundreds position.

200 eights $(200 \cdot 8 = 1600)$

---

* An algorithm is a process or pattern of steps to be followed in working with numbers.

$$\begin{array}{r} 25 \\ 8\,\overline{)\,2076} \\ -1600 \\ \hline 476 \\ -400 \\ \hline 76 \end{array}$$

STEP 2:   Write 5 in tens position.

50 eights ($50 \cdot 8 = 400$)

$$\begin{array}{r} 259 \\ 8\,\overline{)\,2076} \\ -1600 \\ \hline 476 \\ -400 \\ \hline 76 \\ -72 \\ \hline 4 \end{array}$$

STEP 3:   Write 9 in units position.

9 eights ($9 \cdot 8 = 72$)

Summary

The process can be shortened by not writing all the 0's and "bringing down" only one digit at a time.

$$\begin{array}{r} 259 \text{ R4} \\ 8\,\overline{)\,2076} \\ 16 \\ \hline 47 \\ 40 \\ \hline 76 \\ 72 \\ \hline 4 \end{array}$$

← "Bring down" the 7 only; then divide 8 into 47.

EXAMPLE 4 $746 \div 32$

STEP 1:
$$\begin{array}{r} 2 \\ 32\,\overline{)\,746} \\ 64 \\ \hline 10 \end{array}$$

Trial divide 30 into 70 or 3 into 7, giving 2 in the tens position. Note that 10 is less than 32.

STEP 2:
$$\begin{array}{r} 23 \text{ R10} \\ 32\,\overline{)\,746} \\ 64 \\ \hline 106 \\ 96 \\ \hline 10 \end{array}$$

Trial divide 30 into 100 or 3 into 10, giving 3 in the units position.

Check

$$
\begin{array}{r}
23 \\
\times\ 32 \\
\hline
46 \\
69 \\
\hline
736
\end{array}
\qquad
\begin{array}{r}
736 \\
+\ 10 \\
\hline
746
\end{array}
$$

EXAMPLE 5 $9325 \div 45$

STEP 1:
$$
\begin{array}{r}
2 \\
45\,\overline{)\,9325} \\
90 \\
\hline
3
\end{array}
$$
Trial divide 40 into 90 or 4 into 9, giving 2 in the hundreds position.

STEP 2:
$$
\begin{array}{r}
20 \\
45\,\overline{)\,9325} \\
90 \\
\hline
32 \\
0 \\
\hline
\end{array}
$$
45 divides into 32 0 times. So, write 0 in the tens column and multiply $0 \cdot 45 = 0$.

STEP 3:
$$
\begin{array}{r}
208 \\
45\,\overline{)\,9325} \\
90 \\
\hline
32 \\
0 \\
\hline
325 \\
360 \\
\hline
\end{array}
$$
Trial divide 45 into 325 or 4 into 32. But the trial quotient 8 is too large since $8 \cdot 45 = 360$ and 360 is larger than 325.

$$
\begin{array}{r}
207\ \text{R}10 \\
45\,\overline{)\,9325} \\
90 \\
\hline
32 \\
0 \\
\hline
325 \\
315 \\
\hline
10
\end{array}
$$
Now the trial divisor is 7. Since $7 \cdot 45 = 315$ and 315 is smaller than 325, 7 is the desired number.

Check

$$
\begin{array}{r}
207 \\
\times\ 45 \\
\hline
1035 \\
828 \\
\hline
9315
\end{array}
\qquad
\begin{array}{r}
9315 \\
+\ 10 \\
\hline
9325
\end{array}
$$

Common Error

In Step 2 of Example 5, we wrote 0 in the quotient because 45 did not divide into 32. Many students fail to write the 0. This error obviously changes the value of the quotient.

Wrong	Right

$$\begin{array}{r} 48 \\ 17\overline{)6938} \\ 68 \\ \hline 138 \\ 136 \\ \hline 2 \end{array}$$ remainder

$$\begin{array}{r} 408 \\ 17\overline{)6938} \\ 68 \\ \hline 13 \\ 0 \\ \hline 138 \\ 136 \\ \hline 2 \end{array}$$ remainder

Writing the 0 in the quotient gives the correct quotient of 408, which is considerably different from 48.

If the remainder is 0, then the following statements are true.

1. Both the divisor and quotient are **factors** of the dividend.

2. We say that both factors **divide exactly** into the dividend.

3. Both factors are called **divisors** of the dividend.

EXAMPLE 6

Show that 17 and 36 are factors (or divisors) of 612.

Solution

$$\begin{array}{r} 36 \\ 17\overline{)612} \\ 51 \\ \hline 102 \\ 102 \\ \hline 0 \end{array}$$

Since 0 is the remainder, both 17 and 36 are factors (or divisors) of 612. Just to double-check, we find 612 ÷ 36.

$$\begin{array}{r} 17 \\ 36\overline{)612} \\ \underline{36} \\ 252 \\ \underline{252} \\ 0 \end{array}$$

Yes, both 17 and 36 divide exactly into 612.

EXAMPLE 7

A plumber purchased 17 special pipe fittings. What was the price of one fitting if the bill was $544 before taxes?

Solution

We need to know how many times 17 goes into 544.

$$\begin{array}{r} 32 \\ 17\overline{)544} \\ \underline{51} \\ 34 \\ \underline{34} \\ 0 \end{array}$$

The price for one fitting was $32.

CLASSROOM PRACTICE

Find the quotient and remainder for each of the following problems.

1. 325 ÷ 7   **2.** 16$\overline{)324}$ **3.** 41$\overline{)24682}$

ANSWERS: **1.** 46 R3   **2.** 20 R4   **3.** 602 R0

Estimating Quotients with Whole Numbers

By rounding off both divisor and dividend, we can estimate the quotient. Estimation can help identify unreasonable answers when the actual value is calculated.

> **To Estimate a Quotient:**
>
> **1.** Round off both the divisor and dividend to the place of the last digit on the left.
>
> **2.** Divide with the rounded-off numbers. (This process is very similar to the trial dividing step in the division algorithm.)

EXAMPLE 8   Estimate the quotient $325 \div 42$ by using rounded-off values; then find the quotient.

Solution

(a) Estimation: $325 \div 42$ $\rightarrow$ $300 \div 40$

$$
\begin{array}{r}
7 \quad \text{estimated quotient} \\
40\,)\overline{300} \\
\underline{280} \\
20
\end{array}
$$

(b) The quotient should be near 7.

$$
\begin{array}{r}
7 \quad \text{quotient} \\
42\,)\overline{325} \\
\underline{294} \\
31 \quad \text{remainder}
\end{array}
$$

In this case, the quotient is the same as the estimated value. The true remainder is different.

EXAMPLE 9   First estimate the quotient $48,062 \div 26$; then find the quotient.

Solution

(a) Estimation: $48,062 \div 26$ $\rightarrow$ $50,000 \div 30$

$$
\begin{array}{r}
1,666 \quad \text{approximate quotient} \\
30\,)\overline{50,000} \\
\underline{30} \\
20\ 0 \\
\underline{18\ 0} \\
2\ 00 \\
\underline{1\ 80} \\
200 \\
\underline{180} \\
20
\end{array}
$$

(b) The quotient should be near 1600 or 1700.

$$
\begin{array}{r}
1{,}848 \\
26\,\overline{)\,48{,}062} \\
26\phantom{{,}062} \\
\overline{22\ 0} \\
20\ 8 \\
\overline{1\ 26} \\
1\ 04 \\
\overline{222} \\
208 \\
\overline{14} \quad \text{remainder}
\end{array}
$$

An Adjustment to the Process of Estimating Answers

The rule for rounding off to the leftmost digit for estimating an answer is flexible. We sometimes adjust the estimating process, because there are times when using two digits gives simpler calculations or more accurate estimates. In Example 9, for instance 3 can be seen to divide into 48, so we use two-digit rounding to make the estimate. (See the discussion below.) Thus, estimating answers does involve some basic understanding and intuitive judgment, and there is no one best way to estimate answers. However, "adjustment" to the leftmost digit rule is more applicable to division than it is to addition, subtraction, or multiplication.

As an example of another technique for estimating answers, we could have proceeded as follows in Example 9, part (a):

Estimation: $48{,}062 \div 26 \quad \rightarrow \quad 48{,}000 \div 30$

Use 48,000 instead of 50,000 because 3 divides evenly into 48. Thus, the estimate could be

$$
\begin{array}{r}
1{,}600 \quad \text{approximate quotient} \\
30\,\overline{)\,48{,}000} \\
30\phantom{{,}000} \\
\overline{18\ 0} \\
18\ 0 \\
\overline{00} \\
0 \\
\overline{00} \\
0 \\
\overline{0}
\end{array}
$$

For completeness, we close this section with two rules about division involving 0. These rules will be discussed in detail in Chapter 3.

Division with 0

1. If a is any nonzero whole number, then

$0 \div a = 0.$

2. If a is any whole number, then

$a \div 0$ is **undefined.**

Exercises 1.6

Please read the text and study the examples before working these exercises.

Find the quotient and remainder for each of the following problems by using the method of repeated subtraction.

1. $240 \div 6$	**2.** $189 \div 9$	**3.** $210 \div 7$	**4.** $140 \div 14$
5. $168 \div 8$	**6.** $70 \div 5$	**7.** $132 \div 11$	**8.** $120 \div 4$
9. $75 \div 15$	**10.** $51 \div 3$	**11.** $52 \div 8$	**12.** $44 \div 6$
13. $600 \div 25$	**14.** $413 \div 20$	**15.** $161 \div 15$	**16.** $182 \div 13$
17. $150 \div 13$	**18.** $500 \div 14$	**19.** $205 \div 5$	**20.** $321 \div 7$

Divide and check using the division algorithm.

21. $6\,)\overline{32}$	**22.** $7\,)\overline{17}$	**23.** $4\,)\overline{25}$	**24.** $5\,)\overline{35}$
25. $8\,)\overline{48}$	**26.** $6\,)\overline{72}$	**27.** $9\,)\overline{81}$	**28.** $2\,)\overline{76}$
29. $3\,)\overline{98}$	**30.** $14\,)\overline{52}$	**31.** $12\,)\overline{108}$	**32.** $11\,)\overline{424}$
33. $16\,)\overline{128}$	**34.** $20\,)\overline{305}$	**35.** $18\,)\overline{206}$	**36.** $30\,)\overline{847}$
37. $10\,)\overline{423}$	**38.** $15\,)\overline{750}$	**39.** $13\,)\overline{260}$	**40.** $17\,)\overline{340}$
41. $12\,)\overline{360}$	**42.** $19\,)\overline{7603}$	**43.** $16\,)\overline{4813}$	**44.** $11\,)\overline{4406}$
45. $13\,)\overline{3917}$	**46.** $73\,)\overline{148}$	**47.** $68\,)\overline{207}$	**48.** $49\,)\overline{993}$
49. $50\,)\overline{3065}$	**50.** $40\,)\overline{2163}$	**51.** $105\,)\overline{210}$	**52.** $116\,)\overline{232}$
53. $213\,)\overline{4760}$	**54.** $716\,)\overline{3056}$	**55.** $630\,)\overline{4768}$	**56.** $414\,)\overline{83276}$
57. $502\,)\overline{98762}$	**58.** $317\,)\overline{70365}$	**59.** $471\,)\overline{50612}$	**60.** $215\,)\overline{64930}$

61. Suppose your income for one year was $30,576 and your income was the same each month. (a) Estimate your monthly income. (b) What was your exact monthly income?   a) $30,000 \div 12 = 2500$
$30,000 \div 10 = 3000$

62. A high school bought 3,075 new textbooks for a total price of $116,850. (a) What was the approximate price of each text? (b) What was the exact price of each text? $a) 100,000 \div 3000 \approx 33$

63. Show that 28 and 36 are both factors of 1008 by using division.

64. Show that 45 and 702 are both factors of 31,590 by using division.

65. The United States has a population of about 248,400,000 people and has a land area of about 3,600,000 square miles. About how many people are there for each square mile? (Do you think that the population per square mile is the same for New York City or for Chicago or for the state of New Mexico as it is for the entire United States? Raw numbers do not always give a true-to-life picture when taken out of context. In many applications, some judgment and numerical skills must be used as well.)

Check Your Number Sense

66. Match each indicated quotient with the closest estimate of that quotient. Perform as many calculations mentally as you can.

Quotient	Estimate
____ (a) $9\overline{)910}$	A. 6
____ (b) $34 \div 5$	B. 10
____ (c) $34\overline{)12,000}$	C. 100
____ (d) $18\overline{)3900}$	D. 200
____ (e) $216 \div 18$	E. 300

67. Match each indicated quotient with the closest estimate of that quotient.

Quotient	Estimate
____ (a) $3\overline{)870}$	A. 30
____ (b) $3\overline{)87,000}$	B. 300
____ (c) $3\overline{)87}$	C. 3000
____ (d) $3\overline{)8700}$	D. 30,000

Writing and Thinking about Mathematics

68. Nothing is said in the text about division being a commutative operation. Do you think that there might be a commutative property for division? Give several examples that help justify your answer.

69. There is no mention in the text of an associative property for division. Do you think that division is associative? Give several examples that help justify your answer.   use: $120 \div 10 \div 2$

Collaborative Learning Exercise

70. Separate the class into teams of two to four students. Each team is to list 10 jobs or work situations in which arithmetic skills are a necessary part of the work. After the lists are complete, each team leader is to read the team's list with classroom discussion to follow.

1.7   Problem Solving with Whole Numbers

OBJECTIVES

1. Develop the reading skills needed for understanding word problems.
2. Be able to analyze a word problem with confidence.
3. Realize that problem solving takes time and that the learning process also involves learning from making mistakes.

Strategy for Solving Word Problems

In this section, problem solving can involve various combinations of the four operations of addition, subtraction, multiplication, and division. The decisions on what operations are to be used are based generally on experience and practice as well as on certain key words.

> **Note:** If you are not exactly sure just what operations to use, at least try something. Even by making errors in technique or judgment you are learning what does not work. If you do nothing, then you learn nothing. Do not be embarrassed by mistakes.

The problems discussed here will come under the following headings: Consumer Items, Checking Account, Geometry, and Average. The steps in the basic strategy listed here will help give an organized approach to problem solving regardless of the type of problem. These steps were developed by George Pōlya, a famous educator and mathematician from Stanford University.

Basic Strategy for Solving Word Problems

1. Read each problem carefully until you understand the problem and know what is being asked for.

2. Draw any type of figure or diagram that might be helpful and decide what operations are needed.

> **3.** Perform these operations.
>
> **4.** Mentally check to see if your answer is reasonable. See if you can think of another more efficient or more interesting way to do the same problem.

Applications Examples

EXAMPLE 1

Consumer Items

Bill bought a car for $9000. The salesman added $540 for taxes and $150 for license fees. If Bill made a down payment of $3500 and financed the rest with his credit union, how much did he finance?

Solution

Find the total cost and subtract the down payment.

$$\begin{array}{r} \$9000 \\ 540 \\ +\ \ 150 \\ \hline \$9690 \end{array} \text{ total cost}$$

$$\begin{array}{r} \$9690 \\ -\ 3500 \\ \hline \$6190 \end{array} \begin{array}{l} \text{total cost} \\ \text{down payment} \\ \text{to be financed} \end{array}$$

Bill financed $6190.

(Mental checking is difficult here. But you might think that if he paid about $10,000 and put down $3500, an answer around $6500 is reasonable.)

EXAMPLE 2

Checking Account

In July, Ms. Smith opened a checking account and deposited $1528. She wrote checks for $132, $425, $196, and $350. What was her balance at the end of July?

Solution

To find the balance, find the sum of the checks and subtract the sum from $1528.

$$\begin{array}{r} \overset{2\ 1}{} \\ \$\ 132 \\ 425 \\ 196 \\ +\ \ 350 \\ \hline \$1103 \end{array} \text{ total checks}$$

$$\begin{array}{r} \$1528 \\ -\ 1103 \\ \hline \$\ 425 \end{array} \text{ balance}$$

The balance was $425 at the end of July.

(Mentally round off each check to hundreds to see if the answer is reasonable.)

EXAMPLE 3 | **Geometry**

A triangle has three sides: a base of 3 feet, a height of 4 feet, and a third side of 5 feet. This triangle is called a right triangle because one angle is 90°. (a) Find the perimeter of (distance around) the triangle. (b) Find the area of the triangle in square feet. (To find the area, multiply base times height and then divide by 2.)

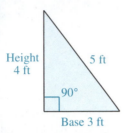

Solution

(a) To find the perimeter, add the lengths of the three sides.

$$\begin{array}{r} 3 \\ 4 \\ + 5 \\ \hline 12 \end{array} \quad \text{feet} \quad \text{(perimeter)}$$

(b) To find the area, multiply the base times the height and divide by 2.

$$\begin{array}{r} 3 \\ \times 4 \\ \hline 12 \end{array} \qquad \begin{array}{r} 6 \\ 2\overline{)12} \\ \underline{12} \\ 0 \end{array} \quad \text{square feet} \quad \text{(area)}$$

The perimeter is 12 feet and the area is 6 square feet.

Average

A topic closely related to addition and division is **average.** Your grade in this course may be based on the average of your exam scores. Newspapers and magazines have information about the Dow Jones stock averages, the average income of American families, the average life expectancy of laboratory rats, and so on. The average of a set of numbers is a kind of "middle number" of the set.* **The average of a set of numbers can be defined as the number found by adding the numbers in the set, then dividing this sum by the number of numbers in the set.** This average is also called the **arithmetic average,** or **mean.**

---

* Such terms as the **average citizen** or **average voter** are not related to numbers and are not as easily defined as the average of a set of numbers.

The average of a set of whole numbers need not be a whole number. However, in this section, the problems will be set up so that the averages will be whole numbers. Other cases will be discussed later in the chapters on fractions and decimals (Chapters 3 and 5).

EXAMPLE 4

Find the average of the three numbers 32, 47, 23.

Solution

$$
\begin{array}{r}
34 \quad \text{(average)} \\
3\,)\,\overline{102} \\
9 \\
\hline
12 \\
12 \\
\hline
0
\end{array}
$$

$$
\begin{array}{r}
32 \\
47 \\
+\ 23 \\
\hline
102
\end{array}
$$

The sum, 102, is divided by 3 because there are three numbers being added.

EXAMPLE 5

Suppose five people had the following incomes for one year: $8,000; $9,000; $10,000; $11,000; $12,000. Find their average income.

Solution

$$
\begin{array}{r}
\$\ 8,000 \\
9,000 \\
10,000 \\
11,000 \\
+\ \ 12,000 \\
\hline
\$50,000
\end{array}
$$

$$
\begin{array}{r}
\$10,000 \\
5\,)\,\overline{50,000} \\
50,000 \\
\hline
0
\end{array}
$$

The average income is $10,000.

EXAMPLE 6

Suppose five people had the following incomes for one year: $1,000; $1,000; $1,000; $1,000; $46,000. Find their average income.

Solution

$$
\begin{array}{r}
\$\ 1,000 \\
1,000 \\
1,000 \\
1,000 \\
+\ \ 46,000 \\
\hline
\$50,000
\end{array}
$$

$$
\begin{array}{r}
\$10,000 \\
5\,)\,\overline{50,000} \\
50,000 \\
\hline
0
\end{array}
$$

The average income is $10,000.

The average of a set of numbers can be very useful, but it can also be misleading. Judging the importance of an average is up to you, the reader of the information.

In Example 5, the average of $10,000 serves well as a "middle score" or "representative" of all the incomes. However, in Example 6, none of the incomes is even close to $10,000. The one large income completely destroys the "representativeness" of the average. Thus, it is useful to see the numbers or at least know something about them before attaching too much importance to an average.

EXAMPLE 7

On an English exam, two students scored 95 points, five students scored 86 points, one student scored 82 points, one student scored 78 points, and six students scored 75 points. What was the mean score of the class?

Solution

95	86	82	78	75
× 2	× 5	× 1	× 1	× 6
190	430	82	78	450

We have multiplied rather than write down all 15 scores. So, when the five products are added, we will divide the sum by 15 because the sum represents 15 scores.

$$
\begin{array}{r}
190 \\
430 \\
82 \\
78 \\
+\ 450 \\
\hline
1230
\end{array}
\qquad
\begin{array}{r}
82 \quad \text{mean score}\\
15\,\overline{)\,1230} \\
120 \\
\hline
30 \\
30 \\
\hline
\end{array}
$$

The class mean is 82 points.

Exercises 1.7

Consumer Items

1. To purchase a new refrigerator for $1200 including tax, Mr. Kline paid $240 down and the remainder in six equal monthly payments. What were his monthly payments?

2. Miguel decided to go shopping for school clothes before college started in the fall. How much did he spend if he bought four pairs of pants for $21 a pair, five shirts for $18 each, three pairs of socks for $4 a pair, and two pairs of shoes for $38 a pair?

3. To purchase a new dining room set for $1200, Mrs. Steel had to pay $72 in sales tax. If she made a deposit of $486, how much did she still owe?

4. Vijay wanted to buy a new car. He could buy a red one for $8500 plus $510 in sales tax and $135 in fees; or he could buy a blue one for $8700 plus $522 in sales tax and $140 in fees. If the manufacturer was giving a $250 rebate on the blue model, which car would be cheaper for Vijay? How much cheaper?

5. Lynn decided to take up surfing. She bought a new surfboard for $675, a wet suit for $130, a beach towel for $12, and a new swimsuit for $57. How much money did she spend? (Sales tax was included in the prices.)

6. Pat needed art supplies for a new course at the local community college. She bought a portfolio for $32, a zinc plate for $44, etching ink for $12, and three sheets of rag paper totaling $6. She received a student discount of $9. What did she spend on art supplies?

Checking Account

7. If you opened a checking account with $875, then wrote checks for $20, $35, $115, $8, and $212, what would be your balance?

8. Your friend has a checking balance of $1250 and writes checks for $375, $52, $83, and $246. What is her balance?

9. On August 1, Matt had a balance of $250. During August, he made deposits of $200, $350, and $236. He wrote checks for $487, $25, $33, and $175. What was his balance on September 1?

10. Melissa deposited $500, $2470, $800, $3562, and $2875 in her checking account during a five-month period. She wrote checks totaling $6742. If her beginning balance was $1400, what was her balance at the end of the five months?

Geometry

11. A rectangle is a four-sided figure with opposite sides equal and all four angles equal. (Each angle is 90°.) (a) Find the perimeter of a rectangle that has a width of 15 meters and a length of 37 meters. (b) Also find its area (multiply length times width) in square meters.

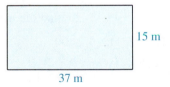

15 m

37 m

12. A regular hexagon is a six-sided figure with all six sides equal and all six angles equal. Find the perimeter of a regular hexagon with one side 19 centimeters.

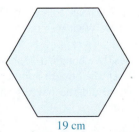

19 cm

13. An isosceles triangle (two sides equal) is placed on top of a square to form a window as shown in the figure below. If each of the two equal sides of the triangle is 18 inches and the square is 28 inches on a side, what is the perimeter of the window?

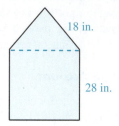

18 in.

28 in.

14. A rectangular picture is mounted in a rectangular frame with a border (called a mat). If the picture is 12 inches by 18 inches and the frame is 16 inches by 24 inches, what is the area of the mat? (Area of a rectangle is length times width.)

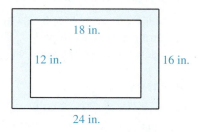

18 in.

12 in.   16 in.

24 in.

Average

In Exercises 15–20, find the average (or mean) of each set of numbers.

15. 102, 113, 97, 100

16. 56, 64, 38, 58

17. 6, 8, 7, 4, 4, 5, 6, 8

18. 5, 4, 5, 6, 5, 8, 9, 6

19. 512, 618, 332, 478

20. 436, 520, 630, 422

21. On an exam in history, two students scored 95, six students scored 90, three students scored 80, and one student scored 50. What was the class average?

22. A salesman sold items from his sales list for $972, $834, $1005, $1050, and $799. What was the average price per item?

23. Ms. Lee bought 100 shares of stock in MicroSoft at $14 per share. Two months later she bought another 200 shares at $20 per share. What average price per share did she pay? If she sold all 300 shares at $22 per share, what was her profit?

24. Three families, each with two children, had incomes of $15,942. Two families, each with four children, had incomes of $18,512. Four families, each with two children, had incomes of $31,111. One family had no children and an income of $20,016. What was the mean income per family?

25. During July, Mr. Rodriguez made deposits in his checking account of $400 and $750 and wrote checks totaling $625. During August, his deposits were $632, $322, and $798, and his checks totaled $978. In September, the deposits were $520, $436, $200, and $376; the checks totaled $836. (a) What was the average monthly difference between his deposits and his withdrawals? (b) What was his bank balance at the end of September if he had a balance of $500 on July 1?

26. In one month (30 days), an airline pilot spent the following number of hours preparing for and flying each of 12 flights: 6, 8, 9, 6, 7, 7, 7, 5, 6, 6, 6, and 11 hours. What was the average (mean) amount of time that the pilot spent per flight?

27. The ten largest cities in South Carolina have the following approximate populations:

Columbia	98,000	Sumter	41,900
Charleston	80,400	Rock Hill	41,600
North Charleston	70,200	Mount Pleasant Town	30,100
Greenville	58,300	Florence	29,800
Spartanburg	44,000	Anderson	26,200

What is the average population of these cities?

28. The five longest rivers in the world are the:

Nile (Africa)	4180 miles
Amazon (South America)	3900 miles
Mississippi-Missouri-Red Rock (North America)	3880 miles
Yangtze (China)	3600 miles
Ob (Russia)	3460 miles

What is the mean length of these rivers?

29. The following list is repeated from Mathematics at Work! at the beginning of this chapter. Tuition, room, and board are given as annual figures (rounded to the nearest ten dollars) for 1992.

| Institution | Enrollment | Tuition | | Room/Board |
		Resident	Nonresident	
Univ. of Alabama	15,940	$ 2,010	$ 5,020	$3,300
Bryn Mawr College	1,180	16,170	16,170	6,150
Univ. of Colorado	10,410	1,970	9,900	3,540
Cornell University	7,600	16,190	16,190	5,410
Duke University	6,020	16,120	16,120	5,240
Harvard and Radcliffe Colleges	6,620	17,670	17,670	5,840
Univ. of Illinois	26,370	3,060	6,800	3,900
Indiana University	25,310	2,370	6,900	3,370
Univ. of Miami	8,640	15,050	15,050	5,910
Univ. of Notre Dame	7,520	13,500	13,500	3,600
Stanford University	6,510	16,540	16,540	6,310
Univ. of Texas	37,000	1,100	4,220	3,400
UCLA	24,370	2,900	10,600	5,410
Vassar College	2,310	17,210	17,210	5,500
Yale University	5,150	17,500	17,500	6,200

For these universities and colleges, find (a) the average enrollment, (b) the average tuition for residents, (c) the average tuition for nonresidents, and (d) the average cost of room and board.

Writing and Thinking about Mathematics

30. If the product of 607 and 93 is divided by 3, what is the quotient? What is the remainder? Is 3 a factor of the product? Explain briefly.

31. If the difference between 347 and 196 is multiplied by 15, what is the product? Is 5 a factor of this product? Explain briefly.

32. In a short paragraph, with complete sentences, discuss one topic in Chapter 1 you found particularly interesting and describe why.

Summary: Chapter 1

Key Terms and Ideas

The **decimal system** is a place value system that depends on three things:

1. the **ten digits:** 0, 1, 2, 3, 4, 5, 6, 7, 8, 9

2. the **placement** of each digit

3. the **value** of each place

Whole numbers are those numbers used for counting and 0.

$$W = \{0, 1, 2, 3, 4, 5, 6, 7, 8, 9, 10, 11, 12, \ldots\}$$

Numbers being added are called **addends,** and the result of the addition is called the **sum.**

Subtraction is a reverse addition, and the missing addend is called the **difference** between the sum and one addend.

Rounding off a given number means to find another number close to the given number.

Estimating an answer means to use rounded-off numbers in a calculation to get some idea of what the size of the actual answer should be.

The result of multiplying two numbers is called the **product,** and the two numbers are called **factors** of the product.

In division:

The number dividing is called the **divisor.**

The number being divided is called the **dividend.**

The result is called the **quotient.**

The number left over is called the **remainder,** and it must be less than the divisor.

An **average** (or **mean**) of a set of numbers is the number found by adding the numbers in the set and then dividing this sum by the number of numbers in the set.

Rules and Properties

Rounding-Off Rule for Whole Numbers

1. Look at the single digit just to the right of the digit that is in the place of desired accuracy.

2. If this digit is 5 or greater, make the digit in the desired place of accuracy one larger and replace all digits to the right with zeros. All digits to the left remain unchanged unless a 9 is made one larger.

3. If this digit is less than 5, leave the digit that is in the place of desired accuracy as it is and replace all digits to the right with zeros. All digits to the left remain unchanged.

Properties of Addition

Commutative property: $a + b = b + a$

Associative property: $a + (b + c) = (a + b) + c$

Identity: $a + 0 = a$

Properties of Multiplication

Commutative property: $a \cdot b = b \cdot a$

Associative property: $a(b \cdot c) = (a \cdot b)c$

Identity: $a \cdot 1 = a$

Zero factor law: $a \cdot 0 = 0$

Distributive property of
multiplication over addition: $a(b + c) = a \cdot b + a \cdot c$

Division with 0

1. If a is any nonzero whole number, then

   $0 \div a = 0$.

2. If a is any whole number, then

   $a \div 0$ is **undefined.**

Procedures

To Estimate an Answer:

1. Round off each number to the place of the leftmost digit.

2. Perform the indicated operation with these rounded-off numbers.

To Multiply Two or More Numbers that End with 0's:

1. Count the ending 0's.

2. Multiply the numbers without the ending 0's.

3. Write the product of Step 2 with the counted number of 0's at the end.

Basic Strategy for Solving Word Problems

1. Read each problem carefully until you understand the problem and know what is being asked for.

2. Draw any type of figure or diagram that might be helpful and decide what operations are needed.

3. Perform these operations.

4. Mentally check to see if your answer is reasonable. See if you can think of another more efficient or more interesting way to do the same problem.

Review Questions: Chapter 1

Write the following decimal numbers in expanded notation and in their English word equivalents.

1. 495   **2.** 1975   **3.** 60,308

Write the following numbers as decimal numbers.

4. four thousand eight hundred fifty-six

5. fifteen million, thirty-two thousand, one hundred ninety-seven

6. six hundred seventy-two million, three hundred forty thousand, eighty-three

Round off as indicated.

7. 625 (to the nearest ten)

8. 14,620 (to the nearest thousand)

9. 749 (to the nearest hundred)

10. 2570 (to the nearest hundred)

State which property of addition or multiplication is illustrated.

11. $17 + 32 = 32 + 17$ **12.** $3(22 \cdot 5) = (3 \cdot 22)5$

13. $28 + (6 + 12) = (28 + 6) + 12$ **14.** $72 \cdot 89 = 89 \cdot 72$

First estimate each sum; then find the sum.

15. 8445
      267
    1351
 +   478

16.   39
   487
   966
+ 182

Subtract.

17. 647
  − 139

18. 7036
  − 4652

19. 5000
  − 2898

Multiply.

20. 0 · 36

21. 70 · 80

22. 90 · 4000

First estimate each product; then find the product.

23. 98
  × 52

24. 8975
  × 436

25. 4837
  × 5000

First estimate each quotient; then find the quotient.

26. 7) 2044

27. 38) 23,028

28. 529) 71,496

29. If the product of 17 and 51 is added to the product of 16 and 12, what is the sum?

30. Find the average of 33, 42, 25, and 40.

31. If the quotient of 546 and 6 is subtracted from 100, what is the difference?

32. Two years ago, Ms. Miller bought five shares of stock at $353 per share. One year ago, she bought another ten shares at $290 per share. Yesterday, she sold all her shares at $410 per share. What was her total profit? What was her average profit per share?

33. On a history exam, two students scored 98 points, five students scored 87 points, one student scored 81 points, and six students scored 75 points. What was the average score in the class?

34. What number should be added to seven hundred forty-three to get a sum of eight hundred thirteen?

35. If you buy a car for $10,000 plus taxes and license fees totaling $1200 and make a down payment of $3500, how much will you finance?

36. Fill in the missing numbers in the chart according to the directions.

Given Number	Add 100	Double	Subtract 200
3	103	206	?
20	120	?	?
15	?	?	?
?	?	?	16

Test: Chapter 1

1. Write 8952 in expanded notation and in its English word equivalent.

2. The number 1 is called the multiplicative _____.

3. Give an example that illustrates the commutative property of multiplication.

4. Round off 997 to the nearest thousand.

5. Round off 135,721 to the nearest ten-thousand.

First estimate each sum; then find the sum.

6.
```
   9586
    345
+  2078
```

7.
```
    37
   486
   493
   162
+  557
```

8.
```
   1,480,900
   2,576,850
   5,200,635
+  4,523,276
```

Subtract.

9.
```
   850
-  362
```

10.
```
   5097
-  3868
```

11.
```
   6000
-   293
```

First estimate each product; then find the product.

12.
```
    34
×   76
```

13.
```
   2593
×    85
```

14.
```
    793
×   266
```

Divide.

15. $25 \overline{)\, 10{,}075}$

16. $462 \overline{)\, 79{,}852}$

17. $603 \overline{)\, 1{,}209{,}015}$

18. Find the average of 82, 96, 49, and 69.

19. If the quotient of 51 and 17 is subtracted from the product of 19 and 3, what is the difference?

20. Robert and his brother were saving money to buy a new TV set for their parents. (a) If Robert saved $23 a week and his brother saved $28 a week, how much did they save in six weeks? (b) What was their average weekly savings? (c) If the set they wanted to buy cost $530 including tax, how much did they still need after the six weeks?

21. Find the area in square inches of a rectangle that is 34 inches wide and 42 inches long. (Area of a rectangle is found by multiplying length times width.)

22. You open a checking account with a deposit of $2500. What is your balance if you write checks for $520, $35, $70, and $230 and make further deposits of $200 and $180?

Prime Numbers

Mathematics at Work!

A tracking and data relay satellite is launched by the space shuttle *Discovery*.

The concepts of multiples and least common multiples are an important part of this chapter. These ideas are used throughout our work with fractions and mixed numbers in Chapters 3 and 4. The following problem is an example of working with the least common multiple (or LCM). (See Section 2.6, Example 7 on page 103).

Suppose three weather satellites A, B, and C orbit the earth in different times: satellite A takes 24 hours for a complete orbit, B takes 18 hours, and C takes 12 hours. If they are directly above each other now (as shown in part (a) of the figure), in how many hours will they again be directly above each other in the position shown in part (a)?

(a) Beginning positions   (b) Positions after 6 hours   (c) Positions after 12 hours

---

What to Expect in Chapter 2

The topics in Chapter 2 form the foundation for all the work in Chapter 3 (Fractions) and Chapter 4 (Mixed Numbers). Exponents are introduced in Section 2.1, and they are used throughout Chapter 2 and in appropriate places in the rest of the text to simplify expressions and represent powers. The rules for order of operations, in Section 2.2, are necessary so that everyone will arrive at the same answer when evaluating an expression involving more than one operation.

Section 2.3 provides a few tests for divisibility by 2, 3, 4, 5, 9, and 10 that indicate divisibility without actually requiring you to divide by any of these numbers. The tests are valuable tools for working with prime factorizations and simplifying fractions. The ideas discussed in the first three sections are carried over and used throughout the remainder of the chapter with prime numbers, prime factorizations, and least common multiple (LCM) in Sections 2.4, 2.5, and 2.6. The concepts of prime numbers and prime factorizations are an integral part of the development of fractions and mixed numbers in Chapters 3 and 4.

2.1   Exponents

OBJECTIVES

1. Know that an exponent is used to indicate repeated multiplication.
2. Know the basic property: For any whole number a, $a = a^1$.
3. Know the basic property: For any nonzero whole number a, $a^0 = 1$.
4. Recognize the squares of the whole numbers from 1 to 20.

Terminology of Exponents

Repeated addition is shortened by using multiplication:

$$2 + 2 + 2 + 2 = 4 \cdot 2 = 8$$

Repeated multiplication can be shortened by using **exponents.** Thus, if 2 is used as a factor three times, we can write

$$2 \cdot 2 \cdot 2 = 2^3 = 8$$

In an expression such as $2^3 = 8$, 2 is called the **base,** 3 is called the **exponent,** and 8 is called the **power.** (Exponents are written slightly to the right and above the base.)

EXAMPLE 1	Repeated Multiplication	Using Exponents
	(a) $7 \cdot 7 = 49$	$7^2 = 49$
	(b) $3 \cdot 3 = 9$	$3^2 = 9$
	(c) $2 \cdot 2 \cdot 2 \cdot 2 = 16$	$2^4 = 16$
	(d) $10 \cdot 10 \cdot 10 = 1000$	$10^3 = 1000$

Definition

An **exponent** is a number that tells how many times its base is to be used as a factor.

Common Error

Do **not** multiply the base times the exponent.

$$10^2 = 10 \cdot 2 = 20 \qquad \text{WRONG}$$

$$6^3 = 6 \cdot 3 = 18 \qquad \text{WRONG}$$

Do multiply the base times itself.

$$10^2 = 10 \cdot 10 = 100 \qquad \text{RIGHT}$$

$$6^3 = 6 \cdot 6 \cdot 6 = 216 \qquad \text{RIGHT}$$

Expressions with exponent 2 are read "squared," with exponent 3 are read "cubed," and with other exponents are read "to the _____ power." For example, we read

$5^2 = 25$ as "five squared is equal to twenty-five."

$4^3 = 64$ as "four cubed is equal to sixty-four."

$3^4 = 81$ as "three to the fourth power is equal to eighty-one."

Special note about the word power: A power is not an exponent. A power is the product indicated by an exponent. Thus, for an equation such as $2^5 = 32$, we must think of the phrase "two to the fifth power" in its entirety, and the corresponding power is the product, 32.

Properties of Exponents

If there is no exponent, the exponent is understood to be 1. Thus

$$8 = 8^1, \qquad 6 = 6^1, \qquad \text{and} \qquad 942 = 942^1.$$

Definition

For any whole number a, $a = a^1$.

When the exponent 0 is used for any base except 0, the value of the power is defined to be 1:

$$2^0 = 1 \qquad 3^0 = 1 \qquad 5^0 = 1 \qquad 46^0 = 1$$

Property of Exponents

For any nonzero whole number a, $a^0 = 1$.

To help in understanding 0 as an exponent, we discuss one of the rules for exponents (which will be studied in algebra) that is related to division. To divide $\frac{2^6}{2^2}$ or $\frac{5^4}{5^3}$, we can write

$$\frac{2^6}{2^2} = \frac{2 \cdot 2 \cdot 2 \cdot 2 \cdot 2 \cdot 2}{2 \cdot 2} = 2 \cdot 2 \cdot 2 \cdot 2 = 2^4 \qquad \text{or} \qquad \frac{2^6}{2^2} = 2^{6-2} = 2^4$$

$$\frac{5^4}{5^3} = \frac{5 \cdot 5 \cdot 5 \cdot 5}{5 \cdot 5 \cdot 5} = 5 \qquad \text{or} \qquad \frac{5^4}{5^3} = 5^{4-3} = 5^1$$

The rule is to **subtract the exponents when dividing numbers with the same base.** So,

$$\frac{3^4}{3^4} = 3^{4-4} = 3^0 \qquad \text{and} \qquad \frac{5^2}{5^2} = 5^{2-2} = 5^0$$

But,

$$\frac{3^4}{3^4} = \frac{81}{81} = 1 \qquad \text{and} \qquad \frac{5^2}{5^2} = \frac{25}{25} = 1$$

For the rules of exponents to make sense, we have $3^0 = 1$ and $5^0 = 1$.

We make special note that the expression 0^0 is not defined.

EXAMPLE 2

(a) $8 = 8^1$

(b) $6^0 = 1$

(c) $6^2 = 36 \qquad (6^2 = 6 \cdot 6)$

(d) $5^3 = 125 \qquad (5^3 = 5 \cdot 5 \cdot 5)$

(e) $3^3 = 27 \qquad (3^3 = 3 \cdot 3 \cdot 3)$

(f) $2^5 = 32 \qquad (2^5 = 2 \cdot 2 \cdot 2 \cdot 2 \cdot 2)$

To improve your speed in factoring (Section 2.5) and working with fractions and algebraic expressions (in your next mathematics course), you should try to memorize all the squares of the numbers from 1 to 20. The following table lists these squares.

Number (N)	1	2	3	4	5	6	7	8	9	10
Square (N^2)	1	4	9	16	25	36	49	64	81	100

Number (N)	11	12	13	14	15	16	17	18	19	20
Square (N^2)	121	144	169	196	225	256	289	324	361	400

These powers and others are listed in the table on the inside back cover of the text.

Exercises 2.1

Study the text and examples carefully before working these exercises.

In each of the following expressions, name (a) the exponent and (b) the base. Also, find each power.

1. 2^3　　　**2.** 2^5　　　**3.** 5^2　　　**4.** 6^2　　　**5.** 7^0

6. 11^2　　**7.** 1^4　　　**8.** 4^3　　　**9.** 4^0　　　**10.** 3^6

11. 3^2　　**12.** 2^4　　**13.** 5^0　　**14.** 1^{50}　　**15.** 62^1

16. 12^2　**17.** 10^2　**18.** 10^3　**19.** 4^2　　**20.** 2^5

21. 10^4　**22.** 5^3　　**23.** 6^3　　**24.** 10^5　**25.** 19^0

Find a base and exponent form for each of the following powers without using the exponent 1. [**Hint:** The table inside the back cover may be helpful.]

26. 4　　　**27.** 25　　　**28.** 16　　　**29.** 27　　　**30.** 32

31. 121　　**32.** 49　　　**33.** 8　　　　**34.** 9　　　　**35.** 36

36. 125　　**37.** 81　　　**38.** 64　　　**39.** 100　　**40.** 1000

41. 10,000　**42.** 216　　**43.** 144　　**44.** 169　　**45.** 243

46. 625　　**47.** 225　　**48.** 196　　**49.** 343　　**50.** 100,000

Rewrite the following products using exponents.

51. $6 \cdot 6 \cdot 6 \cdot 6 \cdot 6$ **52.** $7 \cdot 7 \cdot 7 \cdot 7$ **53.** $2 \cdot 2 \cdot 7 \cdot 7$

54. $5 \cdot 5 \cdot 9 \cdot 9 \cdot 9$ **55.** $2 \cdot 2 \cdot 3 \cdot 3 \cdot 3$ **56.** $3 \cdot 3 \cdot 5 \cdot 5 \cdot 5$

57. $7 \cdot 7 \cdot 13$ **58.** $11 \cdot 11 \cdot 11$ **59.** $2 \cdot 3 \cdot 3 \cdot 11 \cdot 11$

60. $5 \cdot 5 \cdot 5 \cdot 11 \cdot 11$

Find the following squares. Write as many of them as you can from memory.

61. 8^2 **62.** 3^2 **63.** 7^2 **64.** 11^2 **65.** 15^2

66. 14^2 **67.** 18^2 **68.** 9^2 **69.** 12^2 **70.** 20^2

71. 10^2 **72.** 16^2 **73.** 30^2 **74.** 40^2 **75.** 50^2

2.2   Order of Operations

OBJECTIVES

1. Know the rules for order of operations.
2. Be able to use the rules for order of operations to simplify numerical expressions.

Rules for Order of Operations

Would you use either of the following procedures to evaluate the expression $5 \cdot 2^2 + 14 \div 2$?

$$5 \cdot 2^2 + 14 \div 2 = 10^2 + 14 \div 2$$
$$= 100 + 14 \div 2$$
$$= 114 \div 2$$
$$= 57$$

OR

$$5 \cdot 2^2 + 14 \div 2 = 5 \cdot 4 + 14 \div 2$$
$$= 5 \cdot 4 + 7$$
$$= 5 \cdot 11$$
$$= 55$$

Both of these answers are **WRONG.** Mathematicians have agreed on a set of rules for simplifying (or evaluating) any numerical expression involving addition, subtraction, multiplication, division, and exponents. Under these rules,

$$5 \cdot 2^2 + 14 \div 2 = 5 \cdot 4 + 14 \div 2$$
$$= 20 + 7$$
$$= 27 \quad \textbf{RIGHT ANSWER}$$

Rules for Order of Operations

1. First, simplify within grouping symbols, such as parentheses ( ), brackets [ ], or braces { }. Start with the innermost grouping.

2. Second, find any powers indicated by exponents.

3. Third, moving from **left to right,** perform any multiplications or divisions in the order in which they appear.

4. Fourth, moving from **left to right,** perform any additions or subtractions in the order in which they appear.

The rules are very explicit. Read them carefully. Note that in Rule 3, neither multiplication nor division has priority over the other. Whichever of these operations occurs first, **moving left to right,** is done first. In Rule 4, addition and subtraction are handled in the same way.

The following examples show how to apply the rules for order of operations. Generally, more than one step is performed at the same time. Use the + and − signs to separate the expressions into various parts. The addition and subtraction occur last unless they are within grouping symbols.

Use the rules for order of operations to find the value of each of the following expressions. [**Note:** In Examples 1–3 the rules are followed in detail. In Examples 4–6 we show how parts separated by a + or − sign can be simplified in the same step.]

EXAMPLE 1

Evaluate $14 \div 7 + 3 \cdot 2 - 5$.

Solution

$$14 \div 7 + 3 \cdot 2 - 5 \quad \text{Divide before multiplying in this case.}$$
$$= \quad 2 \quad + 3 \cdot 2 - 5 \quad \text{Multiply before adding or subtracting.}$$
$$= \quad 2 \quad + \quad 6 \quad - 5 \quad \text{Add before subtracting in this case.}$$
$$= \quad 8 \quad - 5 \quad \text{Subtract.}$$
$$= \quad 3$$

EXAMPLE 2

Evaluate $3 \cdot 6 \div 9 - 1 + 4 \cdot 7$.

Solution

$$3 \cdot 6 \div 9 - 1 + 4 \cdot 7 \qquad \text{Multiply.}$$
$$= \quad 18 \div 9 - 1 + 4 \cdot 7 \qquad \text{Divide.}$$
$$= \quad\quad 2 \quad - 1 + 4 \cdot 7 \qquad \text{Multiply.}$$
$$= \quad\quad 2 \quad - 1 + 28 \qquad \text{Subtract and add, left to right.}$$
$$= \quad\quad\quad\quad 1 \quad + 28$$
$$= \quad\quad\quad\quad\quad 29$$

EXAMPLE 3

Evaluate $(6 + 2) + (8 + 1) \div 9$.

Solution

$$(6 + 2) + (8 + 1) \div 9 \qquad \text{Operate within parentheses.}$$
$$= \quad 8 \quad + \quad 9 \quad \div 9 \qquad \text{Divide.}$$
$$= \quad 8 \quad + \quad\quad 1$$
$$= \quad\quad\quad 9$$

EXAMPLE 4

Evaluate $30 \div 3 \cdot 2 + 3(6 - 21 \div 7)$.

Solution

$$30 \div 3 \cdot 2 + 3(6 - 21 \div 7) \qquad \text{Operate within parentheses.}$$
$$= 30 \div 3 \cdot 2 + 3(6 - 3) \qquad \text{Divide and simplify within parentheses since the two parts are separated by a + sign. Do \textbf{not} multiply by 2 yet.}$$
$$= \quad 10 \cdot 2 + 3(3) \qquad \text{Multiply.}$$
$$= \quad\quad 20 + 9$$
$$= \quad\quad\quad 29$$

EXAMPLE 5

Evaluate $2 \cdot 3^2 + 18 \div 3^2$.

Solution

$$2 \cdot 3^2 + 18 \div 3^2 \qquad \text{Find powers.}$$
$$= 2 \cdot 9 + 18 \div 9 \qquad \text{Multiply and divide since the two parts are separated by a + sign.}$$
$$= \quad 18 + 2$$
$$= \quad\quad 20$$

EXAMPLE 6 | Evaluate $[(5 + 2^2) \div 3 + 8](14 - 10)$.

Solution

$$[(5 + 2^2) \div 3 + 8](14 - 10) \quad \text{Operate within parentheses (twice).}$$

$$= [(5 + 4) \div 3 + 8] \cdot (4) \quad \text{Operate within parentheses.}$$

$$= [9 \div 3 + 8](4) \quad \text{Divide within brackets.}$$

$$= [3 + 8](4) \quad \text{Add within brackets.}$$

$$= [11](4) \quad \text{Multiply.}$$

$$= 44$$

CLASSROOM PRACTICE

Find the value for each of the following expressions by using the rules for order of operations.

1. $15 \div 15 + 10 \cdot 2$

2. $3 \cdot 2^3 - 12 - 3 \cdot 2^2$

3. $4 \div 2^2 + 3 \cdot 2^2$

4. $(5 + 7) \div 3 + 1$

5. $19 - 5(3 - 1)$

ANSWERS: **1.** 21   **2.** 0   **3.** 13   **4.** 5   **5.** 9

Exercises 2.2

Be sure to study the examples before you begin these exercises.

Copy the problem just as you see it here and fill in each blank with the word (or words) that indicates which operation (or operations) is performed at each step in evaluating the expression.

1. $5 + 6 \cdot 4 \div 2$ _____ (a)

$= 5 + 24 \div 2$ _____ (b)

$= 5 + 12$ _____ (c)

$= 17$

2. $(9 - 2) - (3 + 2) \div 5$ _____ (a)

$= 7 - 5 \div 5$ _____ (b)

$= 7 - 1$ _____ (c)

$= 6$

3. $(22 \div 2 + 3) \div 7$ (a) _____

   $= \quad (11 \quad + 3) \div 7$ (b) _____

   $= \qquad 14 \quad \div \ 7$ (c) _____

   $= \qquad\qquad 2$

4. $3 \cdot 2^2 - 6 \div 2 + 5 \cdot 7$ (a) _____

   $= 3 \cdot 4 \ - \ 6 \div 2 + 5 \cdot 7$ (b) _____

   $= \quad 12 \quad - \quad 3 \quad + \ 35$ (c) _____

   $= \qquad\quad 9 \quad + \ 35$ (d) _____

   $= \qquad\qquad 44$

Find the value of each of the following expressions by using the rules for order of operations.

5. $4 \div 2 + 7 - 3 \cdot 2$

6. $8 \cdot 3 \div 12 + 13$

7. $6 + 3 \cdot 2 - 10 \div 2$

8. $14 \cdot 3 \div 7 \div 2 + 6$

9. $6 \div 2 \cdot 3 - 1 + 2 \cdot 7$

10. $5 \cdot 1 \cdot 3 - 4 \div 2 + 6 \cdot 3$

11. $72 \div 4 \div 9 - 2 + 3$

12. $14 + 63 \div 3 - 35$

13. $(2 + 3 \cdot 4) \div 7 + 3$

14. $(2 + 3) \cdot 4 \div 5 + 3 \cdot 2$

15. $(7 - 3) + (2 + 5) \div 7$

16. $16(2 + 4) - 90 - 3 \cdot 2$

17. $35 \div (6 - 1) - 5 + 6 \div 2$

18. $22 - 11 \cdot 2 + 15 - 5 \cdot 3$

19. $(42 - 2 \div 2 \cdot 3) \div 13$

20. $18 + 18 \div 2 \div 3 - 3 \cdot 1$

21. $4(7 - 2) \div 10 + 5$

22. $(33 - 2 \cdot 6) \div 7 + 3 - 6$

23. $72 \div 8 + 3 \cdot 4 - 105 \div 5$

24. $6(14 - 6 \div 2 - 11)$

25. $48 \div 12 \div 4 - 1 + 6$

26. $5 - 1 \cdot 2 + 4(6 - 18 \div 3)$

27. $8 - 1 \cdot 5 + 6(13 - 39 \div 3)$

28. $(21 \div 7 - 3)42 + 6$

29. $16 - 16 \div 2 - 2 + 7 \cdot 3$

30. $(135 \div 3 + 21 \div 7) \div 12 - 4$

31. $(13 - 5) \div 4 + 12 \cdot 4 \div 3 - 72 \div 18 \cdot 2 + 16$

32. $15 \div 3 + 2 - 6 + (3)(2)(18)(0)(5)$

33. $100 \div 10 \div 10 + 1000 \div 10 \div 10 \div 10 - 2$

34. $[(85 + 5) \div 3 \cdot 2 + 15] \div 15$

35. $2 \cdot 5^2 - 4 \div 2 + 3 \cdot 7$

36. $16 \div 2^4 - 9 \div 3^2$

37. $(4^2 - 7) \cdot 2^3 - 8 \cdot 5 \div 10$

38. $4^2 - 2^4 + 5 \cdot 6^2 - 10^2$

39. $(2^5 + 1) \div 11 - 3 + 7(3^3 - 7)$

40. $(6 + 8^2 - 10 \div 2) \div 5 + 5 \cdot 3^2$

41. $(5 + 7) \div 4 + 2$

42. $(2^3 + 2) \div 5 + (7^2 \div 7)$

43. $(5^2 + 7) \div 8 - (14 \div 7 \cdot 2)$

44. $(3 \cdot 2^2 - 5 \cdot 2 + 2) - (1 \cdot 2^2 + 5 \cdot 2 - 10)$

45. $2^3 \cdot 3^2 \div 24 - 3 + 6^2 \div 4$

46. $2 \cdot 3^2 + 5 \cdot 3^2 + 15^2 - (21 \cdot 3^2 + 6)$

47. $3 \cdot 2^3 - 2^2 + 4 \cdot 2 - 2^4$

48. $2 \cdot 5^2 - 4(21 \div 3 - 7) + 10^3 - 1000$

49. $(4 + 3)^2 - (2 + 3)^2$

50. $40 \div 2 \cdot 5 + 1 \cdot 3^2 \cdot 2$

51. $20 - 2(3 - 1) + 6^2 \div 2 \cdot 3$

52. $8 \div 2 \cdot 4 + 16 \div 4 \cdot 2 + 3 \cdot 2^2$

53. $50 \cdot 10 \div 2 - 2^2 \cdot 5 + 14 - 2 \cdot 7$

54. $(20 \div 2^2 \cdot 5) + (51 \div 17)^2$

55. $[(2 + 3)(5 - 1) \div 2](10 + 1)$

56. $3[4 + (6 \div 3 \cdot 2)]$

57. $5[3^2 + (8 + 2^3)] - 15$

58. $(2^4 - 16)[13 - (5^2 - 20)]$

59. $100 + 2[(7^2 - 9)(5 + 1)^2]$

60. $(3 + 5)^2[(2 + 1)^2 - 14 \div 2]$

Tests for Divisibility (2, 3, 4, 5, 9, and 10)

OBJECTIVES

1. Understand the terms **exactly divisible, divisible,** and **divides.**
2. Know the tests for easily checking divisibility by 2, 3, 4, 5, 9, and 10.

Rules of Tests for Divisibility

In our work with factoring (Section 2.5) and fractions (Chapter 3), we will need to be able to divide quickly by small numbers. We will want to know if a number is **exactly divisible** (remainder 0) by some number **before** we divide.

Definition

If a number can be divided by another number so that the remainder is 0, then we say:

1. The first number is **exactly divisible by** (or is **divisible by**) the second

OR

2. The second number **divides** the first.

There are simple tests we can use to determine whether a number is divisible by 2, 3, 4, 5, 9, or 10 **without actually dividing.** For example, can you tell (without dividing) if 6495 is divisible by 2? By 3? The answer is that 6495 is divisible by 3 but not by 2.

$$
\begin{array}{r}
2165 \\
3\,)\overline{6495} \\
\underline{6} \\
04 \\
\underline{3} \\
19 \\
\underline{18} \\
15 \\
\underline{15} \\
0 \quad \text{Remainder}
\end{array}
\qquad
\begin{array}{r}
3247 \\
2\,)\overline{6495} \\
\underline{6} \\
04 \\
\underline{4} \\
09 \\
\underline{8} \\
15 \\
\underline{14} \\
1 \quad \text{Remainder (remainder not 0)}
\end{array}
$$

The following list of rules explains how to test for divisibility by 2, 3, 4, 5, 9, and 10. There are other tests for other numbers such as 6, 7, 8, and 15, but the rules given here are sufficient for our purposes.

Tests for Divisibility by 2, 3, 4, 5, 9, and 10

For 2:   If the last digit (units digit) of a whole number is 0, 2, 4, 6, or 8, then the whole number is divisible by 2.

For 3:   If the sum of the digits of a whole number is divisible by 3, then the number is divisible by 3.

For 4:   If the last two digits of a whole number form a number that is divisible by 4, then the number is divisible by 4. (00 is considered to be divisible by 4.)

For 5:   If the last digit of a whole number is 0 or 5, then the number is divisible by 5.

For 9:   If the sum of the digits of a whole number is divisible by 9, then the number is divisible by 9.

For 10:   If the last digit of a whole number is 0, then the number is divisible by 10.

Even and Odd Whole Numbers

Even whole numbers are divisible by 2.
(If a whole number is divided by 2 and the remainder is 0, then the whole number is even.)

Odd whole numbers are not divisible by 2.
(If a whole number is divided by 2 and the remainder is 1, then the whole number is odd.)

EXAMPLE 1

(a) 356 is divisible by 2 since the last digit is 6.

(b) 6801 is divisible by 3 since $6 + 8 + 0 + 1 = 15$ and 15 is divisible by 3.

(c) 9036 is divisible by 4 since 36 (last 2 digits) is divisible by 4.

(d) 1365 is divisible by 5 since 5 is the last digit.

(e) 9657 is divisible by 9 since $9 + 6 + 5 + 7 = 27$ and 27 is divisible by 9.

(f) 3590 is divisible by 10 since 0 is the last digit.

If a number is divisible by 9, then it will be divisible by 3. In Example 1(e), $9 + 6 + 5 + 7 = 27$ and 27 is divisible by 9 and by 3, so 9657 is divisible by 9 and by 3. We also say that 9 and 3 **divide into** (or **divide exactly into**) 9657 with the implication that the remainder is 0.

But, a number that is divisible by 3 may not be divisible by 9. In Example 1(b), $6 + 8 + 0 + 1 = 15$ and 15 is **not** divisible by 9, so 6801 is divisible by 3 but not by 9.

Similarly, any number divisible by 10 is also divisible by 5, but a number that is divisible by 5 might not be divisible by 10. The number 2580 is divisible by 10 and also by 5, but 4365 is divisible by 5 and not by 10.

Many numbers are divisible by more than one of the numbers 2, 3, 4, 5, 9, and 10. These numbers will satisfy more than one of the six tests. For example, 4365 is divisible by 5 (last digit is 5) and by 3 ($4 + 3 + 6 + 5 = 18$) and by 9 ($4 + 3 + 6 + 5 = 18$).

Use all six tests to determine which of the numbers 2, 3, 4, 5, 9, and 10 will divide into the numbers in Examples 2–4.

EXAMPLE 2

5712

(a) divisible by 2 (last digit is 2, an even digit)

(b) divisible by 3 ($5 + 7 + 1 + 2 = 15$ and 15 is divisible by 3)

(c) divisible by 4 (12 is divisible by 4)

(d) not divisible by 5 (last digit is not 0 or 5)

(e) not divisible by 9 ($5 + 7 + 1 + 2 = 15$ and 15 is not divisible by 9)

(f) not divisible by 10 (last digit is not 0)

EXAMPLE 3   2530

(a) divisible by 2 (last digit is 0, an even digit)

(b) not divisible by 3 (2 + 5 + 3 + 0 = 10 and 10 is not divisible by 3)

(c) not divisible by 4 (30 is not divisible by 4)

(d) divisible by 5 (last digit is 0)

(e) not divisible by 9 (2 + 5 + 3 + 0 = 10 and 10 is not divisible by 9)

(f) divisible by 10 (last digit is 0)

EXAMPLE 4   3401

(a) not divisible by 2 (last digit is 1, an odd digit)

(b) not divisible by 3 (3 + 4 + 0 + 1 = 8 and 8 is not divisible by 3)

(c) not divisible by 4 (1 is not divisible by 4)

(d) not divisible by 5 (last digit is not 0 or 5)

(e) not divisible by 9 (3 + 4 + 0 + 1 = 8 and 8 is not divisible by 9)

(f) not divisible by 10 (last digit is not 0)

Therefore, 3401 is not divisible by any numbers in this list.

Checking Divisibility of Products

To emphasize the relationships among the concepts of multiplication, factors, and divisibility, we now discuss how these are related to given products. We will use these concepts as the basis of our discussion of the least common multiple (LCM) in Section 2.6 and with common denominators of fractions in Chapter 3.

Consider the fact that $3 \cdot 4 \cdot 5 \cdot 5 = 300$. This result means that if any one factor (or the product of two or more factors) is divided into 300, the quotient will be the product of the remaining factors. For example, if 300 is divided by 3, then the product $4 \cdot 5 \cdot 5 = 100$ will be the quotient. Similarly, various groupings give

$$3 \cdot 4 \cdot 5 \cdot 5 = (3 \cdot 4)(5 \cdot 5) = 12 \cdot 25 = 300$$

$$3 \cdot 4 \cdot 5 \cdot 5 = (4 \cdot 5)(3 \cdot 5) = 20 \cdot 15 = 300$$

and

$$3 \cdot 4 \cdot 5 \cdot 5 = (3 \cdot 4 \cdot 5)(5) = 60 \cdot 5 = 300$$

Thus, we can make the following statements:

12 divides 300 25 times   and   25 divides 300 12 times
(or, $300 \div 12 = 25$ and $300 \div 25 = 12$)

20 divides 300 15 times   and   15 divides 300 20 times
(or, 300 ÷ 20 = 15   and   300 ÷ 15 = 20)

60 divides 300 5 times   and   5 divides 300 60 times
(or, 300 ÷ 60 = 5   and   300 ÷ 5 = 60)

EXAMPLE 5

Does 36 divide the product $4 \cdot 5 \cdot 9 \cdot 3 \cdot 7$? If so, how many times?

Solution

Since $36 = 4 \cdot 9$, we have

$$4 \cdot 5 \cdot 9 \cdot 3 \cdot 7 = (4 \cdot 9)(5 \cdot 3 \cdot 7)$$
$$= 36 \cdot 105$$

Thus, 36 does divide the product, and it divides the product 105 times.

EXAMPLE 6

Does 15 divide the product $5 \cdot 7 \cdot 2 \cdot 3 \cdot 2$? If so, how many times?

Solution

Since $15 = 3 \cdot 5$, we have

$$5 \cdot 7 \cdot 2 \cdot 3 \cdot 2 = (3 \cdot 5)(7 \cdot 2 \cdot 2)$$
$$= (15)(28)$$

Thus, 15 does divide the product, and it divides the product 28 times.

EXAMPLE 7

Does 35 divide the product $3 \cdot 4 \cdot 5 \cdot 11$?

Solution

We know that $35 = 5 \cdot 7$ and even though 5 is a factor of the product, 7 is not. Therefore, 35 does not divide the product $3 \cdot 4 \cdot 5 \cdot 11$. In other words, $3 \cdot 4 \cdot 5 \cdot 11 = 660$; 660 is not divisible by 35.

CLASSROOM PRACTICE

Using the techniques of this section, determine which of the numbers 2, 3, 4, 5, 9, and 10 (if any) will divide exactly into each of the following numbers.

1. 842   **2.** 9030   **3.** 4031

4. Does 16 divide the product $3 \cdot 5 \cdot 4 \cdot 7 \cdot 4$? If so, how many times?

ANSWERS: **1.** 2   **2.** 2, 3, 5, 10   **3.** None   **4.** Yes, 105 times.

Exercises 2.3

Be sure to study the text and the examples before working these exercises.

Using the techniques of this section, determine which of the numbers 2, 3, 4, 5, 9, and 10 (if any) will divide exactly into each of the following numbers.

1. 72	**2.** 81	**3.** 105	**4.** 333	**5.** 150
6. 471	**7.** 664	**8.** 154	**9.** 372	**10.** 375
11. 443	**12.** 173	**13.** 567	**14.** 480	**15.** 331
16. 370	**17.** 571	**18.** 466	**19.** 897	**20.** 695
21. 795	**22.** 777	**23.** 45,000	**24.** 885	**25.** 4422
26. 1234	**27.** 4321	**28.** 8765	**29.** 5678	**30.** 402
31. 705	**32.** 732	**33.** 441	**34.** 555	**35.** 666
36. 9000	**37.** 10,000	**38.** 576	**39.** 549	**40.** 792
41. 5700	**42.** 4391	**43.** 5476	**44.** 6930	**45.** 4380
46. 510	**47.** 8805	**48.** 7155	**49.** 8377	**50.** 2222
51. 35,622	**52.** 75,495	**53.** 12,324	**54.** 55,555	
55. 632,448	**56.** 578,400	**57.** 9,737,001	**58.** 17,158,514	
59. 36,762,252	**60.** 20,498,105			

Determine whether each of the given numbers divides (or is a factor of) the given product. If it does divide the product, tell how many times. Find each product and make a written statement concerning the divisibility of the product by the given number.

61. $6; 2 \cdot 3 \cdot 3 \cdot 5$

62. $10; 2 \cdot 3 \cdot 3 \cdot 5$

63. $14; 2 \cdot 3 \cdot 5 \cdot 7$

64. $20; 3 \cdot 4 \cdot 5 \cdot 11$

65. $10; 3 \cdot 3 \cdot 5 \cdot 7$

66. $25; 2 \cdot 3 \cdot 5 \cdot 7 \cdot 11$

67. $25; 2 \cdot 2 \cdot 3 \cdot 5 \cdot 5$

68. $35; 3 \cdot 4 \cdot 5 \cdot 7 \cdot 10$

69. $21; 3 \cdot 3 \cdot 5 \cdot 7 \cdot 11$

70. $30; 2 \cdot 3 \cdot 4 \cdot 5 \cdot 13$

Writing and Thinking about Mathematics

71. If a number is divisible by both 2 and 9, must it be divisible by 18? Explain your reasoning and give several examples to support your answer.

72. If a number is divisible by both 3 and 9, must it be divisible by 27? Explain your reasoning and give several examples to support your answer.

73. With your understanding of factors and divisibility, make up a rule for divisibility by 6 and a rule for divisibility by 15.

♻ The Recycle Bin   (from Section 1.4)

Round off as indicated.

To the nearest ten: **1.** 847   **2.** 1931

To the nearest hundred: **3.** 439   **4.** 2563

To the nearest thousand: **5.** 13,612   **6.** 20,500

First estimate the answers using rounded-off numbers; then find the following sums and differences.

7.   485
           93
       + 115

8.   661
         1730
       + 2059

9.   8752
       − 3527

10. 74,605
       − 46,083

2.4 Prime Numbers and Composite Numbers

OBJECTIVES

1. Know the definition of a prime number.
2. Know the definition of a composite number.
3. Be able to list all the prime numbers less than 50.
4. Be able to determine whether a number is prime or composite.

Prime Numbers and Composite Numbers

Every counting number, except 1, has at least two factors, as the following list shows.

Counting Numbers	Factors
18 $\longrightarrow$	1, 2, 3, 6, 9, 18
14 $\longrightarrow$	1, 2, 7, 14
23 $\longrightarrow$	1, 23
17 $\longrightarrow$	1, 17
21 $\longrightarrow$	1, 3, 7, 21
36 $\longrightarrow$	1, 2, 3, 4, 6, 9, 12, 18, 36
3 $\longrightarrow$	1, 3

We are particularly interested in those numbers that have exactly two different factors. In the list just given, 23, 17, and 3 fall into that category. They are called **prime numbers.**

Definition

A **prime number** is a counting number greater than 1 that has only 1 and itself as factors.

OR

A **prime number** is a counting number with exactly two different factors (or divisors).

Definition

A **composite number** is a counting number with more than two different factors (or divisors).

Thus, in the list discussed, 18, 14, 21, and 36 are **composite numbers.**

Note: 1 is **neither** a prime nor a composite number. $1 = 1 \cdot 1$, and 1 is the only factor of 1. 1 does not have **exactly** two **different** factors, and it does not have more than two different factors.

EXAMPLE 1   Some prime numbers:

2	2 has exactly two different factors, 1 and 2.
7	7 has exactly two different factors, 1 and 7.
11	11 has exactly two different factors, 1 and 11.
29	29 has exactly two different factors, 1 and 29.

EXAMPLE 2   Some composite numbers:

12    $1 \cdot 12 = 12, \quad 2 \cdot 6 = 12, \quad$ and $\quad 3 \cdot 4 = 12$
So, 1, 2, 3, 4, 6, and 12 are all factors of 12; and 12 has more than two different factors.

33    $1 \cdot 33 = 33$ and $3 \cdot 11 = 33$
So, 1, 3, 11, and 33 are all factors of 33; and 33 has more than two different factors.

The Sieve of Eratosthenes

There is no formula to help us find all the prime numbers. However, there is a technique developed by a Greek mathematician named Eratosthenes. He used the concept of **multiples.** To find the multiples of any counting number, multiply each of the counting numbers by that number.

Counting Numbers	**1, 2, 3, 4, 5, 6, 7, 8, . . .**
multiples of 8	8, 16, 24, 32, 40, 48, 56, 64, . . .
multiples of 2	2, 4, 6, 8, 10, 12, 14, 16, . . .
multiples of 3	3, 6, 9, 12, 15, 18, 21, 24, . . .
multiples of 10	10, 20, 30, 40, 50, 60, 70, 80, . . .

None of the multiples of a number, except possibly the number itself, can be prime since they all have that number as a factor. To sift out the prime numbers according to the **Sieve of Eratosthenes,** we proceed by eliminating multiples as the following steps describe.

1. To find the prime numbers from 1 to 50, list all the counting numbers from 1 to 50 in rows of ten.

1	2	3	4	5	6	7	8	9	10
11	12	13	14	15	16	17	18	19	20
21	22	23	24	25	26	27	28	29	30
31	32	33	34	35	36	37	38	39	40
41	42	43	44	45	46	47	48	49	50

2. Start by crossing out 1 (since 1 is not a prime number). Next, circle 2 and cross out all the other multiples of 2; that is, cross out every second number.

1	2	3	4	5	6	7	8	9	10
11	12	13	14	15	16	17	18	19	20
21	22	23	24	25	26	27	28	29	30
31	32	33	34	35	36	37	38	39	40
41	42	43	44	45	46	47	48	49	50

3. The first number after 2 not crossed out is 3. Circle 3 and cross out all multiples of 3 that are not already crossed out; that is, after 3, every third number should be crossed out.

1	2	3	4	5	6	7	8	9	10
11	12	13	14	15	16	17	18	19	20
21	22	23	24	25	26	27	28	29	30
31	32	33	34	35	36	37	38	39	40
41	42	43	44	45	46	47	48	49	50

4. The next number not crossed out is 5. Circle 5 and cross out all multiples of 5 that are not already crossed out. If we proceed this way, we will have the prime numbers circled and the composite numbers crossed out.

The final table shows that the prime numbers less than 50 are:

2, 3, 5, 7, 11, 13, 17, 19, 23, 29, 31, 37, 41, 43, 47

You should also note that

(a) 2 is the only even prime number.

(b) All other prime numbers are odd, but not all odd numbers are prime.

1	2	3	4	5	6	7	8	9	10
11	12	13	14	15	16	17	18	19	20
21	22	23	24	25	26	27	28	29	30
31	32	33	34	35	36	37	38	39	40
41	42	43	44	45	46	47	48	49	50

Determining Prime Numbers

Computers can be used to determine whether or not very large numbers are prime. The following procedure of dividing by prime numbers can be used to determine whether or not relatively small numbers are prime. If a prime number smaller than the given number is found to be a factor (or divisor), then the given number is composite.

To Determine Whether a Number is Prime:

Divide the number by progressively larger **prime numbers** (2, 3, 5, 7, 11, and so forth) until one of the following occurs:

1. You find a remainder 0 (meaning that the prime number is a factor and the given number is composite).

2. You find a quotient smaller than the prime divisor (meaning the given number has no smaller prime factors and is therefore prime itself).

NOTE: We divide only by prime numbers. There is no need to divide by a composite number because if a composite number was a factor, then one of its prime factors would have been found to be a factor in an earlier division.

EXAMPLE 3

Is 103 prime?

Tests for 2, 3, and 5 fail. (103 is not even; $1 + 0 + 3 = 4$ and 4 is not divisible by 3; the last digit is not 0 or 5.)

Divide by 7:

$$
\begin{array}{r}
14 \\
7 \overline{)103} \\
\underline{7} \\
33 \\
\underline{28} \\
5
\end{array}
$$

14   quotient greater than divisor

5   remainder not 0

Divide by 11:

$$
\begin{array}{r}
9 \\
11 \overline{)103} \\
\underline{99} \\
4
\end{array}
$$

9   quotient is less than divisor

So 103 is prime.

EXAMPLE 4

Is 221 prime?

Tests for 2, 3, and 5 fail.

Divide by 7:

$$\begin{array}{r} 31 \\ 7\overline{)221} \\ \underline{21} \\ 11 \\ \underline{7} \\ 4 \end{array}$$ quotient greater than divisor

remainder not 0

Divide by 11:

$$\begin{array}{r} 20 \\ 11\overline{)221} \\ \underline{22} \\ 01 \\ \underline{0} \\ 1 \end{array}$$ quotient greater than divisor

remainder not 0

Divide by 13:

$$\begin{array}{r} 17 \\ 13\overline{)221} \\ \underline{13} \\ 91 \\ \underline{91} \\ 0 \end{array}$$ remainder is 0

So, 221 is composite and not prime. [**Note:** 221 = 13 · 17.]

EXAMPLE 5

One interesting application of factors of counting numbers that is very useful in beginning algebra involves finding two factors whose sum is some specified number. As an example, find two factors of 70 such that their product is 70 and their sum is 19.

Solution

The factors of 70 are 1, 2, 5, 7, 10, 14, 35, and 70, and the pairs whose products are 70 are

$$1 \cdot 70 = 70 \qquad 2 \cdot 35 = 70 \qquad 5 \cdot 14 = 70 \qquad 7 \cdot 10 = 70$$

Thus, the numbers we are looking for are 5 and 14 because

$$5 \cdot 14 = 70 \qquad \text{and} \qquad 5 + 14 = 19$$

Exercises 2.4 _____

Be sure you know what prime numbers are and what multiples are before you begin these exercises.

List the multiples for each of the following numbers.

1. 5   **2.** 7   **3.** 11   **4.** 13   **5.** 12

6. 9   **7.** 20   **8.** 17   **9.** 16   **10.** 25

11. Construct a Sieve of Eratosthenes for the numbers from 1 to 100. List the prime numbers from 1 to 100.

Decide whether each of the following numbers is prime or composite. If the number is composite, find at least three factors for the number.

12. 17   **13.** 19   **14.** 28   **15.** 32   **16.** 47

17. 59   **18.** 16   **19.** 63   **20.** 14   **21.** 51

22. 67   **23.** 89   **24.** 73   **25.** 61   **26.** 52

27. 57   **28.** 98   **29.** 86   **30.** 53   **31.** 37

Two numbers are given. Find two factors of the first number such that their product is the first number and their sum is the second number.

EXAMPLE: 12, 8
            Two factors of 12 whose sum is 8 are 6 and 2 since $6 \cdot 2 = 12$ and
            $6 + 2 = 8$.

32. 24, 10   **33.** 12, 7   **34.** 16, 10   **35.** 12, 13   **36.** 14, 9

37. 50, 27   **38.** 20, 9   **39.** 24, 11   **40.** 48, 19   **41.** 36, 15

42. 7, 8   **43.** 63, 24   **44.** 51, 20   **45.** 25, 10   **46.** 16, 8

47. 60, 17   **48.** 52, 17   **49.** 27, 12   **50.** 72, 22

**Writing and Thinking
about Mathematics**

51. Find the set of all prime numbers less than 1000 that are not odd.

52. Are all odd numbers also prime numbers? Explain your answer.

53. Explain why the number 1 is not prime and not composite.

54. The number 1001 is composite. Find all the factors of 1001.

**Collaborative
Learning Exercises**

55. In teams of 2 to 4 students, each team is to try to find the largest prime number it can within fifteen minutes. The team leader is to explain the team's reasoning to the class.

56. Mathematicians have been interested since ancient times in a search for perfect numbers. A **perfect number** is a counting number that is equal to the sum of its *proper divisors* (divisors not including itself). For example, the first perfect number is 6. The proper divisors of 6 are 1, 2, and 3 and $1 + 2 + 3 = 6$. The teams formed for Exercise 55 are to try to find the second and third perfect numbers. (**Hint:** The second perfect number is between 20 and 30, and the third perfect number is between 450 and 500.)

 ### The Recycle Bin   (from Section 1.5)

Tell what property of multiplication is illustrated.

1. $8 \cdot 7 = 7 \cdot 8$　　　　　　　**2.** $5(2 \cdot 6) = (5 \cdot 2)6$

Use the technique of multiplying by powers of 10 to find the following products mentally.

3. $40 \cdot 400$　　　　**4.** $300 \cdot 500$　　　　**5.** $120 \cdot 7000$

Find the following products.

6.　314　　　　　**7.**　182　　　　　**8.**　105
　　　$\times\ 73$　　　　　　　$\times\ 466$　　　　　　$\times\ 1700$

2.5　Prime Factorizations

OBJECTIVES

1. Know the Fundamental Theorem of Arithmetic.
2. Know the meaning of the term **prime factorization.**
3. Be able to find the prime factorization of a composite number.

Finding a Prime Factorization

To find common denominators for fractions, which will be discussed in Chapter 3, finding **all** the prime factors of a number will be very useful. For example, $28 = 4 \cdot 7$ and 7 is a prime factor, but 4 is not prime. We can write $28 = 4 \cdot 7$

= 2 · 2 · 7 and this product (2 · 2 · 7) contains all prime factors. It is called the **prime factorization** of 28.

The **prime factorization** of a number is the product of all the prime factors of that number, including repeated factors.

Two methods, I and II, for finding prime factorizations are discussed here. Both methods give the same prime factorization for a composite number. Method I is easier to apply in working with fractions, and Method II is easier to follow as an introductory method for some students.

The tests for divisibility by 2, 3, 4, 5, 9, and 10 discussed in Section 2.3 are very useful here. Review them now if you do not remember them. Many times they provide a beginning factor.

Method I

To find the **prime factorization** of a composite number:

1. Factor the composite number into **any** two factors.

2. If either or both factors are not prime, factor each of these.

3. Continue this process until all factors are prime.

Using Method I, find the prime factorization for each of the following numbers.

EXAMPLE 1

60

$$60 = 6 \cdot 10 \quad \text{Since the last digit is 0, we know 10 is a factor.}$$
$$= 2 \cdot 3 \cdot 2 \cdot 5 \quad \text{6 and 10 can both be factored so that each factor is a prime number. This is the prime factorization of 60.}$$

or,

$$60 = 3 \cdot 20 \quad \text{3 is prime, but 20 is not.}$$
$$= 3 \cdot 4 \cdot 5 \quad \text{4 is not prime.}$$
$$= 3 \cdot 2 \cdot 2 \cdot 5 \quad \text{All factors are prime.}$$

Since multiplication is commutative, the order of the factors is not important. What is important is that **all the factors must be prime numbers.** With the factors in order, from smallest to largest, we can write the prime factorization of 60 as 2 · 2 · 3 · 5 or, with exponents, $2^2 \cdot 3 \cdot 5$.

EXAMPLE 2 | 85

$85 = 5 \cdot 17$ 5 is a factor since the last digit is 5. You can find 17 either by dividing mentally or by writing out the long division.

EXAMPLE 3 | 72

$72 = 8 \cdot 9$ 9 is a factor since $7 + 2 = 9$.
$= 2 \cdot 4 \cdot 3 \cdot 3$
$= 2 \cdot 2 \cdot 2 \cdot 3 \cdot 3 = 2^3 \cdot 3^2$

or

$72 = 2 \cdot 36$ 2 is a factor since the last digit is even.
$= 2 \cdot 6 \cdot 6$
$= 2 \cdot 2 \cdot 3 \cdot 2 \cdot 3 = 2^3 \cdot 3^2$

We can form an arrangement called a **factor tree** in which the lines lead to prime factors. Two factor trees for 72 are shown here.

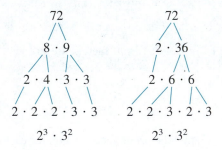

We can form an arrangement called a **factor tree** in which the lines lead to prime factors. Two factor trees for 72 are shown here.

EXAMPLE 4 | 264

Two factor trees for 264 are shown here. Note that the prime factorization is the same regardless of which two factors are chosen first.

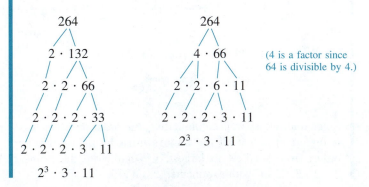

(4 is a factor since 64 is divisible by 4.)

> **Method II**
>
> To find the **prime factorization** of a composite number:
>
> **1.** Divide the composite number by **any** prime number that will divide into it.
>
> **2.** Continue to divide the **quotient** by prime numbers until the quotient is prime.
>
> **3.** The prime factorization is the product of all the prime divisors and the last quotient, which is a prime number.

Using Method II, find the prime factorization for each of the following numbers.

EXAMPLE 5 70

$$\overset{35}{2\overline{)70}} \qquad \overset{7}{5\overline{)35}}$$

So, $70 = 2 \cdot 5 \cdot 7$.

EXAMPLE 6 245

$$\overset{49}{5\overline{)245}} \qquad \overset{7}{7\overline{)49}}$$

So, $245 = 5 \cdot 7 \cdot 7 = 5 \cdot 7^2$.

EXAMPLE 7 168

$$\overset{84}{2\overline{)168}} \qquad \overset{42}{2\overline{)84}} \qquad \overset{21}{2\overline{)42}} \qquad \overset{3}{7\overline{)21}}$$

So, $168 = 2 \cdot 2 \cdot 2 \cdot 7 \cdot 3 = 2^3 \cdot 3 \cdot 7$.

Finding Factors of Composite Numbers

Regardless of what factors you start with or what method you use, **there is only one prime factorization for any composite number.** This fact is so important that it is called the **Fundamental Theorem of Arithmetic.**

> **Fundamental Theorem of Arithmetic**
>
> Every composite number has exactly one prime factorization.

We can use prime factorizations to find all the factors (or divisors) of a composite number.

The only factors (or divisors) of a composite number are

(a) 1 and the number itself

(b) each prime factor

(c) products of various combinations of prime factors

EXAMPLE 8

Find all the factors of 30.

Since $30 = 2 \cdot 3 \cdot 5$, the factors are

(a) 1 and 30

(b) 2, 3, and 5

(c) $2 \cdot 3$, $2 \cdot 5$, and $3 \cdot 5$

The factors are 1, 30, 2, 3, 5, 6, 10, and 15. These are the only factors of 30.

EXAMPLE 9

Find all the factors of 196.

$$196 = 2 \cdot 98$$
$$= 2 \cdot 2 \cdot 49$$
$$= 2 \cdot 2 \cdot 7 \cdot 7$$

The factors are

(a) 1 and 196

(b) 2 and 7

(c) $2 \cdot 2$, $2 \cdot 7$, $7 \cdot 7$, $2 \cdot 2 \cdot 7$, and $2 \cdot 7 \cdot 7$

The factors are 1, 196, 2, 7, 4, 14, 49, 28, and 98. These are the only factors (or divisors) of 196.

CLASSROOM PRACTICE

Find the prime factorization of each of the following numbers.

1. 42   **2.** 56   **3.** 230

4. Using the prime factorization of 63, find all the factors of 63.

ANSWERS: **1.** $2 \cdot 3 \cdot 7$ **2.** $2^3 \cdot 7$ **3.** $2 \cdot 5 \cdot 23$ **4.** 1, 63, 3, 7, 9, 21

Exercises 2.5

Study the text carefully before you start these exercises.

Find the prime factorization for each of the following numbers. Use the tests for divisibility given in Section 2.3 whenever you need them to help you get started.

1. 24	**2.** 28	**3.** 27	**4.** 16	**5.** 36
6. 60	**7.** 72	**8.** 90	**9.** 81	**10.** 105
11. 125	**12.** 160	**13.** 75	**14.** 150	**15.** 210
16. 40	**17.** 250	**18.** 93	**19.** 168	**20.** 360
21. 126	**22.** 48	**23.** 17	**24.** 47	**25.** 51
26. 144	**27.** 121	**28.** 169	**29.** 225	**30.** 52
31. 32	**32.** 98	**33.** 108	**34** 103	**35.** 101
36. 202	**37.** 78	**38.** 500	**39.** 10,000	**40.** 100,000

Using the prime factorization of each number, find all the factors (or divisors) of each number.

41. 12	**42.** 18	**43.** 28	**44.** 98	**45.** 121
46. 45	**47.** 105	**48.** 54	**49.** 97	**50.** 144

Collaborative Learning Exercises

51. The product of the counting numbers from 1 to a given number is called the **factorial** of that number. For example, the symbol 5! is read "five factorial," and the value is $5! = 1 \cdot 2 \cdot 3 \cdot 4 \cdot 5$.

In teams of 2 to 4 students, analyze the following questions and statements and discuss your conclusions in class.

(a) What are the meanings of the symbols 6!, 7!, 8!, and 10!?

(b) Write the prime factorization of each of these factorials.

(c) Determine whether or not each of the following statements is true or false:

    1. 24 divides 6!.   **2.** 28 divides 6!.
    3. 75 divides 8!.   **4.** 75 divides 10!.

52. The following expressions are part of formulas used in probability and statistics. For example, the answer to part (a) indicates the number of committees that can be formed with two people from a group of six people.

(a) $\dfrac{6!}{2! \, (4!)}$ (b) $\dfrac{10!}{2! \, (8!)}$ (c) $\dfrac{20!}{6! \, (14!)}$

In teams of 2 to 4 students, evaluate each of these expressions and discuss how you arrived at these results in class. Did you find any shortcuts in the calculations? What are possible "committee problems" related to parts (b) and (c)?

The Recycle Bin   (from Section 1.7)

1. Find the difference between the product of 42 and 38 and the quotient of 1470 and 70.

2. Chris bought a car for $16,200. The salesperson added $972 for taxes and $520 for license fees. If he made a down payment of $4750 and financed the rest with his bank, how much did he finance?

3. Find the average (or mean) of the numbers 103, 114, 98, and 101.

4. A right triangle (one angle is 90°) has three sides: a base of 12 meters, a height of 5 meters, and a third side of 13 meters. (See diagram.)

   (a) Find the perimeter (distance around) of the triangle.

   (b) Find the area of the triangle in square meters. (To find the area, multiply the base times height, then divide by 2.)

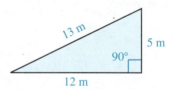

2.6   Least Common Multiple (LCM) with Applications

OBJECTIVES

1. Understand the meaning of the term **multiple.**
2. Use prime factorizations to find the least common multiple of a set of numbers.
3. Recognize the application of the LCM concept in a word problem.

Finding the LCM of a Set of Counting Numbers

The techniques discussed in this section are used throughout Chapter 3 on fractions. Study these ideas thoroughly because they will make your work with fractions much easier.

Remember that the multiples of a number are the products of that number with the counting numbers. The first multiple of a number is that number, and all other multiples are larger than the number.

Counting Numbers   1,   2,   3,   4,   5,   6,   7,   8,   9,   10, . . .

multiples of 6   6, 12, 18, 24, ⟨30,⟩ 36, 42, 48, 54, ⟨60,⟩ . . .

multiples of 10   10, 20, ⟨30,⟩ 40, 50, ⟨60,⟩ 70, 80, ⟨90,⟩100, . . .

The common multiples for 6 and 10 are 30, 60, 90, 120, ...
The **Least Common Multiple** is 30.

Definition

The **Least Common Multiple (LCM)** of a set of counting numbers is the smallest number common to all the sets of multiples of the given numbers.

Except for the number itself, **factors** (or **divisors**) of the number are smaller than the number, and **multiples** are larger than the number. For example, 6 and 10 are factors of 30, and both are smaller than 30. 30 is a multiple of both 6 and 10 and is larger than either 6 or 10.

Listing all the multiples as we did for 6 and 10 and then choosing the least common multiple (LCM) is not very efficient. Either of two other techniques, one involving prime factorizations and the other involving division by prime factors, is easier to apply.

Method I

To find the LCM of a set of counting numbers:

1. Find the prime factorization of each number.

2. Find the prime factors that appear in **any one** of the prime factorizations.

3. Form the product of these primes using each prime the most number of times it appears in **any one** of the prime factorizations.

EXAMPLE 1

Find the LCM for 18, 30, and 45.

(a) Prime factorizations:

$18 = 2 \cdot 9 = 2 \cdot 3 \cdot 3$ One 2, two 3's.
$30 = 6 \cdot 5 = 2 \cdot 3 \cdot 5$ One 2, one 3, one 5.
$45 = 9 \cdot 5 = 3 \cdot 3 \cdot 5$ Two 3's, one 5.

(b) 2, 3, and 5 are the only prime factors.

(c) Most of each factor in any one factorization:

one 2 (in 18 and in 30)
two 3's (in 18 and in 45)
one 5 (in 30 and in 45)

So LCM $= 2 \cdot 3 \cdot 3 \cdot 5 = 90$

90 is the smallest number divisible by all the numbers 18, 30, and 45. (Note also that the LCM, 90, contains all the factors of each of the numbers 18, 30, and 45.)

EXAMPLE 2

Find the LCM for 36, 24, and 48.

(a) Prime factorizations:

$$\left. \begin{array}{l} 36 = 4 \cdot 9 = 2^2 \cdot 3^2 \\ 24 = 8 \cdot 3 = 2^3 \cdot 3 \\ 48 = 16 \cdot 3 = 2^4 \cdot 3 \end{array} \right\} \quad \begin{array}{l} \text{LCM} = 2^4 \cdot 3^2 \\ \qquad = 144 \end{array}$$

(b) The only prime factors that appear are 2 and 3.

(c) There are four 2's in 48 and two 3's in 36.

Thus, the product $2^4 \cdot 3^2 = 144$ is the least common multiple.

EXAMPLE 3

Find the LCM for 27, 30, and 42.

$$\left. \begin{array}{l} 27 = 3 \cdot 9 = 3^3 \\ 30 = 6 \cdot 5 = 2 \cdot 3 \cdot 5 \\ 42 = 2 \cdot 21 = 2 \cdot 3 \cdot 7 \end{array} \right\} \quad \begin{array}{l} \text{LCM} = 2 \cdot 3^3 \cdot 5 \cdot 7 \\ \qquad = 1890 \end{array}$$

The second technique for finding the LCM involves division.

Method II

To find the LCM of a set of counting numbers:

1. Write the numbers horizontally and find a prime number that will divide into more than one number, if possible.

> **2.** Divide by that prime and write the quotients beneath the dividends. Rewrite any numbers not divided beneath themselves.
>
> **3.** Continue the process until no two numbers have a common prime divisor.
>
> **4.** The LCM is the product of all the prime divisors and the last set of quotients.

EXAMPLE 4

Find the LCM for 20, 25, 18, and 6.

$$
\begin{array}{r|rrrr}
5 & 20 & 25 & 18 & 6 \\
\hline
2 & 4 & 5 & 18 & 6 \\
\hline
3 & 2 & 5 & 9 & 3 \\
\hline
& 2 & 5 & 3 & 1
\end{array}
$$

$$
\begin{aligned}
\text{LCM} &= 5 \cdot 2 \cdot 3 \cdot 2 \cdot 5 \cdot 3 \cdot 1 \\
&= 2^2 \cdot 3^2 \cdot 5^2 \\
&= 900
\end{aligned}
$$

Finding How Many Times Each Number Divides into the LCM

In Example 3, by using prime factorizations we found that for 27, 30, and 42 the LCM is 1890. We know that 1890 is a multiple of 27. That is, 27 will divide evenly into 1890. How many times will 27 divide into 1890? You could perform long division:

$$
\begin{array}{r}
70 \\
27 \overline{)\,1890} \\
189 \\
\hline
00 \\
0 \\
\hline
0
\end{array}
$$

Thus, 27 divides 70 times into 1890.

However, we do not need to divide. We can find the same result much more quickly by looking at the prime factorization of 1890 (which we have as a result of finding the LCM). First, find the prime factors for 27. Then the remaining prime factors will have 70 as their product.

$$
\begin{aligned}
1890 &= 2 \cdot 3^3 \cdot 5 \cdot 7 = \underbrace{(3 \cdot 3 \cdot 3)}_{27} \cdot \underbrace{(2 \cdot 5 \cdot 7)}_{70}
\end{aligned}
$$

The following examples illustrate how to use prime factors to tell how many times each number in a set divides into the LCM for that set of numbers. **This technique is very useful in working with fractions.**

EXAMPLE 5

Find the LCM for 27, 30, and 42, and tell how many times each number divides into the LCM.

$$27 = 3^3$$
$$30 = 2 \cdot 3 \cdot 5 \left.\vphantom{\begin{matrix}a\\b\\c\end{matrix}}\right\} \quad \text{LCM} = 2 \cdot 3^3 \cdot 5 \cdot 7 = 1890$$
$$42 = 2 \cdot 3 \cdot 7$$

$$1890 = 2 \cdot 3 \cdot 3 \cdot 3 \cdot 5 \cdot 7 = (3 \cdot 3 \cdot 3) \cdot (2 \cdot 5 \cdot 7)$$
$$= \quad 27 \quad \cdot \quad 70$$

$$1890 = 2 \cdot 3 \cdot 3 \cdot 3 \cdot 5 \cdot 7 = (2 \cdot 3 \cdot 5) \cdot (3 \cdot 3 \cdot 7)$$
$$= \quad 30 \quad \cdot \quad 63$$

$$1890 = 2 \cdot 3 \cdot 3 \cdot 3 \cdot 5 \cdot 7 = (2 \cdot 3 \cdot 7) \cdot (3 \cdot 3 \cdot 5)$$
$$= \quad 42 \quad \cdot \quad 45$$

So, 27 divides into 1890 70 times;
    30 divides into 1890 63 times;
    42 divides into 1890 45 times.

EXAMPLE 6

Find the LCM for 12, 18, and 66, and tell how many times each number divides into the LCM.

$$12 = 2^2 \cdot 3$$
$$18 = 2 \cdot 3^2 \left.\vphantom{\begin{matrix}a\\b\\c\end{matrix}}\right\} \quad \text{LCM} = 2^2 \cdot 3^2 \cdot 11 = 396$$
$$66 = 2 \cdot 3 \cdot 11$$

$$396 = 2 \cdot 2 \cdot 3 \cdot 3 \cdot 11 = (2 \cdot 2 \cdot 3) \cdot (3 \cdot 11)$$
$$= \quad 12 \quad \cdot \quad 33$$

$$396 = 2 \cdot 2 \cdot 3 \cdot 3 \cdot 11 = (2 \cdot 3 \cdot 3) \cdot (2 \cdot 11)$$
$$= \quad 18 \quad \cdot \quad 22$$

$$396 = 2 \cdot 2 \cdot 3 \cdot 3 \cdot 11 = (2 \cdot 3 \cdot 11) \cdot (2 \cdot 3)$$
$$= \quad 66 \quad \cdot \quad 6$$

So, 12 divides into 396 33 times;
    18 divides into 396 22 times;
    66 divides into 396 6 times.

An Application

Whenever various events occur at regular intervals of time, one question that arises is, "How frequently will these events occur at the same time?" The answer can be stated in terms of the LCM, as illustrated in Example 7.

EXAMPLE 7   Suppose three weather satellites A, B, and C orbit the Earth in different times: A takes 18 hours, B takes 14 hours, and C takes 10 hours. If they are directly above each other now (as shown in part (a) of the figure), in how many hours will they again be directly above each other in exactly the same position as shown in part (a)?

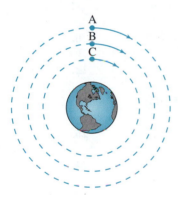

(a) Beginning positions

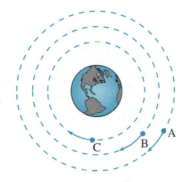

(b) Positions after 5 hours

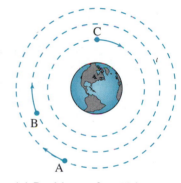

(c) Positions after 10 hours

Solution

When the three satellites are again in the position shown, each will have made some number of complete orbits. Since satellite A takes 18 hours to make one complete orbit of the Earth, the solution must be a multiple of 18. Similarly, the solution must be a multiple of 14 and a multiple of 10 to account for the complete orbits of satellites B and C.

The solution is the LCM of 18, 14, and 10.

$$\left. \begin{array}{l} 18 = 2 \cdot 3^2 \\ 14 = 2 \cdot 7 \\ 10 = 2 \cdot 5 \end{array} \right\} \quad \text{LCM} = 2 \cdot 3^2 \cdot 5 \cdot 7 = 630$$

Thus, the satellites will align again in the position shown in 630 hours (or 26 days and 6 hours).

CLASSROOM PRACTICE

Find the LCM for each of the following sets of numbers.

1. 30, 40, 50   **2.** 28, 70   **3.** 168, 140

ANSWERS: **1.** LCM = 600   **2.** LCM = 140   **3.** LCM = 840

Exercises 2.6

Study the procedures and examples in this section carefully before you begin these exercises.

Find the LCM for each of the following sets of numbers.

1. 8, 12   **2.** 3, 5, 7   **3.** 4, 6, 9   **4.** 3, 5, 9

5. 2, 5, 11   **6.** 4, 14, 18   **7.** 6, 10, 12   **8.** 6, 8, 27

9. 25, 40   **10.** 40, 75   **11.** 28, 98   **12.** 30, 75

13. 30, 80   **14.** 16, 28   **15.** 25, 100   **16.** 20, 50

17. 35, 100   **18.** 144, 216   **19.** 36, 42   **20.** 40, 100

21. 2, 4, 8   **22.** 10, 15, 35   **23.** 8, 13, 15   **24.** 25, 35, 49

25. 6, 12, 15   **26.** 8, 10, 120   **27.** 6, 15, 80   **28.** 13, 26, 169

29. 45, 125, 150   **30.** 34, 51, 54   **31.** 33, 66, 121

32. 36, 54, 72   **33.** 45, 145, 290   **34.** 54, 81, 108

35. 45, 75, 135   **36.** 35, 40, 72   **37.** 10, 20, 30, 40

38. 15, 25, 30, 40   **39.** 24, 40, 48, 56   **40.** 169, 637, 845

Find the LCM and tell how many times each number divides into the LCM.

41. 8, 10, 15   **42.** 6, 15, 30   **43.** 10, 15, 24

44. 8, 10, 120   **45.** 6, 18, 27, 45   **46.** 12, 95, 228

47. 45, 63, 98   **48.** 40, 56, 196   **49.** 99, 143, 363

50. 125, 135, 225

51. Two long-distance joggers are running on the same course in the same direction. They meet at the water fountain and say, "Hi." One jogger goes around the course in 10 minutes and the other goes around the course in 14 minutes. They continue to jog until they meet again at the water fountain. (a) How many minutes elapsed between meetings at the fountain? (b) How many times will each have jogged around the course?

52. Three night watchmen meet at a doughnut shop before they walk around inspecting buildings at a shopping center. The watchmen take 9, 12, and 14 minutes, respectively, for the inspection trip. (a) If they start at the same time, in how many minutes will they meet again at the doughnut shop? (b) How many inspection trips will each watchman have made?

53. Two astronauts miss connections at their first rendezvous in space. (a) If one astronaut circles the earth every 12 hours and the other every 16 hours, in how many hours will they rendezvous again at the same place? (b) How many orbits will each astronaut have made before the second rendezvous?

54. Three truck drivers eat lunch together whenever all three are at the routing station at the same time. The first driver's route takes 5 days, the second driver's takes 15 days, and the third driver's takes 6 days. (a) How often do the three drivers eat lunch together? (b) If the first driver's route was changed to take 6 days, how often would they eat lunch together?

55. Four book salespersons leave the home office the same day. They take 10 days, 12 days, 15 days, and 18 days, respectively, to travel their own sales regions. (a) In how many days will they all meet again at the home office? (b) How many sales trips will each have made?

Writing and Thinking about Mathematics

56. From studying Example 7 and working Exercises 51–55, write out a general statement about the types of word problems (or characteristics of these problems) in which the least common multiple is a related idea.

Collaborative Learning Exercise

57. With the class separated into teams of two to four students, each team is to write one or two paragraphs explaining what topic in this chapter they found to be the most interesting and why. Each team leader is to read the paragraph(s), with classroom discussion to follow.

Summary: Chapter 2

Key Terms and Ideas

An **exponent** is a number that tells how many times its base is to be used as a factor.

A **prime number** is a counting number with exactly two different factors (or divisors). A **composite number** is a counting number with more than two different factors (or divisors). The number 1 is neither prime nor composite.

The only factors (or divisors) of a composite number are:

(a) 1 and the number itself

(b) each prime factor

(c) products of various combinations of prime factors

The **least common multiple (LCM)** of a set of natural numbers is the smallest number common to all the sets of multiples of the given numbers.

Rules and Properties

For any whole number a, $a = a^1$.
For any nonzero number a, $a^0 = 1$.

Rules for Order of Operations

1. First, simplify within grouping symbols, such as parentheses ( ), brackets [ ], or braces { }. Start with the innermost grouping.

2. Second, find any power indicated by exponents.

3. Third, moving from **left to right,** perform any multiplications or divisions in the order in which they appear.

4. Fourth, moving from **left to right,** perform any additions or subtractions in the order in which they appear.

Tests for Divisibility by 2, 3, 4, 5, 9, and 10

For 2:   If the last digit (units digit) of a whole number is 0, 2, 4, 6, or 8, then the whole number is divisible by 2. (In other words, even whole numbers are divisible by 2; odd whole numbers are not divisible by 2.)

For 3:   If the sum of the digits of a whole number is divisible by 3, then the number is divisible by 3.

For 4:   If the last two digits of a whole number form a number that is divisible by 4, then the number is divisible by 4. (00 is considered to be divisible by 4.)

For 5:   If the last digit of a whole number is 0 or 5, then the number is divisible by 5.

For 9:   If the sum of the digits of a whole number is divisible by 9, then the number is divisible by 9.

For 10:   If the last digit of a whole number is 0, then the number is divisible by 10.

The **Fundamental Theorem of Arithmetic** states that every composite number has a unique prime factorization.

Procedures

To Find the Prime Factorization of a Composite Number:

Method I

1. Factor the composite number into **any** two factors.

2. If either or both factors are not prime, factor each of these.

3. Continue this process until all factors are prime.

Method II

1. Divide the composite number by **any** prime number that will divide into it.

2. Continue to divide the **quotient** by prime numbers until the quotient is prime.

3. The prime factorization is the product of all the prime divisors and the last quotient, which is a prime number.

To Find the LCM of a Set of Counting Numbers:

Method I

1. Find the prime factorization of each number.

2. Find the prime factors that appear in **any one** of the prime factorizations.

3. Form the product of these primes using each prime the most number of times it appears in **any one** of the prime factorizations.

> **Method II**
>
> **1.** Write the numbers horizontally and find a prime number that will divide into more than one number, if possible.
>
> **2.** Divide by that prime and write the quotients beneath the dividends. Rewrite any numbers not divided beneath themselves.
>
> **3.** Continue the process until no two numbers have a common prime divisor.
>
> **4.** The LCM is the product of all the prime divisors and the last set of quotients.

Review Questions: Chapter 2

1. In the expression $3^4 = 81$, 3 is called the _____, 4 is called the _____, and 81 is called the _____.

2. A prime number is a counting number with exactly two _____.

3. Every composite number has a unique _____ factorization.

4. Since 36 has more than two different factors, 36 is a _____ number.

Find each power.

5. 2^4 **6.** 3^3 **7.** 13^2 **8.** 21^2

Evaluate each of the following expressions.

9. $7 + 3 \cdot 2 - 1 + 9 \div 3$

10. $3 \cdot 2^5 - 2 \cdot 5^2$

11. $14 \div 2 + 2 \cdot 8 + 30 \div 5 \cdot 2$

12. $(16 \div 2^2 + 6) \div 2 + 8$

13. $(75 - 3 \cdot 5) \div 10 - 4$

14. $(7^2 \cdot 2 + 2) \div 10 \div (2 + 3)$

Determine which of the numbers 2, 3, 4, 5, 9, and 10 (if any) will divide exactly into each of the following numbers.

15. 45   **16.** 72   **17.** 479

18. 5040   **19.** 8836   **20.** 575,493

21. List the multiples of 3. Are any of these multiples prime numbers?

22. Is 223 a prime number? Show your work.

23. List the prime numbers less than 60.

24. Find two factors of 24 whose product is 24 and whose sum is 10.

25. Find two factors of 60 whose product is 60 and whose sum is 17.

Find the prime factorization for each of the following numbers.

26. 150   **27.** 65   **28.** 84   **29.** 92

Find the LCM for each of the following sets of numbers.

30. 8, 14, 24   **31.** 8, 12, 25, 36   **32.** 27, 54, 135

33. Find the LCM for the numbers 18, 39, and 63 and tell how many times each number divides into the LCM.

34. One racing car goes around the track every 30 seconds and the other every 35 seconds. If both cars start a race from the same point, in how many seconds will the first car be exactly one lap ahead of the second car? How many laps will each car have made when the first car is two laps ahead of the second?

Test: Chapter 2

1. Name (a) the exponent and (b) the base, and (c) find the power of 7^3.

2. List the squares of the prime numbers between 5 and 20.

Evaluate each of the following expressions.

3. $15 \div 3 + 6 \cdot 2^3$ **4.** $7 \cdot 3^2 + 18 \div 6$

5. $4 + (75 \div 5^2 + 5) + 16 \div 2^3$ **6.** $12 \div 3 \cdot 2 - 5 + 2^3 \div 4$

Using the tests for divisibility, tell which of the numbers 2, 3, 4, 5, 9, and 10 (if any) will divide the following numbers.

7. 216   **8.** 221   **9.** 180

10. List the multiples of 19 that are less than 100.

11. List all of the composite numbers found in the following set:

1, 2, 6, 9, 13, 15, 17, 27, 39, 41, 51, 59

Find the prime factorization for each of the following numbers.

12. 70   **13.** 234   **14.** 750

15. List the factors of 45.

Find the LCM for each of the following sets of numbers.

16. 15 and 24   **17.** 12, 27, and 36

18. 15, 25, and 75   **19.** 7, 9, and 16

20. Two long-distance runners practice on the same track and start at the same point. One takes 54 seconds and the other takes 63 seconds to go around the track once. (a) If each continues at the same pace, in how many seconds will they meet again at the starting point? (b) How many laps will each have run by the time they meet again at the starting point?

1. Write 40,586 in expanded notation and in its English word equivalent.

Name each property illustrated in Exercises 2–4.

2. $18 \cdot 3 = 3 \cdot 18$

3. $5 + (6 + 10) = (5 + 6) + 10$

4. $72 \cdot 1 = 72$

Perform the indicated operations.

5.
$$
\begin{array}{r}
72 \\
857 \\
+\ 348 \\
\hline
\end{array}
$$

6.
$$
\begin{array}{r}
5004 \\
-\ 897 \\
\hline
\end{array}
$$

7.
$$
\begin{array}{r}
35 \\
\times\ 65 \\
\hline
\end{array}
$$

8. Round off 39,842 to the nearest thousand.

9. Estimate the product 83×52; then find the product.

10. Estimate the quotient $2210 \div 17$; then find the quotient.

11. Divide 10,840 by 35 and indicate any remainder with capital **R**.

12. Multiply 507(3000).

13. Evaluate $5 + (11 \cdot 6 + 3^2) \div 3 \cdot 5$.

14. Without actually dividing, determine which of the numbers 2, 3, 4, 5, 9, and 10 will divide into 72,450.

15. (True or False)   Two is the only even prime number.

16. Find the prime factorization of 375.

17. Find the LCM of the numbers 12, 45, and 50.

18. Find the average of 72, 106, 59, and 79.

19. If you bought two textbooks for a price of $34 each, one textbook for $43, and other supplies for $22, how much did you spend at the bookstore (before taxes)?

20. A painting is in a rectangular frame that is 24 inches wide and 36 inches high. How many square inches of wall space is covered when the painting is hanging?

Fractions

Mathematics at Work!

Young voters meet at a local polling place.

Politics and the political process affect everyone in some way. In local or state or national elections, registered voters make decisions about who will represent them and make choices about various ballot measures.

In major issues at the state and national level, pollsters use mathematics (in particular, statistics and statistical methods) to indicate attitudes and to predict, within certain percentages, how the voters will vote. When there is an important election in your area, read the papers and magazines and listen to the television reports for mathematically related statements predicting the outcome.

Consider the following situation: There are 8000 registered voters in Brownsville, and $\frac{3}{8}$ of these voters live in neighborhoods on the north side of town. A survey indicates that $\frac{4}{5}$ of these north-side voters are in favor of a bond measure for constructing a new recreation facility that would largely benefit their neighborhoods. Also, $\frac{7}{10}$ of the registered voters from all other parts of town are in favor of the measure. We might then want to know:

(a) How many north-side voters favor the bond measure?

(b) How many voters in all favor the bond measure?

(See Example 10, Section 3.1.)

What to Expect in Chapter 3

Chapter 3 deals with understanding fractions (called *rational numbers*) and operations with fractions. Each section has some word problems to reinforce the concepts as they are developed. The term *improper fraction* is introduced in Section 3.1, and then used throughout the chapter. Despite the use of the term "improper," you should understand that improper fractions are perfectly good and as useful as any other type of number.

The commutative and associative properties for multiplication (Section 3.2) and addition (Section 3.4) are shown to apply with fractions just as they do with whole numbers. The prime factorization techniques developed in Chapter 2 are used to reduce fractions to lower terms and to raise fractions to higher terms and form the basis for the development of the operations of multiplication, division (Section 3.3), addition, and subtraction (Section 3.5). For many students this prime factoring method provides valuable insight into the nature of fractions and helps to build a new confidence in working with fractions.

The rules for order of operations (Section 3.6) are the same for evaluating expressions that contain fractions as they are for evaluating expressions with whole numbers.

3.1 Basic Multiplication

OBJECTIVES

1. Know the terms **numerator** and **denominator**.
2. Learn two meanings of a fraction.
   (a) To indicate equal parts of a whole
   (b) To indicate division
3. Learn how to multiply fractions.
4. Understand why no denominator can be 0.

Rational Numbers (or Fractions)

In this chapter we will deal only with fractions in which the **numerator** (top number) and **denominator** (bottom number) are whole numbers, and the denominator is not 0. Such fractions have the technical name of **rational numbers.**

Examples of rational numbers are $\frac{1}{2}, \frac{3}{4}, \frac{9}{10}$, and $\frac{17}{3}$.

Figure 3.1 shows how a whole may be separated into equal parts. We see that the rational numbers $\frac{1}{2}, \frac{2}{4}, \frac{4}{8}$, and $\frac{6}{12}$ all represent the same amount of the whole. These numbers are **equivalent,** or **equal.**

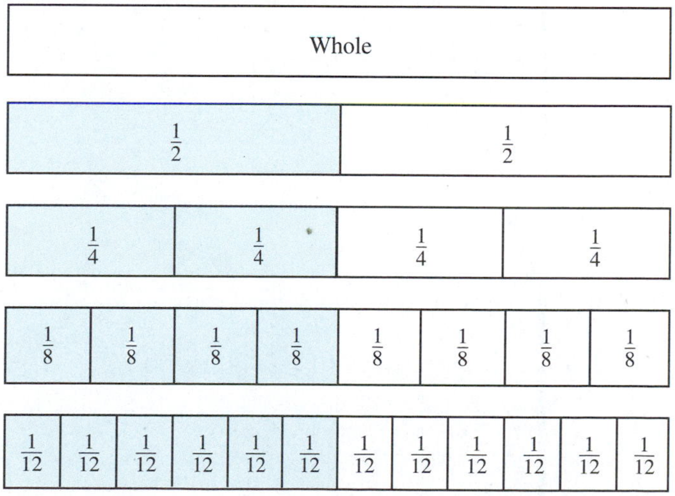

Whole

$\frac{1}{2}$	$\frac{1}{2}$

$\frac{1}{4}$	$\frac{1}{4}$	$\frac{1}{4}$	$\frac{1}{4}$

$\frac{1}{8}$	$\frac{1}{8}$	$\frac{1}{8}$	$\frac{1}{8}$	$\frac{1}{8}$	$\frac{1}{8}$	$\frac{1}{8}$	$\frac{1}{8}$

$\frac{1}{12}$	$\frac{1}{12}$	$\frac{1}{12}$	$\frac{1}{12}$	$\frac{1}{12}$	$\frac{1}{12}$	$\frac{1}{12}$	$\frac{1}{12}$	$\frac{1}{12}$	$\frac{1}{12}$	$\frac{1}{12}$	$\frac{1}{12}$

The shaded regions indicate equivalent (or equal) parts of the whole.

Figure 3.1

Definition

A **rational number** is a number that can be written in the form $\frac{a}{b}$, where a is a whole number and b is a nonzero whole number.

$$\frac{a}{b} \quad \begin{array}{l} \leftarrow \text{numerator} \\ \leftarrow \text{denominator} \end{array}$$

Note: The numerator a can be 0, but the denominator b cannot be 0.

Until otherwise stated, we will use the terms **fraction** and **rational number** to mean the same thing.

Fractions can be used to indicate:

1. Equal parts of a whole.

2. Division.

EXAMPLE 1 $\frac{1}{2}$ can indicate 1 of 2 equal parts.

$\frac{1}{2}$ shaded

EXAMPLE 2 $\frac{3}{4}$ can indicate 3 of 4 equal parts.

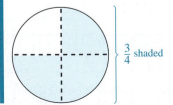

$\frac{3}{4}$ shaded

EXAMPLE 3 $\frac{2}{3}$ can indicate 2 of 3 equal parts.

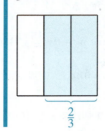

$\frac{2}{3}$

Improper fractions, such as $\frac{5}{3}, \frac{13}{6}$, and $\frac{18}{10}$, are fractions in which the numerator is larger than the denominator. The term "improper fraction" is misleading because there is nothing improper about such fractions. In fact, in algebra and other courses in mathematics, improper fractions are sometimes preferred over mixed numbers such as $2\frac{3}{4}$ and $5\frac{1}{8}$. Mixed numbers will be discussed in Chapter 4.

EXAMPLE 4 $\frac{5}{3}$ can indicate 5 of 3 equal parts (more than a whole).

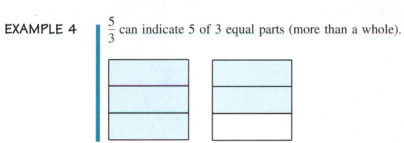

We stated earlier that, in a fraction $\frac{a}{b}$, b is nonzero. That is, $b \neq 0$ ($\neq$ means "is not equal to"). Under the division meaning of fractions, $\frac{21}{7}$ is the same as $21 \div 7$. By the meaning of division related to multiplication,

$$\frac{21}{7} = 3 \qquad \text{because} \qquad 21 = 7 \cdot 3.$$

Similarly,

$$\frac{0}{4} = 0 \qquad \text{because} \qquad 0 = 4 \cdot 0.$$

In general, if $b \neq 0$,

$$\frac{0}{b} = 0 \qquad \text{because} \qquad 0 = b \cdot 0.$$

However, if the denominator is 0, then the fraction has no meaning. We say that **division by 0 is undefined.** The following discussion explains the reasoning.

Division by 0 Is Undefined. No Denominator Can Be 0.

1. Consider $\frac{5}{0} = \boxed{?}$. Then, we must have $5 = 0 \cdot \boxed{?}$. But this is impossible because $0 \cdot \boxed{?} = 0$ and $5 \neq 0$. Therefore, $\frac{5}{0}$ is undefined.

2. Next, consider $\frac{0}{0} = \boxed{?}$. Then, we must have $0 = 0 \cdot \boxed{?}$. But $0 = 0 \cdot \boxed{?}$ is true for any value of $\boxed{?}$, and in arithmetic, an operation such as division cannot give more than one answer. So, we agree that $\frac{0}{0}$ is undefined.

Thus, $\dfrac{a}{0}$ is undefined for any value of a.

EXAMPLE 5 $\dfrac{13}{0}$ is undefined.

EXAMPLE 6 $\dfrac{0}{20} = 0$

Multiplying Rational Numbers (or Fractions)

Now we state the rule for multiplying two fractions and illustrate the rule with diagrams.

To Multiply Fractions:

1. Multiply the numerators.

2. Multiply the denominators.

$$\frac{a}{b} \cdot \frac{c}{d} = \frac{a \cdot c}{b \cdot d}$$

Finding the product of two fractions can be thought of as finding one fractional part **of** another fraction. Thus, when we multiply $\frac{1}{3} \cdot \frac{1}{2}$, we are finding $\frac{1}{3}$ **of** $\frac{1}{2}$.

We use a square to illustrate the product $\frac{2}{3} \cdot \frac{4}{5}$ $\left(\text{or to find } \frac{2}{3} \text{ of } \frac{4}{5}\right)$ in Figure 3.2.

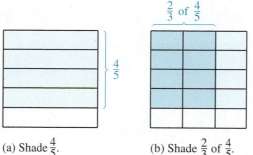

(a) Shade $\frac{4}{5}$.   (b) Shade $\frac{2}{3}$ of $\frac{4}{5}$.   (c) Shaded region is

$\frac{2}{3}$ of $\frac{4}{5}$ or $\frac{2}{3} \cdot \frac{4}{5} = \frac{8}{15}$.

Figure 3.2

EXAMPLE 7   Find the product of $\frac{1}{3}$ and $\frac{2}{5}$ and illustrate the product by shading parts of a square.

Solution

$$\frac{1}{3} \cdot \frac{2}{5} = \frac{1 \cdot 2}{3 \cdot 5} = \frac{2}{15}$$

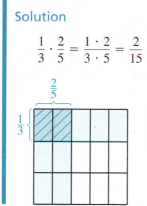

The square is separated into thirds in one direction and into fifths in the other direction. There are 15 equal regions. The overlapping shaded region represents

$$\frac{1}{3} \text{ of } \frac{2}{5} \quad \text{or} \quad \frac{1}{3} \cdot \frac{2}{5} = \frac{2}{15}.$$

EXAMPLE 8   Find $\frac{3}{4}$ of $\frac{5}{7}$.

Solution

$$\frac{3}{4} \cdot \frac{5}{7} = \frac{3 \cdot 5}{4 \cdot 7} = \frac{15}{28}$$

EXAMPLE 9   Find $\frac{7}{6}$ of $\frac{11}{2}$.

Solution

$$\frac{7}{6} \cdot \frac{11}{2} = \frac{7 \cdot 11}{6 \cdot 2} = \frac{77}{12}$$

EXAMPLE 10

In a certain voting district, $\frac{3}{5}$ of the eligible voters are actually registered to vote.

Of these registered voters, $\frac{2}{7}$ are registered as independents with no party affiliation. What fraction of the eligible voters are registered as having no party affiliation?

Solution

Since the unaffiliated voters are a fraction **of** the eligible voters, we multiply:

$$\frac{2}{7} \cdot \frac{3}{5} = \frac{2 \cdot 3}{7 \cdot 5} = \frac{6}{35}$$

Thus, $\frac{6}{35}$ of the eligible voters are registered as independent, with no party affiliation.

Whole numbers can be considered to be fractions with denominator 1. We say that the whole numbers are **equivalent** to their corresponding fractions. Thus,

$$0 = \frac{0}{1}, \quad 1 = \frac{1}{1}, \quad 2 = \frac{2}{1}, \quad 3 = \frac{3}{1}, \quad 4 = \frac{4}{1}, \quad 5 = \frac{5}{1}, \quad \text{and so on.}$$

So, in multiplying fractions and whole numbers, rewrite the whole number as a fraction. Also, the rule for multiplying fractions can be extended to multiplication with more than two fractions.

EXAMPLE 11

$$\frac{3}{14} \cdot 5 = \frac{3}{14} \cdot \frac{5}{1} = \frac{3 \cdot 5}{14 \cdot 1} = \frac{15}{14}$$

EXAMPLE 12

$$\frac{7}{8} \cdot 0 = \frac{7}{8} \cdot \frac{0}{1} = \frac{7 \cdot 0}{8 \cdot 1} = \frac{0}{8} = 0$$

EXAMPLE 13

$$\frac{1}{4} \cdot \frac{3}{5} \cdot \frac{7}{2} = \frac{1 \cdot 3 \cdot 7}{4 \cdot 5 \cdot 2} = \frac{21}{40}$$

EXAMPLE 14

$$\frac{5}{11} \cdot \frac{7}{8} \cdot 3 \cdot \frac{1}{4} = \frac{5}{11} \cdot \frac{7}{8} \cdot \frac{3}{1} \cdot \frac{1}{4} = \frac{5 \cdot 7 \cdot 3 \cdot 1}{11 \cdot 8 \cdot 1 \cdot 4} = \frac{105}{352}$$

Both the commutative property and the associative property of multiplication apply to fractions.

Commutative Property of Multiplication

If $\frac{a}{b}$ and $\frac{c}{d}$ are fractions, then

$$\frac{a}{b} \cdot \frac{c}{d} = \frac{c}{d} \cdot \frac{a}{b}$$

Associative Property of Multiplication

If $\frac{a}{b}, \frac{c}{d},$ and $\frac{e}{f}$ are fractions, then

$$\frac{a}{b} \cdot \frac{c}{d} \cdot \frac{e}{f} = \frac{a}{b}\left(\frac{c}{d} \cdot \frac{e}{f}\right) = \left(\frac{a}{b} \cdot \frac{c}{d}\right)\frac{e}{f}$$

EXAMPLE 15

$\frac{3}{4} \cdot \frac{1}{7} = \frac{1}{7} \cdot \frac{3}{4}$ Illustrates the commutative property of multiplication.

$\frac{3}{4} \cdot \frac{1}{7} = \frac{3 \cdot 1}{4 \cdot 7} = \frac{3}{28}$ and $\frac{1}{7} \cdot \frac{3}{4} = \frac{1 \cdot 3}{7 \cdot 4} = \frac{3}{28}$

EXAMPLE 16

$\left(\frac{5}{8} \cdot \frac{3}{2}\right) \cdot \frac{3}{11} = \frac{5}{8} \cdot \left(\frac{3}{2} \cdot \frac{3}{11}\right)$ Illustrates the associative property of multiplication.

$\left(\frac{5}{8} \cdot \frac{3}{2}\right) \cdot \frac{3}{11} = \left(\frac{5 \cdot 3}{8 \cdot 2}\right) \cdot \frac{3}{11} = \frac{15}{16} \cdot \frac{3}{11} = \frac{15 \cdot 3}{16 \cdot 11} = \frac{45}{176}$

and $\frac{5}{8} \cdot \left(\frac{3}{2} \cdot \frac{3}{11}\right) = \frac{5}{8} \cdot \left(\frac{3 \cdot 3}{2 \cdot 11}\right) = \frac{5}{8} \cdot \frac{9}{22} = \frac{5 \cdot 9}{8 \cdot 22} = \frac{45}{176}$

CLASSROOM PRACTICE

Find the products.

1. $\frac{1}{4} \cdot \frac{1}{4}$ **2.** $\frac{3}{5} \cdot \frac{4}{7}$ **3.** $0 \cdot \frac{5}{6} \cdot \frac{7}{8}$ **4.** $\frac{1}{2} \cdot \frac{3}{7} \cdot \frac{9}{2}$

5. Find $\frac{1}{2}$ of $\frac{1}{6}$.

ANSWERS: **1.** $\frac{1}{16}$ **2.** $\frac{12}{35}$ **3.** 0 **4.** $\frac{27}{28}$ **5.** $\frac{1}{12}$

Exercises 3.1

1. Each of the following products has the same value. What is that value?

(a) $\frac{0}{4} \cdot \frac{5}{6}$ (b) $\frac{3}{10} \cdot \frac{0}{8}$ (c) $\frac{1}{10} \cdot \frac{0}{10}$ (d) $\frac{0}{6} \cdot \frac{0}{52}$

2. What is the value, if any, of each of the following numbers?

(a) $\dfrac{7}{0}$ 　　　(b) $\dfrac{16}{0}$ 　　　(c) $\dfrac{75}{0} \cdot \dfrac{2}{0}$ 　　　(d) $\dfrac{1}{0} \cdot \dfrac{0}{2}$

Write a fraction that represents the shaded parts in each of the following diagrams.

3.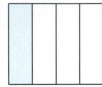

4.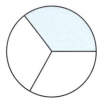

5.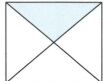

6.

7.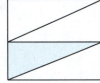

8.

Draw a square with appropriate shading to illustrate each of the following products. (See Example 7.)

9. $\dfrac{1}{5} \cdot \dfrac{3}{4}$ 　　　**10.** $\dfrac{5}{6} \cdot \dfrac{5}{6}$ 　　　**11.** $\dfrac{2}{3} \cdot \dfrac{4}{7}$ 　　　**12.** $\dfrac{3}{4} \cdot \dfrac{3}{4}$

Find the fractional parts as indicated.

13. Find $\dfrac{1}{5}$ of $\dfrac{3}{5}$. 　　　　　　**14.** Find $\dfrac{1}{4}$ of $\dfrac{1}{4}$.

15. Find $\dfrac{1}{9}$ of $\dfrac{2}{3}$. 　　　　　　**16.** Find $\dfrac{3}{4}$ of $\dfrac{2}{5}$.

17. Find $\dfrac{1}{2}$ of $\dfrac{1}{2}$. 　　　　　　**18.** Find $\dfrac{1}{3}$ of $\dfrac{1}{3}$.

Find the following products.

19. $\dfrac{3}{4} \cdot \dfrac{3}{4}$ 　　　**20.** $\dfrac{5}{6} \cdot \dfrac{5}{6}$ 　　　**21.** $\dfrac{1}{9} \cdot \dfrac{4}{9}$ 　　　**22.** $\dfrac{4}{7} \cdot \dfrac{3}{5}$

23. $\dfrac{5}{16} \cdot \dfrac{1}{2}$ 　　　**24.** $\dfrac{7}{16} \cdot \dfrac{1}{4}$ 　　　**25.** $\dfrac{0}{5} \cdot \dfrac{7}{6}$ 　　　**26.** $\dfrac{0}{3} \cdot \dfrac{5}{7}$

27. $\dfrac{3}{5} \cdot \dfrac{3}{5}$ 　　　**28.** $\dfrac{4}{9} \cdot \dfrac{4}{9}$ 　　　**29.** $\dfrac{3}{10} \cdot \dfrac{3}{10}$ 　　　**30.** $\left(\dfrac{4}{1}\right)\left(\dfrac{3}{1}\right)$

31. $\left(\dfrac{2}{1}\right)\left(\dfrac{5}{1}\right)$ **32.** $\dfrac{9}{10}\left(\dfrac{1}{5}\right)$ **33.** $\dfrac{7}{10}\left(\dfrac{1}{5}\right)$ **34.** $\left(\dfrac{3}{5}\right)\dfrac{1}{10}$

35. $\left(\dfrac{2}{3}\right)\dfrac{2}{7}$ **36.** $\dfrac{5}{6}\cdot\dfrac{11}{3}$ **37.** $\dfrac{9}{4}\cdot\dfrac{11}{5}$ **38.** $\dfrac{1}{5}\cdot\dfrac{2}{7}\cdot\dfrac{3}{11}$

39. $\dfrac{4}{13}\cdot\dfrac{2}{5}\cdot\dfrac{6}{7}$ **40.** $\dfrac{7}{8}\cdot\dfrac{7}{9}\cdot\dfrac{7}{3}$ **41.** $\dfrac{1}{6}\cdot\dfrac{1}{10}\cdot\dfrac{1}{6}$

42. $\dfrac{9}{100}\cdot 1\cdot 3$ **43.** $\dfrac{5}{1}\cdot\dfrac{12}{1}\cdot\dfrac{14}{1}$ **44.** $\dfrac{1}{3}\cdot\dfrac{1}{10}\cdot\dfrac{7}{3}$

45. $\dfrac{3}{11}\cdot\dfrac{0}{8}\cdot\dfrac{6}{7}$ **46.** $\dfrac{0}{4}\cdot\dfrac{3}{8}\cdot\dfrac{1}{5}$ **47.** $\dfrac{5}{6}\cdot\dfrac{5}{11}\cdot\dfrac{5}{7}\cdot\dfrac{0}{3}$

48. $\left(\dfrac{3}{4}\right)\left(\dfrac{5}{8}\right)\left(\dfrac{11}{13}\right)$ **49.** $\left(\dfrac{7}{5}\right)\left(\dfrac{8}{3}\right)\left(\dfrac{13}{3}\right)$ **50.** $\left(\dfrac{1}{10}\right)\left(\dfrac{3}{5}\right)\left(\dfrac{3}{10}\right)$

Tell which property of multiplication is illustrated in each problem.

51. $\dfrac{5}{9}\cdot\dfrac{11}{14}=\dfrac{11}{14}\cdot\dfrac{5}{9}$ **52.** $\dfrac{3}{5}\cdot\left(\dfrac{7}{8}\cdot\dfrac{1}{2}\right)=\left(\dfrac{3}{5}\cdot\dfrac{7}{8}\right)\cdot\dfrac{1}{2}$

53. $\left(\dfrac{6}{11}\cdot\dfrac{13}{5}\right)\cdot\dfrac{7}{5}=\dfrac{6}{11}\cdot\left(\dfrac{13}{5}\cdot\dfrac{7}{5}\right)$ **54.** $\dfrac{2}{3}\cdot 4=4\cdot\dfrac{2}{3}$

55. Explain in your own words why division by 0 is undefined. (You might show your explanation to a friend to see if it is easily understood.)

56. If you have $10 and you spend $3 for a sandwich and drink, what fraction of your money did you spend for food? What fraction do you still have?

57. In a class of 35 students, 6 students received A's on a mathematics test. What fraction of the class did not receive an A?

58. A glass is 4 inches tall. If the glass is five-eighths full with water, what is the height of the water in the glass?

59. A recipe calls for $\dfrac{3}{4}$ cup of flour. How much flour should be used if only half the recipe is to be made?

60. One of Michelle's birthday presents was a box of candy. If half the candy was chocolate-covered and one-fourth of the chocolate-covered candy had cherries inside, what fraction of the candy was chocolate-covered cherries?

3.2 Multiplication and Reducing

OBJECTIVES

1. Learn how to reduce a fraction to lowest terms.
2. Know when a fraction is in lowest terms.
3. Be able to multiply fractions and reduce at the same time.

Raising Fractions to Higher Terms

We know that 1 is called the **multiplicative identity** for whole numbers; that is, $a \cdot 1 = a$. The number 1 is also the **multiplicative identity** for fractions since

$$\frac{a}{b} \cdot 1 = \frac{a}{b} \cdot \frac{1}{1} = \frac{a \cdot 1}{b \cdot 1} = \frac{a}{b}$$

Multiplicative Identity

For any fraction $\frac{a}{b}$, where $b \neq 0$,

$$\frac{a}{b} \cdot 1 = \frac{a}{b}$$

Now, the number 1 can be written as a fraction with the same counting number as numerator and denominator. In general,

$$1 = \frac{1}{1} = \frac{2}{2} = \frac{3}{3} = \frac{4}{4} = \frac{5}{5} = \cdots = \frac{k}{k} = \cdots$$

where k is any counting number.

Therefore, we have the following important property of fractions.

$$\frac{a}{b} = \frac{a}{b} \cdot 1 = \frac{a}{b} \cdot \frac{k}{k} = \frac{a \cdot k}{b \cdot k} \qquad \text{where } k \neq 0$$

This means that a fraction is unchanged if the numerator and denominator are multiplied by the same nonzero number.

If we find a fraction equal to (or equivalent to) a given fraction with a larger denominator, we say that the given fraction has been **raised to higher terms.** So, **to raise a fraction to higher terms,** multiply the numerator and denominator by the same counting number.

EXAMPLE 1

$$\frac{3}{4} = \frac{?}{28}$$

Solution

We want to raise fourths to twenty-eighths. Try multiplying by $\frac{5}{5}$.

$$\frac{3}{4} = \frac{3}{4} \cdot \frac{5}{5} = \frac{15}{20} \quad \text{WRONG FRACTION}$$

The fraction $\frac{15}{20}$ does equal $\frac{3}{4}$, but $\frac{15}{20}$ has the **wrong denominator.** We want an equal fraction with denominator 28. Since $4 \cdot 7 = 28$, we should use $\frac{7}{7}$ instead of $\frac{5}{5}$.

$$\frac{3}{4} = \frac{3}{4} \cdot \frac{7}{7} = \frac{21}{28} \quad \text{RIGHT FRACTION}$$

Just multiplying by 1 was not good enough. The right form of 1, namely $\frac{7}{7}$, was needed.

EXAMPLE 2

$$\frac{9}{10} = \frac{?}{30}$$

Solution

Now we are smarter. Since $10 \cdot 3 = 30$, multiply $\frac{9}{10}$ by $\frac{3}{3}$ to get an equal fraction with denominator 30.

$$\frac{9}{10} = \frac{9}{10} \cdot \frac{3}{3} = \frac{9 \cdot 3}{10 \cdot 3} = \frac{27}{30}$$

REMEMBER: The numerator and denominator must be multiplied by the same number.

EXAMPLE 3

$$\frac{11}{8} = \frac{?}{40}$$

Solution

Use $\dfrac{k}{k} = \dfrac{5}{5}$ since $8 \cdot 5 = 40$.

$$\frac{11}{8} = \frac{11}{8} \cdot \frac{5}{5} = \frac{11 \cdot 5}{8 \cdot 5} = \frac{55}{40}$$

Reducing Fractions

Changing a fraction to lower terms is the same as **reducing** the fraction. **A fraction is reduced if the numerator and denominator have no common factor other than 1.**

To Reduce a Fraction to Lowest Terms:

1. Factor the numerator and denominator into prime factors.

2. Use the fact that $\dfrac{k}{k} = 1$ and "cancel" or "divide out" common factors.

Note: Reduced fractions might be improper fractions. We will discuss mixed numbers in Chapter 4.

Reduce each fraction to lowest terms.

EXAMPLE 4 $\dfrac{21}{35}$

$$\frac{21}{35} = \frac{3 \cdot 7}{5 \cdot 7} = \frac{3}{5} \cdot \frac{7}{7} = \frac{3}{5} \cdot 1 = \frac{3}{5}$$

or $\dfrac{21}{35} = \dfrac{3 \cdot \cancel{7}}{5 \cdot \cancel{7}} = \dfrac{3}{5}$ Here we "cancel" or "divide out" the 7's with the understanding that $\dfrac{7}{7} = 1$.

EXAMPLE 5 $\dfrac{72}{90}$

$$\frac{72}{90} = \frac{6 \cdot 12}{9 \cdot 10} = \frac{2 \cdot 3 \cdot 2 \cdot 2 \cdot 3}{3 \cdot 3 \cdot 2 \cdot 5} = \frac{2}{2} \cdot \frac{3}{3} \cdot \frac{3}{3} \cdot \frac{2 \cdot 2}{5} = 1 \cdot 1 \cdot 1 \cdot \frac{4}{5} = \frac{4}{5}$$

or $\dfrac{72}{90} = \dfrac{\cancel{2} \cdot \cancel{3} \cdot 2 \cdot 2 \cdot \cancel{3}}{\cancel{3} \cdot \cancel{3} \cdot \cancel{2} \cdot 5} = \dfrac{4}{5}$

EXAMPLE 6 $\dfrac{36}{27}$

$$\frac{36}{27} = \frac{2 \cdot 2 \cdot \cancel{3} \cdot \cancel{3}}{3 \cdot \cancel{3} \cdot \cancel{3}} = \frac{4}{3}$$

It is **not necessary** for you to use prime factors. If you "see" any common factor, you can reduce using that factor. However, using prime factors allows you to see all the factors and gives you confidence that the fraction is completely reduced.

$$\frac{36}{27} = \frac{4 \cdot \cancel{9}}{3 \cdot \cancel{9}} = \frac{4}{3}$$

EXAMPLE 7 $\dfrac{15}{75}$

$$\frac{15}{75} = \frac{\cancel{3} \cdot \cancel{5} \cdot 1}{\cancel{3} \cdot \cancel{5} \cdot 5} = \frac{1}{5}$$ If all the factors in the numerator (or denominator) cancel, then 1 **must** be used as a factor.

or $$\frac{15}{75} = \frac{\cancel{15} \cdot 1}{\cancel{15} \cdot 5} = \frac{1}{5}$$ The numerator is a factor of the denominator.

Multiplying and Reducing Fractions at the Same Time

Now we can multiply fractions and reduce all in one step by using prime factors. Find each product and reduce.

EXAMPLE 8 $\dfrac{18}{35} \cdot \dfrac{7}{9} = \dfrac{18 \cdot 7}{35 \cdot 9} = \dfrac{2 \cdot \cancel{3} \cdot \cancel{3} \cdot \cancel{7}}{5 \cdot \cancel{7} \cdot \cancel{3} \cdot \cancel{3}} = \dfrac{2}{5}$

EXAMPLE 9 $\dfrac{9}{10} \cdot \dfrac{25}{24} \cdot \dfrac{44}{55} = \dfrac{9 \cdot 25 \cdot 44}{10 \cdot 24 \cdot 55} = \dfrac{3 \cdot \cancel{3} \cdot \cancel{5} \cdot \cancel{5} \cdot \cancel{2} \cdot \cancel{2} \cdot \cancel{11}}{2 \cdot \cancel{5} \cdot 2 \cdot \cancel{2} \cdot \cancel{2} \cdot \cancel{3} \cdot \cancel{5} \cdot \cancel{11}} = \dfrac{3}{2 \cdot 2} = \dfrac{3}{4}$

EXAMPLE 10 $\dfrac{17}{100} \cdot \dfrac{25}{34} \cdot 8 = \dfrac{17 \cdot 25 \cdot 8}{100 \cdot 34 \cdot 1} = \dfrac{\cancel{17} \cdot \cancel{25} \cdot \cancel{4} \cdot \cancel{2} \cdot 1}{\cancel{25} \cdot \cancel{4} \cdot \cancel{2} \cdot \cancel{17} \cdot 1} = \dfrac{1}{1} = 1$

It is **not necessary** for you to use prime factors if you notice other common factors. However, if you have any difficulty understanding how to multiply and reduce, use prime factors. Then you will be sure that your answer is reduced.

Another technique frequently used is to divide the numerators and denominators by common factors whether they are prime or not. If these factors are easily seen, then this technique is probably faster. But you must be careful and organized. Examples 8, 9, and 10 are shown again using the division technique.

EXAMPLE 8'

$$\frac{\overset{2}{\cancel{18}}}{\underset{5}{\cancel{35}}} \cdot \frac{\overset{1}{\cancel{7}}}{\underset{1}{\cancel{9}}} = \frac{2 \cdot 1}{5 \cdot 1} = \frac{2}{5}$$

9 is divided into both 9 and 18.
7 is divided into both 7 and 35.

EXAMPLE 9'

$$\frac{\overset{3}{\cancel{9}}}{\underset{2}{\cancel{10}}} \cdot \frac{\overset{1}{\cancel{25}}}{\underset{8}{\cancel{24}}} \cdot \frac{\overset{1}{\cancel{44}}}{\underset{8}{\cancel{55}}} = \frac{3}{4}$$

11 is divided into both 44 and 55.
5 is divided into both 5 and 25.
5 is divided into both 5 and 10.
3 is divided into both 9 and 24.
4 is divided into both 4 and 8.

EXAMPLE 10'

$$\frac{\overset{1}{\cancel{17}}}{\underset{1}{\cancel{100}}} \cdot \frac{\overset{1}{\cancel{25}}}{\underset{1}{\cancel{34}}} \cdot \frac{\overset{2}{\cancel{8}}}{1} = \frac{1}{1} = 1$$

17 is divided into both 17 and 34.
25 is divided into both 25 and 100.
4 is divided into both 4 and 8.
2 is divided into both 2 and 2.

Use the technique that suits your own abilities and understanding.

CLASSROOM PRACTICE

Raise to higher terms as indicated.

1. $\dfrac{3}{5} = \dfrac{?}{100}$ **2.** $\dfrac{7}{2} = \dfrac{?}{10}$

Reduce to lowest terms.

3. $\dfrac{25}{55}$ **4.** $\dfrac{34}{51}$

Multiply.   [Hint: Write 6 as $\dfrac{6}{1}$ and factor before multiplying.]

5. $\dfrac{17}{100} \cdot \dfrac{27}{34} \cdot \dfrac{25}{9} \cdot 6$

ANSWERS: **1.** $\dfrac{60}{100}$ **2.** $\dfrac{35}{10}$ **3.** $\dfrac{5}{11}$ **4.** $\dfrac{2}{3}$ **5.** $\dfrac{9}{4}$

Exercises 3.2

Raise each fraction to higher terms as indicated.

1. $\dfrac{3}{4} = \dfrac{3}{4} \cdot \dfrac{?}{?} = \dfrac{?}{12}$ **2.** $\dfrac{3}{5} = \dfrac{3}{5} \cdot \dfrac{?}{?} = \dfrac{?}{10}$ **3.** $\dfrac{2}{3} = \dfrac{2}{3} \cdot \dfrac{?}{?} = \dfrac{?}{12}$

4. $\dfrac{1}{2} = \dfrac{1}{2} \cdot \dfrac{?}{?} = \dfrac{?}{6}$ **5.** $\dfrac{6}{7} = \dfrac{6}{7} \cdot \dfrac{?}{?} = \dfrac{?}{14}$ **6.** $\dfrac{5}{8} = \dfrac{5}{8} \cdot \dfrac{?}{?} = \dfrac{?}{16}$

7. $\dfrac{5}{9} = \dfrac{5}{9} \cdot \dfrac{?}{?} = \dfrac{?}{27}$ **8.** $\dfrac{3}{4} = \dfrac{3}{4} \cdot \dfrac{?}{?} = \dfrac{?}{20}$ **9.** $\dfrac{1}{2} = \dfrac{1}{2} \cdot \dfrac{?}{?} = \dfrac{?}{20}$

10. $\dfrac{1}{4} = \dfrac{1}{4} \cdot \dfrac{?}{?} = \dfrac{?}{40}$ **11.** $\dfrac{3}{8} = \dfrac{3}{8} \cdot \dfrac{?}{?} = \dfrac{?}{40}$ **12.** $\dfrac{1}{6} = \dfrac{1}{6} \cdot \dfrac{?}{?} = \dfrac{?}{30}$

13. $\dfrac{2}{3} = \dfrac{2}{3} \cdot \dfrac{?}{?} = \dfrac{?}{15}$ **14.** $\dfrac{1}{6} = \dfrac{1}{6} \cdot \dfrac{?}{?} = \dfrac{?}{12}$ **15.** $\dfrac{3}{5} = \dfrac{3}{5} \cdot \dfrac{?}{?} = \dfrac{?}{45}$

16. $\dfrac{2}{3} = \dfrac{2}{3} \cdot \dfrac{?}{?} = \dfrac{?}{9}$ **17.** $\dfrac{10}{11} = \dfrac{10}{11} \cdot \dfrac{?}{?} = \dfrac{?}{44}$ **18.** $\dfrac{5}{11} = \dfrac{5}{11} \cdot \dfrac{?}{?} = \dfrac{?}{33}$

19. $\dfrac{5}{21} = \dfrac{5}{21} \cdot \dfrac{?}{?} = \dfrac{?}{42}$ **20.** $\dfrac{3}{7} = \dfrac{3}{7} \cdot \dfrac{?}{?} = \dfrac{?}{42}$ **21.** $\dfrac{5}{7} = \dfrac{5}{7} \cdot \dfrac{?}{?} = \dfrac{?}{42}$

22. $\dfrac{3}{8} = \dfrac{3}{8} \cdot \dfrac{?}{?} = \dfrac{?}{24}$ **23.** $\dfrac{3}{16} = \dfrac{3}{16} \cdot \dfrac{?}{?} = \dfrac{?}{80}$ **24.** $\dfrac{1}{16} = \dfrac{1}{16} \cdot \dfrac{?}{?} = \dfrac{?}{80}$

25. $\dfrac{7}{2} = \dfrac{7}{2} \cdot \dfrac{?}{?} = \dfrac{?}{20}$ **26.** $\dfrac{14}{3} = \dfrac{14}{3} \cdot \dfrac{?}{?} = \dfrac{?}{9}$ **27.** $\dfrac{5}{2} = \dfrac{5}{2} \cdot \dfrac{?}{?} = \dfrac{?}{20}$

28. $\dfrac{12}{5} = \dfrac{12}{5} \cdot \dfrac{?}{?} = \dfrac{?}{15}$ **29.** $\dfrac{3}{1} = \dfrac{3}{1} \cdot \dfrac{?}{?} = \dfrac{?}{5}$ **30.** $\dfrac{4}{1} = \dfrac{4}{1} \cdot \dfrac{?}{?} = \dfrac{?}{8}$

31. $\dfrac{8}{10} = \dfrac{8}{10} \cdot \dfrac{?}{?} = \dfrac{?}{100}$ **32.** $\dfrac{9}{10} = \dfrac{9}{10} \cdot \dfrac{?}{?} = \dfrac{?}{100}$

33. $\dfrac{6}{10} = \dfrac{6}{10} \cdot \dfrac{?}{?} = \dfrac{?}{100}$ **34.** $\dfrac{7}{10} = \dfrac{7}{10} \cdot \dfrac{?}{?} = \dfrac{?}{100}$

35. $\dfrac{11}{10} = \dfrac{11}{10} \cdot \dfrac{?}{?} = \dfrac{?}{60}$ **36.** $\dfrac{3}{100} = \dfrac{3}{100} \cdot \dfrac{?}{?} = \dfrac{?}{1000}$

37. $\dfrac{9}{100} = \dfrac{9}{100} \cdot \dfrac{?}{?} = \dfrac{?}{1000}$ **38.** $\dfrac{9}{10} = \dfrac{9}{10} \cdot \dfrac{?}{?} = \dfrac{?}{1000}$

39. $\dfrac{3}{10} = \dfrac{3}{10} \cdot \dfrac{?}{?} = \dfrac{?}{1000}$ **40.** $\dfrac{5}{10} = \dfrac{5}{10} \cdot \dfrac{?}{?} = \dfrac{?}{1000}$

Reduce each of the following fractions to lowest terms. Just rewrite the fraction if it is already in lowest terms.

41. $\dfrac{3}{9}$ **42.** $\dfrac{16}{24}$ **43.** $\dfrac{9}{12}$ **44.** $\dfrac{6}{20}$ **45.** $\dfrac{16}{40}$

46. $\dfrac{24}{30}$

47. $\dfrac{14}{36}$

48. $\dfrac{5}{11}$

49. $\dfrac{0}{25}$

50. $\dfrac{75}{100}$

51. $\dfrac{22}{55}$

52. $\dfrac{60}{75}$

53. $\dfrac{30}{36}$

54. $\dfrac{7}{28}$

55. $\dfrac{26}{39}$

56. $\dfrac{27}{56}$

57. $\dfrac{34}{51}$

58. $\dfrac{36}{48}$

59. $\dfrac{24}{100}$

60. $\dfrac{16}{32}$

61. $\dfrac{30}{45}$

62. $\dfrac{28}{42}$

63. $\dfrac{12}{35}$

64. $\dfrac{66}{84}$

65. $\dfrac{14}{63}$

66. $\dfrac{30}{70}$

67. $\dfrac{25}{76}$

68. $\dfrac{70}{84}$

69. $\dfrac{50}{100}$

70. $\dfrac{48}{12}$

71. $\dfrac{54}{9}$

72. $\dfrac{51}{6}$

73. $\dfrac{6}{51}$

74. $\dfrac{27}{72}$

75. $\dfrac{18}{40}$

76. $\dfrac{144}{156}$

77. $\dfrac{150}{135}$

78. $\dfrac{121}{165}$

79. $\dfrac{140}{112}$

80. $\dfrac{96}{108}$

Multiply and reduce each product. [HINT: Factor first so you can reduce as you are multiplying.]

81. $\dfrac{23}{36} \cdot \dfrac{20}{46}$

82. $\dfrac{7}{8} \cdot \dfrac{4}{21}$

83. $\dfrac{5}{15} \cdot \dfrac{18}{24}$

84. $\dfrac{20}{32} \cdot \dfrac{9}{13} \cdot \dfrac{26}{7}$

85. $\dfrac{69}{15} \cdot \dfrac{30}{8} \cdot \dfrac{14}{46}$

86. $\dfrac{42}{52} \cdot \dfrac{27}{22} \cdot \dfrac{33}{9}$

87. $\dfrac{3}{4} \cdot 18 \cdot \dfrac{7}{2} \cdot \dfrac{22}{54}$

88. $\dfrac{9}{10} \cdot \dfrac{35}{40} \cdot \dfrac{65}{15}$

89. $\dfrac{66}{84} \cdot \dfrac{12}{5} \cdot \dfrac{28}{33}$

90. $\dfrac{24}{100} \cdot \dfrac{36}{48} \cdot \dfrac{15}{9}$

91. $\dfrac{17}{10} \cdot \dfrac{5}{42} \cdot \dfrac{18}{51} \cdot 4$

92. $\dfrac{75}{8} \cdot \dfrac{16}{36} \cdot 9 \cdot \dfrac{7}{25}$

93. $5 \cdot \dfrac{3}{10} \cdot \dfrac{8}{9} \cdot \dfrac{1}{2}$

94. $\dfrac{2}{5} \cdot \dfrac{3}{4} \cdot \dfrac{15}{20} \cdot \dfrac{1}{18}$

95. $6 \cdot \dfrac{2}{3} \cdot \dfrac{5}{8} \cdot \dfrac{9}{10} \cdot \dfrac{7}{12}$

96. Suppose that a ball is dropped from a height of 20 feet. If the ball bounces back to five-eighths the height it was dropped, how high will it bounce on its third bounce?

97. A study showed that one-fifth of the students in an elementary school were left-handed. If the school had an enrollment of 550 students, how many were left-handed?

98. For her sprinkler system, Patricia needs to cut a 20-foot piece of plastic pipe into four pieces. If each new piece of pipe is to be half the length of the previous piece, how long will each of the four pieces be?

99. If you go on a bicycle trip of 100 miles in the mountains and one-fourth of the trip is downhill, how many miles are uphill? (Assume no level ground.)

100. A pizza pie is to be cut into fourths. Each of these fourths is to be cut into thirds. What fraction of the pie is each of the final pieces?

101. Major league baseball teams play 162 games each season. If a team has played $\frac{4}{9}$ of its games by the All-Star break (around mid-season), how many games has it played by that time?

102. Venus orbits the sun in $\frac{45}{73}$ of the time that Earth takes to orbit the sun. If we assume that one "earth-year" is 365 days long, how long is a "Venus-year" in terms of "earth-days"?

Check Your Number Sense

103. Consider the following statement:

"If a fraction is less than 1 and greater than 0, the product of this fraction with another number (fraction or whole number) must be less than this other number."

Do you agree with this statement? To help your understanding of this concept, answer the following questions involving the fraction $\frac{3}{4}$.

Is $\frac{3}{4}$ of 12 less than 12?   Is $\frac{3}{4}$ of $\frac{2}{3}$ less than $\frac{2}{3}$?

Is $\frac{3}{4}$ of $\frac{1}{2}$ less than $\frac{1}{2}$?   Is $\frac{3}{4}$ of $\frac{4}{5}$ less than $\frac{4}{5}$?

Answer these same questions using $\frac{2}{3}$ and $\frac{1}{2}$ in place of $\frac{3}{4}$.

♻ The Recycle Bin   (from Section 2.5)

Find the prime factorization for each of the following numbers. Use the tests for divisibility for 2, 3, 4, 5, 9, and 10 whenever they help to find the beginning factors.

1. 45   **2.** 78   **3.** 130

4. 460   **5.** 1,000,000

Find all the factors (or divisors) of each number.

6. 30   **7.** 48   **8.** 63

3.3   Division

Reciprocals

If the product of two fractions is 1, then the fractions are called **reciprocals** of each other. For example,

$$\frac{2}{3} \quad \text{and} \quad \frac{3}{2} \quad \text{are reciprocals because} \quad \frac{2}{3} \cdot \frac{3}{2} = \frac{6}{6} = 1$$

$$\frac{5}{6} \quad \text{and} \quad \frac{6}{5} \quad \text{are reciprocals because} \quad \frac{5}{6} \cdot \frac{6}{5} = \frac{30}{30} = 1$$

Definition

The **reciprocal** of $\dfrac{a}{b}$ is $\dfrac{b}{a}$ ($a \neq 0$ and $b \neq 0$).

The product of a nonzero number and its reciprocal is always 1.

$$\frac{a}{b} \cdot \frac{b}{a} = 1$$

Note: If $a = 0$, then $\dfrac{0}{b}$ has no reciprocal because $\dfrac{b}{0}$ is undefined. That is, **the number 0 has no reciprocal.**

Division with Fractions

To understand how to divide with fractions, remember that division is related to multiplication. For example, with whole numbers we know that

$$18 \div 3 = 6 \quad \text{because} \quad 18 = 6 \cdot 3$$

With fractions, the same reasoning gives

$$\frac{2}{3} \div \frac{7}{8} = \square \qquad \text{because} \qquad \frac{2}{3} = \square \cdot \frac{7}{8}$$

We need to find out what goes in the box. Since division can be indicated with a fraction, we can write

$$\frac{2}{3} \div \frac{7}{8} = \frac{\frac{2}{3}}{\frac{7}{8}}$$

Now, multiplying the numerator (a fraction) and the denominator (another fraction) by the same number will not change the value of the fraction. Multiply by $\frac{8}{7}$, the reciprocal of $\frac{7}{8}$, and see what happens.

$$\frac{2}{3} \div \frac{7}{8} = \frac{\frac{2}{3}}{\frac{7}{8}} = \frac{\frac{2}{3} \cdot \frac{8}{7}}{\frac{7}{8} \cdot \frac{8}{7}} = \frac{\frac{2}{3} \cdot \frac{8}{7}}{1} = \frac{2}{3} \cdot \frac{8}{7}$$

$$\left(\text{Note that here } \frac{k}{k} = \frac{\frac{8}{7}}{\frac{8}{7}} = 1. \right)$$

A division problem has been changed into a multiplication problem. Thus,

$$\frac{2}{3} \div \frac{7}{8} = \frac{2}{3} \cdot \frac{8}{7} = \frac{16}{21}$$

In general,

$$\frac{a}{b} \div \frac{c}{d} = \frac{a}{b} \cdot \frac{d}{c} \qquad \text{where } b, c, d \neq 0$$

In words:
To divide by any number (except 0), multiply by its reciprocal.

Find each of the following quotients. Reduce to lowest terms whenever possible.

EXAMPLE 1

$$\frac{3}{4} \div \frac{8}{5} = \frac{3}{4} \cdot \frac{5}{8} = \frac{15}{32}$$

The divisor is $\frac{8}{5}$, and we multiply by its reciprocal, $\frac{5}{8}$.

EXAMPLE 2 $\dfrac{2}{3} \div \dfrac{1}{2} = \dfrac{2}{3} \cdot \dfrac{2}{1} = \dfrac{4}{3}$ The reciprocal of $\dfrac{1}{2}$ is $\dfrac{2}{1}$.

EXAMPLE 3 $\dfrac{16}{27} \div \dfrac{8}{9} = \dfrac{16}{27} \cdot \dfrac{9}{8} = \dfrac{8 \cdot 2 \cdot 9}{3 \cdot 9 \cdot 8} = \dfrac{2}{3}$ The reciprocal of $\dfrac{8}{9}$ is $\dfrac{9}{8}$. By factoring, the quotient is reduced to $\dfrac{2}{3}$.

EXAMPLE 4 $\dfrac{5}{16} \div 5 = \dfrac{5}{16} \div \dfrac{5}{1} = \dfrac{\overset{1}{\cancel{5}}}{16} \cdot \dfrac{1}{\underset{1}{\cancel{5}}} = \dfrac{1}{16}$ The reciprocal of 5 is $\dfrac{1}{5}$.

EXAMPLE 5 $\dfrac{9}{10} \div \dfrac{9}{10} = \dfrac{\overset{1}{\cancel{9}}}{\underset{1}{\cancel{10}}} \cdot \dfrac{\overset{1}{\cancel{10}}}{\underset{1}{\cancel{9}}} = \dfrac{1}{1} = 1$ Here we get the expected result that a number divided by itself is 1.

EXAMPLE 6 $\dfrac{7}{5} \div \dfrac{5}{7} = \dfrac{7}{5} \cdot \dfrac{7}{5} = \dfrac{49}{25}$ We are not dividing a number by itself, and the quotient is not 1.

The next two examples are given to reinforce your understanding of the use of 0 in division.

EXAMPLE 7 $\dfrac{3}{5} \div 0$ is undefined.

As we discussed in Section 3.1, division by 0 is not allowed under any circumstances.

EXAMPLE 8 $0 \div \dfrac{7}{8} = 0 \cdot \dfrac{8}{7} = \dfrac{0}{1} \cdot \dfrac{8}{7} = \dfrac{0}{7} = 0$

If 0 is divided by any nonzero number, the quotient is always 0.

If you know the product of two numbers and you know one of the numbers, how do you find the other number? The answer is: Divide. For example, with whole numbers, suppose the product of two numbers is 24 and one number is 6. To find the other number, divide 24 by 6:

$$24 \div 6 = 4 \qquad \text{or} \qquad 6\overline{)24}$$
$$\phantom{24 \div 6 = 4 \qquad \text{or} \qquad 6)}\underline{24}$$
$$\phantom{24 \div 6 = 4 \qquad \text{or} \qquad 6)2}0$$

The other number is 4.

The following examples show that the same technique applies with fractions.

EXAMPLE 9

The result of multiplying two numbers is $\frac{7}{8}$. If one of the numbers is $\frac{3}{4}$, what is the other number?

Solution

Divide $\frac{7}{8}$ by $\frac{3}{4}$. This is similar to dividing 24 by 6 to find the other factor as we just discussed.

$$\frac{7}{8} \div \frac{3}{4} = \frac{7}{8} \cdot \frac{4}{3} = \frac{7 \cdot \overset{1}{\cancel{4}}}{\underset{2}{\cancel{8}} \cdot 3} = \frac{7}{6}$$

The other number is $\frac{7}{6}$.

Check by multiplying:

$$\frac{7}{6} \cdot \frac{3}{4} = \frac{7 \cdot \overset{1}{\cancel{3}}}{\underset{2}{\cancel{6}} \cdot 4} = \frac{7}{8}$$

EXAMPLE 10

If the product of $\frac{3}{2}$ with another number is $\frac{5}{18}$, what is the other number?

Solution

We know the product of two numbers. So, divide the product by the given number to find the other number.

$$\frac{5}{18} \div \frac{3}{2} = \frac{5}{18} \cdot \frac{2}{3} = \frac{5 \cdot \cancel{2}}{\cancel{2} \cdot 9 \cdot 3} = \frac{5}{27}$$

$\frac{5}{27}$ is the other number.

Check by multiplying:

$$\frac{3}{2} \cdot \frac{5}{27} = \frac{\cancel{3} \cdot 5}{2 \cdot \cancel{3} \cdot 9} = \frac{5}{18}$$

(In rethinking the problem, remember that $\frac{5}{18}$ is the product of two numbers.)

CLASSROOM PRACTICE

Find each of the following quotients. Reduce whenever possible.

1. $\frac{5}{8} \div \frac{5}{8}$ 2. $\frac{3}{4} \div \frac{1}{4}$ 3. $\frac{3}{4} \div 4$ 4. $\frac{8}{25} \div \frac{2}{15}$ 5. $\frac{6}{7} \div 0$

ANSWERS: 1. 1 2. 3 3. $\frac{3}{16}$ 4. $\frac{12}{5}$ 5. undefined

Exercises 3.3

1. What is the reciprocal of $\frac{5}{7}$?

2. To divide by any nonzero number, multiply by its _____.

3. The quotient $\frac{2}{3} \div 0$ is _____.

4. The quotient $0 \div \frac{2}{3}$ is equal to _____.

Find each of the following quotients. Reduce whenever possible.

5. $\frac{3}{4} \div \frac{2}{3}$ 6. $\frac{5}{8} \div \frac{3}{5}$ 7. $\frac{1}{3} \div \frac{1}{5}$ 8. $\frac{2}{11} \div \frac{1}{5}$

9. $\frac{2}{7} \div \frac{1}{2}$ 10. $\frac{7}{16} \div \frac{3}{4}$ 11. $\frac{9}{16} \div \frac{3}{4}$ 12. $\frac{5}{12} \div \frac{5}{3}$

13. $\frac{3}{5} \div \frac{27}{25}$ 14. $\frac{7}{8} \div \frac{2}{3}$ 15. $\frac{7}{8} \div \frac{3}{2}$ 16. $\frac{4}{5} \div \frac{2}{3}$

17. $\frac{5}{7} \div \frac{1}{7}$ 18. $\frac{3}{8} \div \frac{1}{8}$ 19. $\frac{7}{10} \div \frac{1}{10}$ 20. $\frac{5}{6} \div \frac{1}{6}$

21. $\frac{2}{15} \div \frac{2}{15}$ 22. $\frac{7}{11} \div \frac{7}{11}$ 23. $\frac{3}{4} \div \frac{4}{3}$ 24. $\frac{9}{10} \div \frac{10}{9}$

25. $\frac{9}{20} \div 3$ 26. $\frac{14}{20} \div 7$ 27. $\frac{18}{20} \div 9$ 28. $\frac{25}{40} \div 10$

29. $\frac{4}{7} \div 0$ 30. $0 \div \frac{6}{11}$ 31. $0 \div \frac{1}{2}$ 32. $\frac{15}{20} \div 0$

33. $\frac{16}{35} \div \frac{1}{7}$ 34. $\frac{15}{27} \div \frac{1}{9}$ 35. $\frac{5}{24} \div \frac{3}{8}$ 36. $\frac{7}{48} \div \frac{14}{16}$

37. $\dfrac{3}{48} \div \dfrac{5}{16}$ 38. $\dfrac{14}{20} \div \dfrac{7}{4}$ 39. $\dfrac{16}{20} \div \dfrac{18}{10}$ 40. $\dfrac{20}{21} \div \dfrac{10}{7}$

41. $\dfrac{13}{4} \div \dfrac{6}{5}$ 42. $\dfrac{12}{13} \div \dfrac{3}{2}$ 43. $\dfrac{16}{35} \div \dfrac{3}{11}$ 44. $\dfrac{19}{24} \div \dfrac{8}{5}$

45. $\dfrac{92}{7} \div \dfrac{46}{11}$ 46. $14 \div \dfrac{1}{7}$ 47. $25 \div \dfrac{1}{5}$ 48. $30 \div \dfrac{1}{10}$

49. $50 \div \dfrac{1}{2}$

50. The product of $\dfrac{9}{10}$ with another number is $\dfrac{5}{3}$. What is the other number?

51. The result of multiplying two numbers is $\dfrac{31}{3}$. If one of the numbers is $\dfrac{43}{6}$, what is the other one?

52. The result of multiplying two numbers is 150. If one of the numbers is $\dfrac{5}{7}$, what is the other number?

53. The product of $\dfrac{4}{5}$ with another number is 180. What is the other number?

54. The product of $\dfrac{7}{8}$ and $\dfrac{5}{10}$ is divided by $\dfrac{21}{20}$. Find the quotient.

55. A small private college has determined that about $\dfrac{44}{100}$ of the students that it accepts will actually enroll. If the college wants 550 freshmen to enroll, how many should it accept?

56. The sea floor of the Atlantic Ocean is spreading apart at an average rate of $\dfrac{3}{50}$ meters per year. About how long does it take for the sea floor to spread 12 meters?

Check Your Number Sense

57. A bus has 45 passengers. This is $\dfrac{3}{4}$ of its capacity. Is the capacity of the bus more or less than 45? What is the capacity of the bus?

58. A computer printer can print 8 pages in one minute. Will the printer print more than or less than 8 pages in 30 seconds? How many pages will the printer print in 15 seconds?

59. If $\dfrac{2}{3}$ of your salary is $1200 per month, is your monthly salary more or less than $1200? If you get a raise of $\dfrac{1}{10}$ of your monthly salary, will your new

monthly salary be more than, less than, or equal to $1320? Why? (Note: $\frac{1}{10}$ of $1200 is $120.)

60. Show that the phrases "6 divided by two" and "6 divided by one-half" do not have the same meaning.

61. Show that the phrases "15 divided by three" and "15 divided by one-third" do not have the same meaning.

62. Show that the phrases "12 divided by three" and "12 times one-third" have the same meaning.

63. Show that the phrases "20 divided by four" and "20 times one-fourth" have the same meaning.

64. Is division a commutative operation? Explain briefly and give three examples using fractions to help justify your answer.

 The Recycle Bin   (from Section 2.6)

Find the least common multiple (LCM) of the following sets of numbers.

1. 30, 65   **2.** 28, 36

3. 10, 20, 50   **4.** 39, 51

5. 15, 25, 100   **6.** 44, 88, 121

For each of the following sets of numbers, (a) find the LCM, (b) state the number of times each number divides into the LCM.

7. 50, 125   **8.** 20, 24, 30

9. 12, 35, 70   **10.** 45, 63, 99

11. Two people meet in the optometrist's office and have a pleasant conversation. They agree to have lunch together the next time they are in the optometrist's office on the same day. If their appointments occur once every 30 days, for one person, and once every 45 days, for the other person, in how many days will they have lunch together?

3.4   Addition

OBJECTIVES

1. Be able to add fractions with the same denominator.
2. Recall how to find the LCM.
3. Know how to add fractions with different denominators.

Addition with Fractions

Figure 3.3 illustrates how the **sum** of the two fractions $\frac{3}{7}$ and $\frac{2}{7}$ might be diagrammed.

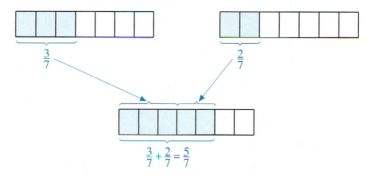

Figure 3.3

The common denominator "names" each fraction. The sum has this common name. Just as 3 *apples* plus 2 *apples* gives a total of 5 *apples,* we have 3 *sevenths* plus 2 *sevenths* giving a total of 5 *sevenths.*

To Add Two (or More) Fractions with the Same Denominator:

1. Add the numerators.

2. Keep the common denominator.

$$\frac{a}{b} + \frac{c}{b} = \frac{a + c}{b}$$

EXAMPLE 1 $\dfrac{1}{5} + \dfrac{3}{5} = \dfrac{1+3}{5} = \dfrac{4}{5}$

EXAMPLE 2 $\dfrac{2}{7} + \dfrac{3}{7} + \dfrac{1}{7} = \dfrac{2+3+1}{7} = \dfrac{6}{7}$

You may be able to reduce after adding.

EXAMPLE 3 $\dfrac{4}{15} + \dfrac{6}{15} = \dfrac{4+6}{15} = \dfrac{10}{15} = \dfrac{2 \cdot \cancel{5}}{3 \cdot \cancel{5}} = \dfrac{2}{3}$

EXAMPLE 4 $\dfrac{1}{8} + \dfrac{2}{8} + \dfrac{7}{8} = \dfrac{1+2+7}{8} = \dfrac{10}{8} = \dfrac{\cancel{2} \cdot 5}{\cancel{2} \cdot 4} = \dfrac{5}{4}$

Of course, fractions to be added will not always have the same denominator. In these cases, the smallest common denominator must be found. **The least common denominator (LCD) is the least common multiple (LCM) of the denominators.**

EXAMPLE 5   Find the LCD for $\dfrac{3}{8}$ and $\dfrac{11}{12}$.

Solution

Using prime factorization:

$$8 = 2 \cdot 2 \cdot 2$$
$$12 = 2 \cdot 2 \cdot 3$$
$$\text{LCD} = 2 \cdot 2 \cdot 2 \cdot 3 = 24$$

EXAMPLE 6   Find the LCD for $\dfrac{5}{21}$ and $\dfrac{5}{28}$.

Solution

Using prime factorization:

$$21 = 3 \cdot 7$$
$$28 = 2 \cdot 2 \cdot 7$$
$$\text{LCD} = 2 \cdot 2 \cdot 3 \cdot 7 = 84$$

To Add Fractions with Different Denominators:

1. Find the least common denominator (LCD).

2. Change each fraction to an equal fraction with that denominator.

3. Add the new fractions.

4. Reduce if possible.

EXAMPLE 7   Find the sum $\dfrac{1}{2} + \dfrac{3}{5}$.

Solution

(a) Find the LCD.

$$\text{LCD} = 2 \cdot 5 = 10$$

(b) Find equal fractions with denominator 10.

$$\frac{1}{2} = \frac{1}{2} \cdot \frac{5}{5} = \frac{5}{10}$$

$$\frac{3}{5} = \frac{3}{5} \cdot \frac{2}{2} = \frac{6}{10}$$

(c) Add.

$$\frac{1}{2} + \frac{3}{5} = \frac{5}{10} + \frac{6}{10} = \frac{5 + 6}{10} = \frac{11}{10}$$

EXAMPLE 8   Find the sum $\dfrac{3}{8} + \dfrac{11}{12}$.

Solution

(a) Find the LCD.

$$\left.\begin{array}{l} 8 = 2 \cdot 2 \cdot 2 \\ 12 = 2 \cdot 2 \cdot 3 \end{array}\right\} \text{LCD} = 2 \cdot 2 \cdot 2 \cdot 3 = 24$$

(b) Find equal fractions with denominator 24.

$$\frac{3}{8} = \frac{3}{8} \cdot \frac{3}{3} = \frac{9}{24} \qquad \text{Multiply by } \tfrac{3}{3} \text{ since } 8 \cdot 3 = 24.$$

$$\frac{11}{12} = \frac{11}{12} \cdot \frac{2}{2} = \frac{22}{24} \qquad \text{Multiply by } \tfrac{2}{2} \text{ since } 12 \cdot 2 = 24.$$

(c) Add.

$$\frac{3}{8} + \frac{11}{12} = \frac{9}{24} + \frac{22}{24} = \frac{9 + 22}{24} = \frac{31}{24}$$

EXAMPLE 9   Add $\dfrac{5}{21} + \dfrac{5}{28}$ and reduce if possible.

Solution

(a) Find the LCD.

$$\left.\begin{array}{l} 21 = 3 \cdot 7 \\ 28 = 2 \cdot 2 \cdot 7 \end{array}\right\} \text{LCD} = 2 \cdot 2 \cdot 3 \cdot 7 = 84$$

(b) Find equal fractions with denominator 84.

$$\frac{5}{21} = \frac{5}{21} \cdot \frac{4}{4} = \frac{20}{84}$$

$$\frac{5}{28} = \frac{5}{28} \cdot \frac{3}{3} = \frac{15}{84}$$

(c) $\dfrac{5}{21} + \dfrac{5}{28} = \dfrac{20}{84} + \dfrac{15}{84} = \dfrac{20 + 15}{84} = \dfrac{35}{84}$

(d) Now reduce.

$$\frac{35}{84} = \frac{\cancel{7} \cdot 5}{2 \cdot 2 \cdot 3 \cdot \cancel{7}} = \frac{5}{12}$$

EXAMPLE 10   Find the sum $\dfrac{2}{3} + \dfrac{1}{6} + \dfrac{5}{12}$.

Solution

(a) LCD = 12   You can simply observe this or use prime factorizations.

(b) Steps (b), (c), and (d) can be written together in one process.

$$\frac{2}{3} + \frac{1}{6} + \frac{5}{12} = \frac{2}{3} \cdot \frac{4}{4} + \frac{1}{6} \cdot \frac{2}{2} + \frac{5}{12}$$

$$= \frac{8}{12} + \frac{2}{12} + \frac{5}{12}$$

$$= \frac{15}{12} = \frac{\cancel{3} \cdot 5}{2 \cdot 2 \cdot \cancel{3}} = \frac{5}{4}$$

Or, the numbers can be written vertically. The process is the same.

$$\frac{2}{3} = \frac{2}{3} \cdot \frac{4}{4} = \frac{8}{12}$$

$$\frac{1}{6} = \frac{1}{6} \cdot \frac{2}{2} = \frac{2}{12}$$

$$\frac{5}{12} = \frac{5}{12} = \frac{5}{12}$$

$$\overline{}$$

$$\frac{15}{12} = \frac{\cancel{3} \cdot 5}{2 \cdot 2 \cdot \cancel{3}} = \frac{5}{4}$$

EXAMPLE 11   Add $5 + \dfrac{7}{10} + \dfrac{3}{1000}$.

Solution

(a) LCD = 1000   All the denominators are powers of 10, and 1000 is the largest. We can write 5 as $\dfrac{5}{1}$.

(b) $5 + \dfrac{7}{10} + \dfrac{3}{1000} = \dfrac{5}{1} \cdot \dfrac{1000}{1000} + \dfrac{7}{10} \cdot \dfrac{100}{100} + \dfrac{3}{1000}$

$$= \dfrac{5000}{1000} + \dfrac{700}{1000} + \dfrac{3}{1000}$$

$$= \dfrac{5703}{1000}$$

Common Error

The following common error must be avoided.

Find the sum $\dfrac{3}{2} + \dfrac{1}{6}$. .

WRONG SOLUTION

$$\dfrac{\overset{1}{\cancel{3}}}{2} + \dfrac{1}{\cancel{6}_{2}} = \dfrac{1}{2} + \dfrac{1}{2} = 1 \quad \textbf{WRONG}$$

You **cannot** cancel across the $+$ sign.

CORRECT SOLUTION

Use LCD $= 6$.

$$\dfrac{3}{2} + \dfrac{1}{6} = \dfrac{3}{2} \cdot \dfrac{3}{3} + \dfrac{1}{6} = \dfrac{9}{6} + \dfrac{1}{6} = \dfrac{10}{6}$$

NOW reduce.

$$\dfrac{10}{6} = \dfrac{5 \cdot \cancel{2}}{3 \cdot \cancel{2}} = \dfrac{5}{3} \qquad \text{2 is a factor in both the numerator and the denominator.}$$

Both the commutative and associative properties of addition apply to fractions.

Commutative Property of Addition

If $\dfrac{a}{b}$ and $\dfrac{c}{d}$ are fractions, then

$$\dfrac{a}{b} + \dfrac{c}{d} = \dfrac{c}{d} + \dfrac{a}{b}$$

> **Associative Property of Addition**
>
> If $\frac{a}{b}, \frac{c}{d}$, and $\frac{e}{f}$ are fractions, then
>
> $$\frac{a}{b} + \frac{c}{d} + \frac{e}{f} = \frac{a}{b} + \left(\frac{c}{d} + \frac{e}{f}\right) = \left(\frac{a}{b} + \frac{c}{d}\right) + \frac{e}{f}$$

CLASSROOM PRACTICE

Find the following sums. Reduce all answers.

1. $\frac{1}{8} + \frac{3}{8} + \frac{2}{8}$ **2.** $\frac{2}{3} + \frac{5}{8} + \frac{1}{6}$ **3.** $\frac{7}{10} + \frac{1}{100} + \frac{5}{1000}$

4. $\quad \dfrac{3}{10}$
$\quad \dfrac{1}{20}$
$+ \dfrac{7}{30}$

5. $\quad \dfrac{17}{100}$
$+ \dfrac{18}{100}$

ANSWERS: **1.** $\frac{3}{4}$ **2.** $\frac{35}{24}$ **3.** $\frac{715}{1000} = \frac{143}{200}$ **4.** $\frac{7}{12}$ **5.** $\frac{7}{20}$

Exercises 3.4

Find the least common denominator (LCD) for the fractions in each exercise.

1. $\frac{3}{8}, \frac{5}{16}$ **2.** $\frac{2}{39}, \frac{1}{3}, \frac{4}{13}$ **3.** $\frac{2}{27}, \frac{5}{18}, \frac{1}{6}$

4. $\frac{5}{8}, \frac{1}{12}, \frac{5}{9}$ **5.** $\frac{3}{10}, \frac{1}{100}, \frac{1}{1000}$

Add the following fractions and reduce all answers.

6. $\frac{6}{10} + \frac{4}{10}$ **7.** $\frac{3}{14} + \frac{2}{14}$ **8.** $\frac{1}{20} + \frac{3}{20}$ **9.** $\frac{3}{4} + \frac{3}{4}$

10. $\frac{5}{6} + \frac{4}{6}$ **11.** $\frac{7}{5} + \frac{3}{5}$ **12.** $\frac{11}{15} + \frac{7}{15}$ **13.** $\frac{7}{9} + \frac{8}{9}$

14. $\frac{3}{25} + \frac{12}{25}$ **15.** $\frac{7}{90} + \frac{37}{90} + \frac{21}{90}$ **16.** $\frac{11}{75} + \frac{12}{75} + \frac{62}{75}$

$\dfrac{57}{95}$

17. $\dfrac{14}{32} + \dfrac{7}{32} + \dfrac{1}{32}$

18. $\dfrac{4}{100} + \dfrac{35}{100} + \dfrac{76}{100}$

19. $\dfrac{21}{95} + \dfrac{33}{95} + \dfrac{3}{95}$

20. $\dfrac{1}{200} + \dfrac{17}{200} + \dfrac{25}{200}$

21. $\dfrac{1}{12} + \dfrac{2}{3} + \dfrac{1}{4}$

22. $\dfrac{3}{8} + \dfrac{5}{16}$

23. $\dfrac{2}{5} + \dfrac{3}{10} + \dfrac{3}{20}$

24. $\dfrac{3}{4} + \dfrac{1}{16} + \dfrac{6}{32}$

25. $\dfrac{2}{7} + \dfrac{4}{21} + \dfrac{1}{3}$

26. $\dfrac{1}{6} + \dfrac{1}{4} + \dfrac{1}{3}$

27. $\dfrac{2}{39} + \dfrac{1}{3} + \dfrac{4}{13}$

28. $\dfrac{1}{2} + \dfrac{3}{10} + \dfrac{4}{5}$

29. $\dfrac{1}{27} + \dfrac{4}{18} + \dfrac{1}{6}$

30. $\dfrac{2}{7} + \dfrac{3}{20} + \dfrac{9}{14}$

31. $\dfrac{1}{8} + \dfrac{1}{12} + \dfrac{1}{9}$

32. $\dfrac{2}{5} + \dfrac{4}{7} + \dfrac{3}{8}$

33. $\dfrac{2}{3} + \dfrac{3}{4} + \dfrac{5}{6}$

34. $\dfrac{1}{5} + \dfrac{7}{30} + \dfrac{1}{6}$

35. $\dfrac{1}{5} + \dfrac{2}{15} + \dfrac{1}{6}$

36. $\dfrac{1}{5} + \dfrac{1}{10} + \dfrac{1}{4}$

37. $\dfrac{1}{5} + \dfrac{7}{40} + \dfrac{1}{4}$

38. $\dfrac{1}{3} + \dfrac{5}{12} + \dfrac{1}{15}$

39. $\dfrac{1}{4} + \dfrac{1}{20} + \dfrac{8}{15}$

40. $\dfrac{7}{10} + \dfrac{3}{25} + \dfrac{3}{4}$

41. $\dfrac{5}{8} + \dfrac{4}{27} + \dfrac{1}{48}$

42. $\dfrac{3}{16} + \dfrac{5}{48} + \dfrac{1}{32}$

43. $\dfrac{72}{105} + \dfrac{2}{45} + \dfrac{15}{21}$

44. $\dfrac{1}{63} + \dfrac{2}{27} + \dfrac{1}{45}$

45. $\dfrac{0}{27} + \dfrac{0}{16} + \dfrac{1}{5}$

46. $\dfrac{5}{6} + \dfrac{0}{100} + \dfrac{0}{70} + \dfrac{1}{3}$

47. $\dfrac{3}{10} + \dfrac{1}{100} + \dfrac{7}{1000}$

48. $\dfrac{11}{100} + \dfrac{15}{10} + \dfrac{1}{10}$

49. $\dfrac{17}{1000} + \dfrac{1}{100} + \dfrac{1}{10,000}$

50. $6 + \dfrac{1}{100} + \dfrac{3}{10}$

51. $8 + \dfrac{1}{10} + \dfrac{9}{100} + \dfrac{1}{1000}$

52. $\dfrac{1}{10} + \dfrac{3}{10} + \dfrac{9}{1000}$

53. $\dfrac{7}{10} + \dfrac{5}{100} + \dfrac{3}{1000}$

54. $\dfrac{1}{2} + \dfrac{3}{4} + \dfrac{1}{100}$

55. $\dfrac{1}{4} + \dfrac{1}{8} + \dfrac{7}{100}$

56. $\dfrac{9}{1000} + \dfrac{7}{1000} + \dfrac{21}{10,000}$

57. $\dfrac{11}{100} + \dfrac{1}{2} + \dfrac{3}{1000}$

58. $\dfrac{3}{4} + \dfrac{17}{1000} + \dfrac{13}{10,000} + 2$

59. $5 + \dfrac{1}{10} + \dfrac{3}{100} + \dfrac{4}{1000}$

60. $\dfrac{13}{10,000} + \dfrac{1}{100,000} + \dfrac{21}{1,000,000}$

61. $\begin{array}{r} \dfrac{3}{4} \\[4pt] \dfrac{1}{2} \\[4pt] +\dfrac{5}{12} \\ \hline \end{array}$

62. $\begin{array}{r} \dfrac{1}{5} \\[4pt] \dfrac{2}{15} \\[4pt] +\dfrac{5}{6} \\ \hline \end{array}$

63. $\begin{array}{r} \dfrac{7}{8} \\[4pt] \dfrac{2}{3} \\[4pt] +\dfrac{1}{9} \\ \hline \end{array}$

64. $\begin{array}{r} \dfrac{1}{27} \\[4pt] \dfrac{1}{18} \\[4pt] +\dfrac{4}{9} \\ \hline \end{array}$

65. $\begin{array}{r} \dfrac{3}{20} \\[4pt] \dfrac{1}{100} \\[4pt] +\dfrac{3}{100} \\ \hline \end{array}$

66. $\begin{array}{r} \dfrac{13}{100} \\[4pt] \dfrac{4}{10} \\[4pt] +\dfrac{1}{1000} \\ \hline \end{array}$

67. $\begin{array}{r} \dfrac{7}{12} \\[4pt] \dfrac{1}{9} \\[4pt] +\dfrac{2}{3} \\ \hline \end{array}$

68. $\begin{array}{r} \dfrac{1}{3} \\[4pt] \dfrac{8}{15} \\[4pt] +\dfrac{7}{10} \\ \hline \end{array}$

69. $\begin{array}{r} \dfrac{9}{16} \\[4pt] \dfrac{5}{48} \\[4pt] +\dfrac{3}{32} \\ \hline \end{array}$

70. $\begin{array}{r} \dfrac{3}{10} \\[4pt] \dfrac{1}{20} \\[4pt] +\dfrac{1}{25} \\ \hline \end{array}$

71. Three letters weigh $\dfrac{1}{2}$ ounce, $\dfrac{1}{5}$ ounce, and $\dfrac{3}{10}$ ounce. What is the total weight of the letters?

72. Using a microscope, a scientist measures the diameters of three hairs to be $\dfrac{1}{1000}$ inch, $\dfrac{3}{1000}$ inch, and $\dfrac{1}{100}$ inch. What is the total of these three diameters?

73. A machinist drills four holes in a straight line. Each hole has a diameter of $\dfrac{1}{10}$ inch and there is $\dfrac{1}{4}$ inch between the holes. What is the distance between the outer edges of the first and last holes?

74. A notebook contains 30 sheets of paper (each $\dfrac{1}{100}$ inch thick), 2 pieces of cardboard (each $\dfrac{1}{16}$ inch thick), and a front and back cover (each $\dfrac{1}{4}$ inch thick). What is the total thickness of the notebook?

75. A rectangular-shaped stereo speaker has a strip of molding around each of its four sides. The strips at the top and bottom each measure $\dfrac{3}{16}$ inch thick and the strips on the sides each measure $\dfrac{1}{8}$ inch thick. What is the total thickness of the molding?

3.5 Subtraction

OBJECTIVES

1. Be able to subtract fractions with the same denominator.
2. Recall how to find the LCM.
3. Know how to subtract fractions with different denominators.

Subtraction with Fractions

Figure 3.4 shows how the **difference** of the two fractions $\frac{7}{8}$ and $\frac{4}{8}$ might be diagrammed.

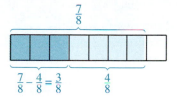

Figure 3.4

From Figure 3.4 we see that $\frac{7}{8} - \frac{4}{8} = \frac{3}{8}$. Just as with addition, the common denominator "names" each fraction. The difference is found by subtracting the numerators and using the common denominator.

> ### To Subtract Fractions that Have the Same Denominator:
>
> 1. Subtract the numerators.
>
> 2. Keep the common denominator.
>
> $$\frac{a}{b} - \frac{c}{b} = \frac{a - c}{b}$$

EXAMPLE 1

$$\frac{5}{6} - \frac{1}{6} = \frac{5 - 1}{6} = \frac{4}{6} = \frac{2 \cdot 2}{2 \cdot 3} = \frac{2}{3}$$

EXAMPLE 2 $\dfrac{9}{10} - \dfrac{7}{10} = \dfrac{9-7}{10} = \dfrac{2}{10} = \dfrac{\cancel{2} \cdot 1}{\cancel{2} \cdot 5} = \dfrac{1}{5}$

EXAMPLE 3 $\dfrac{19}{10} - \dfrac{11}{10} = \dfrac{19-11}{10} = \dfrac{8}{10} = \dfrac{\cancel{2} \cdot 4}{\cancel{2} \cdot 5} = \dfrac{4}{5}$

EXAMPLE 4 $\dfrac{7}{8} - \dfrac{3}{8} = \dfrac{7-3}{8} = \dfrac{4}{8} = \dfrac{\cancel{4} \cdot 1}{\cancel{4} \cdot 2} = \dfrac{1}{2}$

To Subtract Fractions with Different Denominators:

1. Find the least common denominator (LCD).

2. Change each fraction to an equal fraction with that denominator.

3. Subtract the new fractions.

4. Reduce if possible.

EXAMPLE 5 $\dfrac{9}{10} - \dfrac{2}{15}$

Solution

(a) Find the LCD.

$\left. \begin{array}{l} 10 = 2 \cdot 5 \\ 15 = 3 \cdot 5 \end{array} \right\}$ LCD $= 2 \cdot 3 \cdot 5$
$ = 30$

(b) Find equal fractions with denominator 30.

$\dfrac{9}{10} = \dfrac{9}{10} \cdot \dfrac{3}{3} = \dfrac{27}{30}$

$\dfrac{2}{15} = \dfrac{2}{15} \cdot \dfrac{2}{2} = \dfrac{4}{30}$

(c) Subtract.

$\dfrac{9}{10} - \dfrac{2}{15} = \dfrac{27}{30} - \dfrac{4}{30} = \dfrac{27-4}{30} = \dfrac{23}{30}$

EXAMPLE 6 $\dfrac{12}{55} - \dfrac{2}{33}$

Solution

(a) Find the LCD.

$\left. \begin{array}{l} 55 = 5 \cdot 11 \\ 33 = 3 \cdot 11 \end{array} \right\}$ LCD $= 3 \cdot 5 \cdot 11 = 165$

(b) Find equal fractions with denominator 165.

$$\frac{12}{55} = \frac{12}{55} \cdot \frac{3}{3} = \frac{36}{165}$$

$$\frac{2}{33} = \frac{2}{33} \cdot \frac{5}{5} = \frac{10}{165}$$

(c) Subtract.

$$\frac{12}{55} - \frac{2}{33} = \frac{36}{165} - \frac{10}{165} = \frac{36 - 10}{165} = \frac{26}{165}$$

(d) Reduce if possible.

$$\frac{26}{165} = \frac{2 \cdot 13}{3 \cdot 5 \cdot 11} = \frac{26}{165}$$ Cannot be reduced because there are no common prime factors in the numerator and denominator.

EXAMPLE 7

$$\frac{7}{12} - \frac{3}{20}$$

Solution

(a) $\left. \begin{array}{l} 12 = 2 \cdot 2 \cdot 3 \\ 20 = 2 \cdot 2 \cdot 5 \end{array} \right\}$ LCD $= 2 \cdot 2 \cdot 3 \cdot 5 = 60$

(b) Steps (b), (c), and (d) of Example 6 can be written as one process.

$$\frac{7}{12} - \frac{3}{20} = \frac{7}{12} \cdot \frac{5}{5} - \frac{3}{20} \cdot \frac{3}{3} = \frac{35}{60} - \frac{9}{60}$$

$$= \frac{35 - 9}{60} = \frac{26}{60} = \frac{2 \cdot 13}{2 \cdot 30} = \frac{13}{30}$$

Or, writing the fractions vertically,

$$\frac{7}{12} = \frac{7}{12} \cdot \frac{5}{5} = \frac{35}{60}$$

$$-\frac{3}{20} = \frac{3}{20} \cdot \frac{3}{3} = \frac{9}{60}$$

$$\frac{26}{60} = \frac{2 \cdot 13}{2 \cdot 30} = \frac{13}{30}$$

EXAMPLE 8

$$1 - \frac{5}{8}$$

Solution

(a) LCD $= 8$ Since 1 can be written $\frac{1}{1}$, the common denominator is $8 \cdot 1 = 8$.

(b) $1 - \dfrac{5}{8} = \dfrac{1}{1} \cdot \dfrac{8}{8} - \dfrac{5}{8} = \dfrac{8}{8} - \dfrac{5}{8} = \dfrac{8 - 5}{8} = \dfrac{3}{8}$

EXAMPLE 9

$$2 - \frac{7}{16}$$

Solution

(a) LCD = 16   Since 2 can be written $\frac{2}{1}$, the common denominator is 16.

(b) $2 - \frac{7}{16} = \frac{2}{1} \cdot \frac{16}{16} - \frac{7}{16} = \frac{32}{16} - \frac{7}{16} = \frac{32 - 7}{16} = \frac{25}{16}$

EXAMPLE 10

The Narragansett Grays baseball team lost 90 games in one season. If $\frac{1}{5}$ of their losses were by 1 or 2 runs and $\frac{4}{9}$ of their losses were by 3 or fewer runs, what fraction of their losses were by exactly 3 runs?

Solution

Before answering the question, we need to analyze the information and decide what arithmetic operation is needed. The losses that were by 3 or fewer runs include those that were by 1 or 2 runs. So, if we subtract the fraction of losses that were by 1 or 2 runs from the fraction of losses that were by 3 or fewer runs, we will be left with the fraction of losses that were by exactly 3 runs. That is, we subtract $\frac{1}{5}$ from $\frac{4}{9}$. (The LCD is 45.)

$$\frac{4}{9} - \frac{1}{5} = \frac{4}{9} \cdot \frac{5}{5} - \frac{1}{5} \cdot \frac{9}{9} = \frac{20}{45} - \frac{9}{45} = \frac{11}{45}$$

Thus, the Grays lost $\frac{11}{45}$ of their games by exactly 3 runs.

CLASSROOM PRACTICE

Find the following differences. Reduce all answers.

1. $\dfrac{5}{9} - \dfrac{1}{9}$ 2. $\dfrac{7}{6} - \dfrac{2}{3}$ 3. $\dfrac{7}{10} - \dfrac{7}{15}$ 4. $1 - \dfrac{3}{4}$

ANSWERS: 1. $\dfrac{4}{9}$ 2. $\dfrac{1}{2}$ 3. $\dfrac{7}{30}$ 4. $\dfrac{1}{4}$

Exercises 3.5

Subtract and reduce if possible.

1. $\dfrac{4}{7} - \dfrac{1}{7}$ **2.** $\dfrac{5}{7} - \dfrac{3}{7}$ **3.** $\dfrac{9}{10} - \dfrac{3}{10}$ **4.** $\dfrac{11}{10} - \dfrac{7}{10}$

5. $\dfrac{5}{8} - \dfrac{1}{8}$ **6.** $\dfrac{7}{8} - \dfrac{5}{8}$ **7.** $\dfrac{11}{12} - \dfrac{7}{12}$ **8.** $\dfrac{7}{12} - \dfrac{3}{12}$

9. $\dfrac{13}{15} - \dfrac{4}{15}$ **10.** $\dfrac{21}{15} - \dfrac{11}{15}$ **11.** $\dfrac{5}{6} - \dfrac{1}{3}$ **12.** $\dfrac{5}{6} - \dfrac{1}{2}$

13. $\dfrac{11}{15} - \dfrac{3}{10}$ **14.** $\dfrac{8}{10} - \dfrac{3}{15}$ **15.** $\dfrac{3}{4} - \dfrac{2}{3}$ **16.** $\dfrac{2}{3} - \dfrac{1}{4}$

17. $\dfrac{15}{16} - \dfrac{21}{32}$ **18.** $\dfrac{3}{8} - \dfrac{1}{16}$ **19.** $\dfrac{5}{4} - \dfrac{3}{5}$ **20.** $\dfrac{5}{12} - \dfrac{1}{6}$

21. $\dfrac{14}{27} - \dfrac{7}{18}$ **22.** $\dfrac{25}{18} - \dfrac{21}{27}$ **23.** $\dfrac{8}{45} - \dfrac{11}{72}$ **24.** $\dfrac{46}{55} - \dfrac{10}{33}$

360

25. $\dfrac{5}{36} - \dfrac{1}{45}$ **26.** $\dfrac{5}{1} - \dfrac{3}{4}$ **27.** $\dfrac{4}{1} - \dfrac{5}{8}$ **28.** $2 - \dfrac{9}{16}$

29. $1 - \dfrac{13}{16}$ **30.** $6 - \dfrac{2}{3}$ **31.** $\dfrac{9}{10} - \dfrac{3}{100}$ **32.** $\dfrac{159}{1000} - \dfrac{1}{10}$

33. $\dfrac{76}{100} - \dfrac{7}{10}$ **34.** $\dfrac{999}{1000} - \dfrac{99}{100}$ **35.** $\dfrac{54}{100} - \dfrac{5}{10}$ **36.** $\dfrac{7}{24} - \dfrac{10}{36}$

37. $\dfrac{31}{40} - \dfrac{5}{8}$ **38.** $\dfrac{14}{35} - \dfrac{12}{30}$ **39.** $\dfrac{20}{35} - \dfrac{24}{42}$ **40.** $\dfrac{3}{10} - \dfrac{298}{1000}$

41. $1 - \dfrac{9}{10}$ **42.** $1 - \dfrac{7}{8}$ **43.** $1 - \dfrac{2}{3}$ **44.** $1 - \dfrac{1}{16}$

45. $1 - \dfrac{3}{20}$ **46.** $1 - \dfrac{4}{9}$

47. $\begin{array}{r} \frac{7}{8} \\ -\frac{2}{3} \end{array}$ **48.** $\begin{array}{r} \frac{14}{15} \\ -\frac{3}{10} \end{array}$ **49.** $\begin{array}{r} \frac{1}{10} \\ -\frac{8}{100} \end{array}$ **50.** $\begin{array}{r} \frac{3}{100} \\ -\frac{1}{1000} \end{array}$

51. Find the sum of $\dfrac{1}{4}$ and $\dfrac{3}{16}$. Subtract $\dfrac{1}{8}$ from the sum. What is the difference?

52. Find the difference between $\frac{2}{3}$ and $\frac{5}{9}$. Add $\frac{1}{12}$ to the difference. What is the sum?

53. Find the sum of $\frac{3}{4}$ and $\frac{5}{8}$. Multiply the sum by $\frac{2}{3}$. What is the product?

54. Find the product of $\frac{9}{10}$ and $\frac{2}{3}$. Divide the product by $\frac{5}{3}$. What is the quotient?

55. Find the quotient of $\frac{3}{4}$ and $\frac{15}{16}$. Add $\frac{3}{10}$ to the quotient. What is the sum?

56. About $\frac{1}{2}$ of all incoming solar radiation is absorbed by the earth, $\frac{1}{5}$ is absorbed by the atmosphere, and $\frac{1}{20}$ is scattered by the atmosphere. The rest is reflected by the earth and clouds. What fraction of solar radiation is reflected?

Collaborative Learning Exercise

57. With the class separated into teams of two to four students, each team is to write one or two paragraphs explaining what topic in this chapter they found to be the most difficult and what techniques they used to learn the related material. Each team leader is to read the paragraph(s) with classroom discussion to follow.

♻ The Recycle Bin (from Sections 2.1 and 2.2)

Find the value of each of the following squares. Write as many as you can from memory.

1. 9^2 **2.** 13^2 **3.** 15^2 **4.** 18^2

Find the value of each of the following expressions by using the rules for order of operations.

5. $8 \div 2 + 6 - 5 \cdot 2$ **6.** $8 \div (2 + 6) + 14 \cdot 2$

7. $10^2 - 9 \cdot 6 \div 2 + 1^3$ **8.** $3(4 + 7) - 4 \cdot 3 - 3 \cdot 7$

9. $2 \cdot 5^2 + 3(16 - 2 \cdot 8) + 5 \cdot 2^2$

10. $(5^2 + 7) \div 4 - (10 \div 5 \cdot 2)$

3.6 Comparisons and Order of Operations

OBJECTIVES

1. Compare fractions by finding a common denominator and comparing the numerators.
2. Follow the rules of order of operations with fractions.
3. Simplify complex fractions.

Comparing Two or More Fractions

Many times we want to compare two (or more) fractions to see which is smaller or larger. Then we can subtract the smaller from the larger, or possibly make some financial decision based on the relative sizes of the fractions. Related word problems will be discussed in detail in later chapters.

> **To Compare Two Fractions (to Find which Is Larger or Smaller):**
>
> **1.** Find the least common denominator (LCD).
>
> **2.** Change each fraction to an equal fraction with that denominator.
>
> **3.** Compare the numerators.

EXAMPLE 1   Which is larger, $\dfrac{5}{6}$ or $\dfrac{7}{8}$? How much larger?

Solution

(a) Find the LCD for 6 and 8.

$$\left.\begin{array}{l} 6 = 2 \cdot 3 \\ 8 = 2 \cdot 2 \cdot 2 \end{array}\right\} \text{LCD} = 2 \cdot 2 \cdot 2 \cdot 3 = 24$$

(b) Find equal fractions with denominator 24.

$$\frac{5}{6} = \frac{5}{6} \cdot \frac{4}{4} = \frac{20}{24} \qquad \text{and} \qquad \frac{7}{8} = \frac{7}{8} \cdot \frac{3}{3} = \frac{21}{24}$$

(c) $\dfrac{7}{8}$ is larger than $\dfrac{5}{6}$, since 21 is larger than 20.

(d) $\dfrac{7}{8} - \dfrac{5}{6} = \dfrac{21}{24} - \dfrac{20}{24} = \dfrac{1}{24}$

$\dfrac{7}{8}$ is larger by $\dfrac{1}{24}$.

EXAMPLE 2   Which is larger, $\dfrac{8}{9}$ or $\dfrac{11}{12}$? How much larger?

Solution

(a) LCD $= 2 \cdot 2 \cdot 3 \cdot 3 = 36$

(b) $\dfrac{8}{9} = \dfrac{8}{9} \cdot \dfrac{4}{4} = \dfrac{32}{36}$ and $\dfrac{11}{12} = \dfrac{11}{12} \cdot \dfrac{3}{3} = \dfrac{33}{36}$

(c) $\dfrac{11}{12}$ is larger than $\dfrac{8}{9}$, since 33 is larger than 32.

(d) $\dfrac{11}{12} - \dfrac{8}{9} = \dfrac{33}{36} - \dfrac{32}{36} = \dfrac{1}{36}$

$\dfrac{11}{12}$ is larger by $\dfrac{1}{36}$.

EXAMPLE 3   Arrange $\dfrac{2}{3}, \dfrac{7}{10}$, and $\dfrac{9}{15}$ in order, smallest to largest.

Solution

(a) LCD $= 30$

(b) $\dfrac{2}{3} = \dfrac{2}{3} \cdot \dfrac{10}{10} = \dfrac{20}{30}; \quad \dfrac{7}{10} = \dfrac{7}{10} \cdot \dfrac{3}{3} = \dfrac{21}{30}; \quad \dfrac{9}{15} = \dfrac{9}{15} \cdot \dfrac{2}{2} = \dfrac{18}{30}$

(c) In order, smallest to largest: $\dfrac{9}{15}, \dfrac{2}{3}, \dfrac{7}{10}$.

Using the Rules for Order of Operations with Fractions

An expression with fractions may involve more than one arithmetic operation. To simplify such expressions, we can use the rules for order of operations just as they were discussed in Chapter 2 for whole numbers. Of course, all the rules for fractions must be followed, too. That is, to add or subtract, you need a common denominator; to divide, you multiply by the reciprocal of the divisor.

> ## Rules for Order of Operations
>
> **1.** First, simplify within grouping symbols, such as parentheses ( ), brackets [ ], or braces { }. Start with the innermost grouping.
>
> **2.** Second, find any powers indicated by exponents.
>
> **3.** Third, moving from **left to right,** perform any multiplications or divisions in the order they appear.
>
> **4.** Fourth, moving from **left to right,** perform any additions or subtractions in the order they appear.

EXAMPLE 4   Evaluate the following expression.

$$\frac{1}{2} \div \frac{3}{4} + \frac{5}{6} \cdot \frac{1}{5}$$

Solution

(a) Divide first.

$$\frac{1}{2} \div \frac{3}{4} + \frac{5}{6} \cdot \frac{1}{5}$$

$$= \frac{1}{\cancel{2}} \cdot \frac{\cancel{4}^{2}}{3} + \frac{5}{6} \cdot \frac{1}{5}$$

(b) Now multiply.

$$= \frac{2}{3} + \frac{\cancel{5}}{6} \cdot \frac{1}{\cancel{5}}$$

(c) Now add.
(LCD = 6)

$$= \frac{2}{3} + \frac{1}{6}$$

$$= \frac{2}{3} \cdot \frac{2}{2} + \frac{1}{6}$$

$$= \frac{4}{6} + \frac{1}{6}$$

$$= \frac{5}{6}$$

EXAMPLE 5   Evaluate the expression $\left(\frac{3}{4} - \frac{5}{8}\right) \div \left(\frac{15}{16} - \frac{1}{2}\right).$

Solution

(a) Work inside the parentheses.

$$\left(\frac{3}{4} - \frac{5}{8}\right) \div \left(\frac{15}{16} - \frac{1}{2}\right)$$

$$= \left(\frac{6}{8} - \frac{5}{8}\right) \div \left(\frac{15}{16} - \frac{8}{16}\right)$$

(b) Now divide.

$$= \left(\frac{1}{8}\right) \div \left(\frac{7}{16}\right)$$

$$= \frac{1}{\cancel{8}} \cdot \frac{\cancel{16}^{\,2}}{7} = \frac{2}{7}$$

EXAMPLE 6   Evaluate the expression $\frac{1}{2} \cdot \frac{5}{6} + \frac{7}{15} \div 2.$

Solution

$$\frac{1}{2} \cdot \frac{5}{6} + \frac{7}{15} \div 2 = \frac{5}{12} + \frac{7}{15} \div 2$$

$$= \frac{5}{12} + \frac{7}{15} \cdot \frac{1}{2}$$

$$= \frac{5}{12} + \frac{7}{30}$$

$$= \frac{5}{12} \cdot \boxed{\frac{5}{5}} + \frac{7}{30} \cdot \boxed{\frac{2}{2}} \qquad \text{(LCD = 60)}$$

$$= \frac{25}{60} + \frac{14}{60}$$

$$= \frac{39}{60} = \frac{\cancel{3} \cdot 13}{\cancel{3} \cdot 20} = \frac{13}{20}$$

EXAMPLE 7   Evaluate the expression $\frac{9}{10} - \left(\frac{1}{4}\right)^2 + \frac{1}{2}.$

Solution

(a) Find the power first.

$$\frac{9}{10} - \left(\frac{1}{4}\right)^2 + \frac{1}{2} = \frac{9}{10} - \frac{1}{16} + \frac{1}{2}$$

(b) Now add and subtract
from left to right.
(LCD = 80)

$$= \frac{9}{10} \cdot \frac{8}{8} - \frac{1}{16} \cdot \frac{5}{5} + \frac{1}{2} \cdot \frac{40}{40}$$

$$= \frac{72}{80} - \frac{5}{80} + \frac{40}{80}$$

$$= \frac{107}{80}$$

Complex Fractions

A **complex fraction** is a form of a fraction in which the numerator and/or denominator are themselves fractions or the sums or differences of fractions. To simplify a complex fraction we follow the rules for order of operations by treating the numerator and denominator as if they were surrounded by parentheses.

To Simplify a Complex Fraction:

1. Simplify the numerator. (Add or subtract as indicated.)

2. Simplify the denominator. (Add or subtract as indicated.)

3. Divide the numerator by the denominator and reduce if possible.

EXAMPLE 8   Simplify the complex fraction $\dfrac{\dfrac{3}{4} + \dfrac{1}{2}}{1 - \dfrac{1}{3}}$.

Solution

Simplifying the numerator gives

$$\frac{3}{4} + \frac{1}{2} = \frac{3}{4} + \frac{2}{4} = \frac{5}{4}.$$

Simplifying the denominator gives

$$1 - \frac{1}{3} = \frac{3}{3} - \frac{1}{3} = \frac{2}{3}.$$

Therefore, $\dfrac{\dfrac{3}{4} + \dfrac{1}{2}}{1 - \dfrac{1}{3}} = \dfrac{\dfrac{5}{4}}{\dfrac{2}{3}} = \dfrac{5}{4} \cdot \dfrac{3}{2} = \dfrac{15}{8}.$

NOTE: We could also write

$$\frac{\frac{3}{4} + \frac{1}{2}}{1 - \frac{1}{3}} = \left(\frac{3}{4} + \frac{1}{2}\right) \div \left(1 - \frac{1}{3}\right),$$

and simplify by following the rules for order of operations.

Exercises 3.6

Find the larger number of each pair and tell how much larger it is.

1. $\frac{2}{3}, \frac{3}{4}$
2. $\frac{5}{6}, \frac{7}{8}$
3. $\frac{4}{5}, \frac{17}{20}$
4. $\frac{4}{10}, \frac{3}{8}$

5. $\frac{13}{20}, \frac{5}{8}$
6. $\frac{13}{16}, \frac{21}{25}$
7. $\frac{14}{35}, \frac{12}{30}$
8. $\frac{10}{36}, \frac{7}{24}$

9. $\frac{17}{80}, \frac{11}{48}$
10. $\frac{37}{100}, \frac{24}{75}$

In each exercise find equivalent fractions with a common denominator. Then write the original numbers in order, smallest to largest.

11. $\frac{2}{3}, \frac{3}{5}, \frac{7}{10}$
12. $\frac{8}{9}, \frac{9}{10}, \frac{11}{12}$

13. $\frac{7}{6}, \frac{11}{12}, \frac{19}{20}$
14. $\frac{1}{3}, \frac{5}{42}, \frac{3}{7}$

15. $\frac{1}{2}, \frac{1}{3}, \frac{1}{4}$
16. $\frac{2}{3}, \frac{3}{4}, \frac{5}{8}$

17. $\frac{7}{9}, \frac{31}{36}, \frac{13}{18}$
18. $\frac{17}{12}, \frac{40}{36}, \frac{31}{24}$

19. $\frac{1}{100}, \frac{3}{1000}, \frac{20}{10,000}$
20. $\frac{32}{100}, \frac{298}{1000}, \frac{3}{10}$

Evaluate each expression using the rules of order of operations.

21. $\frac{1}{2} \div \frac{7}{8} + \frac{1}{7} \cdot \frac{2}{3}$
22. $\frac{3}{5} \cdot \frac{1}{6} + \frac{1}{5} \div 2$

23. $\frac{1}{2} \div \frac{1}{2} + \frac{2}{3} \cdot \frac{2}{3}$
24. $5 - \frac{3}{4} \div 3$

25. $6 - \frac{5}{8} \div 4$
26. $\frac{2}{15} \cdot \frac{1}{4} \div \frac{3}{5} + \frac{1}{25}$

27. $\dfrac{5}{8} \cdot \dfrac{1}{10} \div \dfrac{3}{4} + \dfrac{1}{6}$

28. $\left(\dfrac{7}{15} + \dfrac{8}{21}\right) \div \dfrac{3}{35}$

29. $\left(\dfrac{1}{2} - \dfrac{1}{3}\right) \div \left(\dfrac{5}{8} + \dfrac{3}{16}\right)$

30. $\left(\dfrac{1}{3} + \dfrac{1}{5}\right) \cdot \left(\dfrac{3}{4} - \dfrac{1}{6}\right)$

31. $\left(\dfrac{1}{2}\right)^2 - \left(\dfrac{1}{4}\right)^3$

32. $\dfrac{2}{3} + \dfrac{3}{4} + \left(\dfrac{1}{2}\right)^2$

33. $\left(\dfrac{1}{3}\right)^2 + \left(\dfrac{1}{6}\right)^2 + \dfrac{2}{3}$

34. $\dfrac{1}{2} \div \dfrac{2}{3} + \left(\dfrac{1}{3}\right)^2$

Simplify the following complex fractions.

35. $\dfrac{\dfrac{3}{4} - \dfrac{1}{2}}{1 + \dfrac{1}{3}}$

36. $\dfrac{\dfrac{1}{8} + \dfrac{1}{2}}{1 - \dfrac{2}{5}}$

37. $\dfrac{\dfrac{1}{5} + \dfrac{1}{6}}{2 + \dfrac{1}{3}}$

38. $\dfrac{\dfrac{5}{6} - \dfrac{1}{3}}{\dfrac{1}{2} + \dfrac{1}{5}}$

39. $\dfrac{\dfrac{2}{3} - \dfrac{1}{4}}{\dfrac{3}{5} - \dfrac{1}{4}}$

40. $\dfrac{\dfrac{5}{6} - \dfrac{2}{3}}{\dfrac{5}{8} - \dfrac{1}{16}}$

41. $\dfrac{\dfrac{7}{8} - \dfrac{3}{16}}{\dfrac{1}{3} - \dfrac{1}{4}}$

42. $\dfrac{\dfrac{3}{5} + \dfrac{4}{7}}{\dfrac{3}{8} + \dfrac{1}{10}}$

Check Your Number Sense

43. (a) If two fractions are less than 1, can their sum be more than 1? Explain.

(b) If two fractions are less than 1, can their product be more than 1? Explain.

44. Consider the fraction $\dfrac{1}{2}$.

(a) If this fraction is divided by 2, will the quotient be more or less than $\dfrac{1}{2}$?

(b) If this fraction is divided by 3, will the quotient be more or less than the quotient in part (a)?

45. Will the quotient always get smaller and smaller when a nonzero number is divided by larger and larger numbers? Can you think of a case in which this is not true? What happens when 0 is divided by larger and larger numbers?

46. Consider any fraction between 0 and 1, not including 0 or 1. If you square this number, will the result be larger or smaller than the original number? Is this always the case? Explain your answer.

Summary: Chapter 3

Key Terms and Ideas

A **rational number** is a number that can be written in the form $\frac{a}{b}$ where a and b are whole numbers and $b \neq 0$.

The terms **fraction** and **rational number** are used to mean the same thing.

Improper fractions are fractions in which the numerator is larger than the denominator.

Fractions can be used to indicate:

1. Equal parts of a whole.

2. Division.

The **reciprocal** of $\frac{a}{b}$ is $\frac{b}{a}$ ($a \neq 0$ and $b \neq 0$).

Rules and Properties

Properties of Addition

Commutative property: $\frac{a}{b} + \frac{c}{d} = \frac{c}{d} + \frac{a}{b}$

Associative property: $\frac{a}{b} + \left(\frac{c}{d} + \frac{e}{f}\right) = \left(\frac{a}{b} + \frac{c}{d}\right) + \frac{e}{f}$

Properties of Multiplication

Commutative property: $\frac{a}{b} \cdot \frac{c}{d} = \frac{c}{d} \cdot \frac{a}{b}$

Associative property: $\frac{a}{b}\left(\frac{c}{d} \cdot \frac{e}{f}\right) = \left(\frac{a}{b} \cdot \frac{c}{d}\right)\frac{e}{f}$

Identity: $\frac{a}{b} \cdot 1 = \frac{a}{b}$

Division by 0 is undefined. No denominator can be 0.

Rules for Order of Operations

1. First, simplify within grouping symbols, such as parentheses ( ), brackets [ ], or braces { }. Start with the innermost grouping.

2. Second, find any powers indicated by exponents.

3. Third, moving from **left to right,** perform any multiplications or divisions in the order they appear.

4. Fourth, moving from **left to right,** perform any additions or subtractions in the order they appear.

Procedures

Operations with fractions

1. Addition: $\dfrac{a}{b} + \dfrac{c}{b} = \dfrac{a+c}{b}$

2. Subtraction: $\dfrac{a}{b} - \dfrac{c}{b} = \dfrac{a-c}{b}$

3. Multiplication: $\dfrac{a}{b} \cdot \dfrac{c}{d} = \dfrac{a \cdot c}{b \cdot d}$

4. Division: $\dfrac{a}{b} \div \dfrac{c}{d} = \dfrac{a}{b} \cdot \dfrac{d}{c}$

For addition (or subtraction) of fractions with different denominators:

1. Find the least common denominator (LCD).

2. Change each fraction to an equal fraction with the LCD as its denominator.

3. Add (or subtract) the new fractions.

4. Reduce if possible.

Review Questions: Chapter 3

1. The denominator of a rational number cannot be _____ .

2. $\dfrac{0}{7} = 0$, but $\dfrac{7}{0}$ is _____ .

3. The reciprocal of $\dfrac{2}{3}$ is _____ , and the reciprocal of $\dfrac{3}{2}$ is _____ .

4. Which property of addition is illustrated by the following statement?

$$\frac{1}{3} + \left(\frac{5}{6} + \frac{1}{2} \right) = \left(\frac{1}{3} + \frac{5}{6} \right) + \frac{1}{2}$$

5. Find $\dfrac{2}{3}$ of $\dfrac{2}{5}$.

Multiply and reduce all answers.

6. $\dfrac{1}{3} \cdot \dfrac{1}{2} \cdot \dfrac{1}{5}$ **7.** $\dfrac{1}{7} \cdot \dfrac{3}{7}$ **8.** $\dfrac{35}{56} \cdot \dfrac{4}{15} \cdot \dfrac{5}{10}$

Fill in the missing terms so that each equation is true.

9. $\dfrac{1}{6} = \dfrac{?}{12}$ **10.** $\dfrac{9}{10} = \dfrac{?}{60}$ **11.** $\dfrac{15}{13} = \dfrac{?}{65}$

Reduce each fraction to lowest terms.

12. $\dfrac{15}{30}$ **13.** $\dfrac{99}{88}$ **14.** $\dfrac{0}{4}$ **15.** $\dfrac{150}{120}$

Add or subtract as indicated and reduce all answers.

16. $\dfrac{3}{7} + \dfrac{2}{7}$ **17.** $\dfrac{5}{6} - \dfrac{1}{6}$ **18.** $\dfrac{5}{8} - \dfrac{3}{8}$

19. $\dfrac{1}{12} + \dfrac{5}{36} + \dfrac{11}{24}$ **20.** $\dfrac{13}{22} - \dfrac{9}{33}$ **21.** $\dfrac{5}{27} + \dfrac{5}{18}$

22. $1 - \dfrac{13}{20}$ **23.** $\dfrac{3}{4} - \dfrac{5}{12}$ **24.** $\begin{array}{r} \dfrac{2}{3} \\[4pt] \dfrac{1}{8} \\[4pt] + \dfrac{1}{12} \\ \hline \end{array}$

Divide and reduce all answers.

25. $\dfrac{2}{3} \div 6$ **26.** $1 \div \dfrac{3}{5}$ **27.** $\dfrac{7}{12} \div \dfrac{7}{12}$

28. $\dfrac{15}{16} \div \dfrac{3}{4}$

29. $\dfrac{3}{4} \div \dfrac{15}{16}$

30. Which is larger, $\dfrac{2}{3}$ or $\dfrac{4}{5}$? How much larger?

31. Arrange the following fractions in order, smallest to largest.

$$\dfrac{7}{12}, \quad \dfrac{5}{9}, \quad \dfrac{11}{20}$$

Evaluate each expression using the rules of order of operations.

32. $\dfrac{5}{8} \cdot \dfrac{3}{10} + \dfrac{1}{14} \div 2$

33. $\left(\dfrac{3}{5} - \dfrac{1}{3}\right) \div \left(\dfrac{1}{6} + \dfrac{7}{8}\right)$

34. $\dfrac{7}{15} + \dfrac{5}{9} \div \dfrac{2}{3} - \dfrac{2}{3}$

35. $\left(\dfrac{2}{3}\right)^2 - \left(\dfrac{1}{3}\right)^2 + \dfrac{1}{18}$

36. Simplify the complex fraction: $\dfrac{\dfrac{3}{8} + \dfrac{1}{2}}{\dfrac{1}{2} - \dfrac{1}{10}}$

Test: Chapter 3

1. What is the reciprocal of $\frac{5}{8}$?

2. Name the property illustrated.

   (a) $\frac{2}{3} + \frac{7}{15} = \frac{7}{15} + \frac{2}{3}$

   (b) $\frac{1}{2}\left(\frac{1}{3} \cdot \frac{1}{4}\right) = \left(\frac{1}{2} \cdot \frac{1}{3}\right)\frac{1}{4}$

Reduce to lowest terms.

3. $\frac{45}{75}$

4. $\frac{216}{264}$

5. $\frac{90}{108}$

Perform the indicated operations. Reduce if possible.

6. $\frac{3}{10} + \frac{1}{10}$

7. $\frac{11}{20} + \frac{5}{20}$

8. $\frac{7}{10} + \frac{4}{15}$

9. $\frac{37}{11} - \frac{15}{11}$

10. $1 - \frac{11}{15}$

11. $\frac{5}{12} - \frac{2}{9}$

12. $\frac{3}{10} + \frac{1}{8} + \frac{5}{6}$

13. $\frac{14}{35} \div \frac{2}{5}$

14. $\frac{21}{26} \cdot \frac{13}{15} \cdot \frac{5}{7}$

15. $\frac{3}{16} \div \frac{5}{12}$

16. $\frac{3}{10} \div \frac{1}{2} + \frac{7}{8} \cdot \frac{2}{5}$

17. $\left(\frac{3}{4} - \frac{1}{5}\right) \div \frac{3}{2}$

18. $\left(\frac{1}{2}\right)^3 - \left(\frac{1}{4}\right)^2$

19. $\frac{9}{10} + \frac{4}{5} \div \frac{2}{7} - \frac{2}{7}$

20. $\frac{3}{5} + \left(\frac{1}{3}\right)^2 - \left(\frac{1}{2} \div \frac{3}{4}\right)$

21. Simplify the complex fraction: $\dfrac{\frac{2}{3} + \frac{3}{4}}{\frac{1}{2} + \frac{2}{5}}$

22. Find $\frac{3}{4}$ of $\frac{20}{21}$.

23. Arrange $\frac{3}{5}, \frac{7}{12}$, and $\frac{5}{8}$ in order, smallest to largest.

24. Find the sum of $\frac{3}{8}$ and $\frac{1}{4}$. Subtract $\frac{3}{16}$ from the sum. What is the difference?

25. The result of multiplying two numbers is $\frac{4}{5}$. If one of the numbers is $\frac{2}{3}$, what is the other number?

26. Find the sum of $\frac{7}{10}$ with the quotient of $\frac{9}{10}$ and $\frac{10}{3}$.

1. Round off 1,283,400 to the nearest hundred thousand.

2. Estimate the product 243 × 65; then find the product.

3. Estimate the quotient 504 ÷ 21; then find the quotient.

4. The equation 3(10 + 7) = 30 + 21 is an example of which property?

5. Find the quotient and remainder: 4570 ÷ 15.

6. Evaluate $(12 \cdot 6 \div 3^2) + 2 \cdot 11$.

7. Find the prime factorization of 375.

8. Find the LCM of the numbers 42, 18, and 36.

9. Multiply 40(3000).

10. Which of the numbers 2, 3, 4, 5, 9, and 10 divide exactly into 1890?

11. Determine whether or not the number 331 is prime.

Perform the indicated operations.

12.
$$\begin{array}{r} 86 \\ 73 \\ + 196 \end{array}$$

13.
$$\begin{array}{r} 85 \\ \times 12 \end{array}$$

14.
$$\begin{array}{r} 9763 \\ - 8495 \end{array}$$

15. $212 \overline{)86{,}425}$

Perform the indicated operations and reduce if possible.

16. $\left(\dfrac{3}{4}\right)^2 + \dfrac{1}{2} \div \dfrac{8}{7}$

17. $\dfrac{3}{40} + \dfrac{5}{8} + \dfrac{2}{15}$

18. $\dfrac{5}{28}\left(\dfrac{3}{20}\right)\left(\dfrac{14}{9}\right)$

19. $\dfrac{29}{34} - \dfrac{1}{51}$

20. Simplify the complex fraction: $\dfrac{\dfrac{7}{8} - \dfrac{3}{4}}{\dfrac{3}{5} + \dfrac{1}{4}}$

21. Find the average of the numbers 96, 84, 75, and 121.

22. Henry opened his checking account with a deposit of $5470. He wrote checks for $85, $275, $86, and $1450 and made another deposit of $250. What was the balance in his account?

23. If the quotient of 102 and 17 is subtracted from the product of 25 and 32, what is the difference?

24. Add $\dfrac{5}{8}$ to the quotient of $\dfrac{1}{2}$ and $\dfrac{2}{3}$. What is the sum?

25. Find the area in square feet of a rectangle with length 125 feet and width 80 feet.

Mixed Numbers

Mathematics at Work!

Everyone knows how to program their VCR, right? Not necessarily. Did you know that VCRs can tape at different speeds, and that depending on what speed you use to tape a program you can get up to 6 hours of programming on one tape? Generally, there are three speeds you can use with your VCR to tape television programs: standard speed gives 2 hours of programming on one tape; long-play speed gives 4 hours of programming on one tape; super-long-play speed gives 6 hours of programming on one tape. The advantage of the faster speed (shorter running time) is that it provides a much better resolution of the picture.

Suppose that you would like to make the most efficient use of a tape; but at the same time, you want your favorite program to be taped at the standard speed so you can view it more clearly. If you have used a tape for 1 hour at standard speed and 1 hour at long-play speed, how many hours of taping do you have left at the super-long-play speed? (Refer to Problem 46 in Section 4.3.)

What to Expect in Chapter 4

In Chapter 4 we will discuss the meaning of mixed numbers and develop techniques for operating with them. All of the concepts and skills developed in Chapter 1 (Whole Numbers), Chapter 2 (Prime Numbers), and Chapter 3 (Fractions) are an integral part of the topics in Chapter 4, and they should be reviewed often, perhaps even on a daily basis.

One interesting fact made clear in Section 4.1 is that a mixed number indicates the sum of a whole number and a fraction. For example, the mixed number $5\frac{1}{3}$ is shorthand for $5 + \frac{1}{3}$. This idea leads to understanding how to change mixed numbers to improper fractions and how to change improper fractions to mixed numbers. Section 4.1 also makes the distinction between the two concepts of reducing an improper fraction and changing an improper fraction to a mixed number.

The importance of the relationship between improper fractions and mixed numbers is emphasized in Section 4.2, where multiplication and division with complex numbers are accomplished by using the numbers in improper fraction form. Sections 4.3 and 4.4 deal with addition and subtraction with mixed numbers. In subtraction, we can sometimes allow the fraction part of a mixed number to be larger than 1. Section 4.5 explains how the rules for order of operations can be applied with mixed numbers.

4.1   Introduction to Mixed Numbers

OBJECTIVES

1. Understand that a mixed number is the sum of a whole number and a fraction.
2. Be able to change a mixed number to the form of an improper fraction.
3. Be able to change an improper fraction to the form of a mixed number.
4. Know the difference between reducing an improper fraction and changing it to a mixed number.

Changing Mixed Numbers to Fraction Form

Most people are familiar with mixed numbers and use them daily, as in "I drive $4\frac{1}{2}$ miles to school" or "This recipe calls for $2\frac{1}{4}$ cups of flour." In this chapter, we will discuss the operations of addition, subtraction, multiplication, and division with mixed numbers, and we will be using many of the ideas related to fractions that we discussed in Chapter 3.

A **mixed number** is the sum of a whole number and a fraction. We usually write the whole number and the fraction side by side without the plus sign. For example,

$$5 + \frac{1}{10} = 5\frac{1}{10} \qquad \text{Read "five and one-tenth."}$$

$$9 + \frac{2}{3} = 9\frac{2}{3} \qquad \text{Read "nine and two-thirds."}$$

We will find that, in many cases, operations with mixed numbers (particularly in multiplication and division) are easily accomplished by first changing each of the mixed numbers to an improper fraction. With this in mind, in this section, we will develop the techniques for relating mixed numbers and improper fractions.

To change a mixed number to an improper fraction, add the whole number and the fraction. Remember, the whole number can be written with 1 as the denominator.

Change each mixed number to an improper fraction.

EXAMPLE 1 $5\dfrac{1}{10} = 5 + \dfrac{1}{10} = \dfrac{5}{1} \cdot \dfrac{10}{10} + \dfrac{1}{10} = \dfrac{50}{10} + \dfrac{1}{10} = \dfrac{51}{10}$

The denominators of $\dfrac{5}{1}$ and $\dfrac{1}{10}$ are 1 and 10, and the LCD = 10.

EXAMPLE 2 $9\dfrac{2}{3} = 9 + \dfrac{2}{3} = \dfrac{9}{1} \cdot \dfrac{3}{3} + \dfrac{2}{3} = \dfrac{27}{3} + \dfrac{2}{3} = \dfrac{29}{3}$

The LCD = 3 since the denominators are 1 and 3.

EXAMPLE 3 $10\dfrac{1}{2} = 10 + \dfrac{1}{2} = \dfrac{10}{1} \cdot \dfrac{2}{2} + \dfrac{1}{2} = \dfrac{20}{2} + \dfrac{1}{2} = \dfrac{21}{2}$

There is a pattern to changing mixed numbers to improper fractions that leads to a familiar shortcut. Since the denominator of the whole number is always 1, the LCD is always the denominator of the fraction part. Thus, in Example 1, the LCD = 10 and we multiplied the whole number 5 times the common denominator 10. Similarly, in Example 2, we multiplied $9 \cdot 3$; and in Example 3, we multiplied $10 \cdot 2$. After each multiplication, we added the numerator of the fraction part, then used the common denominator. This process is summarized as follows.

> **Shortcut to Changing Mixed Numbers to Fraction Form**
>
> **1.** Multiply the denominator of the fraction part by the whole number.
>
> **2.** Add the numerator of the fraction part to this product.
>
> **3.** Write this sum over the denominator of the fraction.

Change each mixed number to an improper fraction using the shortcut method.

EXAMPLE 4 $3\dfrac{2}{5}$

Solution

(a) Multiply $5 \cdot 3 = 15$ and add 2: $15 + 2 = 17$.

(b) Write 17 over 5:

$$3\frac{2}{5} = \frac{17}{5}$$

EXAMPLE 5 $9\frac{1}{2}$

Solution

(a) Multiply $2 \cdot 9 = 18$ and add 1: $18 + 1 = 19$.

(b) Write 19 over 2:

$$9\frac{1}{2} = \frac{19}{2}$$

EXAMPLE 6 $7\frac{3}{10}$

Solution

(a) Multiply $10 \cdot 7 = 70$ and add 3: $70 + 3 = 73$.

(b) Write 73 over 10:

$$7\frac{3}{10} = \frac{73}{10}$$

We can diagram the procedure as follows:

$$7\frac{3}{10} = \frac{10 \times 7 + 3}{10} = \frac{73}{10}$$

Changing Improper Fractions to Mixed Numbers

To reverse the process (that is, to change an improper fraction to a mixed number), we use the fact that a fraction can indicate division.

> **To Change an Improper Fraction to a Mixed Number:**
>
> 1. Divide the numerator by the denominator.
>
> 2. Write the remainder over the denominator as the fraction part of the mixed number.

EXAMPLE 7   Change $\dfrac{29}{4}$ to a mixed number.

Solution

$$4\overline{)29} \quad \frac{29}{4} = 7 + \frac{1}{4} = 7\frac{1}{4}$$
$$\phantom{4\overline{)}}\begin{array}{r} 7 \\ \hline 29 \\ 28 \\ \hline 1 \end{array}$$

EXAMPLE 8   Change $\dfrac{59}{3}$ to a mixed number.

Solution

$$3\overline{)59} \quad \frac{59}{3} = 19 + \frac{2}{3} = 19\frac{2}{3}$$
$$\begin{array}{r} 19 \\ \hline 59 \\ 3 \\ \hline 29 \\ 27 \\ \hline 2 \end{array}$$

Distinction Between Reducing and Changing to a Mixed Number

Changing an improper fraction to a mixed number is not the same as reducing it. **Reducing** involves finding common factors in the numerator and denominator. **Changing to a mixed number** involves division of the numerator by the denominator. Common factors are not involved.

The fraction part of a mixed number should be reduced. You can proceed in either of the following ways:

1. Reduce the improper fraction first and then change it to a mixed number.

OR

2. Change the improper fraction to a mixed number and then reduce the fraction part.

EXAMPLE 9

Change $\dfrac{24}{10}$ to a mixed number with the fraction part reduced.

Solution

(a) Reducing first;   then changing to a mixed number:

$$\frac{24}{10} = \frac{2 \cdot 12}{2 \cdot 5} = \frac{12}{5} \qquad 5\overline{)12} \qquad \frac{12}{5} = 2\frac{2}{5}$$
$$\begin{array}{r} 2 \\ 5\overline{)12} \\ \underline{10} \\ 2 \end{array}$$

(b) Changing to a mixed number; then reducing:

$$\begin{array}{r} 2 \\ 10\overline{)24} \\ \underline{20} \\ 4 \end{array} \qquad \frac{24}{10} = 2\frac{4}{10} = 2\,\frac{2 \cdot 2}{2 \cdot 5} = 2\frac{2}{5}$$

Both procedures give the same result: $2\dfrac{2}{5}$.

EXAMPLE 10

A customer at a supermarket deli ordered the following amounts of sliced meats: $\dfrac{1}{3}$ pound of roast beef, $\dfrac{3}{4}$ pound of turkey, $\dfrac{3}{8}$ pound of salami, and $\dfrac{1}{2}$ pound of boiled ham. What was the total amount of meats purchased? (Express the answer as a mixed number.)

Solution

To find the total amount of meat purchased, we **add** the individual amounts. From Section 3.4, we know that to add fractions, we need a common denominator. The LCD is the least common multiple of the numbers 3, 4, 8, and 2: LCD = $2 \cdot 2 \cdot 2 \cdot 3 = 24$.

$$\frac{1}{3} + \frac{3}{4} + \frac{3}{8} + \frac{1}{2} = \frac{1}{3} \cdot \frac{8}{8} + \frac{3}{4} \cdot \frac{6}{6} + \frac{3}{8} \cdot \frac{3}{3} + \frac{1}{2} \cdot \frac{12}{12}$$

$$= \frac{8}{24} + \frac{18}{24} + \frac{9}{24} + \frac{12}{24}$$

$$= \frac{47}{24} = 1\frac{23}{24}$$

The total purchase was $1\dfrac{23}{24}$ pounds of meat.

Reduce to lowest terms.

1. $\dfrac{18}{16}$ **2.** $\dfrac{35}{15}$

Change to a mixed number.

3. $\dfrac{51}{34}$ **4.** $\dfrac{35}{15}$

Change to an improper fraction.

5. $6\dfrac{2}{3}$ **6.** $7\dfrac{1}{100}$

ANSWERS: **1.** $\dfrac{9}{8}$ **2.** $\dfrac{7}{3}$ **3.** $1\dfrac{1}{2}$ **4.** $2\dfrac{1}{3}$ **5.** $\dfrac{20}{3}$ **6.** $\dfrac{701}{100}$

Exercises 4.1

Reduce to lowest terms.

1. $\dfrac{24}{18}$ **2.** $\dfrac{25}{10}$ **3.** $\dfrac{16}{12}$ **4.** $\dfrac{10}{8}$ **5.** $\dfrac{39}{26}$

6. $\dfrac{48}{32}$ **7.** $\dfrac{35}{25}$ **8.** $\dfrac{18}{16}$ **9.** $\dfrac{80}{64}$ **10.** $\dfrac{75}{60}$

Change to mixed numbers with the fraction part reduced to lowest terms.

11. $\dfrac{100}{24}$ **12.** $\dfrac{25}{10}$ **13.** $\dfrac{16}{12}$ **14.** $\dfrac{10}{8}$ **15.** $\dfrac{39}{26}$

16. $\dfrac{42}{8}$ **17.** $\dfrac{43}{7}$ **18.** $\dfrac{34}{16}$ **19.** $\dfrac{45}{6}$ **20.** $\dfrac{75}{12}$

21. $\dfrac{56}{18}$ **22.** $\dfrac{31}{15}$ **23.** $\dfrac{36}{12}$ **24.** $\dfrac{48}{16}$ **25.** $\dfrac{72}{16}$

26. $\dfrac{70}{34}$ **27.** $\dfrac{45}{15}$ **28.** $\dfrac{60}{36}$ **29.** $\dfrac{35}{20}$ **30.** $\dfrac{185}{100}$

Change to fraction form.

31. $4\dfrac{5}{8}$ **32.** $3\dfrac{3}{4}$ **33.** $5\dfrac{1}{15}$ **34.** $1\dfrac{3}{5}$ **35.** $4\dfrac{2}{11}$

36. $2\dfrac{11}{44}$ **37.** $2\dfrac{9}{27}$ **38.** $4\dfrac{6}{7}$ **39.** $10\dfrac{8}{12}$ **40.** $11\dfrac{3}{8}$

41. $6\dfrac{8}{10}$ **42.** $14\dfrac{1}{5}$ **43.** $16\dfrac{2}{3}$ **44.** $12\dfrac{4}{8}$ **45.** $20\dfrac{3}{15}$

46. $9\dfrac{4}{10}$ **47.** $13\dfrac{1}{7}$ **48.** $49\dfrac{0}{12}$ **49.** $17\dfrac{0}{3}$ **50.** $3\dfrac{1}{50}$

51. $6\dfrac{3}{100}$ **52.** $4\dfrac{237}{1000}$ **53.** $8\dfrac{921}{1000}$ **54.** $3\dfrac{477}{1000}$ **55.** $2\dfrac{631}{1000}$

In each of the following problems, write the answer in the form of a mixed number.

56. A tree (in Yosemite National Forest) grew $\dfrac{2}{3}$ foot, $\dfrac{3}{4}$ foot, $\dfrac{7}{8}$ foot, and $\dfrac{1}{2}$ foot in four consecutive years. How many feet did the tree grow during these four years?

57. A baby grew $\dfrac{1}{4}$ inch, $\dfrac{3}{8}$ inch, $\dfrac{3}{4}$ inch, and $\dfrac{9}{16}$ inch over a four-month period. How many inches did the baby grow over this time period?

58. During five days in one week the price of Xerox stock rose $\dfrac{1}{4}$ of a dollar, rose $\dfrac{7}{8}$ of a dollar, rose $\dfrac{3}{4}$ of a dollar, fell $\dfrac{1}{2}$ of a dollar, and rose $\dfrac{3}{8}$ of a dollar. What was the net gain (or loss) in the price of this stock over these five days?

59. After typing the manuscript for this textbook, Dr. Wright measured the height (or thickness) of each of the stack of pages for each of the first four chapters as $\dfrac{7}{8}$ inch, $\dfrac{1}{2}$ inch, $\dfrac{3}{4}$ inch, and $\dfrac{5}{8}$ inch. What was the total height of the first four chapters in typed form?

Writing and Thinking about Mathematics

60. You were probably familiar with the shortcut for changing a mixed number to an improper fraction (See page 168.) before you read this section. Explain, in your own words, how this method works and why it does indeed give the correct fraction every time.

61. Explain, in your own words, why the use of the word "improper" is somewhat misleading when referring to improper fractions. Can you think of some other word used in a special context, other than in its normal English usage, that could be misleading to a person not familiar with the new meaning?

The Recycle Bin   (from Sections 3.2 and 3.3)

Reduce each fraction to lowest terms.

1. $\dfrac{35}{45}$

2. $\dfrac{128}{320}$

3. $\dfrac{300}{100}$

4. $\dfrac{102}{221}$

Find each of the following products reduced to lowest terms.

5. $\dfrac{2}{3} \cdot \dfrac{15}{16} \cdot \dfrac{8}{25}$

6. $\dfrac{13}{24} \cdot \dfrac{15}{39} \cdot \dfrac{18}{20}$

Find each of the following quotients reduced to lowest terms.

7. $\dfrac{25}{36} \div \dfrac{35}{28}$

8. $\dfrac{65}{22} \div \dfrac{91}{33}$

4.2   Multiplication and Division

OBJECTIVES

1. Learn how to multiply with mixed numbers.
2. Learn how to divide with mixed numbers.

Multiplication with Mixed Numbers

In Sections 4.3 and 4.4, we will see that addition and subtraction with mixed numbers both rely on the fact that (as we discussed in Section 4.1) a mixed number is the sum of a whole number and a fraction. However, in this section, we will see that multiplication and division are easier to accomplish by changing mixed numbers to the form of improper fractions rather than by treating them as sums. Thus, multiplication and division with mixed numbers is the same as multiplication and division with fractions. Just as with fractions, we use prime factorizations and reduce as we multiply and/or divide.

> **To Multiply Mixed Numbers:**
>
> **1** Change each number to fraction form.
>
> **2.** Multiply by factoring numerators and denominators; then reduce.
>
> **3.** Change the answer to a mixed number or leave it in fraction form. (The form of the answer sometimes depends on what use is to be made of the result.)

EXAMPLE 1 Find the product: $\dfrac{5}{6} \cdot 3\dfrac{3}{10}$.

Solution

$$\frac{5}{6} \cdot 3\frac{3}{10} = \frac{5}{6} \cdot \frac{33}{10} = \frac{\cancel{5} \cdot \cancel{3} \cdot 11}{2 \cdot \cancel{3} \cdot 2 \cdot \cancel{5}} = \frac{11}{4} \quad \text{or} \quad 2\frac{3}{4}$$

EXAMPLE 2 Multiply and reduce to lowest terms: $4\dfrac{1}{2} \cdot 1\dfrac{1}{6} \cdot 3\dfrac{1}{3}$.

Solution

$$4\frac{1}{2} \cdot 1\frac{1}{6} \cdot 3\frac{1}{3} = \frac{9}{2} \cdot \frac{7}{6} \cdot \frac{10}{3} = \frac{\cancel{3} \cdot \cancel{3} \cdot 7 \cdot \cancel{2} \cdot 5}{\cancel{2} \cdot 2 \cdot \cancel{3} \cdot \cancel{3}} = \frac{35}{2} \quad \text{or} \quad 17\frac{1}{2}$$

Large mixed numbers can be multiplied in the same way as shown in Examples 1 and 2. The numerators will be relatively large numbers and their products will also be large; but, the technique of changing to improper fractions and then multiplying and reducing is the same.

EXAMPLE 3 Linda is framing an antique circus poster (in the shape of a rectangle) that measures $24\dfrac{3}{8}$ inches wide by $45\dfrac{1}{4}$ inches long. What is the area of the glass needed to cover the poster?

Solution

We find the area (measured in square inches) of the rectangle by multiplying the width times the length.

(a) To multiply $24\frac{3}{8} \cdot 45\frac{1}{4}$, we first change both numbers to improper fractions.

$$
\left.\begin{array}{r}
24 \\
\times\ 8 \\
\hline
192
\end{array}\quad
\begin{array}{r}
192 \\
+\ \ 3 \\
\hline
195
\end{array}\right\}
\quad \text{so } 24\frac{3}{8} = \frac{195}{8}
$$

$$
\left.\begin{array}{r}
45 \\
\times\ 4 \\
\hline
180
\end{array}\quad
\begin{array}{r}
180 \\
+\ \ 1 \\
\hline
181
\end{array}\right\}
\quad \text{so } 45\frac{1}{4} = \frac{181}{4}
$$

(b) Now multiply the improper fractions and change the product back to a mixed number.

$$
24\frac{3}{8} \cdot 45\frac{1}{4} = \frac{195}{8} \cdot \frac{181}{4} = \frac{35{,}295}{32} = 1102\frac{31}{32}
$$

$$
\begin{array}{r}
195 \\
\times\ 181 \\
\hline
195 \\
15\ 60 \\
19\ 5\ \ \\
\hline
35{,}295
\end{array}
\qquad
\begin{array}{r}
1102\frac{31}{32} \\
\hline
32\,)\,\overline{35{,}295} \\
32 \\
\hline
3\ 2 \\
3\ 2 \\
\hline
09 \\
0 \\
\hline
95 \\
64 \\
\hline
31
\end{array}
$$

Thus, the area of the glass is $1102\frac{31}{32}$ square inches.

The product in Example 3 with large mixed numbers can be found by writing the numbers one under the other and then multiplying. But this procedure is not simple because four products must be calculated and then added. The procedure is shown here for comparison, but it is **not** recommended.

$$24\frac{3}{8}$$

$$45\frac{1}{4}$$

$$\frac{3}{32} \quad \leftarrow \frac{1}{4} \cdot \frac{3}{8} = \frac{3}{32}$$

$$6 \quad \leftarrow \frac{1}{4} \cdot 24 = 6$$

$$16\frac{7}{8} \quad \leftarrow 45 \cdot \frac{3}{8} = \frac{135}{8} = 16\frac{7}{8}$$

$$120$$

$$960 \quad \leftarrow 45 \cdot 24$$

$$1102\frac{31}{32} \quad \left(\frac{3}{32} + \frac{7}{8} = \frac{3}{32} + \frac{28}{32} = \frac{31}{32}\right)$$

In Section 3.1 we discussed the idea that finding a fractional part **of** another number indicates multiplication. The key word is **of,** and we will find that this same idea is related to decimals and percents in later chapters. In a phrase such as "$\frac{3}{4}$ of 80," the implication is that you are to **multiply** $\frac{3}{4} \cdot 80$.

> A fraction **of** a number means to **multiply** the number by the fraction.

EXAMPLE 4   Find $\frac{3}{4}$ of 80.

Solution

$$\frac{3}{4} \cdot 80 = \frac{3}{\overset{}{\underset{1}{4}}} \cdot \frac{\overset{20}{80}}{1} = 60$$

EXAMPLE 5   Find $\frac{2}{3}$ of $\frac{9}{10}$.

Solution

$$\frac{2}{3} \cdot \frac{9}{10} = \frac{\cancel{2} \cdot \cancel{3} \cdot 3}{\cancel{3} \cdot \cancel{2} \cdot 5} = \frac{3}{5}$$

Note: Whenever you find a fraction **of** a number (and the fraction is between 0 and 1), the result should always be less than the number. If you get a larger number for an answer, you have made a mistake.

Find the indicated products.

1. $4\frac{1}{3} \cdot \frac{2}{13}$

2. $5\frac{1}{2} \cdot 3\frac{1}{3}$

3. $\frac{5}{8} \cdot 2\frac{3}{5} \cdot 5\frac{1}{3}$

4. $16\frac{1}{2} \cdot 20\frac{2}{3}$

5. Find $\frac{9}{10}$ of 70.

ANSWERS: **1.** $\frac{2}{3}$ **2.** $\frac{55}{3}$ or $18\frac{1}{3}$ **3.** $\frac{26}{3}$ or $8\frac{2}{3}$ **4.** 341   **5.** 63

Division with Mixed Numbers

Division with mixed numbers is the same as division with fractions, as discussed in Section 3.3. Simply change each mixed number to an improper fraction before dividing.

As review,

the **reciprocal** of $\quad \frac{c}{d} \quad$ is $\quad \frac{d}{c}$

and $\quad \frac{a}{b} \div \frac{c}{d} = \frac{a}{b} \cdot \frac{d}{c}.$

That is, **to divide by any number (except 0), multiply by its reciprocal.**

To Divide Mixed Numbers:

1. Change each number to fraction form.

2. Write the reciprocal of the divisor.

3. Multiply by factoring numerators and denominators; then reduce.

EXAMPLE 6

$$3\frac{1}{4} \div 7\frac{4}{5} = \frac{13}{4} \div \frac{39}{5} = \frac{13}{4} \cdot \boxed{\frac{5}{39}}$$

Note that the divisor is $\frac{39}{5}$, and we multiply by its reciprocal, $\frac{5}{39}$.

$$= \frac{\cancel{13} \cdot 5}{4 \cdot \cancel{13} \cdot 3} = \frac{5}{12}$$

EXAMPLE 7

$$\frac{2\frac{1}{4}}{4\frac{1}{2}} = \frac{\frac{9}{4}}{\frac{9}{2}} = \frac{\cancel{9}^{1}}{\cancel{4}_{2}} \cdot \boxed{\frac{\cancel{2}^{1}}{\cancel{9}_{1}}} = \frac{1}{2}$$

Note that the divisor is $\frac{9}{2}$, and we multiply by its reciprocal, $\frac{2}{9}$.

EXAMPLE 8

$$6 \div 7\frac{7}{8} = 6 \div \frac{63}{8} = \frac{6}{1} \cdot \boxed{\frac{8}{63}}$$

$$= \frac{\cancel{3} \cdot 2 \cdot 8}{\cancel{3} \cdot 3 \cdot 7} = \frac{16}{21}$$

EXAMPLE 9

$$7\frac{7}{8} \div 6 = \frac{63}{8} \div \frac{6}{1} = \frac{63}{8} \cdot \boxed{\frac{1}{6}}$$

$$= \frac{\cancel{3} \cdot 3 \cdot 7}{8 \cdot 2 \cdot \cancel{3}} = \frac{21}{16} \quad \text{or} \quad 1\frac{5}{16}$$

If we know two numbers, say 2 and 7, we can find their product, 14, by multiplying the two numbers. Now suppose we know the product of two numbers is 24 and one of the numbers is 3. How do we find the other number? Recall from our work with whole numbers in Chapter 1, that the answer is found by dividing the product by the known number. Thus, the missing number is $24 \div 3 = 8$. Example 10 illustrates this same idea (of finding a number by dividing the product of two numbers by the known number) with mixed numbers.

EXAMPLE 10

The product of $2\frac{1}{3}$ with another number is $5\frac{1}{6}$. What is the other number?

Solution

$$2\frac{1}{3} \cdot \; ? \; = 5\frac{1}{6}$$

To find the missing number *divide* $5\frac{1}{6}$ by $2\frac{1}{3}$.

$$5\frac{1}{6} \div 2\frac{1}{3} = \frac{31}{6} \div \frac{7}{3} = \frac{31}{\overset{}{\underset{2}{6}}} \cdot \frac{\overset{1}{\cancel{3}}}{7} = \frac{31}{14} \quad \text{or} \quad 2\frac{3}{14}$$

The other number is $2\frac{3}{14}$.

CHECK

$$2\frac{1}{3} \cdot 2\frac{3}{14} = \frac{\overset{1}{\cancel{7}}}{3} \cdot \frac{31}{\underset{2}{\cancel{14}}} = \frac{31}{6} = 5\frac{1}{6}$$

CLASSROOM PRACTICE

Find the indicated quotients.

1. $2\frac{1}{3} \div \frac{7}{3}$ **2.** $5\frac{1}{4} \div 3\frac{1}{2}$ **3.** $12 \div \frac{3}{4}$ **4.** $6\frac{3}{5} \div 11$

ANSWERS: **1.** 1 **2.** $\frac{3}{2}$ or $1\frac{1}{2}$ **3.** 16 **4.** $\frac{3}{5}$

Exercises 4.2

Find the indicated products.

1. $\left(2\frac{1}{3}\right)\left(3\frac{1}{4}\right)$

2. $\left(1\frac{1}{5}\right)\left(1\frac{1}{7}\right)$

3. $4\frac{1}{2}\left(2\frac{1}{3}\right)$

4. $3\frac{1}{3}\left(2\frac{1}{5}\right)$

5. $6\frac{1}{4}\left(3\frac{3}{5}\right)$

6. $5\frac{1}{3}\left(2\frac{1}{4}\right)$

7. $\left(8\frac{1}{2}\right)\left(3\frac{2}{3}\right)$

8. $\left(9\frac{1}{3}\right)2\frac{1}{7}$

9. $\left(6\frac{2}{7}\right)1\frac{3}{11}$

10. $\left(11\frac{1}{4}\right)1\frac{1}{15}$

11. $6\frac{2}{3} \cdot 4\frac{1}{2}$

12. $4\frac{3}{8} \cdot 2\frac{2}{7}$

13. $9\frac{3}{4} \cdot 2\frac{6}{26}$ **14.** $7\frac{1}{2} \cdot \frac{2}{15}$ **15.** $\frac{3}{4} \cdot 1\frac{1}{3}$

16. $3\frac{4}{5} \cdot 2\frac{1}{7}$ **17.** $12\frac{1}{2} \cdot 2\frac{1}{5}$ **18.** $9\frac{3}{5} \cdot 1\frac{1}{16}$

19. $6\frac{1}{8} \cdot 3\frac{1}{7}$ **20.** $5\frac{1}{4} \cdot 11\frac{1}{3}$ **21.** $\frac{1}{4} \cdot \frac{2}{3} \cdot \frac{6}{7}$

22. $\frac{7}{8} \cdot \frac{24}{25} \cdot \frac{5}{21}$ **23.** $\frac{3}{16} \cdot \frac{8}{9} \cdot \frac{3}{5}$ **24.** $\frac{2}{5} \cdot \frac{1}{5} \cdot \frac{4}{7}$

25. $\frac{6}{7} \cdot \frac{2}{11} \cdot \frac{3}{5}$ **26.** $\left(3\frac{1}{2}\right)\left(2\frac{1}{7}\right)\left(5\frac{1}{4}\right)$ **27.** $\left(4\frac{3}{8}\right)\left(2\frac{1}{5}\right)\left(1\frac{1}{7}\right)$

28. $42\frac{5}{6} \cdot 30\frac{1}{7}$ **29.** $75\frac{1}{3} \cdot 40\frac{1}{25}$ **30.** $36\frac{3}{4} \cdot 17\frac{5}{12}$

31. Find $\frac{2}{3}$ of 60.   **32.** Find $\frac{1}{4}$ of 80.   **33.** Find $\frac{1}{5}$ of 100.

34. Find $\frac{3}{5}$ of 100.   **35.** Find $\frac{1}{2}$ of $2\frac{5}{8}$.   **36.** Find $\frac{1}{6}$ of $1\frac{3}{4}$.

37. Find $\frac{9}{10}$ of $3\frac{5}{7}$.   **38.** Find $\frac{7}{8}$ of $6\frac{4}{5}$.

Find the indicated quotients.

39. $\frac{2}{21} \div \frac{2}{7}$ **40.** $\frac{9}{32} \div \frac{5}{8}$ **41.** $\frac{5}{12} \div \frac{3}{4}$

42. $\frac{6}{17} \div \frac{6}{17}$ **43.** $\frac{5}{6} \div 3\frac{1}{4}$ **44.** $\frac{7}{8} \div 7\frac{1}{2}$

45. $\frac{29}{50} \div 3\frac{1}{10}$ **46.** $4\frac{1}{5} \div 3\frac{1}{3}$ **47.** $2\frac{1}{17} \div 1\frac{1}{4}$

48. $5\frac{1}{6} \div 3\frac{1}{4}$ **49.** $2\frac{2}{49} \div 3\frac{1}{14}$ **50.** $6\frac{5}{6} \div 2$

51. $4\frac{1}{5} \div 3$ **52.** $4\frac{5}{8} \div 4$ **53.** $6\frac{5}{6} \div \frac{1}{2}$

54. $4\frac{5}{8} \div \frac{1}{4}$ **55.** $4\frac{1}{5} \div \frac{1}{3}$ **56.** $1\frac{1}{32} \div 3\frac{2}{3}$

57. $7\dfrac{5}{11} \div 4\dfrac{1}{10}$ **58.** $13\dfrac{1}{7} \div 4\dfrac{2}{11}$

59. A telephone pole is 32 feet long. If $\dfrac{5}{16}$ of the pole must be underground and $\dfrac{11}{16}$ of the pole aboveground, how much of the pole is underground? How much is aboveground?

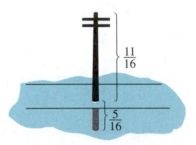

$20\frac{1}{4}$

60. The total distance around a square (its perimeter) is found by multiplying the length of one side by 4. Find the perimeter of a square if the length of one side is $5\dfrac{1}{16}$ inches.

$5\frac{1}{16}$ in.

61. A man driving to work drives $17\dfrac{7}{10}$ miles one way five days a week. How many miles does he drive each week going to and from work?

62. A length of pipe is $27\dfrac{3}{4}$ feet. What would be the total length if $36\dfrac{1}{2}$ of these pipes were laid end to end?

63. A woman reads $\dfrac{1}{6}$ of a book in 3 hours. If the book contains 540 pages, how

$\frac{1}{6}$ of 540? → many pages will she read in 3 hours? How long will she take to read the entire book?

64. Three towns (A, B, and C) are located on the same highway (assume a straight section of highway). Towns A and B are 53 kilometers apart. Town C is

$45\dfrac{9}{10}$ kilometers from town B. How far apart are towns A and C? (The sketch shows that there are two possible situations to consider.)

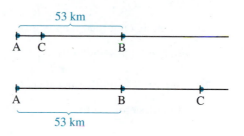

65. The result of multiplying two numbers is $10\dfrac{1}{3}$. If one of the numbers is $7\dfrac{1}{6}$, what is the other number?

66. An airplane is carrying 150 passengers. This is $\dfrac{6}{7}$ of its capacity. What is the capacity of the airplane?

67. The sale price of a coat is $135. This is $\dfrac{3}{4}$ of the original price. What was the original price?

68. If the product of $5\dfrac{1}{2}$ and $2\dfrac{1}{4}$ is added to the quotient of $\dfrac{9}{10}$ and $\dfrac{3}{4}$, what is the sum?

69. If the quotient of $\dfrac{5}{8}$ and $\dfrac{1}{2}$ is subtracted from the product of $2\dfrac{1}{4}$ and $3\dfrac{1}{5}$, what is the difference?

70. If $\dfrac{9}{10}$ of 70 is divided by $\dfrac{3}{4}$ of 10, what is the quotient?

71. The sale price of a new computer system, including a printer, is $2400. This is $\dfrac{3}{4}$ of the original price. What was the original price?

72. A used car is advertised as a special for $3500. If this sale price is $\dfrac{4}{5}$ of the original price, what was the original price before the sale?

73. Your car averages $26\dfrac{3}{10}$ miles per gallon of gas. (a) How many miles can your car travel on $17\dfrac{1}{2}$ gallons of gas? (b) If each gallon costs $132\dfrac{9}{10}$¢, what would you pay (in cents) for $17\dfrac{1}{2}$ gallons?

74. An equilateral triangle is one in which all three sides have the same length. What is the perimeter of an equilateral triangle with sides of length $15\frac{7}{10}$ centimeters?

75. The scale on a road map is given so that one inch equals 50 miles. How many miles are represented by $3\frac{1}{4}$ inches?

76. Butter has 36 milligrams of cholesterol in one tablespoon. How many milligrams of cholesterol are there in $2\frac{3}{4}$ tablespoons of butter?

77. In one second, the velocity of a free falling object increases by $9\frac{4}{5}$ meters per second. By how much has the velocity of a falling object increased after $5\frac{1}{2}$ seconds?

Check Your Number Sense

78. Suppose that the product of $5\frac{7}{10}$ with some other number is $10\frac{1}{2}$. Answer the following questions without doing any calculations:

    (a) Do you think that this other number is more than 1 or less than 1? Why?

    (b) Do you think that this other number is more than 2 or less than 2? Why?

    Find the other number.

The Recycle Bin   (from Sections 3.4 and 3.5)

Find each of the following sums and reduce to lowest terms.

1. $\dfrac{2}{3} + \dfrac{1}{7}$ 　　　　　　　　　　 **2.** $\dfrac{5}{8} + \dfrac{9}{10}$

3. $\dfrac{3}{5} + \dfrac{7}{20} + \dfrac{1}{12}$ 　　　　　 **4.** $\dfrac{11}{18} + \dfrac{2}{5} + \dfrac{3}{10}$

Find each of the following differences and reduce to lowest terms.

5. $\dfrac{7}{8} - \dfrac{9}{20}$ 　　　　　　　　　　 **6.** $\dfrac{19}{21} - \dfrac{3}{28}$

7. $\dfrac{17}{32} - \dfrac{17}{64}$ 　　　　　　　　　 **8.** $\dfrac{20}{39} - \dfrac{9}{26}$

4.3   Addition

OBJECTIVES

1. Know how to add mixed numbers when the fraction parts have the same denominator.
2. Know how to add mixed numbers when the fraction parts have different denominators.
3. Be able to write the sum of mixed numbers in the form of a mixed number with the fraction part less than 1.

Addition with Mixed Numbers

Since a mixed number itself represents addition, two or more mixed numbers can be added by adding the whole numbers and the fraction parts separately.

> **To Add Mixed Numbers:**
>
> **1.** Add the fraction parts.
>
> **2.** Add the whole numbers.
>
> **3.** Write the mixed number so that the fraction part is less than 1.

Examples 1–6 illustrate addition with mixed numbers.

EXAMPLE 1

$4\frac{2}{7} + 6\frac{3}{7}$

Solution

$$4\frac{2}{7} + 6\frac{3}{7} = 4 + \frac{2}{7} + 6 + \frac{3}{7}$$

$$= (4 + 6) + \left(\frac{2}{7} + \frac{3}{7}\right)$$

$$= 10 + \frac{5}{7}$$

$$= 10\frac{5}{7}$$

Or, vertically,

$$4\frac{2}{7}$$
$$+\ 6\frac{3}{7}$$
$$\overline{10\frac{5}{7}}$$

EXAMPLE 2 $25\dfrac{1}{6} + 3\dfrac{7}{18}$

Solution

$$25\frac{1}{6} + 3\frac{7}{18} = 25 + \frac{1}{6} + 3 + \frac{7}{18}$$

$$= (25 + 3) + \left(\frac{1}{6} + \frac{7}{18}\right)$$

$$= 28 + \left(\frac{1}{6} \cdot \frac{3}{3} + \frac{7}{18}\right) \qquad \text{(The LCD, 18, is needed to add the fraction parts.)}$$

$$= 28 + \left(\frac{3}{18} + \frac{7}{18}\right)$$

$$= 28\frac{10}{18} = 28\frac{5}{9}$$

EXAMPLE 3 $7\dfrac{2}{3} + 9\dfrac{4}{5}$

Solution

The LCD is 15.

$$7\frac{2}{3} = \quad 7\frac{2}{3} \cdot \frac{5}{5} = \quad 7\frac{10}{15}$$

$$+\ 9\frac{4}{5} = +\ 9\frac{4}{5} \cdot \frac{3}{3} = \quad 9\frac{12}{15}$$

$$\overline{\qquad\qquad\qquad\qquad 16\frac{22}{15} = 16 + 1\frac{7}{15} = 17\frac{7}{15}}$$

$$\uparrow \qquad\qquad\qquad \uparrow$$

Fraction part is   Change it to a
greater than 1.   mixed number.

EXAMPLE 4 $3 + 6\frac{1}{2}$

Solution

$$3 + 6\frac{1}{2} = 3 + 6 + \frac{1}{2}$$

$$= 9 + \frac{1}{2}$$

$$= 9\frac{1}{2}$$

Or, vertically,

$$\begin{array}{r} 3 \\ + \, 6\dfrac{1}{2} \\ \hline 9\dfrac{1}{2} \end{array}$$

The whole number 3 has no fraction part. Be sure to align the whole numbers.

EXAMPLE 5 $4\frac{3}{4} + 5 + 7\frac{3}{10}$

Solution

The vertical arrangement is generally easier to use for adding mixed numbers. The LCD is 20.

$$\begin{array}{rcccl} 4\dfrac{3}{4} & = & 4\,\dfrac{3}{4} \cdot \dfrac{5}{5} & = & 4\dfrac{15}{20} \\ 5 & = & 5 & = & 5 \\ + \, 7\dfrac{3}{10} & = & + \, 7\,\dfrac{3}{10} \cdot \dfrac{2}{2} & = & 7\dfrac{6}{20} \\ \hline & & & & 16\dfrac{21}{20} = 16 + 1\dfrac{1}{20} = 17\dfrac{1}{20} \end{array}$$

↑ Fraction part is greater than 1.

↑ Change it to a mixed number.

EXAMPLE 6

A triangle has sides measuring $25\frac{1}{2}$ meters, $32\frac{3}{10}$ meters, and $41\frac{7}{10}$ meters. Find the perimeter (total distance around) of the triangle.

Solution

We find the perimeter by adding the lengths of the three sides.

$$25\frac{1}{2} = 25\frac{5}{10}$$

$$32\frac{3}{10} = 32\frac{3}{10}$$

$$41\frac{7}{10} = 41\frac{7}{10}$$

$$98\frac{15}{10} = 98 + 1\frac{5}{10} = 99\frac{1}{2} \text{ meters}$$

The perimeter is $99\frac{1}{2}$ meters.

Estimating with Mixed Numbers

Sometimes you can concentrate so hard on remembering the rules for operating with fractions that you can overlook the fact that the fraction parts of mixed numbers have much less effect on the answer than do the whole number parts. Thus, to help avoid some errors when operating with mixed numbers, you can simply estimate the answer by using only the whole number parts and ignoring the fraction parts. We say that the fraction parts have been **truncated** (cut off) from the mixed numbers. A **truncated mixed number,** then, consists of the whole number part only.

> **To Estimate an Answer with Mixed Numbers:**
>
> **1.** Perform the related operations with truncated mixed numbers.
>
> **2.** If one of the numbers is a fraction only, replace this fraction with the whole number 1. (This replacement can be used for estimating as long as no denominator or divisor is 0.)

Note: This technique is illustrated here in Example 7 with addition. However, it applies to all the operations of addition, subtraction, multiplication, and division.

EXAMPLE 7

Estimate the value of the following expression by using truncated mixed numbers:

$$14\frac{3}{5} + 6\frac{1}{3} + 2\frac{1}{4}$$

Solution

Truncate $14\frac{3}{5}$ to 14.

Truncate $6\frac{1}{3}$ to 6.

Truncate $2\frac{1}{4}$ to 2.

Thus, before any work with finding a common denominator for the fraction parts in the mixed numbers, we can estimate the sum

$$14\frac{3}{5} + 6\frac{1}{3} + 2\frac{1}{4}$$

as

$$14 + 6 + 2 = 22$$

Exercises 4.3

First (a) estimate each sum by using truncated mixed numbers, then, (b) find each sum.

1. $\begin{aligned} 10 \\ + \ 4\frac{5}{7} \end{aligned}$

2. $\begin{aligned} 6\frac{1}{2} \\ + 3\frac{1}{2} \end{aligned}$

3. $\begin{aligned} 5\frac{1}{3} \\ + 2\frac{2}{3} \end{aligned}$

4. $\begin{aligned} 8 \\ + 7\frac{3}{11} \end{aligned}$

5. $\begin{aligned} 12 \\ + \ \ \frac{3}{4} \end{aligned}$

6. $\begin{aligned} 15 \\ + \ \ \frac{9}{10} \end{aligned}$

7. $\begin{aligned} 3\frac{1}{4} \\ + 5\frac{5}{12} \end{aligned}$

8. $\begin{aligned} 5\frac{11}{12} \\ + 3\frac{5}{6} \end{aligned}$

9. $21\dfrac{1}{3}$
$+\ 13\dfrac{1}{18}$

10. $9\dfrac{2}{3}$
$+\ \dfrac{5}{6}$

11. $4\dfrac{1}{2} + 3\dfrac{1}{6}$

12. $3\dfrac{1}{4} + 7\dfrac{1}{8}$

13. $25\dfrac{1}{10} + 17\dfrac{1}{4}$

14. $5\dfrac{1}{7} + 3\dfrac{1}{3}$

15. $6\dfrac{5}{12} + 4\dfrac{1}{3}$

16. $5\dfrac{3}{10} + 2\dfrac{1}{14}$

17. $8\dfrac{2}{9} + 4\dfrac{1}{27}$

18. $11\dfrac{3}{4} + 2\dfrac{5}{16}$

19. $6\dfrac{4}{9} + 12\dfrac{1}{15}$

20. $4\dfrac{1}{6} + 13\dfrac{9}{10}$

21. $21\dfrac{3}{4} + 6\dfrac{3}{4}$

22. $3\dfrac{5}{8} + 3\dfrac{5}{8}$

23. $7\dfrac{3}{5} + 2\dfrac{1}{8}$

24. $9\dfrac{1}{8} + 3\dfrac{7}{12}$

25. $3\dfrac{1}{3} + 4\dfrac{1}{4} + 5\dfrac{1}{5}$

26. $\dfrac{3}{7} + 2\dfrac{1}{14} + 2\dfrac{1}{6}$

27. $20\dfrac{5}{8} + 42\dfrac{5}{6}$

28. $25\dfrac{2}{3} + 1\dfrac{1}{16}$

29. $32\dfrac{1}{64} + 4\dfrac{1}{24} + 17\dfrac{3}{8}$

30. $3\dfrac{1}{20} + 7\dfrac{1}{15} + 2\dfrac{3}{10}$

31. $4\dfrac{1}{3}$
$8\dfrac{3}{8}$
$+\ 6\dfrac{1}{6}$

32. $3\dfrac{2}{3}$
$14\dfrac{1}{10}$
$+\ 5\dfrac{1}{5}$

33. $13\dfrac{5}{8}$
$13\dfrac{1}{12}$
$+\ 10\dfrac{1}{4}$

34. $5\dfrac{1}{8}$
$1\dfrac{1}{5}$
$+\ 3\dfrac{1}{40}$

35. $27\dfrac{2}{3}$
$30\dfrac{5}{8}$
$+\ 31\dfrac{5}{6}$

36. A bus trip is made in three parts. The first part takes $2\dfrac{1}{3}$ hours, the second part takes $2\dfrac{1}{2}$ hours, and the third part takes $3\dfrac{3}{4}$ hours. How many hours does the trip take?

37. A construction company built three sections of highway. One section was $20\dfrac{7}{10}$ kilometers, the second section was $3\dfrac{4}{10}$ kilometers, and the third section was $11\dfrac{6}{10}$ kilometers. What was the total length of highway built?

38. A triangle (three-sided figure) has sides of $42\dfrac{3}{4}$ feet, $23\dfrac{1}{2}$ feet, and $22\dfrac{7}{8}$ feet. What is the perimeter (total distance around) of the triangle?

39. A quadrilateral (four-sided figure) has sides of $3\frac{1}{2}$ inches, $2\frac{1}{4}$ inches, $3\frac{5}{8}$ inches, and $2\frac{3}{4}$ inches. What is the perimeter (total distance around) of the quadrilateral?

40. During one week, Sam played three rounds of golf. The first round took $4\frac{1}{2}$ hours, the second round took $3\frac{3}{4}$ hours, and the third round took $4\frac{3}{4}$ hours. How much time did Sam spend playing golf that week?

41. On five different days, a test driver drove a new car $5\frac{1}{2}$ miles, $4\frac{3}{10}$ miles, $2\frac{3}{4}$ miles, $7\frac{1}{2}$ miles, and $25\frac{7}{10}$ miles. How far did he drive that particular car?

42. During four days of training, a marathon runner ran 34 miles, $25\frac{1}{2}$ miles, $50\frac{3}{10}$ miles, and $42\frac{7}{10}$ miles. What total distance did she run on those four days?

43. A pentagon (five-sided figure) has sides of $2\frac{3}{8}$ cm, $5\frac{1}{2}$ cm, $6\frac{1}{4}$ cm, $8\frac{1}{10}$ cm, and $3\frac{7}{8}$ cm. What is the perimeter in centimeters of the pentagon?

44. The United States Postal Service sells bulk rate postage stamps with denominations of $7\frac{3}{5}$¢, $13\frac{1}{5}$¢, $16\frac{7}{10}$¢, $17\frac{1}{2}$¢, and $24\frac{1}{10}$¢. If a company bought 1000 of each of these stamps, what would be the total amount paid?

45. Among the top 50 movie rentals in 1990, "The Hunt for Red October" earned $58\frac{1}{2}$ million, "Presumed Innocent" earned $43\frac{4}{5}$ million, "Robocop 2" earned $22\frac{3}{10}$ million, and "Goodfellas" earned $18\frac{1}{5}$ million. What were the total rental earnings of these four films in 1990?

46. A video cassette tape has a running time of 2 hours at the standard speed, 4 hours at the long-play speed, and 6 hours at the super-long-play speed. If $\frac{1}{2}$ of the tape is recorded at the standard speed, $\frac{1}{4}$ is recorded at the long play speed, and $\frac{1}{4}$ is recorded at the super-long-play speed, what is the total playing time of the entire recording?

Check Your Number Sense

47. Using the idea of truncated mixed numbers, estimate each of the following products mentally.

(a) $2\frac{1}{7} \cdot 3\frac{5}{8}$ (b) $20\frac{1}{7} \cdot 30\frac{5}{8}$ (c) $200\frac{1}{7} \cdot 300\frac{5}{8}$

48. Of your three estimated answers in Exercise 47, which one do you think is closest to the actual product? Why?

4.4 Subtraction

OBJECTIVES

1. Know how to subtract mixed numbers when the fraction part of the number being subtracted is smaller than or equal to the fraction part of the other number.
2. Know how to subtract mixed numbers when the fraction part of the number being subtracted is larger than the fraction part of the other number.

Subtraction with Mixed Numbers

Subtraction with mixed numbers involves working with the whole numbers and fraction parts separately in a manner similar to addition. With subtraction, though, we have another concern related to the sizes of the fraction parts. We will see that when the fraction part being subtracted is larger than the other fraction part, the smaller fraction part must be changed to an improper fraction by "borrowing" 1 from its whole number.

To Subtract Mixed Numbers:

1. Subtract the fraction parts.

2. Subtract the whole numbers.

Examples 1–3 illustrate subtraction with mixed numbers.

EXAMPLE 1 $4\frac{3}{5} - 1\frac{2}{5}$

Solution

$$4\frac{3}{5} - 1\frac{2}{5} = (4 - 1) + \left(\frac{3}{5} - \frac{2}{5}\right) \qquad \text{Or, vertically,}$$

$$= 3 + \frac{1}{5} \qquad\qquad 4\frac{3}{5}$$

$$= 3\frac{1}{5} \qquad\qquad \frac{- 1\frac{2}{5}}{3\frac{1}{5}}$$

EXAMPLE 2 $5\frac{6}{7} - 3\frac{2}{7}$

Solution

$$5\frac{6}{7}$$
$$\frac{- 3\frac{2}{7}}{2\frac{4}{7}}$$

EXAMPLE 3 $10\frac{3}{5} - 6\frac{3}{20}$

Solution

$$10\frac{3}{5} \;=\; 10\frac{12}{20}$$
$$\frac{- 6\frac{3}{20} \;=\; 6\frac{3}{20}}{4\frac{9}{20}}$$

Sometimes the **fraction part** of the number being subtracted is larger than the fraction part of the first number. By "borrowing" a whole 1 from the whole number, we write the first number as a whole number with an improper fraction part. Then the subtraction of the fraction parts is possible.

> **If the Fraction Part Being Subtracted Is Larger than the First Fraction:**
>
> **1.** "Borrow" a whole 1 from the first whole number.
>
> **2.** Add this 1 to the first fraction. (The result will always be an improper fraction that is larger than the fraction being subtracted.)
>
> **3.** Now subtract.

Examples 4–9 illustrate subtraction with "borrowing."

EXAMPLE 4

$$7\frac{2}{5} - 4\frac{3}{5}$$

Solution

$$7\frac{2}{5}$$
$$-\,4\frac{3}{5}$$

$\frac{3}{5}$ is larger than $\frac{2}{5}$, so "borrow" 1 from 7.

$$7\frac{2}{5} = 6 + 1\frac{2}{5} = 6\frac{7}{5} \quad \leftarrow \left(7\frac{2}{5} = 6 + 1 + \frac{2}{5} = 6 + 1\frac{2}{5} = 6\frac{7}{5}\right)$$
$$-\,4\frac{3}{5} = 4 + \frac{3}{5} = 4\frac{3}{5}$$
$$\rule{2cm}{0.4pt}$$
$$2\frac{4}{5}$$

EXAMPLE 5

$$19\frac{2}{3} - 5\frac{3}{4}$$

Solution

$$19\frac{2}{3} = 19\frac{8}{12} = 18\frac{20}{12} \quad \leftarrow \left(1 + \frac{8}{12} = \frac{12}{12} + \frac{8}{12} = \frac{20}{12}\right)$$
$$-\,5\frac{3}{4} = 5\frac{9}{12} = 5\frac{9}{12}$$
$$\rule{3cm}{0.4pt}$$
$$13\frac{11}{12}$$

EXAMPLE 6

$$10 - 6\frac{5}{8}$$ (Note: The whole number 10 has no fraction part. So, the fraction part is understood to be 0.)

Solution

$$10 \quad = 9\frac{8}{8} \quad \leftarrow \left(10 = 9 + 1 = 9 + \frac{8}{8}\right)$$

$$\underline{- \ 6\frac{5}{8} = 6\frac{5}{8}}$$

$$3\frac{3}{8}$$

EXAMPLE 7

$$14\frac{3}{4} \ = 14\frac{15}{20} = 13\frac{35}{20} \quad \leftarrow \left(1 + \frac{15}{20} = \frac{20}{20} + \frac{15}{20} = \frac{35}{20}\right)$$

$$\underline{- \ 5\frac{9}{10} = \ 5\frac{18}{20} = \ 5\frac{18}{20}}$$

$$8\frac{17}{20}$$

EXAMPLE 8

$$406\frac{5}{12} \ = 406\frac{25}{60} = 405\frac{85}{60} \qquad \left.\begin{array}{l} 12 = 2^2 \cdot 3 \\ 20 = 2^2 \cdot 5 \end{array}\right\} \ \text{LCD} = 2^2 \cdot 3 \cdot 5 = 60$$

$$\underline{- \ 168\frac{13}{20} = 168\frac{39}{60} = 168\frac{39}{60}}$$

$$237\frac{46}{60} = 237\frac{23}{30}$$

EXAMPLE 9

When he was first using his new word processor on his computer, Karl found that he was taking about $25\frac{1}{2}$ minutes to type 4 pages of a term paper. Now he takes about $18\frac{3}{5}$ minutes to type 4 pages. By how many minutes has he improved in typing 4 pages? In typing 1 page?

Solution

(a) To find how many minutes he has improved in typing 4 pages, we simply find the difference between the two times.

$$25\frac{1}{2} = 25\frac{5}{10} = 24\frac{15}{10}$$

$$\underline{18\frac{3}{5} = 18\frac{6}{10} = 18\frac{6}{10}}$$

$$6\frac{9}{10}$$

So, in typing 4 pages, Karl has improved by about $6\frac{9}{10}$ minutes. (**Note:** Using truncated numbers, we could have estimated an improvement of approximately $25 - 18 = 7$ minutes.)

(b) To find his improvement in typing 1 page, we divide the answer in part (a) by 4.

$$6\frac{9}{10} \div 4 = \frac{69}{10} \div \frac{4}{1} = \frac{69}{10} \cdot \frac{1}{4} = \frac{69}{40} = 1\frac{29}{40}$$

He has improved by about $1\frac{29}{40}$ minutes per page.

Exercises 4.4

Find the differences.

1. $5\frac{1}{2}$ **2.** $7\frac{3}{4}$ **3.** $4\frac{5}{12}$ **4.** $3\frac{5}{8}$
  $-\ 1$ $-\ 2$ $-\ 3$ $-\ 2$

5. $6\frac{1}{2}$ **6.** $9\frac{1}{4}$ **7.** $5\frac{3}{4}$ **8.** $7\frac{9}{10}$
  $-\ 2\frac{1}{2}$ $-\ 5\frac{1}{4}$ $-\ 2\frac{1}{4}$ $-\ 3\frac{3}{10}$

9. $14\frac{5}{8}$ **10.** $20\frac{7}{16}$ **11.** $4\frac{7}{8}$ **12.** $9\frac{5}{16}$
  $-\ 11\frac{3}{8}$ $-\ 15\frac{5}{16}$ $-\ 1\frac{1}{4}$ $-\ 2\frac{1}{4}$

13. $5\frac{11}{12}$ **14.** $10\frac{5}{6}$ **15.** $8\frac{5}{6}$ **16.** $15\frac{5}{8}$
  $-\ 1\frac{1}{4}$ $-\ 4\frac{2}{3}$ $-\ 2\frac{1}{4}$ $-\ 11\frac{3}{4}$

17. $14\frac{6}{10}$ **18.** $8\frac{3}{32}$ **19.** $12\frac{3}{4}$ **20.** $8\frac{11}{12}$
  $-\ 3\frac{4}{5}$ $-\ 4\frac{3}{16}$ $-\ 7\frac{1}{6}$ $-\ 5\frac{9}{10}$

21. $4\frac{7}{16} - 3$ **22.** $5\frac{9}{10} - 2$ **23.** $7 - 6\frac{2}{3}$

24. $12 - 4\frac{1}{5}$ **25.** $2 - 1\frac{3}{8}$ **26.** $75 - 17\frac{5}{6}$

27. $4\frac{9}{16} - 2\frac{7}{8}$ **28.** $3\frac{7}{10} - 2\frac{5}{6}$ **29.** $15\frac{11}{16} - 13\frac{7}{8}$

30. $20\frac{3}{6} - 3\frac{4}{8}$ **31.** $17\frac{3}{12} - 12\frac{2}{8}$ **32.** $18\frac{2}{7} - 4\frac{1}{3}$

33. $13\dfrac{5}{8} - 6\dfrac{11}{20}$

34. $18\dfrac{7}{8} - 2\dfrac{2}{3}$

35. $10\dfrac{3}{10} - 2\dfrac{1}{2}$

36. $1 - \dfrac{2}{3}$

37. $1 - \dfrac{5}{8}$

38. $1 - \dfrac{3}{4}$

39. $1 - \dfrac{9}{10}$

40. $1 - \dfrac{7}{16}$

41. $1 - \dfrac{12}{15}$

42. $1 - \dfrac{2}{5}$

43. $1 - \dfrac{1}{6}$

44. $2 - \dfrac{1}{2}$

45. $2 - \dfrac{3}{4}$

46. $93\dfrac{5}{12} - 36\dfrac{7}{8}$

47. $206\dfrac{5}{12} - 69\dfrac{13}{20}$

48. $182\dfrac{3}{20} - 104\dfrac{7}{15}$

49. $250\dfrac{1}{8} - 84\dfrac{7}{12}$

50. A swimming pool contains $500\dfrac{2}{3}$ gallons of water. If $23\dfrac{4}{5}$ gallons evaporate, how many gallons of water are left in the pool?

51. A box holds 5 pounds of candy. If $3\dfrac{1}{4}$ pounds are eaten, how much candy remains in the box?

52. Sara can paint a room in $3\dfrac{3}{5}$ hours, and Emily can paint a room of the same size in $4\dfrac{1}{5}$ hours. How many hours are saved by having Sara paint the room? How many minutes are saved?

53. A teacher graded two sets of test papers. The first set took $3\dfrac{3}{4}$ hours to grade, and the second set took $2\dfrac{3}{5}$ hours. How much faster did the teacher grade the second set?

54. Mike takes $1\dfrac{1}{2}$ hours to clean a pool, and Tom takes $2\dfrac{1}{3}$ hours to clean the same pool. How much longer does Tom take?

55. A long-distance runner was in training. He ran ten miles in $50\dfrac{3}{10}$ minutes. Three months later, he ran the same ten miles in $43\dfrac{7}{10}$ minutes. By how much did his time improve?

56. Certain shares of stock were selling for $43\dfrac{7}{8}$ dollars per share. One month later, the same stock was selling for $48\dfrac{1}{2}$ dollars per share. By how much did the stock increase in price?

57. You want to lose 10 pounds in weight. If you weigh 180 pounds now and you lose $3\frac{1}{4}$ pounds in one week and $3\frac{1}{2}$ pounds during the second week, how much weight do you still need to lose?

58. Mr. Johnson originally weighed 240 pounds. During each week of six weeks of dieting, he lost $5\frac{1}{2}$ pounds, $2\frac{3}{4}$ pounds, $4\frac{5}{16}$ pounds, $1\frac{3}{4}$ pounds, $2\frac{5}{8}$ pounds, and $3\frac{1}{4}$ pounds. If he was 35 years old, what did he weigh at the end of the six weeks?

59. A board is 10 feet long. If a piece $6\frac{3}{4}$ feet is cut off, what is the length of the remaining piece?

60. A salesman drove $5\frac{3}{4}$ hours one day and then $6\frac{1}{2}$ hours ·the next day. How much more time did he spend driving on the second day?

61. On average, the air that we inhale includes $1\frac{1}{4}$ parts water, and the air that we exhale includes $5\frac{9}{10}$ parts water. How many more parts water are in exhaled air than in inhaled air?

62. A person who is running will burn about $14\frac{7}{10}$ calories each minute and a person who is walking will burn about $5\frac{1}{2}$ calories each minute. How many more calories does a runner burn in a minute than a walker?

Writing and Thinking about Mathematics

63. In your own words, describe the technique for estimating answers when mixed numbers are involved in calculations.

64. Using the rule you just described in Exercise 63, mentally estimate each of the following differences.

(a) $15\frac{6}{7} - 11\frac{3}{4}$ (b) $136\frac{17}{40} - 125\frac{23}{30}$ (c) $945\frac{1}{10} - 845\frac{3}{100}$

65. Consider the following problem: The product of two numbers is $16\frac{1}{2}$. If one of the numbers is $2\frac{3}{4}$, what is the other number?

(a) Rewrite the problem using truncated mixed numbers and solve this new problem.

(b) Does this technique help you to better understand the original problem? Why or why not?

(c) What is the solution to the original problem?

The Recycle Bin   (from Sections 2.2 and 3.6)

Use the rules for order of operations to simplify each of the following expressions.

1. $6 + 2(13 - 5 \cdot 2) + 12 \div 2$ **2.** $5^3 + 6^2 - 3 \cdot 5 \cdot 2^2$

3. $\left(\dfrac{1}{3}\right)^2 + \dfrac{5}{8} \div \dfrac{1}{4}$ **4.** $\dfrac{3}{4} \div \dfrac{1}{2} \cdot 6 - \dfrac{1}{5}(45)$

5. $\dfrac{7}{8} + \dfrac{2}{3} \cdot \dfrac{1}{4} + \left(\dfrac{1}{2}\right)^3 \div 3$

4.5   Complex Fractions and Order of Operations

OBJECTIVES

1. Recognize and know how to simplify complex fractions.
2. Be able to follow the rules of order of operations with mixed numbers.

Simplifying Complex Fractions

In Section 3.6 we defined a **complex fraction** as a form of a fraction in which the numerator and/or denominator are themselves fractions or the sums or differences of fractions. Since a mixed number is the sum of a whole number and a fraction and whole numbers can be written in fraction form, we can now adjust the definition of a complex fraction to include mixed numbers. Thus, we have the following more complete definition of a complex fraction.

Definition

A **complex fraction** is a fraction in which the numerator and/or denominator are themselves fractions or mixed numbers (or are the sums or differences of fractions or mixed numbers).

To Simplify a Complex Fraction:

1. Simplify the numerator so that it is a single fraction, possibly an improper fraction.

2. Simplify the denominator so that it also is a single fraction, possibly an improper fraction.

3. Divide the numerator by the denominator and reduce if possible.

EXAMPLE 1

Simplify the complex fraction $\dfrac{3\frac{2}{5}}{\frac{1}{4} + \frac{3}{5}}$.

Solution

$$3\frac{2}{5} = \frac{17}{5} \qquad \text{Change the mixed number to an improper fraction.}$$

$$\frac{1}{4} + \frac{3}{5} = \frac{5}{20} + \frac{12}{20} = \frac{17}{20}$$

So,

$$\frac{3\frac{2}{5}}{\frac{1}{4} + \frac{3}{5}} = \frac{\frac{17}{5}}{\frac{17}{20}} = \frac{17}{5} \div \frac{17}{20} = \frac{17}{5} \cdot \frac{20}{17} = \frac{17 \cdot 4 \cdot 5}{5 \cdot 17} = \frac{4}{1} = 4$$

EXAMPLE 2

Simplify the complex fraction $\dfrac{5\frac{2}{3} - 2\frac{1}{3}}{1\frac{1}{2} + \frac{1}{2}}$.

Solution

$$5\frac{2}{3} - 2\frac{1}{3} = 3\frac{1}{3} = \frac{10}{3} \qquad \text{and} \qquad 1\frac{1}{2} + \frac{1}{2} = 2$$

So,

$$\frac{5\frac{2}{3} - 2\frac{1}{3}}{1\frac{1}{2} + \frac{1}{2}} = \frac{\frac{10}{3}}{\frac{2}{1}} = \frac{\frac{10}{3}}{\frac{2}{1}} = \frac{10}{3} \div \frac{2}{1} = \frac{10}{3} \cdot \frac{1}{2} = \frac{5 \cdot 2 \cdot 1}{3 \cdot 2} = \frac{5}{3} = 1\frac{2}{3}$$

Order of Operations

Working with complex fractions, we know to simplify the numerator and denominator separately. That is, **we treat the fraction bar as a grouping symbol.** For expressions that involve more than one operation, we evaluate (or simplify) them by using the rules for order of operations. These rules were discussed in Sections 2.2 and 3.6 and are listed again here for easy reference.

Rules for Order of Operations

1. First, simplify within grouping symbols, such as parentheses ( ), brackets [ ], or braces { }. Start with the innermost grouping.

2. Second, find any powers indicated by exponents.

3. Third, moving from **left to right,** perform any multiplications or divisions in the order in which they appear.

4. Fourth, moving from **left to right,** perform any additions or subtractions in the order in which they appear.

EXAMPLE 3

Use the rules for order of operations to simplify the following expression:

$$2\frac{1}{2} \cdot 1\frac{1}{6} + 7 \div \frac{3}{4}.$$

Solution

$$2\frac{1}{2} \cdot 1\frac{1}{6} + 7 \div \frac{3}{4} = \frac{5}{2} \cdot \frac{7}{6} + \frac{7}{1} \cdot \frac{4}{3} \qquad \text{Multiply and divide from left to right.}$$

$$= \frac{35}{12} + \frac{28}{3} \qquad \text{Now add.}$$

$$= \frac{35}{12} + \frac{28}{3} \cdot \frac{4}{4}$$

$$= \frac{35}{12} + \frac{112}{12}$$

$$= \frac{147}{12} = 12\frac{3}{12} = 12\frac{1}{4}$$

Or, you could write:

$$2\frac{1}{2} \cdot 1\frac{1}{6} = \frac{5}{2} \cdot \frac{7}{6} = \frac{35}{12} = 2\frac{11}{12}$$ multiplying

$$7 \div \frac{3}{4} = \frac{7}{1} \cdot \frac{4}{3} = \frac{28}{3} = 9\frac{1}{3}$$ dividing

$$
\begin{array}{rcl}
2\dfrac{11}{12} & = & 2\dfrac{11}{12} \\[2mm]
+\,9\dfrac{1}{3} & = & 9\dfrac{4}{12} \\[2mm]
\hline
11\dfrac{15}{12} & = & 12\dfrac{3}{12} = 12\dfrac{1}{4}
\end{array}
$$ adding the results

EXAMPLE 4   Evaluate the following expression by using the rules for order of operations:

$$\left(2\frac{1}{2}\right)^2 + \frac{1}{5} \cdot \frac{1}{6} \div \frac{1}{15}.$$

Solution

$$\left(2\frac{1}{2}\right)^2 + \frac{1}{5} \cdot \frac{1}{6} \div \frac{1}{15} = \left(\frac{5}{2}\right)^2 + \frac{1}{5} \cdot \frac{1}{6} \div \frac{1}{15}$$

$$= \frac{25}{4} + \frac{1}{30} \cdot \frac{15}{1}$$

$$= \frac{25}{4} + \frac{1}{2}$$

$$= \frac{25}{4} + \frac{2}{4}$$

$$= \frac{27}{4}$$

$$= 6\frac{3}{4}$$

EXAMPLE 5   Find the average of the mixed numbers

$$1\frac{1}{2}, \quad 2\frac{3}{4}, \quad \text{and} \quad 3\frac{5}{8}.$$

> **Solution**
>
> We find the sum first and then divide by 3.
>
> $$1\frac{1}{2} = 1\frac{4}{8}$$
>
> $$2\frac{3}{4} = 2\frac{6}{8}$$
>
> $$+\ 3\frac{5}{8} = 3\frac{5}{8}$$
>
> $$6\frac{15}{8} = 7\frac{7}{8}$$
>
> $$7\frac{7}{8} \div 3 = \frac{63}{8} \cdot \frac{1}{3} = \frac{3 \cdot 3 \cdot 7}{8 \cdot 3} = \frac{21}{8} = 2\frac{5}{8}$$
>
> Therefore, the average is $2\frac{5}{8}$.

Exercises 4.5

Simplify each of the following complex fractions.

1. $\dfrac{2 + \dfrac{1}{5}}{1 + \dfrac{1}{4}}$ **2.** $\dfrac{\dfrac{2}{3} + \dfrac{1}{5}}{4\dfrac{1}{2}}$ **3.** $\dfrac{4 + \dfrac{1}{3}}{6 + \dfrac{1}{4}}$

Ans: $2\frac{1}{2}$

4. $\dfrac{2 - \dfrac{1}{3}}{1 - \dfrac{1}{3}}$ **5.** $\dfrac{\dfrac{2}{3} - \dfrac{1}{4}}{\dfrac{3}{5} - \dfrac{1}{4}}$ **6.** $\dfrac{\dfrac{5}{6} - \dfrac{2}{3}}{\dfrac{5}{8} - \dfrac{1}{16}}$

7. $\dfrac{1}{\dfrac{1}{12} + \dfrac{1}{6}}$ *Ans:* 3 **8.** $\dfrac{1}{\dfrac{1}{5} + \dfrac{2}{15}}$ **9.** $\dfrac{\dfrac{3}{10} + \dfrac{1}{6}}{3}$

10. $\dfrac{\dfrac{7}{12} + \dfrac{1}{15}}{5}$ **11.** $\dfrac{7\dfrac{1}{3} + 2\dfrac{1}{5}}{6\dfrac{1}{9} + 2}$ *Ans:* $\frac{27}{40}$ **12.** $\dfrac{5\dfrac{2}{3} - 1\dfrac{1}{6}}{3\dfrac{1}{2} + 3\dfrac{1}{6}}$

13. $\dfrac{6\dfrac{1}{100} + 5\dfrac{3}{100}}{2\dfrac{1}{2} + 3\dfrac{1}{10}}$ **14.** $\dfrac{4\dfrac{7}{10} - 2\dfrac{9}{10}}{5\dfrac{1}{100}}$ $= \frac{60}{167}$

Evaluate the following expressions using the rules for order of operations.

15. $\dfrac{3}{5} \cdot \dfrac{1}{6} + \dfrac{1}{5} \div 2$ Ansi 0

16. $\dfrac{1}{2} \div \dfrac{1}{2} + 1 - \dfrac{2}{3} \cdot 3$

17. $3\dfrac{1}{2} \cdot 5\dfrac{1}{3} + \dfrac{5}{12} \div \dfrac{15}{16}$

18. $2\dfrac{1}{4} + 1\dfrac{1}{5} + 2 \div \dfrac{20}{21}$

19. $\dfrac{5}{8} - \dfrac{1}{3} \cdot \dfrac{2}{5} + 6\dfrac{1}{10}$ Ans $1\frac{1}{2}$

20. $1\dfrac{1}{6} \cdot 1\dfrac{2}{19} \div \dfrac{7}{8} + \dfrac{1}{38}$

21. $\dfrac{3}{10} + \dfrac{5}{6} \div \dfrac{1}{4} \cdot \dfrac{1}{8} - \dfrac{7}{60}$

22. $5\dfrac{1}{7} \div (2 + 1)$

23. $\left(2 - \dfrac{1}{3}\right) \div \left(1 - \dfrac{1}{3}\right)$ Ans $3\frac{6}{19}$

24. $\left(2\dfrac{4}{9} + 1\dfrac{1}{18}\right) \div \left(1\dfrac{2}{9} - \dfrac{1}{6}\right)$

25. If the product of $5\dfrac{1}{2}$ and $2\dfrac{1}{4}$ is added to the quotient of $\dfrac{9}{10}$ and $\dfrac{3}{4}$, what is the sum?

26. If the quotient of $\dfrac{5}{8}$ and $\dfrac{1}{2}$ is subtracted from the product of $2\dfrac{1}{4}$ and $3\dfrac{1}{5}$, what is the difference?

27. If $\dfrac{9}{10}$ of 70 is divided by $\dfrac{3}{4}$ of 10, what is the quotient?

28. If $\dfrac{2}{3}$ of $4\dfrac{1}{4}$ is added to $\dfrac{5}{8}$ of $6\dfrac{1}{3}$, what is the sum?

29. Find the average of the numbers $\dfrac{7}{8}, \dfrac{9}{10},$ and $1\dfrac{3}{4}$.

30. Find the average of the numbers $\dfrac{5}{6}, \dfrac{1}{15},$ and $\dfrac{17}{30}$.

31. Find the average of the numbers $5\dfrac{1}{8}, 7\dfrac{1}{2}, 4\dfrac{3}{4},$ and $10\dfrac{1}{2}$.

32. Find the average of the numbers $4\dfrac{7}{10}, 3\dfrac{9}{10}, 5\dfrac{1}{100},$ and $11\dfrac{3}{20}$.

Collaborative Learning Exercise

33. After the class has separated into teams of two to four students, each team is to write one or two paragraphs on the following two topics. The team leader is to read the paragraphs with classroom discussion to follow.

(a) In the text so far, what topic have you found to be the most interesting? Why?

(b) In the text so far, what topic have you found to be the most useful? Why?

Summary: Chapter 4

Key Terms and Ideas

A **mixed number** is the sum of a whole number and a fraction. A fraction **of** a number means to **multiply** the number by the fraction. If the fraction is between 0 and 1, then the product will be less than the number.

Rules and Properties

To change a mixed number to an improper fraction, add the whole number and the fraction.

Shortcut to Changing Mixed Numbers to Fraction Form

1. Multiply the denominator of the fraction part by the whole number.

2. Add the numerator of the fraction part to this product.

3. Write this sum over the denominator of the fraction.

To Change an Improper Fraction to a Mixed Number:

1. Divide the numerator by the denominator.

2. Write the remainder over the denominator as the fraction part of the mixed number.

Rules for Order of Operations

1. First, simplify within grouping symbols, such as parentheses ( ), brackets [ ], or braces { }. Start with the innermost grouping.

2. Second, find any powers indicated by exponents.

3. Third, moving from **left to right,** perform any multiplications or divisions in the order they appear.

4. Fourth, moving from **left to right,** perform any additions or subtractions in the order they appear.

Procedures

FOR MIXED NUMBERS

To add:

1. Add the fraction parts.

2. Add the whole numbers.

3. Write the mixed number so that the fraction part is less than 1.

To subtract:

1. Subtract the fraction parts.

2. Subtract the whole numbers.

If the fraction part being subtracted is larger than the first fraction, "borrow" a whole 1 from the first whole number; then subtract.

To multiply:

1. Change each number to fraction form.

2. Multiply by factoring numerators and denominators; then reduce.

3. Change the answer to a mixed number or leave it in fraction form.

To divide:

1. Change each number to fraction form.

2. Write the reciprocal of the divisor.

3. Multiply by factoring numerators and denominators; then reduce.

Review Questions: Chapter 4

Change to mixed numbers with the fraction part reduced.

1. $\dfrac{53}{8}$ **2.** $\dfrac{91}{13}$ **3.** $\dfrac{342}{100}$

Change to fraction form.

4. $5\dfrac{1}{10}$ **5.** $2\dfrac{11}{12}$ **6.** $13\dfrac{2}{5}$

Perform the indicated operations. Reduce all fractions to lowest terms.

7. $27\dfrac{1}{4} + 3\dfrac{1}{2}$ **8.** $5\dfrac{2}{5} - 4\dfrac{2}{3}$ **9.** $4\dfrac{5}{7} \cdot 2\dfrac{6}{11}$

10. $\dfrac{5}{6} \div 3\dfrac{3}{4}$

11. $4\dfrac{5}{8} + 2\dfrac{3}{4}$

12. $\dfrac{5}{12}\left(6\dfrac{3}{10}\right)\left(7\dfrac{1}{9}\right)$

13. $7\dfrac{1}{11}\left(2\dfrac{3}{4}\right)\left(5\dfrac{1}{3}\right)$

14. $2\dfrac{4}{5} \div 4$

15. $12\dfrac{5}{6} - 6\dfrac{1}{4}$

16. $4\dfrac{7}{8} - 3\dfrac{1}{4}$

17. $\begin{aligned} &14\dfrac{5}{12} \\ -\ &3\dfrac{3}{4} \\ \hline \end{aligned}$

18. $\begin{aligned} &42\dfrac{7}{9} \\ &53\dfrac{4}{15} \\ +\ &24\dfrac{9}{10} \\ \hline \end{aligned}$

19. $4\dfrac{2}{7} \div 3\dfrac{3}{5} + 4\dfrac{1}{6} \cdot 2\dfrac{4}{5}$

20. $2\dfrac{3}{10} - 5\dfrac{3}{5} \div 4\dfrac{2}{3} + 2\dfrac{1}{6}$

Simplify.

21. $\dfrac{\dfrac{1}{5} + \dfrac{1}{6}}{2\dfrac{1}{3}}$

22. $\dfrac{\dfrac{7}{8} - \dfrac{3}{16}}{\dfrac{1}{3} - \dfrac{1}{4}}$

23. Find the average of $\dfrac{3}{4}, \dfrac{5}{8}$, and $\dfrac{9}{10}$.

24. Find the average of $1\dfrac{3}{10}, 2\dfrac{1}{5}$, and $1\dfrac{3}{4}$.

25. Find $\dfrac{2}{3}$ of 96.

26. Find $\dfrac{2}{5}$ of $6\dfrac{1}{4}$.

27. The product of $20\dfrac{2}{3}$ with some other number is $6\dfrac{1}{2}$. What is the other number?

28. While dieting, Ken lost $4\dfrac{1}{3}$ pounds the first week, $1\dfrac{3}{4}$ pounds the second week, and $2\dfrac{1}{6}$ pounds the third week. What was his average weekly weight loss?

29. The sale price of a television is \$375. This is $\dfrac{3}{5}$ of the original price. What was the original price?

30. The sum of $12\dfrac{5}{8}$ and some other number is $20\dfrac{1}{4}$. What is the other number?

31. Rachel wants to make matching dresses for her three daughters. The pattern requires $1\dfrac{7}{8}$ yards of material for a size 2 dress, $2\dfrac{1}{4}$ yards for a size 5, and

$3\frac{1}{2}$ yards for a size 8. What is the least amount of material Rachel should buy if she makes one dress of each size?

32. Jason won 7500 dollars in the state lottery. If $\frac{1}{5}$ of his prize money was withheld for taxes, how much money did he actually receive?

33. Michael is building a fence $4\frac{1}{2}$ feet high. It is recommended that $\frac{1}{4}$ of the length of each fence post should be underground. How long must each fence post be?

34. An English class has 42 students enrolled. If $\frac{6}{7}$ of the students enrolled are present in class, how many are absent?

35. A bicyclist rode his bicycle $37\frac{9}{10}$ miles each day for five days. How far did he ride his bicycle during those five days?

Test: Chapter 4

Reduce all fractions to lowest terms.

1. Change $\frac{34}{7}$ to a mixed number.

2. Change $7\frac{3}{8}$ to fraction form.

3. Find the sum of $19\frac{3}{7}$ and $23\frac{9}{14}$.

4. Find the difference between $18\frac{2}{5}$ and $7\frac{7}{10}$.

5. Find the product of $4\frac{3}{8}$ and $2\frac{2}{5}$.

6. Find the quotient of 3 and $4\frac{1}{5}$.

Perform the indicated operations.

7. $\begin{array}{r} 6\frac{5}{14} \\ + \frac{16}{21} \\ \hline \end{array}$

8. $\begin{array}{r} 9 \\ -7\frac{9}{11} \\ \hline \end{array}$

9. $4\frac{2}{3} + 3\frac{3}{4}$

10. $4\frac{3}{8} - 2\frac{5}{6}$

11. $4\frac{1}{5} + 2\frac{7}{8} + 7\frac{1}{20}$

12. $7\frac{2}{9} - 6\frac{3}{4}$

13. $\left(2\dfrac{2}{11}\right)\left(4\dfrac{1}{8}\right)$ **14.** $3\dfrac{1}{9} \div 5\dfrac{1}{4}$ **15.** $\left(1\dfrac{1}{6}\right)\left(2\dfrac{3}{7}\right)\left(5\dfrac{5}{8}\right)$

16. $2\dfrac{2}{7} \cdot 5\dfrac{3}{5} + \dfrac{3}{5} \div 2$

17. Simplify $\dfrac{1\dfrac{1}{9} + \dfrac{5}{18}}{4 - 2\dfrac{2}{3}}$ **18.** Find $\dfrac{4}{9}$ of $6\dfrac{3}{7}$.

19. Find the average of $2\dfrac{3}{4}$, $3\dfrac{2}{5}$, and $3\dfrac{3}{10}$.

20. The number $5\dfrac{1}{4}$ is the product of $\dfrac{2}{3}$ with some other number. What is the other number?

21. During three days of practice, a high school cross country team member ran $22\dfrac{3}{10}$ miles, $14\dfrac{1}{5}$ miles, and $20\dfrac{1}{4}$ miles. (a) How far did he run during those three days? (b) What was his average distance per day?

22. Each day during a five-day period, a woman drove $8\dfrac{1}{2}$ miles from home to work (taking her children to school along the way), and then after work she drove $6\dfrac{3}{4}$ miles to her home. How many total miles did the woman drive to and from work during those five days?

23. (a) Find the perimeter of (distance around) a rectangle that is 8 inches wide and $11\dfrac{1}{2}$ inches long. (b) Find the area of the rectangle. (The area of a rectangle is found by multiplying length times width.)

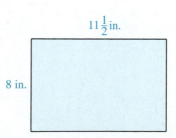

$11\dfrac{1}{2}$ in.

8 in.

1. Write 43,570 in expanded notation and in its English word equivalent.

2. Write six thousand four hundred seventy-five in decimal form.

3. Round off 175,800 to the nearest ten thousand.

4. What property of addition is illustrated?

   $17 + 25 = 25 + 17$

5. Match each expression with its best estimate.

   ____ (a) $175 + 82 + 96 + 131$ A. 200

   ____ (b) $8452 \div 41$ B. 480

   ____ (c) $32 \cdot 47$ C. 1000

   ____ (d) $5476 - 4391$ D. 1500

6. Multiply $80(900)$.

7. Evaluate $5^2 + (12 \cdot 5 \div 2 \cdot 3) - 6 \cdot 4$.

8. List all the prime numbers less than 20.

9. Find the prime factorizations of 45 and 105.

10. The reciprocal of 10 is _____.

11. The value of $0 \div 18$ is _____, while $18 \div 0$ is _____.

12. Find the average of 45, 36, 54, and 41.

Perform the following operations. Reduce all fractions to lowest terms.

13. $\begin{array}{r} 7586 \\ + 3752 \\ \hline \end{array}$

14. $\begin{array}{r} 893 \\ - 576 \\ \hline \end{array}$

15. $\begin{array}{r} 741 \\ \times 35 \\ \hline \end{array}$

16. $14\overline{)2856}$

17. $\dfrac{3}{8} \div 6$

18. $\dfrac{5}{18}\left(\dfrac{3}{10}\right)\left(\dfrac{2}{3}\right)$

19. $\dfrac{5}{8} - \dfrac{5}{12}$

20. $\dfrac{7}{10} + \dfrac{3}{5} + \dfrac{5}{6}$

21. $\begin{array}{r} 37\frac{4}{5} \\ + 10\frac{7}{12} \\ \hline \end{array}$

22. $\begin{array}{r} 3\frac{4}{5} \\ - 2\frac{5}{7} \\ \hline \end{array}$

23. $3\dfrac{2}{3}\left(2\dfrac{6}{11}\right)$

24. $4\dfrac{3}{8} \div 2\dfrac{1}{2}$

25. $4\left(5\dfrac{2}{5}\right)\left(4\dfrac{1}{6}\right)$

26. $\dfrac{1\frac{3}{10} + 2\frac{3}{5}}{2\frac{4}{5} - 1\frac{1}{2}}$

27. $1\dfrac{1}{3} + 6\dfrac{4}{5} \div 2\dfrac{1}{8} - 1\dfrac{7}{10}$

28. Arrange the following fractions in order from smallest to largest.

$$\frac{7}{10}, \frac{3}{4}, \frac{5}{6}$$

29. Find $\frac{1}{2}$ of $2\frac{3}{4}$.

30. On Monday morning, Kirk had $342 in his checking account. That day, he deposited $600 in his account and wrote checks for the following amounts: $96, $150, and $475. What is the new balance in his checking account on Tuesday morning?

31. Vikki bought two pairs of shoes for $17 per pair, a blouse for $28, a pair of pants for $19, and a sweater for $24. What was the average cost per item of clothing?

32. If a car has an 18-gallon gas tank, how many gallons of gas will it take to fill the tank when it is $\frac{3}{8}$ full?

33. David needed to gain weight to join the football team. When he began his training, he weighed $135\frac{3}{4}$ pounds. Six weeks later, he weighed in at $157\frac{1}{2}$ pounds. What was his average weekly weight gain (in pounds)?

34. Find the perimeter of a triangle with sides of length $4\frac{5}{8}$ in., $2\frac{2}{3}$ in., and $5\frac{7}{12}$ in.

35. The discount price of a new refrigerator is $760. If this price is $\frac{4}{5}$ of the original price, what was the original price?

Decimal Numbers

Mathematics at Work!

When ordering a deli item by the fraction of a pound, we will usually see that the scale displays the decimal equivalent form.

Today, decimal numbers are in use almost everywhere we turn: in weights, prices, sizes, speeds, and measures of tools and equipment. Now that the displays on gas pumps are digitized, the prices are given to the thousandths. For example, the price of a gallon of gasoline is now given as $1.259 instead of the previous form of a mixed number as 1.25\frac{9}{10}$. Even in the deli department of the supermarket the scales are set to measure hundredths of a pound in decimal form rather than the previous fraction or mixed number form.

As a common application of decimal numbers, consider the following situation. A local supermarket showed these recent prices:

Low cholesterol macaroni salad $2.29 per pound
Pasta salad $2.89 per pound
Fruit delight $3.99 per pound

(a) If you wanted to buy 0.5 pound of each of these salads, about how much would you expect to pay?

(b) As the clerk weighed each container, the actual weights were: 0.56 pound of macaroni salad, 0.47 pound of pasta salad, and 0.6 pound of fruit delight. How much would you pay for these amounts?

What to Expect in Chapter 5

Chapter 5 develops an understanding of decimal numbers and deals with the basic operations with decimal numbers. Section 5.1 discusses techniques for reading and writing decimal numbers and use of the word **and** to indicate placement of the decimal point. Section 5.2 shows how to round off decimal numbers, and Sections 5.3 through 5.5 discuss the operations of addition, subtraction, multiplication, and division with decimal numbers and the use of rounded-off numbers to estimate answers.

Because of the current emphasis on working with calculators, scientific notation is introduced in Section 5.6 so that you can read results on your calculator when very large or very small decimal numbers are involved or if your calculator is set in scientific mode. As discussed in Section 5.7, fractions and decimals are closely related and can be used together by changing all the numbers in a problem to one form or the other.

5.1 Reading and Writing Decimal Numbers

OBJECTIVES

1. Learn to read and write decimal numbers.
2. Realize the importance of the word **and** in reading and writing decimal numbers.
3. Understand that **th** at the end of a word indicates a fraction part.

Classifications of Decimal Numbers

In Section 1.1, we discussed whole numbers and their representation using a **place-value system** along with the ten digits 0, 1, 2, 3, 4, 5, 6, 7, 8, and 9. The value of each place (for whole numbers) in the system is a power of ten:

$$1, \quad 10, \quad 100, \quad 1000, \quad 10,000, \quad 100,000 \quad \text{and so on.}$$

In this section, we show how this **decimal system** can be extended to include fractions as well as whole numbers. We will need the following two concepts:

1. A **finite** amount is an amount that can be counted.

2. An **infinite** amount is an amount that cannot be counted.

> ### Definition
>
> A **finite decimal number** (or **terminating decimal number**) is any rational number (fraction or mixed number) with a power of ten in the denominator of the fraction.

While we generally operate only with finite decimal numbers in arithmetic, you should be aware that there are three classifications of decimal numbers:

1. finite (or terminating) decimals

2. infinite repeating decimals

3. infinite nonrepeating decimals

Infinite repeating decimals and infinite nonrepeating decimals will be discussed in Sections 5.6 and 5.7. **In Sections 5.1–5.5, we use the term decimal number to refer only to finite (or terminating) decimals.**

Note that the whole numbers can also be classified as decimal numbers since any whole number can be written in fraction form with denominator 1. Examples of decimal numbers are

$$\frac{3}{10}, \quad \frac{75}{100}, \quad \frac{349}{1000}, \quad \frac{7}{1}, \quad 8\frac{1}{10}, \quad 21\frac{16}{100}$$

The common **decimal notation** uses a decimal point, with whole numbers written to the left of the decimal point and fractions to the right of the decimal point. The values of several places in this decimal system are shown in Figure 5.1.

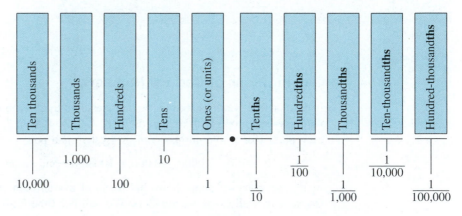

Figure 5.1

> **Note: th** is used to indicate the fraction parts. You should memorize the value of each position.

Reading and Writing Decimal Numbers

To Read or Write a Decimal Number:

1. Read (or write) the whole number as before.

2. Read (or write) **and** in place of the decimal point.

3. Read (or write) the fraction part as a whole number with the name of the place of the last digit on the right.

EXAMPLE 1

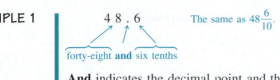

4 8 . 6   The same as $48\frac{6}{10}$.

forty-eight **and** six tenths

And indicates the decimal point and the dight 6 is in the tenths position.

EXAMPLE 2

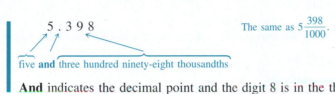

5 . 3 9 8   The same as $5\frac{398}{1000}$.

five **and** three hundred ninety-eight thousandths

And indicates the decimal point and the digit 8 is in the thousandths position.

EXAMPLE 3

Two 0's must be inserted as placeholders.

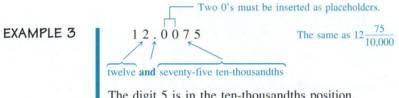

1 2 . 0 0 7 5   The same as $12\frac{75}{10,000}$

twelve **and** seventy-five ten-thousandths

The digit 5 is in the ten-thousandths position.

EXAMPLE 4

In reading a fraction such as $\frac{183}{1000}$, you read the numerator as a whole number (one hundred eighty-three), then attach the name of the denominator (thousandths). Remember to follow the same procedure with decimal numbers.

0.183   is read   "one hundred eighty-three thousandths"

as a numerator   as a denominator

Special Notes:

1. The **th** at the end of a word indicates a fraction part (a part to the right of the decimal point).

eight hundred = 800
eight hundred**ths** = 0.08

2. The hyphen (-) indicates one word.

eight hundred thousand = 800,000
eight hundred-thousand**ths** = 0.00008

EXAMPLE 5

Write $13\frac{506}{1000}$ in decimal notation.

13.506

EXAMPLE 6   Write fifteen thousandths in decimal notation.

┌── 0 is inserted as a placeholder.

0.015

Note that the digit 5 is in the thousandths position.

EXAMPLE 7   Write sixty-two and forty-three hundredths in decimal notation.

62.43

EXAMPLE 8   Write six hundred and five thousandths in decimal notation.

┌── Two 0's are inserted as placeholders.

600.005

EXAMPLE 9   Write six hundred five thousandths in decimal notation.

0.605

Note carefully how the use of **and** in Example 8 gives a completely different result from the result in Example 9.

> **Note:** As a safety measure when writing a check, the amount is written in number form and in word form to avoid problems of spelling and poor penmanship. If there is a discrepancy between the number and the written form, the bank will honor the written form. (See Example 10.)

EXAMPLE 10

MARIA SANCHEZ   No. **125**

123 ELM STREET 123-4569
FRANKLIN, TEXAS 70001

70–1876
711

January 15 19 *95*

Pay to the
order of *Kinderlings Day Care* $ *95.00*

Ninety-five and ⁿᵒ/100   Dollars

SURBURBAN BANK
FRANKLIN, TEXAS
70001

Maria Sanchez

⑈035446⑈⑊021⑊⑉679⑈⑊

CLASSROOM PRACTICE

1. Write 20.7 in words.

2. Write 18.051 in words.

3. Write $4\dfrac{6}{100}$ in decimal notation.

4. Write eight hundred and three tenths in decimal notation.

5. Write eight hundred three thousandths in decimal notation.

ANSWERS: **1.** twenty and seven tenths   **2.** eighteen and fifty-one thousandths   **3.** 4.06
4. 800.3   **5.** 0.803

Exercises 5.1

Write the following mixed numbers in decimal notation.

1. $37\dfrac{498}{1000}$ **2.** $18\dfrac{76}{100}$ **3.** $4\dfrac{11}{100}$ **4.** $56\dfrac{3}{100}$

5. $87\dfrac{3}{1000}$ **6.** $95\dfrac{2}{10}$ **7.** $62\dfrac{7}{10}$ **8.** $100\dfrac{25}{100}$

9. $100\dfrac{38}{100}$ **10.** $250\dfrac{623}{1000}$

Write the following decimal numbers in mixed number form.

11. 82.56   **12.** 93.07   **13.** 10.576   **14.** 100.6

15. 65.003

Write the following numbers in decimal notation.

16. three tenths

17. fourteen thousandths

18. seventeen hundredths

19. six and twenty-eight hundredths

20. sixty and twenty-eight thousandths

21. seventy-two and three hundred ninety-two thousandths

22. eight hundred fifty and thirty-six ten-thousandths

23. seven hundred and seventy-seven hundredths

24. eight thousand four hundred ninety-two and two hundred sixty-three thousandths

25. six hundred thousand, five hundred and four hundred two thousandths

Write the following decimal numbers in words.

26. 0.5	**27.** 0.93	**28.** 5.06	**29.** 32.58
30. 71.06	**31.** 35.078	**32.** 7.003	**33.** 18.102
34. 50.008	**35.** 607.607	**36.** 593.86	**37.** 593.860
38. 4700.617	**39.** 5000.005	**40.** 603.0065	**41.** 900.4638

Write sample checks for the amounts indicated.

42. $356.45

43. $651.50

MARIA SANCHEZ   No. **127**
123 ELM STREET 123-4569
FRANKLIN, TEXAS 70001   **70–1876**
     711

February 9 19 **95**

Pay to the
order of *McCann's Furniture* $ _____

_____ Dollars

SURBURBAN BANK
FRANKLIN, TEXAS
70001

Maria Sanchez

⑈035446⑈021⑈679⑈

44. $2506.64

```
┌─────────────────────────────────────────────────────────────────┐
│   MARIA SANCHEZ                                   No. 128         │
│   123 ELM STREET 123-4569                                         │
│   FRANKLIN, TEXAS 70001                             70–1876       │
│                                                     ───────       │
│                              February 15  19  95       711        │
│   Pay to the        _____        │
│   order of     ___Eastern Securities, Inc.__ $ _____       │
│                                                                   │
│   _____ Dollars        │
│                                                                   │
│          SURBURBAN BANK                                           │
│       ✳    FRANKLIN, TEXAS                                        │
│            70001                          Maria Sanchez           │
│                                          _____     │
│   ⑆035446⑉021⑉679⑈                                               │
└─────────────────────────────────────────────────────────────────┘
```

In each of the following exercises, write the decimal numbers that are not whole numbers in words.

45. One inch is equal to 2.54 centimeters.

46. One gallon of water weighs 8.33 pounds.

47. The winning car in the 1991 Indianapolis 500 race averaged 176.457 mph.

48. In 1990, the U. S. dollar was worth 1.3892 Swiss francs.

49. The number π is approximately equal to 3.14159.

50. The number e (used in higher-level mathematics) is approximately equal to 2.71828.

51. One acre is approximately equal to 0.405 hectares.

52. One liter is approximately equal to 0.264 gallons.

Writing and Thinking about Mathematics

53. In your own words, state why the word **and** is so commonly misused when numbers are spoken and/or written. Bring an example of this from a newspaper, magazine, or television show to class for discussion.

54. Why are hyphens used when words are written? Give four examples of the use of hyphens in writing decimal numbers in word form.

> ## ♻ The Recycle Bin (from Sections 1.2 and 1.3)
>
> First, (a) estimate each sum; then (b) find each sum.
>
> **1.** 2093
>   + 7014
>
> **2.** 875
>   573
>   + 107
>
> **3.** 127,484
>   382,567
>   + 52,661
>
> First, (a) estimate each difference; then (b) find each difference.
>
> **4.** 357
>   − 169
>
> **5.** 7844
>   − 3843
>
> **6.** 90,006
>   − 5,115
>
> **7.** What number should be added to 850 to get a sum of 1200?

5.2 Rounding Off Decimal Numbers

OBJECTIVE

1. Know how to round off decimal numbers to indicated places of accuracy.

Rounding Off Decimal Numbers

Measuring devices made by humans give only approximate measurements. (See Figure 5.2.) The units can be large, such as miles and kilometers, or small, such as inches and centimeters. But there are always smaller units, such as eighths of an inch and millimeters. If a recipe calls for 1.5 cups of flour and you put in 1.52 cups of flour or 1.47 cups of flour, is this acceptable? Will the result be tasty?

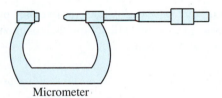

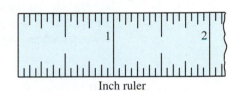

Micrometer   Inch ruler

(a) The micrometer is marked to give   (b) The ruler is marked to give approximate
    approximate measures of small   measures of lengths in fourths, eighths,
    objects.   and sixteenths of an inch.

Figure 5.2

Rounding Off

Rounding off a given number means to find another number close to the given number. The desired place of accuracy must be stated.

EXAMPLE 1

Round off 5.76 to the nearest tenth.

Solution

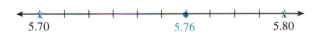

5.70   5.76   5.80

We can see from the number line that 5.76 is closer to 5.80. So, 5.76 rounds off to 5.80 or 5.8 (to the nearest tenth).

EXAMPLE 2

Round off 0.153 to the nearest hundredth.

Solution

0.153 is between 0.150 and 0.160.

0.150   0.153   0.160

There are several rules for rounding off decimal numbers. The IRS, for example, allows rounding off to the nearest dollar on income tax forms. Sometimes, in cases when many numbers are involved, rounding is done to the nearest even digit at some particular place value. The rule chosen depends on the use of the numbers and whether there might be some sort of penalty for an error.

In this text, we will use the following rules for rounding off with one exception: In the case of dollars and cents, we will follow the common business practice of always rounding up to the next higher cent.

> ## Rules for Rounding Off Decimal Numbers
>
> **1.** Look at the single digit just to the right of the place of desired accuracy.
>
> **2.** If this digit is 5 or greater, make the digit in the desired place of accuracy one larger and replace all digits to the right with 0's. If a 9 is made one larger, then the next digit to the left is increased by 1. Otherwise, all digits to the left remain unchanged.
>
> **3.** If this digit is less than 5, leave the digit in the desired place of accuracy as it is and replace all digits to the right with 0's. All digits to the left remain unchanged.
>
> **Note:** Trailing 0's to the right of the decimal point must be dropped so that the place of accuracy is clearly understood. If a rounded-off number has a 0 in the desired place of accuracy, then that 0 remains. In whole numbers, all 0's must remain.

EXAMPLE 3

Round off 6.749 to the nearest tenth.

Solution

(a)   6 . 7 4 9

7 is in the tenths position.   The next digit is 4.

(b) Since 4 is less than 5, leave the 7 and replace 4 and 9 with 0's.

(c) So 6.749 rounds off to **6.7** to the nearest tenth.

Note that even though 6.7 = 6.700, the trailing 0's in 6.700 are dropped to indicate the position of accuracy.

EXAMPLE 4

Round off 13.73962 to the nearest thousandth.

Solution

(a)   1 3 . 7 3 9 6 2

9 is in the thousandths position.   The next digit is 6.

(b) Since 6 is greater than 5, make 9 one larger and replace 6 and 2 with 0's. (Making 9 one larger gives 10, which affects the digit 3 as well as the 9.)

(c) So 13.73962 rounds off to **13.740** to the nearest thousandth. (Note that in 13.74000 only two trailing 0's are dropped. The remaining 0 is written to indicate that the accuracy is to the thousandths place.)

EXAMPLE 5   Round off 8.00241 to the nearest ten-thousandth.

Solution

(a) The digit in the ten-thousandth position is 4.

(b) The next digit on the right is 1.

(c) Since 1 is less than 5, leave 4 as it is and replace 1 with a 0.

(d) So 8.00241 rounds off to **8.0024** to the nearest ten-thousandth.

EXAMPLE 6   Round off 7361 to the nearest hundred.

Solution

(a) The decimal point is understood to be to the right of 1.

(b) The digit in the hundreds position is 3.

(c) The next digit on the right is 6.

(d) Since 6 is greater than 5, change the 3 to 4 and replace 6 and 1 with 0's.

(e) So 7361 rounds off to **7400** to the nearest hundred.

CLASSROOM PRACTICE

Round off as indicated.

1. 572.3 (to the nearest ten)   **2.** 6.749 (to the nearest tenth)

3. 7558 (to the nearest thousand)   **4.** 0.07961 (to the nearest thousandth)

ANSWERS: **1.** 570   **2.** 6.7   **3.** 8000   **4.** 0.080

Exercises 5.2

Fill in the blanks to correctly complete each statement.

1. Round off 34.76 to the nearest tenth.

   (a) The digit in the tenths position is _____.

   (b) The next digit is _____.

(c) Since _____ is greater than 5, change _____ to _____ and replace 6 with 0.

(d) So 34.76 rounds off to _____ to the nearest tenth.

2. Round off 6.832 to the nearest hundredth.

   (a) The digit in the hundredths position is _____.

   (b) The next digit is _____.

   (c) Since _____ is less than 5, leave _____ as it is and replace _____ with 0.

   (d) 6.832 rounds off to _____ to the nearest hundredth.

3. Round off 1.00643 to the nearest ten-thousandth.

   (a) The digit in the ten-thousandths position is _____.

   (b) The next digit is _____.

   (c) Since _____ is less than 5, leave _____ as it is and replace _____ with 0.

   (d) 1.00643 rounds off to _____ to the nearest _____.

4. Round off 6275.38 to the nearest ten.

   (a) The digit in the tens position is _____.

   (b) The next digit is _____.

   (c) Since _____ is equal to 5, change _____ to _____ and replace 5, 3, and 8 with _____.

   (d) 6275.38 rounds off to _____ to the nearest ten.

Round off each of the following decimal numbers as indicated. To the nearest tenth:

5. 89.015   **6.** 7.555   **7.** 18.009   **8.** 37.666

9. 14.3338   **10.** 0.036

To the nearest hundredth:

11. 0.385   **12.** 0.296   **13.** 5.722   **14.** 8.987

15. 6.996   **16.** 13.1346   **17.** 0.0782   **18.** 6.0035

19. 5.7092   **20.** 2.8347

To the nearest thousandth:

21. 0.0672	**22.** 0.05550	**23.** 0.6338	**24.** 7.6666
25. 32.4785	**26.** 9.4302	**27.** 17.36371	**28.** 4.44449
29. 0.00191	**30.** 20.76962		

To the nearest whole number (or nearest unit):

31. 479.23	**32.** 6.8	**33.** 17.5	**34.** 19.999
35. 382.48	**36.** 649.66	**37.** 439.78	**38.** 701.413
39. 6333.11	**40.** 8122.825		

To the nearest ten:

41. 5163.	**42.** 6475	**43.** 495	**44.** 572.5
45. 998.5	**46.** 378.92	**47.** 5476.2	**48.** 76,523.1
49. 92,540.9	**50.** 7007.7		

To the nearest thousand:

51. 7398	**52.** 62,275	**53.** 47,823.4	**54.** 103,499
55. 217,480.2	**56.** 9872.5	**57.** 379,500	**58.** 4,500,762
59. 7,305,438	**60.** 573,333.3		

61. 0.0005783 (to the nearest hundred-thousandth)

62. 0.5449 (to the nearest hundredth)

63. 473.8 (to the nearest ten)

64. 5.00632 (to the nearest thousandth)

65. 473.8 (to the nearest hundred)

66. 5750 (to the nearest thousand)

67. 3.2296 (to the nearest thousandth)

68. 15.548 (to the nearest tenth)

69. 78,419 (to the nearest ten thousand)

70. 78,419 (to the nearest ten)

5.3   Addition and Subtraction (and Estimating)

OBJECTIVES

1. Be able to add with decimal numbers.
2. Be able to subtract with decimal numbers.
3. Know how to estimate sums and differences with rounded-off decimal numbers.

Addition with Decimal Numbers

We can add decimal numbers by writing each number in expanded notation and adding the whole numbers and fractions separately. Thus, we can emphasize that fractions with common denominators are added. For example,

$$6.15 + 3.42 = 6 + \frac{1}{10} + \frac{5}{100} + 3 + \frac{4}{10} + \frac{2}{100}$$

$$= (6 + 3) + \left(\frac{1}{10} + \frac{4}{10}\right) + \left(\frac{5}{100} + \frac{2}{100}\right)$$

$$= 9 + \frac{5}{10} + \frac{7}{100}$$

$$= 9 + \frac{50}{100} + \frac{7}{100}$$

$$= 9 + \frac{57}{100}$$

$$= 9.57$$

Of course, addition of decimal numbers can be accomplished in a much easier way by writing the decimal numbers one under the other and keeping the decimal points aligned vertically. In this way, whole numbers will be added to whole numbers, tenths added to tenths, hundredths added to hundredths, and so on (as we did with the fraction forms in the example just given). The decimal point in the sum is in line with the decimal points in the addends. Thus,

Decimal points are aligned vertically.

$$\begin{array}{r} 6.15 \\ + 3.42 \\ \hline 9.57 \end{array}$$

Any number of 0's may be written to the right of the last digit in the fraction part of a number to help keep the digits in the correct alignment. This will not change the value of any number or the sum.

To Add Decimal Numbers:

1. Write the addends in a vertical column.

2. Keep the decimal points aligned vertically.

3. Keep digits with the same position value aligned. (Zeros may be filled in as aids in keeping the digits aligned properly.)

4. Add the numbers, keeping the decimal point in the sum aligned with the other decimal points.

EXAMPLE 1   Find the sum 6.3 + 5.42 + 14.07.

Solution

$$
\begin{array}{r}
6.30 \\
5.42 \\
+14.07 \\
\hline
25.79
\end{array}
$$

← 0 may be filled in to help keep the digits in line.

EXAMPLE 2   Find the sum 9 + 4.86 + 37.479 + 0.6.

Solution

$$
\begin{array}{r}
9.000 \\
4.860 \\
37.479 \\
+\ \ 0.600 \\
\hline
51.939
\end{array}
$$

The decimal point is understood to be to the right of 9, as in 9.0 or 9.00 or 9.000.

0's are filled in to help keep the digits in line properly.

EXAMPLE 3   Add:

$$
\begin{array}{r}
56.2 \\
85.75 \\
+29.001
\end{array}
$$

Solution

You can write

$$
\begin{array}{r}
56.200 \\
85.750 \\
+29.001 \\
\hline
170.951
\end{array}
$$

0's are filled in to keep the digits in line.

Subtraction with Decimal Numbers

> **To Subtract Decimal Numbers:**
>
> **1.** Write the numbers in a vertical column.
>
> **2.** Keep the decimal points aligned vertically.
>
> **3.** Keep digits with the same position value aligned. (Zeros may be filled in as aids.)
>
> **4.** Subtract, keeping the decimal point in the difference aligned with the other decimal points.

EXAMPLE 4   Find the difference $16.715 - 4.823$.

Solution

$$
\begin{array}{r}
16.715 \\
-\ 4.823 \\
\hline
11.892
\end{array}
$$

EXAMPLE 5   Find the difference $21.2 - 13.716$.

Solution

$$
\begin{array}{r}
2\,1\,.\,2\,0\,0 \\
-\,1\,3\,.\,7\,1\,6 \\
\hline
7\,.\,4\,8\,4
\end{array}
$$
← Fill in 0's.

EXAMPLE 6   Find the difference $17 - 0.5618$.

Solution

$$
\begin{array}{r}
1\,7\,.\,0\,0\,0\,0 \\
-\ \ 0\,.\,5\,6\,1\,8 \\
\hline
1\,6\,.\,4\,3\,8\,2
\end{array}
$$
← Fill in 0's.

EXAMPLE 7   Joe decided he needed some new fishing equipment. He bought a new rod for $55, a rod (on sale) for $22.50, and some fishing line for $2.70. If tax totaled $4.82, how much change did he receive from a $100 bill?

Solution

(a) Find the total of his expenses including tax.

$$
\begin{array}{r}
\$55.00 \\
22.50 \\
2.70 \\
+\ \ \ 4.82 \\
\hline
\$85.02
\end{array}
$$

(b) Subtract the answer in part (a) from $100.

$$
\begin{array}{r}
\$100.00 \\
-\ \ \ 85.02 \\
\hline
\$\ 14.98
\end{array}
$$
   His change was $14.98.

CLASSROOM PRACTICE

Find each indicated sum or difference.

1. $46.2 + 3.07 + 2.6$ **2.** $9 + 5.6 + 0.58$

3. $6.4 - 3.7$ **4.** $18 - 0.4384$

ANSWERS: **1.** 51.87   **2.** 15.18   **3.** 2.7   **4.** 17.5616

Estimating Sums and Differences

We can estimate a sum (or difference) by rounding off each number to the place of the **leftmost nonzero digit** and then adding (or subtracting) these rounded-off numbers. This technique of estimating answers is especially helpful when working with decimal numbers, where the placement of the decimal point is so important.

Note that, depending on the position of the leftmost nonzero digit, the rounded-off numbers may be whole numbers or decimal fractions. Also, different numbers within one problem might be rounded off to different places of accuracy.

EXAMPLE 8

First (a) estimate the sum $74 + 3.529 + 52.61$; then (b) find the actual sum.

Solution

(a) Estimate by adding rounded-off numbers.

(b) Find the actual sum.

$$
\begin{array}{lll}
74 & \text{rounds to} \rightarrow & 70 \\
3.529 & \text{rounds to} \rightarrow & 4 \\
52.61 & \text{rounds to} \rightarrow & +50 \\
\hline
& & 124 \quad \leftarrow \text{estimate}
\end{array}
$$

$$
\begin{array}{r}
74.000 \\
3.529 \\
+52.610 \\
\hline
130.139 \quad \leftarrow \text{actual sum}
\end{array}
$$

EXAMPLE 9

(a) Estimate the difference $22.418 - 17.526$; then (b) find the difference.

Solution

(a) Estimate the difference.
In this case, both numbers round to 20.

$$
\begin{array}{r}
20 \\
-\,20 \\
\hline
0
\end{array}
$$

← estimate

(b) Find the difference.

$$
\begin{array}{r}
\overset{\overset{1}{1}\ \overset{1}{1}\ \overset{3}{ }\ 1}{2\!\!\!/\,2\!\!\!/\,.\,4\,1\,8} \\
-\,1\,7\,.\,5\,2\,6 \\
\hline
4\,.\,8\,9\,2
\end{array}
$$

← actual difference

We find that the actual difference is less than 5, and that is the reason the estimated difference was 0.

EXAMPLE 10

Mrs. Finn went to the local store and bought a pair of shoes for $42.50, a blouse for $25.60, and a skirt for $37.55. How much did she spend? (Tax was included in the prices.) Estimate her expenses mentally before calculating her actual expenses.

Solution

$$
\begin{array}{r}
\$42.50 \\
25.60 \\
+\quad 37.55 \\
\hline
\$105.65
\end{array}
$$

She spent $105.65. (Did you estimate $110?)

Exercises 5.3

Find each of the indicated sums. Estimate your answers, either mentally or on paper, before doing the actual calculations. Check to see that your sums are close to the estimated values.

1. $0.6 + 0.4 + 1.3$

2. $5 + 6.1 + 0.4$

3. $0.59 + 6.91 + 0.05$

4. $3.488 + 16.593 + 25.002$

5. $37.02 + 25 + 6.4 + 3.89$

6. $4.0086 + 0.034 + 0.6 + 0.05$

7. $43.766 + 9.33 + 17 + 206$

8. $52.3 + 6 + 21.01 + 4.005$

9. $2.051 + 0.2006 + 5.4 + 37$

10. $5 + 2.37 + 463 + 10.88$

11.
$$
\begin{array}{r}
47.3 \\
42.03 \\
+\,29.003 \\
\hline
\end{array}
$$

12.
$$
\begin{array}{r}
1.007 \\
20.063 \\
+\ 0.49 \\
\hline
\end{array}
$$

13.
$$
\begin{array}{r}
4.128 \\
0.02 \\
+\,3. \\
\hline
\end{array}
$$

14.
$$
\begin{array}{r}
5.0015 \\
2.443 \\
+\,0.0469 \\
\hline
\end{array}
$$

15. 75.2
    3.682
 + 14.995

16. 107.39
    5.061
   23.54
 + 64.9801

17. 34.967
   50.6
   8.562
 + 9.3

18. 4.156
   3.7
  25.682
 + 13.405

19. 74.
  3.529
 52.62
 + 7.001

20. 983.4
  47.518
 805.411
 + 300.766

Find each of the indicated differences. First estimate the differences mentally.

21. $5.2 - 3.76$

22. $17.83 - 8.9$

23. $29.5 - 13.61$

24. $1.0057 - 0.03$

25. $78.015 - 13.068$

26. 22.418
 − 17.523

27. 4.8
 − 0.0026

28. 31.009
 − 0.534

29. 4.
 − 1.0566

30. 40.718
 − 6.532

31. Theresa got a haircut for $30.00 and a manicure for $10.50. If she tipped the stylist $5, how much change did she receive from a $50 bill?

32. The inside radius of a pipe is 2.38 inches and the outside radius is 2.63 inches, as shown in the figure. What is the thickness of the pipe? (**Note:** The radius of a circle is the distance from the center of the circle to a point on the circle.)

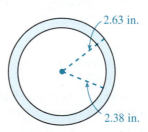

2.63 in.

2.38 in.

33. Mr. Johnson bought the following items at a department store: slacks, $32.50; shoes, $43.75; shirt, $18.60. (a) How much did he spend? (b) What was his change if he gave the clerk a $100 bill? (Tax was included in the prices.)

34. An architect's scale drawing shows a rectangular lot 2.38 inches on one side and 3.76 inches on the other side. What is the perimeter (distance around) of the rectangle in the drawing?

2.38 in.

3.76 in.

35. Mrs. Johnson bought the following items at a department store: dress, $47.25; shoes, $35.75; purse, $12.50. (a) How much did she spend? (b) What was her change if she gave the clerk a $100 bill? (Tax was included in the prices.)

36. Suppose your checking account shows a balance of $280.35 at the beginning of the month. During the month, you make deposits of $310.50, $200, and $185.50, and you write checks for $85.50, $210.20, and $600. (a) Estimate the balance at the end of the month. (b) Find the balance at the end of the month.

37. Your income for one month was $1800. You paid $570 for rent, $40 for electricity, $20 for water, and $35 for gas. **Estimate** what remained of your income after you paid those bills.

38. Martin wants to buy a car for $15,000. His credit union will loan him $10,500, but he must pay $280 for a license fee and $900 for taxes. **Estimate** the amount of cash he needs to buy the car.

39. In 1991 U.S. farmers produced 242.526 million bushels of oats, 464.495 million bushels of barley, and 1980.704 million bushels of wheat. What was the total combined production for these grains in 1991?

40. In 1990 Procter & Gamble spent $2,284.5 million on advertising. In the same year, Johnson & Johnson spent $653.7 million and Bristol-Myers spent $428.7 million on advertising. How much more did Procter & Gamble spend on advertising than the other two companies combined?

41. The eccentricity of a planet's orbit is a measure of how much the orbit varies from a perfectly circular pattern. Mercury's orbit has an eccentricity of 0.205630 while Earth's orbit has an eccentricity of 0.016711. How much greater is Mercury's eccentricity of orbit than Earth's?

42. In a recent year, Boston, MA received 42.25 inches of rain and 23.7 inches of snow, while Burlington, VT received 32.52 inches of rain and 42.1 inches of snow. How much more rain was there in Boston than in Burlington? How much more snow was there in Burlington than in Boston?

43. Suppose that you were to add two decimal numbers with 0 as their whole number parts. Can their sum possibly be (a) more than 2? (b) more than 1? (c) less than 1? Explain briefly.

44. Suppose that you were to subtract two decimal numbers with 0 as their whole number parts. Can their difference possibly be (a) more than 1? (b) less than 1? (c) equal to 1? Explain briefly.

no   *no*   *yes*

 The Recycle Bin   (from Section 1.5)

Find each of the following products.

1. 50 · 7000   **2.** 35   **3.** 195
                              × 15   × 36

4. First estimate the following product; then find the product.

   4571
× 375

5. Find the sum of seventy-five and ninety-three and the difference between one hundred sixty-four and eighty-five. Then, find the product of the sum and the difference.

5.4 Multiplication (and Estimating)

OBJECTIVES

1. Be able to multiply decimal numbers and place the decimal point correctly in the product.
2. Be able to mentally multiply decimal numbers by powers of 10.
3. Know how to estimate products with rounded-off decimal numbers.

Multiplying Decimal Numbers

The method for multiplying decimal numbers is similar to that for multiplying whole numbers. For decimal numbers, however, we need to determine how to place the decimal point in the product. To illustrate how to place the decimal point in a product, several products are shown here in both fraction form and decimal form. The denominators in the fractions are all powers of ten.

Products in Fraction Form

$$\frac{1}{10} \cdot \frac{1}{100} = \frac{1}{1000}$$

$$\frac{3}{10} \cdot \frac{5}{100} = \frac{15}{1000}$$

$$\frac{6}{100} \cdot \frac{4}{1000} = \frac{24}{100,000}$$

Products in Decimal Form

$$
\begin{array}{rl}
.1 & \leftarrow \text{1 place} \\
.01 & \leftarrow \text{2 places} \\
\hline
.001 &
\end{array}
$$ total of 3 places

$$
\begin{array}{rl}
.3 & \leftarrow \text{1 place} \\
.05 & \leftarrow \text{2 places} \\
\hline
.015 &
\end{array}
$$ total of 3 places

$$
\begin{array}{rl}
.004 & \leftarrow \text{3 places} \\
.06 & \leftarrow \text{2 places} \\
\hline
.00024 &
\end{array}
$$ total of 5 places

As the products in decimal form show, **there is no need to keep the decimal points lined up for multiplication.** The following rule states how to multiply two decimal numbers and place the decimal point in the product.

To Multiply Decimal Numbers:

1. Multiply the two numbers as if they were whole numbers.

2. Count the total number of places to the right of the decimal points in both numbers being multiplied.

3. Place the decimal point in the product so that the number of places to the right is the same as that found in Step 2.

EXAMPLE 1   Multiply 2.432×5.1.

Solution

$$
\begin{array}{r}
2.432 \quad \leftarrow \text{3 places} \\
\times \quad\; 5.1 \quad \leftarrow \text{1 place} \\
\hline
2432 \\
12\,160 \\
\hline
12.4032 \quad \leftarrow \text{4 places in the product}
\end{array}
$$

total of 4 places

EXAMPLE 2   Multiply 4.35×12.6.

Solution

$$
\begin{array}{r}
4.35 \quad \leftarrow 2 \text{ places} \\
\times \quad 12.6 \quad \leftarrow 1 \text{ place} \\
\hline
2\ 610 \\
8\ 70 \\
43\ 5 \\
\hline
54.810 \quad \leftarrow 3 \text{ places in the product}
\end{array}
$$

$\left.\begin{array}{c} \leftarrow 2 \text{ places} \\ \leftarrow 1 \text{ place} \end{array}\right\}$ total of 3 places

EXAMPLE 3   Multiply $(0.046)(0.007)$.

Solution

$$
\begin{array}{r}
0.046 \quad \leftarrow 3 \text{ places} \\
\times \quad 0.007 \quad \leftarrow 3 \text{ places} \\
\hline
0.000322 \quad \leftarrow 6 \text{ places in the product}
\end{array}
$$

$\left.\begin{array}{c} \leftarrow 3 \text{ places} \\ \leftarrow 3 \text{ places} \end{array}\right\}$ total of 6 places

Three 0's had to be inserted between the 3 and the decimal point.

Multiplying Decimal Numbers by Powers of 10

In Section 1.5, we discussed how to multiply whole numbers by powers of ten by placing 0's to the right of the number. Now that decimal numbers have been introduced, we can see that inserting 0's in a whole number has the effect of moving the decimal point to the right.

Note: Even though we do not always write the decimal point in a whole number, it is understood to be in place to the right of the rightmost digit.

The following more general guidelines can be used to multiply any decimal number by a power of ten.

> **To Multiply a Decimal Number by a Power of 10:**
>
> **1.** Move the decimal point to the right.
>
> **2.** Move it the same number of places as the number of 0's in the power of 10.
>
> Multiplication by **10** moves the decimal point **one** place **to the right.**
>
> Multiplication by **100** moves the decimal point **two** places **to the right.**
>
> Multiplication by **1000** moves the decimal point **three** places **to the right.**
>
> And so on.

EXAMPLE 4

The following products illustrate multiplication by powers of 10.

(a) $10(2.68) = 26.8$ Move the decimal point 1 place to the right.

(b) $100(2.68) = 268. = 268$ Move the decimal point 2 places to the right.

(c) $1000(0.9653) = 965.3$ Move the decimal point 3 places to the right.

(d) $1000(7.2) = 7200. = 7200$ Move the decimal point 3 places to the right.

(e) $10^2(3.5149) = 351.49$ Move the decimal point 2 places to the right. The exponent tells how many places to move the decimal point.

In the metric system of measure, units of length are set up so that there are ten of one unit in the next longer unit. For example there are 10 millimeters (mm) in 1 centimeter (cm), and there are 10 cm in 1 decimeter (dm), and there are 10 dm in 1 meter (m). Therefore, to change from any number of a particular unit of length in the metric system to a smaller unit of measure, we multiply by some power of 10 (or simply move the decimal point to the right). Table 5.1 illustrates some basic relationships between units of length in the metric system.

Note: There are other units of measure in the metric system for weight, area, and volume.

Table 5.1 Basic Units of Measure of Length in the Metric System

1 m = 10 dm	1 cm = 10 mm
1 m = 100 cm	1 dm = 10 cm
1 m = 10 dm = 100 cm = 1000 mm	1 m = 1000 mm
1 km = 1000 m	

Changing Metric Measures of Length

To change to a measure that is   **Example**

(a) one unit smaller, multiply by 10.   3 cm = 30 mm

(b) two units smaller, multiply by 100.   5 m = 500 cm

(c) three units smaller, multiply by 1000.   14 m = 14 000 mm

And so on.

EXAMPLE 5

The following examples illustrate changing from larger to smaller units of length in the metric system. (Refer to Table 5.1 if you need help.)

(a) 4.32 m = 100(4.32) cm = 432 cm

(b) 4.32 m = 1000(4.32) mm = 4320 mm

(c) 14.6 cm = 10(14.6) mm = 146 mm

(d) 3.51 km = 1000(3.51) m = 3510 m

As an aid in understanding the relative lengths in the metric system, we show some comparisons to the U.S. customary system here.

1 meter is about 39.36 inches.   1 meter is about 3.28 feet.

1 centimeter is about 0.394 inches.   1 kilometer is about 0.62 miles.

Estimating Products of Decimal Numbers

Estimating products can be done by rounding off each number to the place of the last nonzero digit on the left and multiplying these rounded-off numbers. This technique is particularly helpful in correctly placing the decimal point in the actual product.

EXAMPLE 6

First (a) estimate the product $(0.356)(6.1)$; then (b) find the product and use the estimation to help place the decimal point.

Solution

(a) Estimate by multiplying rounded-off numbers.

$$
\begin{array}{r}
0.4 \quad \text{(0.356 rounded off)} \\
\times\ \ 6 \quad \text{(6.1 rounded off)} \\
\hline
2.4 \quad \leftarrow \text{estimate}
\end{array}
$$

(b) Find the actual product.

$$
\begin{array}{r}
0.356 \\
\times \quad 6.1 \\
\hline
356 \\
2\;136 \\
\hline
2.1716
\end{array}
$$
← actual product

The estimated product 2.4 helps place the decimal point correctly in the product 2.1716. Thus, an answer of 0.21716 or 21.716 would indicate an error in the placement of the decimal point since the answer should be near 2.4 (or between 2 and 3).

EXAMPLE 7   First (a) find the product (19.9)(2.3); then (b) estimate the product and use this estimation as a check.

Solution

(a) Find the actual product.

$$
\begin{array}{r}
19.9 \\
\times \quad 2.3 \\
\hline
5\;97 \\
39\;8 \\
\hline
45.77
\end{array}
$$
← actual product

(b) Estimate the product and use this estimation to check that the actual product is reasonable.

$$
\begin{array}{r}
20.0 \\
\times \quad 2 \\
\hline
40.0
\end{array}
$$
20.0 (19.9 rounded off)
× 2 (2.3 rounded off)
40.0 ← estimated product

The estimated product of 40.0 indicates that the actual product is reasonable and the placement of the decimal point is correct.

Note that even though estimated products (or estimated sums or estimated differences) are close to the actual value, they do not guarantee the absolute accuracy of this actual value. The estimates help only in placing decimal points and checking the reasonableness of answers. Experience and understanding are needed so that you can judge whether or not a particular answer is reasonably close to an estimate.

Some word problems may involve several operations with decimal numbers. The words do not usually say directly to add, subtract, or multiply. You need to use your experience and reasoning abilities to decide which operation (if any) to perform with the given numbers. In Example 8 we illustrate a problem that involves several steps, and we show how estimating can provide a check for a reasonable answer.

EXAMPLE 8

You can buy a car for $7500 cash or you can make a down payment of $1875 and then pay $546.67 each month for 12 months. How much can you save by paying cash?

Solution

(a) Find the amount paid in monthly payments by **multiplying** the amount of each payment by 12. In this case, judgment dictates that we do not want to lose two full monthly payments in our estimate, so we use 12 and do not round off to 10. (See Exercise 80 to understand what happens if 10 is used in the estimated calculations.)

Estimate	Actual Amount
$500	$546.67
× 12	× 12
1000	1093 34
500	5466 7
$6000 estimated monthly payments	$6560.04 actual monthly payments

(b) Find the total amount paid by **adding** the down payment to the answer in part (a).

Estimate	Actual Amount
$2000 down payment	$1875.00 down payment
+ 6000 monthly payments	+ 6560.04 monthly payments
$8000 estimated total	$8435.04 total paid

(c) Find the savings by **subtracting** $7500 (the cash price) from the answer in part (b).

Estimate	Actual Amount
$8000 estimated total	$8435.04 total paid
− 7500 cash price	− 7500.00 cash price
$ 500 estimated savings	$ 935.04 savings by paying cash

The $500 estimate in savings by paying cash is reasonably close to the actual savings of $935.04.

Exercises 5.4

1. Match each indicated product with the best estimate of that product.

    ____ (a) (0.7)(0.8)   A. 0.06

    ____ (b) (34.5)(0.11)   B. 1.0

    ____ (c) (0.63)(9.81)   C. 3.0

    ____ (d) (0.34)(0.18)   D. 5.0

    ____ (e) (4.6)(1.2)   E. 6.0

2. Match each indicated product with the best estimate of that product.

    ____ (a) 1.75(0.04)   A. 0.008

    ____ (b) 1.75(0.004)   B. 0.08

    ____ (c) 17.5(0.04)   C. 0.8

    ____ (d) 1.75(4)   D. 8.0

Find each of the indicated products.

3. (0.6)(0.7)   **4.** (0.3)(0.8)   **5.** (0.2)(0.2)

6. (0.3)(0.3)   **7.** 8(2.7)   **8.** 4(9.6)

9. 1.4(0.3)   **10.** 1.5(0.6)   **11.** (0.2)(0.02)

12. (0.3)(0.03)   **13.** 5.4(0.02)   **14.** 7.3(0.01)

15. 0.23 × 0.12   **16.** 0.15 × 0.15   **17.** 8.1 × 0.006

18. 7.1 × 0.008   **19.** 0.06 × 0.01   **20.** 0.25 × 0.01

21. 3(0.125)   **22.** 4(0.375)   **23.** 1.6(0.875)

24. 5.3(0.75)   **25.** 6.9(0.25)   **26.** 4.8(0.25)

27. 0.83(6.1)   **28.** 0.27(0.24)   **29.** 0.16(0.5)

30. 0.28(0.5)

Find each of the following products mentally by using your knowledge of multiplication by powers of ten.

31. 100(3.46)   **32.** 100(20.57)   **33.** 100(7.82)

34. 100(6.93)   **35.** 100(16.1)   **36.** 100(38.2)

37. 10(0.435)   **38.** 10(0.719)   **39.** 10(1.86)

40. 1000(4.1782)   **41.** 1000(0.38)   **42.** 1000(0.47)

43. 10,000 × 0.005   **44.** 10,000 × 0.00615 **45.** 10,000 × 7.4

Find the equivalent measures in the metric system. (Refer to Table 5.1 for help.)

46. 5 cm = _____ mm   **47.** 13 cm = _____ mm

48. 3 m = _____ cm   **49.** 15 m = _____ cm

50. 3.2 m = _____ mm   **51.** 6.17 m = _____ mm

52. 6.5 km = _____ m   **53.** 16 km = _____ m

54. 0.5 km = _____ m

55. 0.6 km = _____ m = _____ cm = _____ mm

56. 2.53 km = _____ m = _____ cm = _____ mm

57. 0.02 km = _____ m = _____ cm = _____ mm

58. 10.7 km = _____ m = _____ cm = _____ mm

In Exercises 59–70, first estimate the product; then find the actual product.

59.   0.106   **60.**   1.07   **61.**   5.08
        × 0.09   × 0.5   × 0.4

62.   0.0106   **63.**   0.0213   **64.**   83.105
        × 0.087   × 0.065   × 0.111

65.   17.002   **66.**   86.1   **67.**   7.83
        × 0.101   ×0.057   ×0.18

68.   95.62   **69.**   6.02   **70.**   8.034
        × 0.57   ×0.57   × 0.29

71. To buy a car, you can pay $2036.50 cash, or put down $400 and make 18 monthly payments of $104.30. How much is saved by paying cash?

72. Suppose a tax assessor figures the tax at 0.07 of the assessed value of a home. (a) If the assessed value is determined at a rate of 0.32 times the market value, what taxes are paid on a home with a market value of $136,500? (b) Estimate these taxes and check to see that the actual taxes are close to the estimate.

73. (a) If the sale price of a new refrigerator is $583 and sales tax is figured at 0.06 times the price, approximately what amount is paid for the refrigerator? (b) What is the exact amount paid for the refrigerator?

74. If you were paid a salary of $350 per week and $13.75 for each hour you worked over 40 hours in a week, how much would you make if you worked 45 hours in one week?

75. The volume of a rectangular solid (a box) is found by multiplying its width times its length times its height. (Volume is measured in cubic units.) Find the volume of a rectangular solid that measures 3.4 in. by 4.5 in. by 6.1 in.

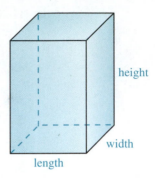

height

width

length

76. Find the volume of a rectangular solid that measures 16.2 cm by 20.5 cm by 30.5 cm. (See Exercise 75.)

77. Find the perimeter and area of a square with sides of length 3 ft.

78. In 1990 New York state led the nation in per capita funding for its public libraries with $30.42 spent for each person. How much funding would have been received that year by a library that served a town of 23,500 people?

79. In 1991 U.S. farmers received an average price of $0.726 per pound for cattle. At this price, what would be the value of cattle with a total weight of 42,500 pounds?

Writing and Thinking about Mathematics

80. In Example 8, we stated the following problem:

You can buy a car for $7500 cash or you can make a down payment of $1875 and then pay $546.67 each month for 12 months. How much can you save by paying cash?

Estimate the savings by using rounded-off numbers. (This includes rounding off 12 months to 10 months.) Explain why this estimated savings does not seem reasonable, and why we must be careful about using rounded-off numbers in practical applications.

81. Suppose you are interested only in a rounded-off answer for a product. Would there be any difference in the products produced by the following two procedures?

(a) First multiply the two numbers as they are and then round off the product to the desired place of accuracy.

(b) First round off each number to the desired place of accuracy and then multiply the rounded-off numbers.

Explain, in your own words, why you think that these two procedures would produce the same result or different results.

82. We stated in the text that one meter is about 39.36 inches. Use each of the techniques discussed in Exercise 81 to find (to the nearest tenth of an inch) how many inches are in 17.523 meters. Discuss why the results are different (or the same).

 The Recycle Bin   (from Section 1.6)

Find the whole number quotients and remainders for each of the following.

1. $15 \overline{)120}$ **2.** $20 \overline{)315}$

3. $50 \overline{)4057}$ **4.** $230 \overline{)46790}$

5. The state of Alaska has a population of about 550,000 people and a land area of about 570,000 square miles (or about 1,480,000 square kilometers). Approximately how many square miles are there in Alaska for each person? About how many square kilometers for each person?

5.5   Division (and Estimating)

OBJECTIVES

1. Be able to divide decimal numbers and place the decimal point correctly in the quotient.
2. Be able to divide decimal numbers mentally by powers of 10.
3. Know how to estimate quotients with rounded-off decimal numbers.

Dividing Decimal Numbers

The process of division (the **division algorithm**) with decimal numbers is, in effect, the same as that for division with whole numbers. As the following example illustrates, division with whole numbers gives a quotient and possibly a remainder.

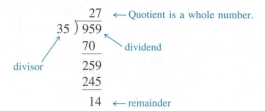

$$
\begin{array}{r}
27 \quad \leftarrow \text{Quotient is a whole number.} \\
35 \overline{)\, 959} \quad\quad\quad\quad \\
70 \quad \searrow \text{dividend} \\
\hline
259 \\
245 \\
\hline
14 \quad \leftarrow \text{remainder}
\end{array}
$$

Now treating the whole numbers in the previous example as decimal numbers, we can continue to divide and get a decimal quotient other than a whole number. Zeros are added on to the dividend if they are needed.

$$
\begin{array}{r}
27.4 \quad \leftarrow \text{Quotient is a decimal number.} \\
35 \overline{)\, 959.0} \quad \leftarrow \text{0 added on} \\
70 \quad\quad \\
\hline
259 \quad\quad \\
245 \quad\quad \\
\hline
14\,0 \\
14\,0 \\
\hline
0
\end{array}
$$

If the divisor is a decimal number other than a whole number, multiply both the divisor and the dividend by a power of 10 so that the divisor is a whole number. For example, we can write

$$4.9 \overline{)\, 51.45} \quad \text{as} \quad \frac{51.45}{4.9} \cdot \frac{10}{10} = \frac{514.5}{49} \quad \text{or} \quad 49 \overline{)\, 514.5}$$

By multiplying both the divisor (4.9) and dividend (51.45) by 10, we have a new divisor (49) that is a whole number. In a similar manner, we can write

$$1.36 \overline{)\, 5.1} \quad \text{as} \quad \frac{5.1}{1.36} \cdot \frac{100}{100} = \frac{510}{136} \quad \text{or} \quad 136 \overline{)\, 510.}$$

In this case, both the divisor (1.36) and the dividend (5.1) are multiplied by 100 so that we have a new whole number divisor (136).

This discussion leads to the following procedure for dividing decimal numbers.

To Divide Decimal Numbers:

1. Move the decimal point in the divisor to the right so that the divisor is a whole number.

2. Move the decimal point in the dividend the same number of places to the right.

3. Place the decimal point in the quotient directly above the new decimal point in the dividend. (**Note: Be sure to do this before dividing.**)

4. Divide just as with whole numbers. (**Note:** 0's may be added in the dividend as needed to be able to continue the division process.)

EXAMPLE 1

Divide 51.45 ÷ 4.9.

Solution

(a) Write the problem as follows:

4.9) 51.45

(b) To move the decimal point so that the divisor is a whole number, move each decimal point one place. This makes the divisor the whole number 49. Place the decimal point in the quotient before dividing.

decimal point in quotient

4.9.) 51.4.5

(c) Divide as with whole numbers.

```
        10.5
49. ) 514.5
      49
      ___
      24
       0
       _
      24 5
      24 5
      ____
         0
```

EXAMPLE 2

Divide 5.1 ÷ 1.36

Solution

(a) Write the problem as follows:

1.36) 5.1

(b) Move the decimal points so the divisor is a whole number. Add 0's in the dividend if needed. Place the decimal point in the quotient before dividing.

decimal point in quotient

1.36.) 5.10.00

Add 0's as needed.

Move each decimal point 2 places.

(c) Divide.

$$
\begin{array}{r}
3.75 \\
136. \overline{)\ 510.00} \\
\underline{408} \\
102\ 0 \\
\underline{95\ 2} \\
6\ 80 \\
\underline{6\ 80} \\
0
\end{array}
$$

EXAMPLE 3

Divide $6.3252 \div 6.3$.

Solution

$$
\begin{array}{r}
1.004 \\
6.3. \overline{)\ 6.3.252} \\
\underline{6\ 3} \\
0\ 2 \\
\underline{0} \\
25 \\
\underline{0} \\
252 \\
\underline{252} \\
0
\end{array}
$$

Note: There **must** be a digit to the right of the decimal point in the quotient above every digit to the right of the decimal point in the dividend.

In each of the Examples 1–3, the remainder was 0. Certainly, this will not always be the case. In general, some place of accuracy for the quotient is agreed upon before the division is performed. If the remainder is not 0 by the time this place of accuracy is reached, then we divide one more place and round off the quotient.

When the Remainder Is Not 0:

1. Decide first how many decimal places are to be in the quotient.

2. Divide until the quotient is **one digit past the place of desired accuracy.**

3. Using this last digit, round off the quotient to the desired place of accuracy.

EXAMPLE 4 Find 8.24 ÷ 2.9 to the nearest tenth.

Solution

Divide until the quotient is in hundredths (one more place than tenths), and then round off to tenths.

$$
\begin{array}{r}
2.84 \approx 2.8 \\
2.9\,\overline{)\,8.2.40} \\
5\,8 \\
\hline
2\,4\,4 \\
2\,3\,2 \\
\hline
1\,20 \\
1\,16 \\
\hline
4
\end{array}
$$

- hundredths
- read **approximately**
- rounded off to tenths

$8.24 \div 2.9 \approx 2.8$ accurate to the nearest tenth

EXAMPLE 5 Find 1.83 ÷ 4.1 to the nearest hundredth.

Solution

Divide until the quotient is in thousandths (one more place than hundredths), and then round off to hundredths.

$$
\begin{array}{r}
0.446 \approx 0.45 \\
4.1\,\overline{)\,1.8.300} \\
1\,6\,4 \\
\hline
1\,90 \\
1\,64 \\
\hline
260 \\
246 \\
\hline
14
\end{array}
$$

- read **approximately**
- Add as many 0's as needed.

$1.83 \div 4.1 \approx 0.45$ accurate to the nearest hundredth

EXAMPLE 6 Find 17 ÷ 3.3 to the nearest thousandth.

Solution

Divide until the quotient is in ten-thousandths, and then round off to thousandths.

$$5.1515 \approx 5.152$$

```
            5.1515 ≈ 5.152
      _____
3.3. ) 17.0.0000
       16 5
          5 0        Add as many 0's as needed.
          3 3
          ___
          1 70
          1 65
          ____
             50
             33
             ___
             170
             165
             ___
               5
```

$17 \div 3.3 \approx 5.152$ accurate to thousandths

Dividing Decimal Numbers by Powers of 10

In Section 5.4, we found that multiplication by powers of 10 can be accomplished by moving the decimal point to the right (giving a number larger than the original number). Division by powers of 10 can be accomplished by moving the decimal point to the left (giving a number smaller than the original number).

> ### To Divide a Decimal Number by a Power of 10:
>
> **1.** Move the decimal point to the left.
>
> **2.** Move it the same number of places as the number of 0's in the power of 10.
>
> Division by **10** moves the decimal point **one** place **to the left.**
>
> Division by **100** moves the decimal point **two** places **to the left.**
>
> Division by **1000** moves the decimal point **three** places **to the left.**
>
> And so on.

Two general guidelines will help you understand work with powers of 10:

1. Multiplication by a power of 10 will make a number larger, so move the decimal point to the right.

2. Division by a power of 10 will make a number smaller, so move the decimal point to the left.

EXAMPLE 7

The following quotients illustrate division by powers of 10.

(a) $4.16 \div 100 = \dfrac{4.16}{100} = 0.0416$ Move the decimal point two places to the left.

(b) $782 \div 10 = \dfrac{782.}{10} = 78.2$ Move the decimal point one place to the left.

(c) $5.933 \div 1000 = \dfrac{5.933}{1000} = 0.005933$ Move the decimal point three places to the left.

As we stated in Section 5.4, the units of length in the metric system are set up so that there are 10 of one unit in the next larger unit. (This relationship is also true for measures of weight and liquid volume.) To change from larger units of length to smaller units of length, we multiply by an appropriate power of 10 (as we did in Section 5.4). Now, to reverse this process (that is, to change from smaller units of length to larger units of length), we divide by an appropriate power of 10. (Refer to Table 5.1 to review the basic relationships of length in the metric system.)

Changing Metric Measures of Length

To change to a measure that is	Example
one unit larger, divide by 10.	4 mm = 0.4 cm
two units larger, divide by 100.	6 mm = 0.06 dm
three units larger, divide by 1000.	8 mm = 0.008 m
And so on.	

EXAMPLE 8

The following examples illustrate changing from smaller to larger units of length in the metric system. (Refer to Table 5.1 if you need help.)

(a) $564 \text{ cm} = \dfrac{564}{100} \text{ m} = 5.64 \text{ m}$

(b) $564 \text{ mm} = \dfrac{564}{1000} \text{ m} = 0.564 \text{ m}$

(c) $1030 \text{ m} = \dfrac{1030}{1000} \text{ km} = 1.03 \text{ km}$

Find each of the indicated quotients.

1. $4 \overline{) 1.83}$ (to the nearest hundredth)

2. $.06 \overline{) 43.721}$ (to the nearest thousandth)

3. $\dfrac{42.31}{1000}$

4. $10^3(42.31)$

Find the equivalent measures in the metric system.

5. 98 mm = _____ cm

6. 3.4 cm = _____ m

ANSWERS: **1.** 0.46   **2.** 728.683   **3.** 0.04231   **4.** 42,310   **5.** 9.8 cm   **6.** 0.034 m

Estimating Quotients of Decimal Numbers

As with addition, subtraction, and multiplication, we can use estimating with division to help in placing the decimal point in the quotient and to verify the reasonableness of the quotient. The technique is to round off both the divisor and the dividend to the place of the last nonzero digit on the left and then divide with these rounded-off values.

EXAMPLE 9

First (a) estimate the quotient $6.2 \div 0.302$, and then (b) find the quotient to the nearest tenth.

Solution

(a) Using $6.2 = 6$ and $0.302 \approx 0.3$, estimate the quotient.

$$
\begin{array}{r}
2\,0. \\
0.3.\,\overline{) 6.0.} \\
\underline{6} \\
0\,0 \\
\underline{0} \\
0
\end{array}
$$

(b) Find the quotient to the nearest tenth.

$$
\begin{array}{r}
20.52 \approx 20.5 \\
0.302.\,\overline{) 6.200.00} \\
\underline{6\,04} \\
160 \\
\underline{0} \\
160\,0 \\
\underline{151\,0} \\
9\,00 \\
\underline{6\,04}
\end{array}
$$

The estimated value 20 is very close to the rounded-off quotient 20.5.

We know from Section 1.7 that the **average** (or **arithmetic average**) of a set of numbers can be found by adding the numbers, then dividing the sum by the number of addends. The term **average** is also used in the sense of an "average speed of 43 miles per hour" or "the average price of a pair of shoes." These averages can also be found by division. If the total amount of a quantity (such as distance, dollars, gallons of gas) and a number of units (such as time, items bought, miles) are known, then we can find the **average amount per unit** by dividing the amount by the number of units.

EXAMPLE 10

The gas tank of a car holds 17 gallons of gasoline. Approximately how many miles per gallon does the car average if it will go 470 miles on one tank of gas?

Solution

The question calls only for an approximate answer. Thus, we can use rounded-off values:

$$17 \approx 20 \text{ gal} \quad \text{and} \quad 470 \approx 500$$

Now divide to approximate the average number of miles per gallon:

$$
\begin{array}{r}
25 \\
20 \overline{)\ 500} \\
40 \\
\hline
100 \\
100 \\
\hline
0
\end{array}
$$
  miles per gallon

The car averages about 25 miles per gallon.

If an average amount per unit is known, then a corresponding total amount can be found by multiplying. For example, if you ride your bicycle at an average speed of 15.2 miles per hour, then the distance you travel can be found by multiplying your average speed by the time you spend riding.

EXAMPLE 11

If you ride your bicycle at an average speed of 15.2 miles per hour, how far will you ride in 3.5 hours?

Solution

Multiply the average speed by the number of hours.

$$
\begin{array}{r}
15.2 \\
\times \quad 3.5 \\
\hline
7\,60 \\
45\,6 \\
\hline
53.20
\end{array}
$$
  miles per hour
  hours

  miles

You will ride 53.2 miles in 3.5 hours.

Exercises 5.5

1. Match the indicated quotient with the best estimate of that quotient.

_B_ (a) $3.1\overline{)6.386}$ A. 0.02

_____ (b) $0.1\overline{)216.5}$ B. 2

_D_ (c) $3.7\overline{)281.6}$ C. 5

_C_ (d) $18.5\overline{)127.9}$ D. 75

_A_ (e) $4.1\overline{)0.0884}$ E. 2000

2. Match the indicated quotient with the best estimate of that quotient.

_____ (a) $27.58 \div 0.003$ A. 10

_____ (b) $27.58 \div 0.03$ B. 100

_____ (c) $27.58 \div 0.3$ C. 1000

_____ (d) $27.58 \div 3$ D. 10,000

Divide.

3. $4.68 \div 2$ **4.** $1.71 \div 3$ **5.** $4.95 \div 5$

6. $1.62 \div 9$ **7.** $0.064 \div 0.8$ **8.** $0.63 \div 0.7$

9. $82.24 \div 0.04$ **10.** $16.02 \div 0.03$ **11.** $48 \div 2.4$

12. $28 \div 5.6$

Find each quotient to the nearest tenth.

13. $8\overline{)455}$ **14.** $4\overline{)263}$ **15.** $9.4\overline{)6.538}$

16. $4.6\overline{)5}$ **17.** $7.05\overline{)0.4977}$ **18.** $0.37\overline{)4.683}$

19. $1.62\overline{)34}$ **20.** $1.33\overline{)75}$

Find each quotient to the nearest hundredth.

21. $24\overline{)0.1463}$ **22.** $1.23\overline{)14.91129}$ **23.** $0.075\overline{)0.42753}$

24. $2.7\overline{)2.583}$ **25.** $23\overline{)62.949}$ **26.** $9\overline{)2}$

27. $13\overline{)65.476}$ **28.** $3.181\overline{)6}$

Find each quotient mentally by using your knowledge of division by powers of ten.

29. $78.4 \div 100$ **30.** $16.4963 \div 100$ **31.** $50.36 \div 100$

32. $45.621 \div 1000$ **33.** $73.85 \div 1000$ **34.** $18.6 \div 1000$

35. $\dfrac{167}{10}$ **36.** $\dfrac{138.1}{10}$ **37.** $\dfrac{7.85}{10}$

38. $\dfrac{1.54}{10,000}$ **39.** $\dfrac{169.9}{10,000}$ **40.** $\dfrac{10.413}{10,000}$

Find the equivalent measures in the metric system.

41. 5 mm = _____ cm

42. 11 mm = _____ cm

43. 83 cm = _____ m

44. 95 cm = _____ m

45. 344 mm = _____ m

46. 255 mm = _____ m

47. 1500 m = _____ km

48. 2400 m = _____ km

49. 97.2 mm = _____ cm

50. 18.5 mm = _____ cm

51. 32 mm = _____ cm = _____ dm = _____ m

52. 560 mm = _____ cm = _____ dm = _____ m

In Exercises 53–60, first estimate each quotient; then find the quotient to the nearest thousandth.

53. $23 \overline{)71}$

54. $69 \overline{)293}$

55. $85.3 \overline{)24.31}$

56. $2.57 \overline{)0.4961}$

57. $13 \overline{)1.029}$

58. $14 \overline{)4.073}$

59. $16.2 \overline{)0.11623}$

60. $25.7 \overline{)6.27}$

Read each problem carefully before you decide what operation (or operations) are required.

61. (a) If a car averages 24.6 miles per gallon, about how far will it go on 18 gallons of gas? (b) Exactly how many miles will it go on 18 gallons of gas?

62. If a motorcycle averages 32.4 miles per gallon, how many miles will it go on 7 gallons of gas?

63. In 1992 the Cleveland Indians had a team batting average of 0.266 and had 1495 base hits. Find the number of team at bats, to the nearest whole number, by dividing base hits by batting average.

64. In a recent year Mike Mussina of the Baltimore Orioles led all American League pitchers with 18 wins and a 0.783 winning percentage. Find Mussina's total number of pitching decisions (games won or lost), to the nearest whole number, by dividing wins by winning percentage.

65. In a recent year Mark Price of the Cleveland Cavaliers basketball team led the NBA with a 0.947 free-throw percentage. He successfully made 270 free throws. Find his number of attempted free throws, to the nearest whole number, by dividing free throws made by free-throw percentage.

66. At Olympic Stadium in Montreal, home of the Expos baseball team, the outfield wall indicates distances from home plate in both feet and meters. The distance down the foul lines is 325 feet. Convert this distance to meters, to the nearest tenth of a meter. (Use 1 m = 3.28 ft.)

67. A marathon footrace is 26.219 miles in length. Convert this distance to kilometers, to the nearest tenth of a kilometer. (Use 1 km = 0.621 mi.)

68. When a textbook is made, it is usually printed and bound in sets of 16 pages called signatures. (Each signature is physically one large sheet of paper with the individual pages laid out in a rectangular arrangement.) Which of the following are possible textbook lengths if only whole signatures are used?

(a) 256 pages

(b) 500 pages

(c) 368 pages

(d) 1264 pages

(e) 648 pages

Explain briefly why you chose the values you did.

69. The public address announcer at Candlestick Park in San Francisco earns $75 for announcing each Giants baseball game. (a) If she announced 72 games in one particular season, about how much did she earn? $3500, $5000, $6500, or $8000? (b) If there are 81 home games in one year, approximately what is the maximum amount that she can earn announcing games? $3500, $5000, $6500, or $8000?

70. Explain in your own words, why, when converting units in the metric system, we *multiply* by a power of 10 to change to a *smaller* unit of measure, and we *divide* by a power of 10 to change to a *larger* unit of measure.

5.6   Scientific Notation

OBJECTIVES

1. Know how to read and write very large numbers in scientific notation.
2. Know how to read and write very small numbers in scientific notation.
3. Be able to read the results on a calculator set in scientific notation mode.

Scientific Notation and Calculators

In scientific applications in fields of study such as physics, chemistry, biology, and astronomy the numbers used are sometimes either very large or very small. Also, any operations with these numbers are likely to yield numbers even larger or smaller.

For example, light travels about 6,000,000,000,000 (6 trillion) miles in one year. (This distance is known as a light-year.) The distance from the earth to the sun is about 93,000,000 (93 million) miles. Due to the large number of 0's, errors can be made when writing down and operating with such numbers. Therefore, mathematicians and scientists have developed a type of shorthand notation for writing such numbers called **scientific notation.**

Also, since hand-held calculators are so widely used now, scientific notation is even more useful because of the limited number of digits on a calculator's display screen. Most calculators display, at most, seven to nine digits. Scientific calculators can be set to display decimal numbers in scientific notation.

> ### Scientific Notation
>
> In **scientific notation,** decimal numbers are written as the product of a decimal number between 1 and 10 and a power of 10. (**Note:** In this section, we will see that the exponent on 10 can be either a positive number or a negative number. Positive and negative numbers will be discussed in detail in Chapter 9.)

To write 93,000,000 in scientific notation, we write the number as follows:

$$93{,}000{,}000 = 9.3 \times 10^7 \quad \text{(distance from Earth to the sun in miles)}$$

between 1 and 10

The exponent 7 tells how many places the decimal point in 9.3 is to be moved to the right so that the product will be equal to 93,000,000.

To write 6,000,000,000,000 in scientific notation, we write the number as follows:

$$6{,}000{,}000{,}000{,}000 = 6.0 \times 10^{12} \quad \text{(a light-year in miles)}$$

between 1 and 10

The exponent 12 tells how many places the decimal point in 6.0 is to be moved to the right so that the product will be equal to 6,000,000,000,000.

Now, if a calculator is used to find the product $93{,}000{,}000 \times 40{,}000$, the display will be similar to one of the following:

$$3.72 \quad 12 \qquad \text{or} \qquad 3.72 \;^{12} \qquad \text{or} \qquad 3.72 \quad \text{E}12$$

In each case 12 is understood to be the exponent on 10 in scientific notation:

$$93{,}000{,}000 \times 40{,}000 = 3{,}720{,}000{,}000{,}000 = 3.72 \times 10^{12}$$

Thus, multiplication by 10^{12} moves the decimal point 12 places to the right:

$$3.72 \times 10^{12} = 3\underbrace{.720,000,000,000}.$$

The decimal point is moved 12 places to the right.

EXAMPLE 1   Each of the following numbers is written in both decimal notation and scientific notation.

Decimal Notation		Scientific Notation
(a) 75,000	=	7.5×10^4
(b) 978,000	=	9.78×10^5
(c) 12,340,000	=	1.234×10^7

Scientific Notation with Negative Exponents

Because **negative numbers** can be used as exponents in scientific notation, we present a brief introduction to the concept of negative numbers here. A more detailed and complete discussion on negative numbers is given in Chapter 9.

Temperature readings below 0 are common in many parts of the world. Mathematically, we can indicate the idea of a reading of **10° below 0** with the number **−10°** (read **negative 10 degrees**). Negative numbers are numbers that are less than 0, and these numbers are indicated with a negative sign (−). Number lines such as that in Figure 5.3 are frequently used to illustrate the idea of negative numbers.

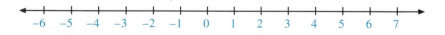

Figure 5.3

At this point, we need only know that **a negative exponent in scientific notation is used to indicate that the decimal point is to be moved to the left.**

Scientific notation makes use of negative exponents in the representations of very small numbers. For example,

$$0.0008 = 8.0 \times 10^{-4} \quad \text{(the diameter of a red blood cell in centimeters)}$$

Or, in another sense,

$$8.0 \times 10^{-4} = 0\underbrace{.0008}.0 = 0.0008$$

The decimal point is moved 4 places to the left.

Similarly,

$$9.234 \times 10^{-6} = 0\underbrace{.000009}.234 = 0.000009234$$

The decimal point is moved 6 places to the left.

EXAMPLE 2   Each of the following numbers is written in both decimal notation and in scientific notation.

Decimal Notation		Scientific Notation
(a) 0.00075	=	7.5×10^{-4}
(b) 0.00632	=	6.32×10^{-3}
(c) 0.0000000004	=	4.0×10^{-10}

CLASSROOM PRACTICE

Write each number in scientific notation.

1. 8,750,000   **2.** 0.000613

Write each number in decimal notation without using exponents.

3. 9.45×10^4 **4.** 7.88×10^{-5}

ANSWERS: **1.** 8.75×10^6 **2.** 6.13×10^{-4} **3.** 94,500 **4.** 0.0000788

Exercises 5.6

Write each of the following numbers in scientific notation.

1. 5,000,000   **2.** 4,300,000   **3.** 750,000 7.5×10^5

4. 890,000   **5.** 67,000,000   **6.** 45,100,000

7. 175,000,000   **8.** 732,000,000,000   **9.** 213,700,000,000

10. 824,500,000,000   **11.** 0.00062   **12.** 0.00057

13. 0.000025   **14.** 0.000034   **15.** 0.0000008

16. 0.0000002   **17.** 0.00000000671   **18.** 0.00000000255

19. 0.00000000000321   **20.** 0.00000000000786

Write each of the following numbers in decimal notation without using exponents.

21. 5.7×10^3 **22.** 6.3×10^4 **23.** 7.54×10^4

24. 1.26×10^7 **25.** 4.72×10^6 **26.** 8.99×10^8

27. 5.7×10^{-3} **28.** 6.3×10^{-4} **29.** 1.84×10^{-5}

30. 3.17×10^{-13} **31.** 5.24×10^{-8} **32.** 2.155×10^{-10}

State whether or not each number given is written in scientific notation. If not, rewrite the number in scientific notation.

33. 56.71×10^3

34. 19.823×10^4

35. 1.44×10^6

36. 1.85×10^5

37. 474.3×10^{-5}

38. 32.1×10^{-3}

39. 17.86×10^{-6}

40. 25.9×10^{-7}

Set your calculator in scientific mode and use it to perform the following operations. Write your answer (a) in the scientific notation you read on the display, and (b) in decimal notation.

41. $120 \div 0.003$

42. $155 \div 0.0005$

43. $57,000 \times 94,000$

44. $125,000 \times 32,000$

45. $88,000 \times 3500$

46. $45,000 \times 2500$

47. $0.00036 \div 1800$

48. $0.00024 \div 600$

49. $5000 \times 65,000 + 7000 \times 2000$

50. $4000 \times 27,000 + 800 \times 30,000$

51. Light travels approximately 3×10^8 meters per second. (a) How fast does light travel in centimeters per second? (b) Write both of these numbers in decimal notation.

52. An atom of gold weighs approximately 3.25×10^{-22} grams. Write this number in decimal notation.

53. The atomic weight of carbon-12 is 1.9926×10^{-26} kilograms. Write this number in decimal notation.

54. Nematode sea worms are the most numerous of all sea or land animals, with an estimated population of 40,000,000,000,000,000,000,000,000. Write this number in scientific notation.

55. The Environmental Protection Agency estimates that, by the year 2010, the amount of municipal solid waste that we produce each year in the United States will reach 2.5×10^8 tons. Write this number in standard decimal notation.

56. A typical slime mold spore measures about 0.000015 meters in diameter. Write this number in scientific notation.

57. In a recent year, the United States consumed 81.17×10^{15} Btu (British thermal units) of energy and produced only 67.47×10^{15} Btu of energy. Write each of these numbers in standard decimal notation.

58. As of 1991, the natural gas reserves of Canada, the United States, and Mexico were estimated at 9.76×10^{13} cubic feet, 1.693×10^{14} cubic feet, and 7.27×10^{13} cubic feet, respectively. Write each of these numbers in standard decimal notation.

5.7   Decimals and Fractions

O B J E C T I V E S

1. Know how to change decimal numbers to fraction and/or mixed number form.

2. Know how to change fractions and mixed numbers to decimal form.

3. Recognize both terminating and nonterminating decimals.

4. Understand that working with decimals and fractions in the same problem may involve rounded-off numbers and approximate answers.

Changing from Decimals to Fractions

In Section 5.1, we discussed the fact that whole numbers, finite decimal numbers, fractions, and mixed numbers are different forms of the same type of number—namely, the **rational numbers.** In this section, we will introduce the concept of **infinite** (or **nonterminating**) **decimals** and show how to add, subtract, multiply, and divide with various combinations of fractions, mixed numbers, and decimal numbers.

Finite (or terminating) decimal numbers can be written in fraction form with denominators that are powers of ten. For example,

$$0.25 = \frac{25}{100} \quad \text{and} \quad 0.025 = \frac{25}{1000}$$

The rightmost digit, 5, is in the hundredths position; so the fraction has 100 in the denominator.

The rightmost digit, 5, is in the thousandths position; so the fraction has 1000 in the denominator.

Changing from Decimals to Fractions

A finite (or terminating) decimal number can be written in fraction form by writing a fraction with

1. a **numerator** that consists of the whole number formed by all the digits of the decimal number

and

2. a **denominator** that is the power of ten that names the position of the rightmost digit

In the following examples, each decimal number is changed to fraction form and then reduced, if possible, by using the factoring techniques discussed in Chapter 3 for reducing fractions.

Change each decimal number to fraction form and reduce if possible.

EXAMPLE 1

$$0.25 = \frac{25}{100} = \frac{\cancel{5} \cdot \cancel{5} \cdot 1}{2 \cdot \cancel{5} \cdot 2 \cdot \cancel{5}} = \frac{1}{4}$$

↑ hundredths

EXAMPLE 2

$$0.32 = \frac{32}{100} = \frac{\cancel{4} \cdot 8}{\cancel{4} \cdot 25} = \frac{8}{25}$$

↑ hundredths

EXAMPLE 3

$$0.131 = \frac{131}{1000}$$

↑ thousandths

EXAMPLE 4

$$0.075 = \frac{75}{1000} = \frac{\cancel{25} \cdot 3}{\cancel{25} \cdot 40} = \frac{3}{40}$$

↑ thousandths

EXAMPLE 5

$$2.6 = \frac{26}{10} = \frac{\cancel{2} \cdot 13}{\cancel{2} \cdot 5} = \frac{13}{5}$$

↑ tenths

or, as a mixed number,

$$2.6 = 2\frac{6}{10} = 2\frac{3}{5}$$

EXAMPLE 6

$$1.42 = \frac{142}{100} = \frac{\cancel{2} \cdot 71}{\cancel{2} \cdot 50} = \frac{71}{50}$$

↑ hundredths

or, as a mixed number,

$$1.42 = 1\frac{42}{100} = 1\frac{21}{50}$$

Changing from Fractions to Decimals

> ### Changing from Fractions to Decimals
>
> A fraction can be written in decimal form by dividing the numerator by the denominator.
>
> **1.** If the remainder is 0, the decimal is said to be **terminating.**
>
> **2.** If the remainder is not 0, the decimal is said to be **nonterminating.**

The following examples illustrate fractions that convert to terminating decimals.

Change each fraction to a decimal number.

EXAMPLE 7

$$\frac{3}{8} \qquad 8\overline{\smash{)}3.000} \quad \begin{array}{r} .375 \end{array} \qquad \frac{3}{8} = .375$$

$$\begin{array}{r}
.375 \\
8\overline{\smash{)}3.000} \\
\underline{2\,4} \\
60 \\
\underline{56} \\
40 \\
\underline{40} \\
0
\end{array}$$

EXAMPLE 8

$$\frac{3}{4}$$

$$\begin{array}{r}
.75 \\
4\overline{\smash{)}3.00} \\
\underline{2\,8} \\
20 \\
\underline{20} \\
0
\end{array} \qquad \frac{3}{4} = .75$$

EXAMPLE 9

$$\frac{4}{5}$$

$$\begin{array}{r}
.8 \\
5\overline{\smash{)}4.0} \\
\underline{4\,0} \\
0
\end{array} \qquad \frac{4}{5} = .8$$

Nonterminating decimals can be **repeating** or **nonrepeating.** A nonterminating repeating decimal has a repeating pattern to its digits. Every fraction with a whole number numerator and nonzero denominator is either a terminating decimal or a repeating decimal. Such numbers are called **rational numbers.** (Nonterminating, nonrepeating decimals are called **irrational numbers** and are discussed later in the text.)

The following examples illustrate fractions that convert to nonterminating repeating decimals.

EXAMPLE 10

$\dfrac{1}{3}$

$$3\overline{)1.000} \quad \begin{array}{r} .333 \\ \end{array}$$

← The 3 will repeat without end.

$$\begin{array}{r} .333 \\ 3\overline{)1.000} \\ \underline{9} \\ 10 \\ \underline{9} \\ 10 \\ \underline{9} \\ 1 \end{array}$$

← Continuing to divide will give a remainder of 1 each time.

We write $\dfrac{1}{3} = 0.333\ \ldots$ The three dots mean "and so on," or to continue without stopping.

EXAMPLE 11

$\dfrac{7}{12}$

$$\begin{array}{r} .5833 \\ 12\overline{)7.0000} \\ \underline{6\,0} \\ 1\,00 \\ \underline{96} \\ 40 \\ \underline{36} \\ 40 \\ \underline{36} \\ 4 \end{array}$$

← The 3 will repeat without end.

Continuing to divide will give a remainder of 4 each time.

We write $\dfrac{7}{12} = 0.5833\ \ldots\ .$

EXAMPLE 12

$\dfrac{1}{7}$

$$\begin{array}{r} .142857 \\ 7\overline{)1.000000} \\ \underline{7} \\ 30 \\ \underline{28} \\ 20 \\ \underline{14} \\ 60 \\ \underline{56} \\ 40 \\ \underline{35} \\ 50 \\ \underline{49} \\ 1 \end{array}$$

← The six digits will repeat in the same pattern without end.

The remainders will repeat in sequence 1, 3, 2, 6, 4, 5, 1, and so on. Therefore, the digits in the quotient will also repeat.

We write $\dfrac{1}{7} = 0.142857142857142857\ \ldots\ .$

Another way to write repeating decimals is to write a **bar** over the repeating digits. Thus, in Examples 10, 11, and 12, we can write

$$\frac{1}{3} = 0.\overline{3} \qquad \text{and} \qquad \frac{7}{12} = 0.58\overline{3} \qquad \text{and} \qquad \frac{1}{7} = 0.\overline{142857}$$

We may choose to round off the quotient to some decimal place just as was done with division in Section 5.5. Perform the division one place past the desired round-off position.

Find the decimal representation of each fraction to the nearest hundredth.

EXAMPLE 13

$$\frac{5}{11}$$

$$\begin{array}{r} .454 \approx .45 \quad \text{(to the nearest hundredth)} \\ 11\overline{)5.000} \\ \underline{4\ 4} \\ 60 \\ \underline{55} \\ 50 \\ \underline{44} \end{array}$$

EXAMPLE 14

$$\frac{5}{6}$$

$$\begin{array}{r} .833 \approx .83 \quad \text{(to the nearest hundredth)} \\ 6\overline{)5.000} \\ \underline{4\ 8} \\ 20 \\ \underline{18} \\ 20 \\ \underline{18} \end{array}$$

Operating with Both Fractions and Decimals

As the following examples illustrate, we can perform operations and comparisons with both fractions and decimal numbers by changing the fractions to decimal form.

Note: In some cases this may involve rounding off the decimal form of a number and settling for an approximate answer. To have a more accurate answer, we may need to change the decimals to fraction form and then perform the operations.

EXAMPLE 15

Find the sum $10\frac{1}{2} + 7.32 + 5\frac{3}{5}$ in decimal form.

Solution

$$10\frac{1}{2} = 10.50 \quad \left(\frac{1}{2} = 0.50\right)$$
$$7.32 = 7.32$$
$$+ 5\frac{3}{5} = 5.60 \quad \left(\frac{3}{5} = 0.60\right)$$
$$\overline{\phantom{+ 5\frac{3}{5} =\ } 23.42}$$

EXAMPLE 16

Determine whether $\frac{3}{16}$ is larger than 0.18 by changing $\frac{3}{16}$ to decimal form and then comparing the two numbers. Find the difference.

Solution

$$
\begin{array}{r}
.1875 \\
16 \overline{)3.0000} \\
1\,6 \\
\hline
1\,40 \\
1\,28 \\
\hline
120 \\
112 \\
\hline
80 \\
80 \\
\hline
0
\end{array}
$$

So $\frac{3}{16} = 0.1875$.

$$
\begin{array}{r}
0.1875 \\
- 0.1800 \\
\hline
0.0075 \quad \text{difference}
\end{array}
$$

Thus, $\frac{3}{16}$ is larger than 0.18 and the difference is 0.0075.

Exercises 5.7

Change each decimal to fraction form. Do not reduce.

1. 0.9	**2.** 0.3	**3.** 0.5	**4.** 0.8
5. 0.62	**6.** 0.38	**7.** 0.57	**8.** 0.41
9. 0.526	**10.** 0.625	**11.** 0.016	**12.** 0.012
13. 5.1	**14.** 7.2	**15.** 8.15	**16.** 6.35

Change each decimal to fraction form (or mixed number form) and reduce if possible.

17. 0.125   **18.** 0.36   **19.** 0.18   **20.** 0.375

21. 0.225   **22.** 0.455   **23.** 0.17   **24.** 0.029

25. 3.2   **26.** 1.25   **27.** 6.25   **28.** 2.75

Change each fraction to decimal form. If the decimal is nonterminating, write it using the bar notation over the repeating pattern of digits.

29. $\dfrac{2}{3}$ **30.** $\dfrac{5}{16}$ **31.** $\dfrac{7}{11}$ **32.** $\dfrac{3}{11}$

33. $\dfrac{11}{16}$ **34.** $\dfrac{9}{16}$ **35.** $\dfrac{3}{7}$ **36.** $\dfrac{5}{7}$

37. $\dfrac{1}{6}$ **38.** $\dfrac{5}{18}$ **39.** $\dfrac{5}{9}$ **40.** $\dfrac{2}{9}$

Change each fraction to decimal form rounded off to the nearest thousandth.

41. $\dfrac{7}{24}$ **42.** $\dfrac{16}{33}$ **43.** $\dfrac{5}{12}$ **44.** $\dfrac{13}{16}$

45. $\dfrac{1}{32}$ **46.** $\dfrac{1}{14}$ **47.** $\dfrac{16}{13}$ **48.** $\dfrac{20}{9}$

49. $\dfrac{30}{21}$ **50.** $\dfrac{40}{3}$

Perform the indicated operations by writing all the numbers in decimal form. Round off to the nearest thousandth, if necessary.

51. $\dfrac{1}{4} + 0.25 + \dfrac{1}{5}$ **52.** $\dfrac{3}{4} + \dfrac{1}{10} + 3.55$

53. $\dfrac{5}{8} + \dfrac{3}{5} + 0.41$ **54.** $6 + 2\dfrac{37}{100} + 3\dfrac{11}{50}$

55. $2\dfrac{53}{100} + 5\dfrac{1}{10} + 7.35$ **56.** $37.02 + 25 + 6\dfrac{2}{5} + 3\dfrac{89}{100}$

57. $1\dfrac{1}{4} - 0.125$ **58.** $2\dfrac{1}{2} - 1.75$

59. $36.71 - 23\dfrac{1}{5}$ **60.** $3.1 - 2\dfrac{1}{100}$

61. $\left(\dfrac{35}{100}\right)^2 (0.73)$ **62.** $\left(5\dfrac{1}{10}\right)^2 (2.25)$

63. $\left(1\dfrac{3}{8}\right)(3.1)(2.6)$

64. $\left(1\dfrac{3}{4}\right)\left(2\dfrac{1}{2}\right)(5.35)$

65. $5\dfrac{54}{100} \div 2.1$

66. $72.16 \div \dfrac{2}{5}$

67. $13.65 \div \dfrac{1}{2}$

68. $91.7 \div \dfrac{1}{4}$

In each of the following exercises, change any fraction to decimal form; then determine which number is larger. Find the difference.

69. $2\dfrac{1}{4}$,   2.3

70. $\dfrac{7}{8}$,   0.878

71. 0.28, $\dfrac{3}{11}$

72. $\dfrac{1}{3}$,   0.3

In Exercises 73–78, change any decimal number (not a yearly date) into fraction or mixed number form.

73. By the 1990 census, there are 70.3 people per square mile in the United States.

74. The average weight for a one-year-old girl is 9.1 kg.

75. The median age for men at first marriage is 26.3 years. The median age for women at first marriage is 24.1 years.

76. The maximum speed of a giant tortoise on land is about 0.17 mph.

77. There are about 22.8 students per teacher in California public schools.

78. The surface gravity on Mars is about 0.38 times the gravity on Earth. The atmospheric pressure on Mars is about 0.01 times the atmospheric pressure on Earth.

In Exercises 79–82, change each fraction or mixed number (not a yearly date) into decimal form.

79. By 1990, one estimation was that $\dfrac{3}{50}$ of U.S. households received 20 or more TV stations.

80. In a recent year, about $\dfrac{8}{57}$ of the advertising budgets in the automotive industry was spent on newspaper ads.

81. Since 1970, the average annual tuition cost for college has increased $5\dfrac{18}{25}$ times.

82. In the 1992 presidential election, Bill Clinton received $\dfrac{1081}{2500}$ of the popular vote.

Summary: Chapter 5

Key Terms and Ideas

A **finite decimal number** (or **terminating decimal number**) is any rational number (fraction or mixed number) with a power of ten in the denominator of the fraction.

To Read or Write a Decimal Number:

1. Read (or write) the whole number as before.

2. Read (or write) **and** in place of the decimal point.

3. Read (or write) the fraction part as a whole number with the name of the place of the last digit on the right.

To **round off** a number means to find another number close to the original number. The desired place of accuracy must be stated.

Estimating can be done by operating with rounded-off numbers.

In **scientific notation,** decimal numbers are written as the product of a decimal number between 1 and 10 and a power of 10.

Rules and Properties

Rules for Rounding Off Decimal Numbers

1. Look at the single digit just to the right of the place of desired accuracy.

2. If this digit is 5 or greater, make the digit in the desired place of accuracy one larger and replace all digits to the right with 0's. If a 9 is made one larger, then the next digit to the left is increased by 1. Otherwise, all digits to the left remain unchanged.

3. If this digit is less than 5, leave the digit in the desired place of accuracy as it is and replace all digits to the right with 0's. All digits to the left remain unchanged.

Note: Trailing 0's to the right of the decimal point must be dropped so that the place of accuracy is clearly understood. If a rounded-off number has a 0 in the desired place of accuracy, then that 0 remains. In whole numbers, all 0's must remain.

Procedures

To Add Decimal Numbers:

1. Write the addends in a vertical column.

2. Keep the decimal points aligned vertically.

3. Keep digits with the same position value aligned. (Zeros may be filled in as aids in keeping the digits aligned properly.)

4. Add the numbers, keeping the decimal point in the sum aligned with the other decimal points.

To Subtract Decimal Numbers:

1. Write the numbers in a vertical column.

2. Keep the decimal points aligned vertically.

3. Keep digits with the same position value aligned. (Zeros may be filled in as aids.)

4. Subtract, keeping the decimal point in the difference aligned with the other decimal points.

To Multiply Decimal Numbers:

1. Multiply the two numbers as if they were whole numbers.

2. Count the total number of places to the right of the decimal points in both numbers being multiplied.

3. Place the decimal point in the product so that the number of places to the right is the same as that found in Step 2.

To Multiply a Decimal Number by a Power of 10:

1. Move the decimal point to the right.

2. Move it the same number of places as the number of 0's in the power of 10.

Multiplication by **10** moves the decimal point **one** place **to the right.**

Multiplication by **100** moves the decimal point **two** places **to the right.**

Multiplication by **1000** moves the decimal point **three** places **to the right.**

And so on.

To Divide Decimal Numbers:

1. Move the decimal point in the divisor to the right so that the divisor is a whole number.

2. Move the decimal point in the dividend the same number of places to the right.

3. Place the decimal point in the quotient directly above the new decimal point in the dividend. (**Note: Be sure to do this before dividing.)**

4. Divide just as with whole numbers. (**Note:** 0's may be added in the dividend as needed to be able to continue the division process.)

To Divide a Decimal Number by a Power of 10:

1. Move the decimal point to the left.

2. Move it the same number of places as the number of 0's in the power of 10.

Division by **10** moves the decimal point **one** place **to the left.**

Division by **100** moves the decimal point **two** places **to the left.**

Division by **1000** moves the decimal point **three** places **to the left.**

And so on.

Changing Metric Measures of Length

To change to a measure that is	Example
one unit smaller, multiply by 10.	3 cm = 30 mm
two units smaller, multiply by 100.	5 m = 500 cm
three units smaller, multiply by 1000.	14 m = 14 000 mm
And so on.	

Changing Metric Measures of Length

To change to a measure that is   **Example**

one unit larger, divide by 10.   4 mm = 0.4 cm

two units larger, divide by 100.   6 mm = 0.06 dm

three units larger, divide by 1000.   8 mm = 0.008 m

And so on.

Changing from Decimals to Fractions

A finite (or terminating) decimal number can be written in fraction form by writing a fraction with

1. a **numerator** that consists of the whole number formed by all the digits of the decimal number

and

2. a **denominator** that is the power of 10 that names the position of the rightmost digit

Changing from Fractions to Decimals

A fraction can be written in decimal form by dividing the numerator by the denominator.

1. If the remainder is 0, the decimal is said to be **terminating.**

2. If the remainder is not 0, the decimal is said to be **nonterminating.**

Review Questions: Chapter 5

1. A _____ number is a mixed number or fraction with a power of 10 in the denominator.

Write the following decimal numbers in words.

2. 0.4   **3.** 7.08   **4.** 92.137   **5.** 18.5526

Write the following decimal numbers in mixed number form.

6. 81.47   **7.** 100.03   **8.** 9.592   **9.** 200.5

Write the following numbers in decimal notation.

10. two and seventeen hundredths

11. eighty-four and seventy-five thousandths

12. three thousand three and three thousandths

Round off as indicated.

13. 5863 (to the nearest hundred)

14. 7.649 (to the nearest tenth)

15. 0.0385 (to the nearest thousandth)

16. 2.069876 (to the nearest hundred-thousandth)

Add or subtract as indicated.

17. $5.4 + 7.34 + 14.08$ **18.** $34.967 + 40.8 + 9.451 + 8.2$

19. $16.92 - 7.9$ **20.** $5 - 1.0377$

21.   78.6
      9.683
   $+ 15.989$

22.   42.008
    $- 19.3$

Multiply as indicated.

23. $(0.8)(0.9)$ **24.** $(0.02)(0.32)$ **25.** $100(2.35)$

26. $10(0.17632)$ **27.** $10^3(5.9641)$ **28.**   1.08
                                            $\times\ 1.6$

29.   36.5
   $\times\ 4.7$

Divide as indicated. (Round off to the nearest hundredth.)

30. $4\overline{)2.83}$ **31.** $0.06\overline{)52.832}$

Divide by using your knowledge of division by powers of 10.

32. $\dfrac{296.1}{100}$ **33.** $\dfrac{5.67}{10^3}$

34. Find the average (to the nearest tenth) of 16.5, 23.4, and 30.7.

Find equivalent measures in the metric system.

35. 182 cm = _____ mm   **36.** 1.35 m = _____ cm

37. 350 mm = _____ cm   **38.** 1800 m = _____ km

Write each of the following decimal numbers in scientific notation.

39. 7,975,000   **40.** 0.0000438

Change each decimal number to fraction (or mixed number) form. Reduce if possible.

41. 0.07   **42.** 2.025   **43.** 0.015

Change each fraction to decimal form. If the decimal is nonterminating, write it using a bar over the repeating digits.

44. $\dfrac{1}{3}$ **45.** $\dfrac{5}{8}$ **46.** $2\dfrac{4}{9}$

Change each fraction to decimal form rounded off to the nearest thousandth.

47. $\dfrac{15}{17}$ **48.** $\dfrac{99}{101}$

49. The buyer for a company purchased 17 cars at a price of $13,450 each. (a) Estimate how much was paid for the cars. (b) Find the exact amount paid for the cars.

50. You are going to make a down payment of $175 on a new stereo and ten equal monthly payments of $48.50. (a) About how much will you pay for the stereo? (b) Exactly how much will you pay for the stereo?

Test: Chapter 5

1. Write the decimal number 30.657 in words.

2. Write the decimal number 0.375 in fraction form reduced to lowest terms.

3. Change $\dfrac{5}{16}$ to decimal form.

In Exercises 4–6, round off as indicated.

4. 203.018 (to the nearest hundredth)

5. 1972.8046 (to the nearest thousand)

6. 0.0693 (to the nearest tenth)

In Exercises 7–16, perform the indicated operations.

7. 85.815 + 17.943   **8.** 95.6 − 93.712

9. 82 + 14.91 + 25.2   **10.** 100.64 − 82.495

11. $13\dfrac{2}{5}$ + 6 + 17.913   **12.** $1\dfrac{1}{1000}$ − 0.09705

13. (0.35)(0.84)   **14.** $\begin{array}{r} 16.31 \\ \times\ 0.785 \\ \hline \end{array}$

15. (1.92)(1000)   **16.** 3.614 ÷ 100

Find each quotient to the nearest hundredth.

17. $82 \div 4.6$

18. $0.13 \overline{)8.617}$

Find the equivalent measures in the metric system.

19. 0.681 km = _____ m = _____ cm = _____ mm

20. 355 mm = _____ cm = _____ m

Write each of the following numbers in scientific notation.

21. 86,500,000

22. 0.000372

23. If you make a down payment of $550.00 on a new refrigerator and make 12 equal payments of $24.80, how much will you pay for the refrigerator?

24. (a) If an automobile gets 18.3 miles per gallon and the tank holds 21.4 gallons, approximately how far can the car travel on a full tank? (b) To the nearest mile, how far can the car actually travel on a full tank?

25. In 1991 conference play (before playoff games), the Washington Redskins professional football team won 14 games for a winning percentage of 0.875. How many conference games did they play?

1. Name the property illustrated.

    (a) $15 + 6 = 6 + 15$

    (b) $2(6 \cdot 7) = (2 \cdot 6)7$

    (c) $51 + 0 = 51$

2. Subtract 503 from the sum of 482 and 367.

3. Mentally, find the product of 70 and 3000.

Evaluate the expressions in Exercises 4–7.

4. $2^2 + 3^2$

5. $18 + 5(2 + 3^2) \div 11 \cdot 2$

6. $24 + [6 + (5^2 \cdot 1 - 3)]$

7. $36 \div 4 + 9 \cdot 2^2$

8. Find the prime factorization of 420.

9. Find the LCM for each set of numbers.

    (a) 28, 70

    (b) 8, 16, 32, 64

In Exercises 10–16, perform the indicated operations.

10. $\dfrac{14}{15} - \dfrac{9}{10}$

11. $4\dfrac{5}{8} + 3\dfrac{9}{10}$

12. $\begin{aligned} & 306\dfrac{3}{8} \\ -\, & 250\dfrac{13}{20} \\ \hline \end{aligned}$

13. $\begin{aligned} & 10\dfrac{1}{2} \\ & 5\dfrac{3}{4} \\ +\, & 16\dfrac{1}{10} \\ \hline \end{aligned}$

14. $\left(\dfrac{3}{4} - \dfrac{1}{8}\right) \div \left(\dfrac{13}{16} - \dfrac{1}{2}\right) = 2$

15. $\left(\dfrac{1}{2}\right)^2\left(\dfrac{14}{15}\right) + \dfrac{4}{15} \div 2$

16. $22\dfrac{1}{2} + 3\dfrac{3}{4}$

17. Simplify the complex fraction: $\dfrac{\dfrac{1}{2} + \dfrac{5}{6}}{1 - \dfrac{5}{8}}$

18. Find $\dfrac{3}{4}$ of 96.

19. Find the quotient: $32.1 \div 1.7$ (to the nearest tenth).

20. Find the product of 21.6 and 0.35 and subtract the sum of 1.375 and 4.

21. Write the following numbers in scientific notation.

   (a) 18,700,000

   (b) 0.000000632

22. If the difference between 29.61 and 17.83 is multiplied by 10^4, what is the product?

23. The business office at a university purchased five vans at a price of $21,540 each, including tax. (a) Approximately how much was paid for the vans? (b) Exactly how much was paid for the vans?

24. The sale price of a washing machine was $240. This was $\frac{4}{5}$ of the original price. What was the original price?

Ratios and Proportions

Mathematics at Work!

Lawn and garden care are perhaps the most routine maintenance tasks faced by every homeowner.

Homeowners are familiar with trips to the store to buy such things as paint, fertilizer, and grass seed for repairs and maintenance. In many cases a homeowner must purchase goods in quantities greater than what is needed at one time for one particular job. The manufacturer's or distributor's packaging units will likely vary from exact needs, and the store will not sell part of a can of paint or part of a bag of fertilizer. A little application of mathematics can help stretch the homeowner's dollar by minimizing any excess amounts that must be bought. Consider the following situation.

You want to treat your lawn with a fungicide, insecticide, fertilizer mix. The local nursery sells bags of this mix for $14 per bag, each bag contains 16 pounds of mix, and the recommended coverage for one bag is 2000 square feet. If your lawn consists of two rectangular shapes, one 30 feet by 100 feet and the other 50 feet by 80 feet, how many pounds of fertilizer mix would cover your lawn? How many bags do you need to buy? How much would you pay (not including tax)? (See Exercises 51–54, Section 6.4.)

What to Expect in Chapter 6

Chapter 6 provides another meaning for fractions (ratios) and introduces the very important topic of solving equations (in the form of proportions). Section 6.1 gives the definition of a ratio and shows three different ways to write a ratio. Also, in Section 6.1, ratios are used in a way familiar to shoppers—price per unit. Proportions (equations stating that two ratios are equal) are introduced in Section 6.2; and then, in Section 6.3, we show how to find the value of one unknown term in a proportion. The unknown term is called a **variable,** and the methods for finding the value of this term provide an introduction to equation solving. (Note: Solving different forms of equations will be discussed in Chapter 9 and in other courses in mathematics.)

The chapter closes with a section on problem solving with the use of proportions.

6.1   Ratios and Price per Unit

OBJECTIVES

1. Understand the meaning of a ratio.
2. Know several notations for ratios.
3. Understand the concept of price per unit.
4. Know how to set up a price per unit as a ratio.
5. Be able to calculate a price per unit by using division.

Understanding Ratios

We know two meanings for fractions.

1. To indicate a part of a whole:

$$\frac{7}{8} \quad \text{means} \quad \frac{7 \text{ pieces of cherry pie}}{8 \text{ pieces in the whole pie}}$$

2. To indicate division:

$$\frac{3}{8} \quad \text{means} \quad 3 \div 8 \quad \text{or} \quad 8 \overline{\smash{)}3.000}$$

$$\begin{array}{r} .375 \\ 8\,\overline{)\,3.000} \\ \underline{2\,4} \\ 60 \\ \underline{56} \\ 40 \\ \underline{40} \\ 0 \end{array}$$

A third use of fractions is to compare two quantities. Such a comparison is called a **ratio.** For example,

$$\frac{3}{4} \quad \text{might mean} \quad \frac{3 \text{ dollars}}{4 \text{ dollars}} \quad \text{or} \quad \frac{3 \text{ hours}}{4 \text{ hours}}$$

> **Definition**
>
> A **ratio** is a comparison of two quantities by division.
>   The ratio of a to b can be written as
>
> $$\frac{a}{b} \quad \text{or} \quad a:b \quad \text{or} \quad a \text{ to } b$$

Ratios have the following characteristics:

1. Ratios can be reduced, just as fractions can be reduced.

2. Whenever the units of the numbers in a ratio are the same, then the ratio has no units. We say that the ratio is an **abstract number.**

3. When the numbers in a ratio have different units, then the numbers must be labeled to clarify what is being compared. Such a ratio is called a **rate.**

   For example, the ratio of 55 miles : 1 hour $\left(\text{or } \dfrac{55 \text{ miles}}{1 \text{ hour}}\right)$ is a rate of 55 miles per hour (or 55 mph).

EXAMPLE 1   Compare the quantities 30 students and 40 chairs as a ratio.

Solution

Since the units (students and chairs) are not the same, the units must be written in the ratio. We can write

(a) $\dfrac{30 \text{ students}}{40 \text{ chairs}}$

(b) 30 students : 40 chairs

(c) 30 students **to** 40 chairs

Furthermore, the same ratio can be simplified by reducing. Since

$$\frac{30}{40} = \frac{3}{4}, \text{ we can write the ratio as } \frac{3 \text{ students}}{4 \text{ chairs}}.$$

The reduced ratio can also be written as

   3 students : 4 chairs   or   3 students **to** 4 chairs.

EXAMPLE 2   Write the comparison of 2 feet to 3 yards as a ratio.

Solution

(a) We can write the ratio as $\dfrac{2 \text{ feet}}{3 \text{ yards}}$.

(b) However, a better procedure is to change to common units. Since 1 yard contains 3 feet, 3 yards = 9 feet. Now we can write the ratio as an abstract number:

$$\frac{2 \text{ feet}}{3 \text{ yards}} = \frac{2 \text{ feet}}{9 \text{ feet}} = \frac{2}{9} \qquad \text{or} \qquad 2:9 \qquad \text{or} \qquad 2 \text{ to } 9$$

EXAMPLE 3

During baseball season, major league players' batting averages are published in the newspapers. Suppose a player has a batting average of .250. What does this indicate?

Solution

A batting average is a ratio (or rate) of hits to times at bat. Thus, a batting average of .250 means

$$.250 = \frac{250 \text{ hits}}{1000 \text{ times at bat}}$$

Reducing gives

$$.250 = \frac{250}{1000} = \frac{250 \cdot 1}{250 \cdot 4} = \frac{1}{4} = \frac{1 \text{ hit}}{4 \text{ times at bat}}$$

This means that we can expect this player to hit successfully at a rate of 1 hit for every 4 times he comes to bat.

EXAMPLE 4

What is the reduced ratio of 300 centimeters (cm) to 2 meters (m)? (In the metric system, there are 100 centimeters in 1 meter.)

Solution

Since 1 m = 100 cm, we have 2 m = 200 cm. Thus the ratio is

$$\frac{300 \text{ cm}}{2 \text{ m}} = \frac{300 \text{ cm}}{200 \text{ cm}} = \frac{3}{2}$$

We can also write the ratio as 3 : 2 or 3 to 2.

Price per Unit

When you buy a new battery for your car, you usually buy just one battery; or, if you buy flashlight batteries, you probably buy two or four. So, in cases such as these, the price for one unit (one battery) is clearly marked or understood. However, when you buy groceries, the same item may be packaged in two (or more) different sizes. Since you want to get the most for your money, you want the better (or best) buy. To calculate the **price per unit** (or **unit price**), you can set up the ratio of price to units and divide.

> **To Find the Price per Unit:**
>
> **1.** Set up a ratio (usually in fraction form) of price to units.
>
> **2.** Divide the price by the number of units.

Note: In the past, many consumers did not understand the concept of price per unit or know how to determine such a number. Now most states have a law that grocery stores must display the price per unit for certain goods they sell so that consumers can be fully informed.

In the following examples, we show the division process as you would write it on paper. However, you may choose to use a calculator to perform these operations. In either case, your first step should be to write each ratio so you can clearly see what you are dividing and that all ratios are comparing the same types of units.

Notice that the comparisons being made in the examples and exercises are with different amounts of the same brand and quality of goods. Any comparison of a relatively expensive brand of high quality with a cheaper brand of lower quality would be meaningless.

EXAMPLE 5

A 12-ounce can of beans is priced at 80¢ while an 18-ounce can of the same beans is $1.10. Which is the better buy?

Solution

We write two ratios of price to units, divide, and compare the results to decide which buy is better. In this problem, we must convert $1.10 to 110¢ so that both ratios will be comparing cents to ounces.

(a) $\dfrac{80¢}{12 \text{ oz}}$

$$
\begin{array}{r}
6.66 \\
12 \overline{)\ 80.00} \\
\underline{72} \\
8\,0 \\
\underline{7\,2} \\
80 \\
\underline{72} \\
\end{array}
$$

or 6.7¢ per ounce

(b) $\dfrac{110¢}{18\text{ oz}}$

$$\begin{array}{r} 6.11 \\ 18\,\overline{)\,110.00} \\ \underline{108} \\ 2\,0 \\ \underline{1\,8} \\ 20 \\ \underline{18} \end{array}$$

or 6.2¢ per ounce

(Note that 6.11 is rounded up to 6.2 because we are dealing with money.)

Thus, the larger can (18 ounces for $1.10) is the better buy since the price per ounce is less.

EXAMPLE 6

Pancake syrup comes in three different sized bottles:

36 fluid ounces for $3.29

24 fluid ounces for $2.49

12 fluid ounces for $1.59

Find the price per fluid ounce for each size of bottle and tell which is the best buy.

Solution

After each ratio is set up, the numerator is changed from dollars and cents to just cents so that the division will yield cents per ounce.

(a) $\dfrac{\$3.29}{36\text{ oz}}\quad \dfrac{329¢}{36\text{ oz}}$

$$\begin{array}{r} 9.11 \\ 36\,\overline{)\,329.00} \\ \underline{324} \\ 5\,0 \\ \underline{3\,6} \\ 40 \\ \underline{36} \end{array}$$

or 9.2¢ per ounce or 9.2¢/oz

(**Note:** "9.2¢/oz" is read "9.2¢ per ounce.")

(b) $\dfrac{\$2.49}{24\text{ oz}}\quad \dfrac{249¢}{24\text{ oz}}$

$$\begin{array}{r} 10.37 \\ 24\,\overline{)\,249.00} \\ \underline{24} \\ 09 \\ \underline{0} \\ 9\,0 \\ \underline{7\,2} \\ 1\,80 \\ \underline{1\,68} \end{array}$$

or 10.4¢/oz

(c) $\dfrac{\$1.59}{12 \text{ oz}}$ $\dfrac{159\cancel{c}}{12 \text{ oz}}$

$$\begin{array}{r} 13.25 \\ 12\overline{)\,159.00} \\ \underline{12} \\ 39 \\ \underline{36} \\ 3\,0 \\ \underline{2\,4} \\ 60 \\ \underline{60} \end{array}$$

or 13.3¢/oz

The largest container (36 fluid ounces) is the best buy.

In each of the examples, the larger (or largest) amount of units was the better (or best) buy. In general, larger amounts are less expensive because the manufacturer wants you to buy more of the product. However, the larger amount is not always the better buy because of other considerations such as packaging or the consumer's individual needs. For example, people who do not use much pancake syrup may want to buy a smaller bottle. Even though they pay more per unit, it's more economical in the long run not to have to throw any away.

Special comment on the term **per:** The student should be aware that the term **per** can be interpreted to mean **divided by.** For example,

cents **per** ounce	means	cents **divided by** ounces
dollars **per** pound	means	dollars **divided by** pounds
miles **per** hour	means	miles **divided by** hours
miles **per** gallon	means	miles **divided by** gallons

Exercises 6.1

Write the following comparisons as ratios reduced to lowest terms. Use common units in the numerator and denominator whenever possible.

1. 1 dime to 4 nickels
2. 5 nickels to 3 quarters
3. 5 dollars to 5 quarters
4. 6 dollars to 50 dimes
5. 250 miles to 5 hours
6. 270 miles to 4.5 hours
7. 50 miles to 2 gallons of gas
8. 60 miles to 5 gallons of gas
9. 30 chairs to 25 people
10. 25 people to 30 chairs
11. 18 inches to 2 feet
12. 36 inches to 2 feet

$\dfrac{18}{24} = \boxed{\dfrac{3}{4}}$ 24 inches

13. 8 days to 1 week

14. 21 days to 4 weeks

15. $200 in profit to $500 invested

16. $200 in profit to $1000 invested

17. 100 centimeters to 1 meter

18. 10 centimeters to 1 millimeter

19. 125 hits to 500 times at bat

20. 100 hits to 500 times at bat

21. About 28 out of every 100 African-Americans have type-A blood. Express this fact as a ratio in lowest terms.

22. A serving of four home-baked chocolate chip cookies weighs 40 grams and contains 12 grams of fat. What is the ratio, in lowest terms, of fat grams to total grams? $12/40 = 3/10$

23. In recent years, 18 out of every 100 students taking the SAT (Scholastic Aptitude Test) have scored 600 or above on the mathematics portion of the test. Write the ratio, in lowest terms, of the number of scores 600 or above to the number of scores below 600.

24. In a recent year, Albany, NY, reported a total of 60 clear days, the rest being cloudy or partly cloudy. For a 365-day year, write the ratio, in lowest terms, of clear days to cloudy or partly cloudy days.   *305 →* $60/305 = 12/61$

Find the unit price of each of the following items and tell which is the better (or best) buy.

25. boxed rice
16 oz (1 lb) at 89¢
32 oz (2 lb) at $1.69

26. noodles
42 oz at $2.99
14 oz at $1.39

27. sour cream
8 oz $\left(\dfrac{1}{2} \text{ pt}\right)$ at 69¢
11 oz at $1.29

28. coffee
48 oz (3 lb) at $7.29
13 oz at $2.45

29. instant coffee
8 oz at $3.49
2 oz at $1.29

30. instant coffee
12 oz at $4.49
8 oz at $3.69

31. frozen orange juice
16 fl oz at $1.69
12 fl oz at 99¢
6 fl oz at 69¢

32. tortilla chips
11 oz at $2.32
16 oz at $2.74
7.5 oz at $1.72

33. boxed dry milk
9.6 oz at $1.49
25.6 oz at $3.59

34. cookies
16 oz at $1.99
20 oz at $2.59

35. saltine crackers
8 oz at 99¢
16 oz at $1.19

36. peanut butter
18 oz at $1.85   10.3 ¢
28 oz at $2.69   9.7 ¢

37. aluminum foil
200 sq ft at $4.19
75 sq ft at $1.59
25 sq ft at 69¢

38. liquid dish soap
32 oz at $2.29
22 oz at $1.69
12 oz at $1.09

39. sliced bologna
8 oz at $1.09
12 oz at $1.59

40. sliced ham
16 oz at $4.69
8 oz at $2.59

41. laundry detergent
9 lb 3 oz (147 oz) at $5.99
72 oz at $3.99
42 oz at $2.39
17 oz at $1.23

42. peanut butter
12 oz at $1.29
18 oz at $1.89
28 oz at $2.79
40 oz at $3.89

43. bottled bleach
1 gal (128 fl oz) at $1.14
$\frac{1}{2}$ gal (64 fl oz) at 85¢
1 qt (32 fl oz) at 59¢

44. jelly
18 oz at $1.29
32 oz at $1.69
48 oz at $2.49

45. boxed doughnuts
14 oz at $1.49
9 oz at $1.09
5 oz at 69¢

46. mustard
8 oz at 65¢
12 oz at $1.09
24 oz at $1.19

47. mayonnaise
16 oz at $1.23
32 oz at $1.69

48. salad dressing
8 oz at $1.09
16 oz at $1.69

49. mustard
8 oz at 69¢
16 oz at $1.09

50. catsup
14 oz at 83¢
32 oz at 99¢

51. chili with beans
40 oz at $2.29
15 oz at 89¢

52. dill pickles
32 oz at $1.59
16 oz at $1.19

53. chili beans
30 oz at 89¢
15.5 oz at 59¢

54. baked beans
31 oz at 79¢
16 oz at 53¢

Writing and Thinking about Mathematics

55. About 8 out of every 100 men exhibit dichromatism, or partial color blindness. What is the ratio of partially color-blind men to men who are unaffected? The ratio of all people affected by dichromatism to those who are not is about 1 to 22. Is this ratio the same as that you have determined for men? What does this tell you about dichromatism in the female population?

The Recycle Bin   (from Sections 3.2, 4.2, 5.4, and 5.5)

Perform the indicated operations and simplify if possible.

1. $3\dfrac{1}{2} \cdot 3\dfrac{1}{7}$

2. $4\dfrac{5}{8} \div 7\dfrac{3}{5}$

3. $\dfrac{14}{32} \cdot \dfrac{12}{28} \cdot \dfrac{4}{15}$

4. $20 \div \dfrac{1}{4}$

5. $(19.2)(7.8)$

6. $14.12(0.0015)$

7. $13.4 \overline{)73.7}$

8. $0.25 \overline{)100}$

6.2   Proportions

OBJECTIVES

1. Understand that a proportion is an equation.
2. Be familiar with the terms **means** and **extremes**.
3. Know that in a true proportion, the product of the means is equal to the product of the extremes.

Understanding Proportions

Consider the equation $\dfrac{3}{6} = \dfrac{4}{8}$. This statement (or equation) says that two ratios are equal. Such an equation is called a **proportion.** As we will see, proportions may be true or false.

Definition

A **proportion** is a statement that two ratios are equal.
In symbols,

$$\frac{a}{b} = \frac{c}{d} \quad \text{is a proportion.}$$

A proportion has four **terms:**

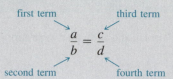

The first and fourth terms (*a* and *d*) are called the **extremes.**

The second and third terms (*b* and *c*) are called the **means.**

To help in remembering which terms are the extremes and which terms are the means, think of a general proportion written with colons as shown here.

With this form, you can see that *a* and *d* are the two end terms and the name **extremes** might seem more reasonable.

EXAMPLE 1   In the proportion $\dfrac{8.4}{4.2} = \dfrac{10.2}{5.1}$, tell which numbers are the extremes and which are the means.

Solution

8.4 and 5.1 are the extremes.

4.2 and 10.2 are the means.

EXAMPLE 2   In the proportion $\dfrac{2\frac{1}{2}}{10} = \dfrac{3\frac{1}{4}}{13}$, tell which numbers are the extremes and which are the means.

Solution

$2\frac{1}{2}$ and 13 are the extremes.

10 and $3\frac{1}{4}$ are the means.

Identifying True Proportions

> **In a true proportion the product of the extremes is equal to the product of the means.**
> In symbols,
>
> $$\frac{a}{b} = \frac{c}{d} \qquad \text{if and only if} \qquad a \cdot d = b \cdot c$$
>
> where $b \neq 0$ and $d \neq 0$.

Note that in a proportion the terms can be any of the types of numbers that we have studied: whole numbers, fractions, mixed numbers, and decimals. Thus, all the related techniques for multiplying these types of numbers should be reviewed at this time.

EXAMPLE 3   Determine whether the proportion $\dfrac{9}{13} = \dfrac{4.5}{6.5}$ is true or false.

Solution

$$\begin{array}{r} 6.5 \\ \times\ 9 \\ \hline 58.5 \end{array} \qquad \begin{array}{r} 4.5 \\ \times\ 13 \\ \hline 13\ 5 \\ 45 \\ \hline 58.5 \end{array}$$

Since $9(6.5) = 13(4.5)$, the proportion is true.

EXAMPLE 4   Is the proportion $\dfrac{5}{8} = \dfrac{7}{10}$ true or false?

Solution

$$5 \cdot 10 = 50 \qquad \text{and} \qquad 8 \cdot 7 = 56$$

Since $50 \neq 56$, the proportion is false.

EXAMPLE 5   Is the proportion $\dfrac{\frac{3}{5}}{\frac{3}{4}} = \dfrac{12}{15}$ true or false?

Solution

The extremes are $\dfrac{3}{5}$ and 15, and their product is

$$\frac{3}{5} \cdot 15 = \frac{3}{5} \cdot \frac{15}{1} = 9$$

The means are $\dfrac{3}{4}$ and 12, and their product is

$$\frac{3}{4} \cdot 12 = \frac{3}{4} \cdot \frac{12}{1} = 9$$

Since the product of the extremes is equal to the product of the means, the proportion is true.

EXAMPLE 6   Is the proportion $\dfrac{1}{4} : \dfrac{2}{3} = 9 : 24$ true or false?

Solution

(a) The extremes are $\dfrac{1}{4}$ and 24.

   The means are $\dfrac{2}{3}$ and 9.

(b) The product of the extremes is

$$\frac{1}{4} \cdot 24 = 6$$

   The product of the means is

$$\frac{2}{3} \cdot 9 = 6$$

(c) The proportion is true, because the products are equal.

Exercises 6.2

1. In the proportion $\dfrac{7}{8} = \dfrac{476}{544}$,

   (a) the extremes are _____ and _____.

   (b) the means are _____ and _____.

2. In the proportion $\dfrac{x}{y} = \dfrac{w}{z}$,

    (a) the extremes are _____ and _____.

    (b) the means are _____ and _____.

    (c) the two terms _____ and _____ cannot be 0.

Determine whether each proportion is true or false by comparing the product of the means with the product of the extremes.

3. $\dfrac{5}{6} = \dfrac{10}{12}$ **4.** $\dfrac{2}{7} = \dfrac{5}{17}$ **5.** $\dfrac{7}{21} = \dfrac{4}{12}$ **6.** $\dfrac{6}{15} = \dfrac{2}{5}$

7. $\dfrac{5}{8} = \dfrac{12}{17}$ **8.** $\dfrac{12}{15} = \dfrac{20}{25}$ **9.** $\dfrac{5}{3} = \dfrac{15}{9}$ **10.** $\dfrac{6}{8} = \dfrac{15}{20}$

11. $\dfrac{2}{5} = \dfrac{4}{10}$ **12.** $\dfrac{3}{5} = \dfrac{60}{100}$ **13.** $\dfrac{125}{1000} = \dfrac{1}{8}$ **14.** $\dfrac{3}{8} = \dfrac{375}{1000}$

15. $\dfrac{1}{4} = \dfrac{25}{100}$ **16.** $\dfrac{7}{8} = \dfrac{875}{1000}$ **17.** $\dfrac{3}{16} = \dfrac{9}{48}$ **18.** $\dfrac{2}{3} = \dfrac{66}{100}$

19. $\dfrac{1}{3} = \dfrac{33}{100}$ **20.** $\dfrac{14}{6} = \dfrac{21}{8}$ **21.** $\dfrac{4}{9} = \dfrac{7}{12}$ **22.** $\dfrac{19}{16} = \dfrac{20}{17}$

23. $\dfrac{3}{6} = \dfrac{4}{8}$ **24.** $\dfrac{12}{18} = \dfrac{14}{21}$ **25.** $\dfrac{5}{6} = \dfrac{7}{8}$ **26.** $\dfrac{7.5}{10} = \dfrac{3}{4}$

27. $\dfrac{6.2}{3.1} = \dfrac{10.2}{5.1}$ **28.** $\dfrac{8\frac{1}{2}}{2\frac{1}{3}} = \dfrac{4\frac{1}{4}}{1\frac{1}{6}}$ **29.** $\dfrac{6\frac{1}{5}}{1\frac{1}{7}} = \dfrac{3\frac{1}{10}}{\frac{8}{14}}$

30. $\dfrac{6}{24} = \dfrac{10}{48}$ **31.** $\dfrac{7}{16} = \dfrac{3\frac{1}{2}}{8}$ **32.** $\dfrac{10}{17} = \dfrac{5}{8\frac{1}{2}}$

33. $3 : 6 = 5 : 10$ **34.** $6 : 16 = 9 : 24$ **35.** $210 : 7 = 20 : \dfrac{2}{3}$

36. $3.75 : 3 = 7.5 : 6$ **37.** $3 : 5 = 60 : 100$ **38.** $12 : 1.09 = 36 : 3.27$

39. $2\frac{1}{2} : \frac{3}{4} = 1\frac{1}{2} : \frac{2}{3}$ **40.** $1\frac{1}{4} : 1\frac{1}{2} = \frac{1}{4} : \frac{1}{2}$ **41.** $6 : 1.56 = 2 : 0.52$

42. $3\frac{1}{5} : 1 = 3\frac{3}{5} : 1\frac{2}{5}$ **43.** $8.5 : 6.5 = 4.5 : 3.5$

44. Consider the proportion $\dfrac{16}{5} = \dfrac{22.4}{7}$. Show why multiplying both sides of the equation by 35 gives the same effect as setting the product of the extremes equal to the product of the means. Do you think that this same technique (of multiplying both sides of the equation by the least common multiple of the denominators) will work with all true proportions? Why or why not?

6.3   Finding the Unknown Term in a Proportion

OBJECTIVES

1. Learn how to find the unknown term in a proportion.
2. Recall the skills in multiplying and dividing with whole numbers, fractions, mixed numbers, and decimals.

Understanding the Meaning of an Unknown Term

Proportions can be used in solving certain types of word problems. In these problems a proportion is set up in which one of the terms in the proportion is not known and the solution to the problem is the value of this unknown term. In this section we will use the fact that in a true proportion the product of the extremes is equal to the product of the means as a method for finding the unknown term in a proportion.

> **Definition**
>
> A **variable** is a symbol (generally a letter of the alphabet) that is used to represent an unknown number or any one of several numbers. The set of possible values for a variable is called its **replacement set.**

As an example of the meaning of a replacement set, suppose that a word problem involves the number of pizza slices you can buy with $5. If you let the variable x represent the possible number of pizza slices, then x can only be a whole number because the restaurant will not sell you a fraction of a slice of pizza. Thus, the set of whole numbers is the replacement set for the variable x.

Finding the Unknown Term in a Proportion

To Find the Unknown Term in a Proportion:

1. In the proportion, the unknown term is represented with a variable (some letter such as x, y, w, A, B, and so forth).

2. Write an equation that sets the product of the extremes equal to the product of the means.

3. Divide both sides of the equation by the number multiplying the variable. (This number is called the **coefficient** of the variable.)

The resulting equation will have a coefficient of 1 for the variable and will give the missing value for the unknown term in the proportion.

Important Notes: As you work through each problem, be sure to write each new equation below the previous equation in the same format shown in the examples. (Arithmetic that cannot be done mentally should be performed to the side with the results written in the next equation.) This format carries over into solving all types of equations at all levels of mathematics. Also, when a number is written next to a variable, such as in $3x$ or $4y$, the meaning is to multiply the number times the value of the variable. That is,

$$3x = 3 \cdot x \text{ and } 4y = 4 \cdot y.$$

The number 3 is the coefficient of x and 4 is the coefficient of y.

EXAMPLE 1

Find the value of x if $\dfrac{3}{6} = \dfrac{5}{x}$.

Solution

In this case you might be able to see that the correct value of x is 10 since

$$\frac{3}{6} \text{ reduces to } \frac{1}{2} \quad \text{and} \quad \frac{1}{2} = \frac{5}{10}.$$

However, not all proportions involve such simple ratios so the following general method of solving for the unknown is important. Follow the steps carefully.

(a) $\dfrac{3}{6} = \dfrac{5}{x}$ Write the proportion.

(b) $3 \cdot x = 6 \cdot 5$ Write the product of the extremes equal to the product of the means.

(c) $\dfrac{3 \cdot x}{3} = \dfrac{30}{3}$ Divide both sides by the coefficient of the variable, 3.

(d) $\dfrac{\cancel{3} \cdot x}{\cancel{3}} = \dfrac{30}{3} = 10$ Reduce both sides to find the solution.

(e) $x = 10$

EXAMPLE 2

Find the value of y if $\dfrac{6}{16} = \dfrac{y}{24}$.

Solution

Note that the variable may appear on the right side of the equation, as shown here in step (b). Or it may appear on the left side of the equation, as shown in Example 1. In either case, we **divide both sides of the equation by the coefficient of the variable.**

(a) $\dfrac{6}{16} = \dfrac{y}{24}$ Write the proportion.

(b) $6 \cdot 24 = 16 \cdot y$ Write the product of the extremes equal to the product of the means.

(c) $\dfrac{6 \cdot 24}{16} = \dfrac{16 \cdot y}{16}$ Divide both sides by 16, the coefficient of y.

(d) $\dfrac{6 \cdot 24}{16} = \dfrac{\cancel{16} \cdot y}{\cancel{16}}$ Reduce both sides to find the value of y.

(e) $9 = y$

Second Solution

Reduce the fraction $\dfrac{6}{16}$ before solving the proportion.

(a) $\dfrac{6}{16} = \dfrac{y}{24}$ Write the proportion.

(b) $\dfrac{3}{8} = \dfrac{y}{24}$ Reduce the fraction: $\dfrac{6}{16} = \dfrac{3}{8}$

(c) $3 \cdot 24 = 8 \cdot y$ Proceed to solve as before.

$\dfrac{3 \cdot 24}{8} = \dfrac{\cancel{8} \cdot y}{\cancel{8}}$

$9 = y$

EXAMPLE 3

Find w if $\dfrac{w}{7} = \dfrac{20}{\frac{2}{3}}$.

Solution

(a) $\quad \dfrac{w}{7} = \dfrac{20}{\frac{2}{3}}$ Write the proportion.

(b) $\quad \dfrac{2}{3} \cdot w = 7 \cdot 20$ Set the product of the extremes equal to the product of the means.

(c) $\quad \dfrac{\frac{2}{3} \cdot w}{\frac{2}{3}} = \dfrac{7 \cdot 20}{\frac{2}{3}}$ Divide each side by the coefficient $\dfrac{2}{3}$.

(d) $\quad w = \dfrac{7}{1} \cdot \dfrac{20}{1} \cdot \dfrac{3}{2} = 210$ Simplify. Remember, to divide by a fraction, multiply by its reciprocal.

EXAMPLE 4

Find A if $\dfrac{A}{9} = \dfrac{7.5}{6}$.

Solution

$$\dfrac{A}{9} = \dfrac{7.5}{6}$$

$$6 \cdot A = 9(7.5)$$

$$\dfrac{6 \cdot A}{6} = \dfrac{67.5}{6}$$

$$A = 11.25$$

$$\begin{array}{r} 11.25 \\ 6 \overline{)\,67.50} \\ \underline{6} \\ 07 \\ \underline{6} \\ 1\,5 \\ \underline{1\,2} \\ 30 \\ \underline{30} \end{array}$$

The unknown can be a decimal or fraction or mixed number as well as a whole number.

Important Note Again: Remember to write one equation under the other, just as in the examples. Even in Example 4, where the division is written to the right, the equations are aligned.

EXAMPLE 5   Find x if $\dfrac{x}{15} = \dfrac{5}{24}$.

Solution

$$\frac{x}{15} = \frac{5}{24}$$

$$24 \cdot x = 5 \cdot 15$$

$$\frac{\cancel{24} \cdot x}{\cancel{24}} = \frac{75}{24}$$

$$x = \frac{\cancel{3} \cdot 5 \cdot 5}{\cancel{3} \cdot 8} = \frac{25}{8}$$

$$x = 3\frac{1}{8}$$

Here the unknown term is written as a mixed number. You might have chosen to divide ($25 \div 8$) and get a decimal answer of 3.125. Both answers are correct.

As illustrated in the following Examples 6 and 7, when the coefficient of the variable is a fraction, we can multiply both sides of the equation by the reciprocal of this coefficient. This is, of course, the same as dividing by the coefficient, but fewer steps are involved.

EXAMPLE 6   Find x if $\dfrac{x}{1\frac{1}{2}} = \dfrac{1\frac{2}{3}}{3\frac{1}{3}}$.

Solution

(a) $\qquad\qquad \dfrac{x}{1\frac{1}{2}} = \dfrac{1\frac{2}{3}}{3\frac{1}{3}}$

(b) $\qquad\qquad \dfrac{10}{3} \cdot x = \dfrac{3}{2} \cdot \dfrac{5}{3}$ $\qquad$ Write each mixed number as an improper fraction.

(c) $\qquad\qquad \dfrac{10}{3} \cdot x = \dfrac{5}{2}$ $\qquad$ Simplify both sides.

(d) $\qquad \dfrac{\cancel{3}}{\cancel{10}} \cdot \dfrac{\cancel{10}}{\cancel{3}} \cdot x = \dfrac{3}{10} \cdot \dfrac{5}{2}$ $\qquad$ Multiply each side by $\dfrac{3}{10}$, the reciprocal of $\dfrac{10}{3}$.

(e) $\qquad\qquad x = \dfrac{3}{\cancel{10}_{2}} \cdot \dfrac{\cancel{5}}{2} = \dfrac{3}{4}$

EXAMPLE 7   Find y if $\dfrac{2\frac{1}{2}}{6} = \dfrac{3}{y}$.

Solution

(a) $\qquad \dfrac{2\frac{1}{2}}{6} = \dfrac{3}{y}$

(b) $\qquad \dfrac{5}{2} \cdot y = 18 \qquad\qquad$ Change $2\frac{1}{2}$ to $\frac{5}{2}$.

(c) $\qquad \dfrac{2}{5} \cdot \dfrac{5}{2} \cdot y = \dfrac{2}{5} \cdot \dfrac{18}{1}$

(d) $\qquad\qquad y = \dfrac{36}{5} \left(\text{or } 7\frac{1}{5}\right)$

CLASSROOM
PRACTICE

Solve the following proportions.

1. $\dfrac{3}{5} = \dfrac{R}{100}$ $\qquad$ **2.** $\dfrac{2\frac{1}{2}}{8} = \dfrac{25}{x}$ $\qquad$ **3.** $\dfrac{3}{2} = \dfrac{B}{4.5}$ $\qquad$ **4.** $\dfrac{x}{1.50} = \dfrac{11}{2.75}$

ANSWERS: **1.** $R = 60$ **2.** $x = 80$ **3.** $B = 6.75$ **4.** $x = 6$

Exercises 6.3

Solve for the variable in each of the following proportions.

1. $\dfrac{3}{6} = \dfrac{6}{x}$ $\qquad$ **2.** $\dfrac{7}{21} = \dfrac{y}{6}$ $\qquad$ **3.** $\dfrac{5}{7} = \dfrac{z}{28}$ $\qquad$ **4.** $\dfrac{4}{10} = \dfrac{5}{x}$

5. $\dfrac{8}{B} = \dfrac{6}{30}$ $\qquad$ **6.** $\dfrac{7}{B} = \dfrac{5}{15}$ $\qquad$ **7.** $\dfrac{1}{2} = \dfrac{x}{100}$ $\qquad$ **8.** $\dfrac{3}{4} = \dfrac{x}{100}$

9. $\dfrac{A}{3} = \dfrac{7}{2}$ $\qquad$ **10.** $\dfrac{x}{100} = \dfrac{1}{20}$ $\qquad$ **11.** $\dfrac{3}{5} = \dfrac{60}{D}$ $\qquad$ **12.** $\dfrac{3}{16} = \dfrac{9}{x}$

13. $\dfrac{\frac{1}{2}}{x} = \dfrac{5}{10}$ $\qquad$ **14.** $\dfrac{\frac{2}{3}}{3} = \dfrac{y}{127}$ $\qquad$ **15.** $\dfrac{\frac{1}{3}}{x} = \dfrac{5}{9}$ $\qquad$ **16.** $\dfrac{\frac{3}{4}}{7} = \dfrac{3}{z}$

17. $\dfrac{\frac{1}{8}}{6} = \dfrac{\frac{1}{2}}{w}$

18. $\dfrac{\frac{1}{6}}{5} = \dfrac{5}{w}$

19. $\dfrac{1}{4} = \dfrac{1\frac{1}{2}}{y}$

20. $\dfrac{1}{5} = \dfrac{x}{2\frac{1}{2}}$

21. $\dfrac{1}{5} = \dfrac{x}{7\frac{1}{2}}$

22. $\dfrac{2}{5} = \dfrac{R}{100}$

23. $\dfrac{3}{5} = \dfrac{R}{100}$

24. $\dfrac{A}{4} = \dfrac{75}{100}$

25. $\dfrac{A}{4} = \dfrac{50}{100}$

26. $\dfrac{20}{B} = \dfrac{1}{4}$

27. $\dfrac{30}{B} = \dfrac{25}{100}$

28. $\dfrac{A}{20} = \dfrac{15}{100}$

29. $\dfrac{1}{3} = \dfrac{R}{100}$

30. $\dfrac{2}{3} = \dfrac{R}{100}$

31. $\dfrac{9}{x} = \dfrac{4\frac{1}{2}}{11}$

32. $\dfrac{y}{6} = \dfrac{2\frac{1}{2}}{12}$

33. $\dfrac{x}{4} = \dfrac{1\frac{1}{4}}{5}$

34. $\dfrac{5}{x} = \dfrac{2\frac{1}{4}}{27}$

35. $\dfrac{x}{3} = \dfrac{16}{3\frac{1}{5}}$

36. $\dfrac{6.2}{5} = \dfrac{x}{15}$

37. $\dfrac{3.5}{2.6} = \dfrac{10.5}{B}$

38. $\dfrac{4.1}{3.2} = \dfrac{x}{6.4}$

39. $\dfrac{7.8}{1.3} = \dfrac{x}{0.26}$

40. $\dfrac{7.2}{y} = \dfrac{4.8}{14.4}$

41. $\dfrac{150}{300} = \dfrac{R}{100}$

42. $\dfrac{19.2}{96} = \dfrac{R}{100}$

43. $\dfrac{12}{B} = \dfrac{25}{100}$

44. $\dfrac{13.5}{B} = \dfrac{15}{100}$

45. $\dfrac{A}{42} = \dfrac{65}{100}$

46. $\dfrac{A}{1000} = \dfrac{18}{100}$

47. $\dfrac{A}{10} = \dfrac{18}{100}$

48. $\dfrac{3}{10} = \dfrac{x}{100}$

49. $\dfrac{x}{10} = \dfrac{33}{100}$

50. $\dfrac{x}{10} = \dfrac{66}{100}$

Check Your Number Sense

Now that you are familiar with proportions and the techniques for finding the unknown term in a proportion, check your general understanding by choosing the answer that seems most reasonable to you in each of the following exercises. (Use mental calculation only.) After you have checked the answers in the back of the text, work out any problem that you missed to help you develop a better understanding of proportions.

51. Given the proportion $\dfrac{x}{100} = \dfrac{1}{4}$, which of the following values seems the most reasonable for x?

(a) 10   (b) 25   (c) 50   (d) 75

52. Given the proportion $\dfrac{x}{200} = \dfrac{1}{10}$, which of the following values seems the most reasonable for x?

(a) 10   (b) 20   (c) 30   (d) 40

53. Given the proportion $\dfrac{3}{5} = \dfrac{60}{x}$, which of the following values seems the most reasonable for x?

(a) 50   (b) 80   (c) 100   (d) 150

54. Given the proportion $\dfrac{4}{10} = \dfrac{20}{x}$, which of the following values seems the most reasonable for x?

(a) 10   (b) 30   (c) 40   (d) 50

55. Given the proportion $\dfrac{1.5}{3} = \dfrac{x}{6}$, which of the following values seems the most reasonable for x?

(a) 1.5   (b) 2.5   (c) 3.0   (d) 4.5

56. Given the proportion $\dfrac{2\frac{1}{3}}{x} = \dfrac{4\frac{2}{3}}{10}$, which of the following values seems the most reasonable for x?

(a) $4\dfrac{2}{3}$ (b) 5   (c) 20   (d) 40

♻ *The Recycle Bin*   (from Section 5.7)

Perform the indicated operations by using decimal forms of the fractions.

1. $\dfrac{3}{4} + \dfrac{3}{5} + 7.16$ **2.** $5\dfrac{1}{2} + 20.3 + 16.8$

3. $27\dfrac{5}{8} - 13.925$ **4.** $8.13\left(2\dfrac{1}{4}\right)$

5. $6\dfrac{7}{10} + 2\dfrac{1}{2}(6.1)$ **6.** $5.8\left(3\dfrac{1}{10}\right) - 3 \div \dfrac{1}{5}$

Problem Solving with Proportions

6.4

OBJECTIVES

1. Learn how to identify the unknown quantity in a problem.
2. Recognize the type of problem that can be solved by using a proportion.
3. Be able to set up a proportion with an unknown term in one of the appropriate patterns that will lead to a solution of the problem.

When to Use Proportions

A proportion is a statement that two ratios are equal. Thus, a problem that involves two ratios can be solved by using proportions. The following procedure is helpful.

> **To Solve a Word Problem by Using Proportions:**
>
> 1. Identify the unknown quantity and use a variable to represent this quantity.
>
> 2. Set up a proportion in which the units are compared as in Pattern A or Pattern B shown here.
>   Suppose that a motorcycle will travel 352 miles on 11 gallons of gas. How far would you expect it to travel on 15 gallons of gas?
>   Let x = the unknown number of miles.
>
> **Pattern A** Each ratio has different units but they are in the same order. For example,
>
> $$\frac{352 \text{ miles}}{11 \text{ gallons}} = \frac{x \text{ miles}}{15 \text{ gallons}}$$
>
> **Pattern B** Each ratio has the same units, the numerators correspond and the denominators correspond. For example,
>
> $$\frac{352 \text{ miles}}{x \text{ miles}} = \frac{11 \text{ gallons}}{15 \text{ gallons}}$$
>
> (352 miles corresponds to 11 gallons and x miles corresponds to 15 gallons.)
>
> 3. Solve the proportion.

Problem Solving with Proportions

EXAMPLE 1

You drove your car 500 miles and used 20 gallons of gasoline. How many miles would you expect to drive using 30 gallons of gasoline?

Solution

(a) Let x represent the number of unknown miles.

(b) Set up a proportion using either Pattern A or Pattern B. Label the numerators and denominators to be sure the units are in the same order.

$$\frac{500 \text{ miles}}{20 \text{ gallons}} = \frac{x \text{ miles}}{30 \text{ gallons}}$$ **Pattern A** Each ratio has different units, but numerators are the same units and denominators are the same units.

(c) Solve the proportion.

$$\frac{500}{20} = \frac{x}{30}$$
$$500 \cdot 30 = 20 \cdot x$$
$$\frac{15{,}000}{20} = \frac{\cancel{20} \cdot x}{\cancel{20}}$$
$$750 = x$$

You would expect to drive 750 miles on 30 gallons of gas.

Note: *Any* of the following proportions would give the same answer.

$$\frac{20 \text{ gallons}}{500 \text{ miles}} = \frac{30 \text{ gallons}}{x \text{ miles}}$$ **Pattern A** Each ratio has different units, but numerators are the same units and denominators are the same units.

$$\frac{500 \text{ miles}}{x \text{ miles}} = \frac{20 \text{ gallons}}{30 \text{ gallons}}$$ **Pattern B** Each ratio has the same units, and numerators correspond and denominators correspond.

$$\frac{x \text{ miles}}{500 \text{ miles}} = \frac{30 \text{ gallons}}{20 \text{ gallons}}$$ **Pattern B** Each ratio has the same units, and numerators correspond and denominators correspond.

EXAMPLE 2

An architect draws the plans for a building using a scale of $\frac{1}{2}$ inch to represent 10 feet. How many feet would 6 inches represent?

Solution

(a) Let y represent the number of unknown feet.

(b) Set up a proportion labeling the numerators and denominators.

$$\frac{\frac{1}{2} \text{ inch}}{6 \text{ inches}} = \frac{y \text{ feet}}{10 \text{ feet}}$$ **WRONG**

This proportion is **WRONG** because $\frac{1}{2}$ inch does *not* correspond to y feet. $\frac{1}{2}$ inch corresponds to 10 feet.

$$\frac{\frac{1}{2} \text{ inch}}{6 \text{ inches}} = \frac{10 \text{ feet}}{y \text{ feet}} \quad \textbf{RIGHT}$$

(c) Solve the **RIGHT** proportion.

$$\frac{\frac{1}{2}}{6} = \frac{10}{y}$$

$$\frac{1}{2} \cdot y = 6 \cdot 10$$

$$\frac{\frac{1}{2} \cdot y}{\frac{1}{2}} = \frac{60}{\frac{1}{2}}$$

$$y = 60 \cdot \frac{2}{1} = 120$$

6 inches would represent 120 feet.

EXAMPLE 3

A recommended mixture of weed killer is 3 cupfuls for 2 gallons of water. How many cupfuls should be mixed with 5 gallons of water?

Solution

(a) Let x = unknown number of capfuls of weed killer.

(b) Set up a proportion.

$$\frac{x \text{ cupfuls}}{5 \text{ gallons}} = \frac{3 \text{ cupfuls}}{2 \text{ gallons}}$$

(c) Solve the proportion.

$$\frac{x}{5} = \frac{3}{2}$$

$$2 \cdot x = 5 \cdot 3$$

$$\frac{2 \cdot x}{2} = \frac{15}{2}$$

$$x = 7\frac{1}{2}$$

$7\frac{1}{2}$ cupfuls of weed killer should be mixed with 5 gallons of water.

Nurses and doctors work with proportions when prescribing medicine and giving injections. Medical texts write proportions in the form

$$2 : 40 :: x : 100$$

instead of

$$\frac{2}{40} = \frac{x}{100}$$

The double colon :: is used in place of an equal ($=$) sign. With either notation, the solution is found by setting the product of the extremes equal to the product of the means and solving the equation for the unknown quantity.

EXAMPLE 4   Solve the proportion

$$2 \text{ ounces} : 40 \text{ grams} :: x \text{ ounces} : 100 \text{ grams}$$

Solution

means

$$2 : 40 :: x : 100$$

extremes

$$2 \cdot 100 = 40 \cdot x$$
$$\frac{200}{40} = \frac{40 \cdot x}{40}$$
$$5 = x$$

The solution is $x = 5$ ounces.

Exercises 6.4

Solve the following word problems using proportions.

1. A map maker uses a scale of 2 inches to represent 30 miles. How many miles are represented by 3 inches?

2. If gasoline sells for \$1.18 per gallon, what will 10 gallons cost?

3. If gasoline sells for \$1.18 per gallon, how many gallons can be bought with \$8.26?

4. If the odds on a horse are \$5 to win on a \$2 bet, how much can be won with a \$5 bet?

5. An investor thinks she should make $12 for every $100 she invests. How much would she expect to make on a $1500 investment?

6. If one dozen (12) eggs cost $1.09, what would three dozen eggs cost?

7. The price of a certain fabric is $1.75 per yard. How many yards can be bought with $35 (not including tax)?

8. An artist figures she can paint 3 portraits every two weeks. At this rate, how long will it take her to paint 18 portraits?

9. Two units of a certain gas weigh 175 grams. What is the weight of 5 units of this gas?

10. A baseball team bought 8 bats for $96. What would they pay for 10 bats?

11. A store owner expects to make a profit of $2 on an item that sells for $10. How much profit will he expect to make on a larger but similar item that sells for $60?

12. A saleswoman makes $8 for every $100 worth of the product she sells. What will she make if she sells $5000 worth of the product?

13. If property taxes are figured at $1.50 for every $100 in evaluation, what taxes will be paid on a home valued at $85,000?

14. A condominium owner pays property taxes of $2000 per year. If taxes are figured at a rate of $1.25 for every $100 in value, what is the value of his condominium?

15. Sales tax is figured at 6¢ for every $1.00 of merchandise purchased. What was the purchase price on an item that had a sales tax of $2.04?

16. An architect drew plans for a city park using a scale of $\frac{1}{4}$ inch to represent 25 feet. How many feet would 2 inches represent?

17. A building 14 stories high casts a shadow of 30 feet at a certain time of day. What is the length of the shadow of a 20-story building at the same time of day in the same city?

18. Two numbers are in the ratio of 4 to 3. The number 10 is in that same ratio to a fourth number. What is the fourth number?

19. A car is traveling at 45 miles per hour. Its speed is increased by 3 miles per hour every 2 seconds. By how much will its speed increase in 5 seconds? How fast will the car be traveling?

20. A truck is traveling at 55 miles per hour and the driver brakes to slow down 2 miles per hour every 3 seconds. How long will it take the truck to slow to a speed of 45 miles per hour?

21. A salesman figured he drove 560 miles every two weeks. How far would he drive in three months (12 weeks)?

22. Driving steadily, a woman made a trip of 200 miles in $4\frac{1}{2}$ hours. How long would she take to drive 500 miles at the same rate of speed?

23. If you can drive 286 miles in $5\frac{1}{2}$ hours, how long will it take you to drive 468 miles at the same rate of speed?

24. What will 21 gallons of gasoline cost if gasoline costs $1.25 per gallon?

25. If diesel fuel costs $1.10 per gallon, how much diesel fuel will $24.53 buy?

26. An electric fan makes 180 revolutions per minute. How many revolutions will the fan make if it runs for 24 hours?

27. An investor made $144 in one year on a $1000 investment. What would she have earned if her investment had been $4500?

28. A typist can type 8 pages of manuscript in 56 minutes. How long will this typist take to type 300 pages?

29. On a map, $1\frac{1}{2}$ inches represent 40 miles. How many inches represent 50 miles?

30. In the metric system, there are 2.54 centimeters in 1 inch. How many centimeters are there in 1 foot?

31. If 40 pounds of fertilizer are used on 2400 square feet of lawn, how many pounds of fertilizer are needed for a lawn of 5400 square feet?

32. An English teacher must read and grade 27 essays. If the teacher takes 20 minutes to read and grade 3 essays, how much time will he need to grade all 27 essays?

33. If 2 cups of flour are needed to make 12 biscuits, how much flour will be needed to make 9 of the same kind of biscuits?

34. If 2 cups of flour are needed to make 12 biscuits, how many of the same kind of biscuits can be made with 3 cups of flour?

35. There are one thousand grams in one kilogram. How many grams are there in four and seven tenths kilograms?

The following exercises are examples of proportions in medicine. Be sure to label your answers. (The abbreviations are from the metric system.)

36. 1 liter : 1000 mL :: x liter : 5000 mL

37. 1 kg : 1000 g :: x kg : 2700 g

38. 1 mg : 1000 mcg :: x mg : 4 mcg

39. 1 g : 1000 mg :: 0.5 g : x mg

40. 1 dram : 60 grains :: x dram : 90 grains

41. $\dfrac{1 \text{ ounce}}{8 \text{ drams}} = \dfrac{0.5 \text{ ounce}}{x}$

42. $\dfrac{1 \text{ dram}}{60 \text{ minims}} = \dfrac{x}{180 \text{ minims}}$

43. $\dfrac{1 \text{ ounce}}{480 \text{ minims}} = \dfrac{4 \text{ ounces}}{x}$

44. $\dfrac{1 \text{ g}}{15 \text{ grains}} = \dfrac{x}{60 \text{ grains}}$

45. $\dfrac{1 \text{ grain}}{60 \text{ mg}} = \dfrac{x}{500 \text{ mg}}$

46. $\dfrac{1 \text{ ounce}}{30 \text{ g}} = \dfrac{2 \text{ ounces}}{x}$

47. $\dfrac{1 \text{ mL}}{15 \text{ minims}} = \dfrac{5 \text{ mL}}{x}$

48. $\dfrac{1 \text{ tsp}}{4 \text{ mL}} = \dfrac{x}{12 \text{ mL}}$

49. $\dfrac{1 \text{ ounce}}{30 \text{ mL}} = \dfrac{x}{15 \text{ mL}}$

50. $\dfrac{1 \text{ pint}}{500 \text{ mL}} = \dfrac{1.5 \text{ pints}}{x}$

51. You want to treat your lawn with a fungicide, insecticide, fertilizer mix. The local nursery sells bags of this mix for \$14 per bag. Each bag contains 16 pounds of mix, and the recommended coverage for one bag is 2000 square feet. (a) If your lawn consists of two rectangular shapes, one 30 feet by 100 feet and the other 50 feet by 80 feet, how many pounds of fertilizer mix would cover your lawn? (b) How many bags do you need to buy? (c) How much would you pay (not including tax)?

52. One bag of Weed & Fertilizer contains 18 pounds of fertilizer and weed treatment with a recommended coverage of 5000 square feet. (a) If your lawn is in the shape of a rectangle 150 feet by 220 feet, how many pounds of Weed & Fertilizer do you need to cover your lawn? (b) If the cost of one bag is \$12, what will you pay for this fertilizer (not including tax)?

53. The local paint store recommends that you use one gallon of paint to cover 400 square feet of properly prepared wall space in your home. You want to paint the walls of three rooms, one room measuring 10 feet by 12 feet, another 12 feet by 13 feet, and the third 16 feet by 15 feet. (a) If each room has 8-foot ceilings, how many gallons of paint do you need to buy? (b) Will you have any paint left over? (c) How much? (Note: For this exercise, ignore window space and door area.)

54. One bag of dichondra lawn food contains 20 pounds of fertilizer and its recommended coverage is 4000 square feet. (a) If you want to cover a lawn that is in the shape of a rectangle 120 feet by 160 feet, how many pounds of lawn food do you need? (b) How many bags of lawn food would you need to buy?

Check Your
Number Sense

In each of the following exercises, use mental calculations and your judgment to choose the best answer. After you have checked your answers in the back of the text, work out any problems that you missed to help reinforce your understanding.

55. A computer microchip manufacturer expects that 3 out of every 100 microchips it produces will be defective. If 5000 microchips are produced in one production run, about how many would be expected to be defective in that run?

(a) 15   (b) 50   (c) 150   (d) 300

56. A professional baseball player has hit 12 home runs in the first 55 games of the season. At this rate, about how many home runs would you expect him to hit over a complete season of 162 games?

(a) 15   (b) 25   (c) 35   (d) 45

57. A food distributor is mailing a survey to shoppers to study their grocery purchasing habits. From experience, the company expects about 6 out of every 100 addressees to return the survey. If the company would like at least 1000 responses, what is the minimum number of surveys that they should mail?

(a) 6000   (b) 10,000   (c) 20,000   (d) 60,000

58. The world's tropical rain forests are being destroyed at a rate of 96,000 acres per day. About how many acres of rain forest are being destroyed every hour?

(a) 4000   (b) 8000   (c) 500,000   (d) 1,000,000

Collaborative
Learning Exercise

With the class separated into teams of two to four, each team is to analyze the following two exercises and decide on the best of the given choices for the answer. The team leader is to discuss the team's choice and the reason for this choice with the class.

59. Marine biologists at a killer whale feeding ground photograph and identify 35 whales during a research trip. The next year they return to the same location and identify 40 whales, 8 of which were ones they had identified the year before. About how many whales would you estimate are in the group that these biologists are studying?

(a) 40   (b) 75   (c) 100   (d) 150

60. Forest rangers are concerned about the population of deer in a certain region. To get a good estimate of the deer population, they find, tranquilize, and tag 50 deer. One month later, they again locate 50 deer and of these 50, 5 are deer that were previously tagged. On the basis of these procedures and results, what do they estimate to be the deer population of that region?

(a) 100   (b) 150   (c) 500   (d) 1000

The Recycle Bin   (from Sections 4.2–4.4)

1. Consider the product $4\frac{1}{2} \cdot 1\frac{1}{6} \cdot 1\frac{1}{5}$. Do you think that this product is (a) close to 6, (b) close to 20, or (c) close to 100. Find the product.

2. Consider the sum $6\frac{1}{3} + 5\frac{3}{10} + 8\frac{1}{15}$. Do you think that this sum is (a) close to 10, (b) close to 20, or (c) close to 100. Find the sum.

3. Consider the quotient $6\frac{2}{3} \div 1\frac{1}{3}$. Do you think that this quotient is (a) more than 6 or (b) less than 6. Find the quotient.

4. Use the rules for order of operations to find the value of the following expression: $\frac{1}{2} \cdot 3\frac{1}{2} + 5\frac{1}{10} \cdot \frac{10}{17} - 3\frac{5}{12}$.

Summary: Chapter 6

Key Terms and Ideas

A **ratio** is a comparison of two quantities by division. The ratio of a to b can be written as

$$\frac{a}{b} \qquad \text{or} \qquad a:b \qquad \text{or} \qquad a \textbf{ to } b$$

Ratios have the following characteristics:

1. Ratios can be reduced, just as fractions can be reduced.

2. Whenever the units of the numbers in a ratio are the same, then the ratio has no units. We say that the ratio is an **abstract number.**

3. When the numbers in a ratio have different units, then the numbers must be labeled to clarify what is being compared. Such a ratio is called a **rate.**

A **proportion** is a statement that two ratios are equal. In symbols,

$$\frac{a}{b} = \frac{c}{d} \qquad \text{is a proportion.}$$

A **variable** is a symbol (generally a letter of the alphabet) that is used to represent an unknown number or any one of several numbers. The set of possible values for a variable is called its **replacement set.**

Rules and Properties

In a true proportion the product of the extremes is equal to the product of the means. In symbols,

$$\frac{a}{b} = \frac{c}{d} \quad \text{if and only if} \quad a \cdot d = b \cdot c,$$

where $b \neq 0$ and $d \neq 0$.

A proportion has four **terms:**

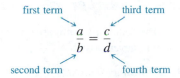

The first and fourth terms (a and d) are called the **extremes.**

The second and third terms (b and c) are called the **means.**

Procedures

To Find the Price per Unit:

1. Set up a ratio (usually in fraction form) of price to units.

2. Divide the price by the number of units.

To Find the Unknown Term in a Proportion:

1. In the proportion, the unknown term is represented with a variable (some letter such as x, y, w, A, B, and so forth).

2. Write an equation that sets the product of the extremes equal to the product of the means.

3. Divide both sides of the equation by the number multiplying the variable. (This number is called the **coefficient** of the variable.)

The resulting equation will have a coefficient of 1 for the variable and will give the missing value for the unknown term in the proportion.

**To Solve a Word Problem
by Using Proportions:**

1. Identify the unknown quantity and use a variable to represent this quantity.

2. Set up a proportion in which the units are compared as in Pattern A or Pattern B (see page 302).

3. Solve the proportion.

Review Questions: Chapter 6

Write the following comparisons as ratios. Reduce to lowest terms and use common units whenever possible.

1. 2 dimes to 5 nickels

2. 20 inches to 1 yard

3. 10.16 centimeters to 4 inches

4. 51 miles to 3 gallons of gas

5. 18 hours to 2 days

6. 26 girls to 39 boys

Find the unit prices and tell which is the better buy.

7. Packaged cheese: 6 oz at 79¢
"Deli" cheese: 1 lb (16 oz) at $1.99

8. 1 qt (32 fl oz) milk at 53¢
1 gal (128 fl oz) milk at $2.02

9. 1 lb 2 oz loaf of bread at 84¢

$1\frac{1}{2}$ lb loaf of bread at $1.15

10. 6 oz bologna at $1.29
8 oz bologna at $1.49

Name the means and the extremes for each proportion.

11. $\dfrac{4}{5} = \dfrac{16}{20}$

12. $\dfrac{\frac{1}{6}}{3} = \dfrac{\frac{1}{9}}{2}$

13. $3 : 7 = 0.75 : 1.75$

Determine whether the following proportions are true or false.

14. $\dfrac{3}{5} = \dfrac{9}{15}$

15. $\dfrac{15}{20} = \dfrac{18}{24}$

16. $\dfrac{6.5}{14} = \dfrac{8}{16}$

Solve the following proportions. Reduce all fractions.

17. $\dfrac{10}{12} = \dfrac{x}{6}$

18. $\dfrac{1.7}{5.1} = \dfrac{100}{y}$

19. $\dfrac{7\frac{1}{2}}{3\frac{1}{3}} = \dfrac{w}{2\frac{1}{4}}$

20. $a : 7 = \dfrac{1}{3} : 5$

21. A motorcycle averages 42.8 miles per gallon of gas. How many miles can the motorcycle travel on 3.5 gallons of gas?

22. Find the difference between sixty-four and five hundred thirty-six ten-thousandths and fifty-nine and three thousand six hundred eighty-one ten-thousandths.

23. On a certain map, 1 inch represents 35.5 miles. What distance is represented by 4.7 inches?

24. A part-time clerk earned $420 the first month on a new job. This was 0.6 of what he had anticipated. How much had he anticipated making?

25. If a machine produces 5000 safety pins in 2 hours, how many will it produce in two 8-hour days?

26. An automobile was slowing down at the rate of 5 miles per hour (mph) for every 3 seconds. If the automobile was going 65 mph when it began to slow down, how fast was it going at the end of 12 seconds?

27. An architect draws house plans using a scale of $\dfrac{3}{4}$ inch to represent 10 feet. How many feet are represented by 2 inches?

28. If you can drive 200 miles in $4\frac{1}{2}$ hours, how far could you drive (at the same rate) in 6 hours?

29. The ratio of kilometers to miles is about 8 to 5. Find the rate in kilometers per hour that is equivalent to 40 miles per hour.

30. In a certain hospital, 55 out of every 100 children born are boys. In a year when 1040 children are born in the hospital, how many are girls?

Test: Chapter 6

1. The ratio 7 to 6 can be expressed as the fraction _____.

2. In the proportion $\dfrac{3}{4} = \dfrac{75}{100}$, 3 and 100 are called the _____.

3. Is the proportion $\dfrac{4}{6} = \dfrac{9}{14}$ true or false? Give a reason for your answer.

Write the following comparisons as ratios reduced to lowest terms. Use common units whenever possible.

4. 3 weeks to 35 days

5. 6 nickels to 3 quarters

6. 220 miles to 4 hours

7. Find the unit prices and tell which is the better buy for a carton of cottage cheese:

8 oz at 54¢   or   32 oz at $2.05

Solve the following proportions.

8. $\dfrac{9}{17} = \dfrac{x}{51}$

9. $\dfrac{3}{5} = \dfrac{10}{y}$

10. $\dfrac{50}{x} = \dfrac{0.5}{0.75}$

11. $\dfrac{2.25}{y} = \dfrac{1.5}{13}$

12. $\dfrac{\frac{3}{8}}{x} = \dfrac{9}{20}$

13. $\dfrac{\frac{1}{3}}{x} = \dfrac{5}{\frac{1}{6}}$

14. $9 : 4 :: y : 2$

15. 9 is to 4 as x is to 16

16. How far (to the nearest tenth of a mile) can you drive on 5 gallons of gas if you can drive 120 miles on 3.5 gallons of gas?

17. On a certain map, 2 inches represents 15 miles. How far apart are two towns that are $3\dfrac{1}{5}$ inches apart on the map?

18. If you can buy 4 tires for $236, what will be the cost of 5 of the same type of tires?

19. If you drive 165 miles in $3\dfrac{2}{3}$ hours, what is your average speed in miles per hour?

20. The ratio of men to women in a certain physical education class is 3 to 4. How many men are in the class if there are 36 women in the class?

21. There are one thousand grams in one kilogram. How many grams are there in four and seven-tenths kilograms?

22. A house painter knows that he can paint three houses every five weeks. At this rate, how long will it take him to paint 15 houses?

23. If 2 units of water weigh 124.8 pounds, what is the weight of 5 of these units of water?

24. A manufacturing company expects to make a profit of $3 on a product that it sells for $8. How much profit does the company expect to make on a similar product that it sells for $20?

1. Write five thousand seven hundred forty in the form of a decimal number.

2. Write the decimal number 3.075 in words.

3. Find the prime factorization of 4510.

Evaluate each of the following expressions.

4. $3 \cdot 5^2 + 20 \div 5$

5. $(80 + 10) \div 5 \cdot 2 - 14 + 2^2$

6. Find the LCM for 30, 35, and 42.

7. Find the average of 50, 54, 60, and 76.

Perform the indicated operations and reduce all answers.

8. $1 - \dfrac{15}{16}$

9. $4\dfrac{5}{8} \div 2$

10. $\left(1\dfrac{2}{3}\right)\left(2\dfrac{4}{5}\right)\left(3\dfrac{1}{7}\right)$

11. $\dfrac{\dfrac{1}{5} + \dfrac{1}{10}}{\dfrac{1}{3} - \dfrac{1}{4}}$

12. $\begin{array}{r} 6 \\ -2\dfrac{3}{4} \\ \hline \end{array}$

13. $\begin{array}{r} 15\dfrac{1}{2} \\ +22\dfrac{3}{4} \\ \hline \end{array}$

14. Find $\dfrac{3}{4}$ of 76.

15. Round off 0.03572 to the nearest thousandth.

16. Write each of the numbers in words.

   (a) 600.006

   (b) 0.606

Perform the indicated operations.

17. $\begin{array}{r} 847.8 \\ 436.92 \\ +354.718 \\ \hline \end{array}$

18. $\begin{array}{r} 32.007 \\ -15.835 \\ \hline \end{array}$

19. $\begin{array}{r} 566.3 \\ \times \quad 7.2 \\ \hline \end{array}$

20. Divide $1.028 \div 1.3$ and find the quotient to the nearest hundredth.

21. Write each of the following numbers in scientific notation.

   (a) 93,500,000

   (b) 0.000000482

Solve each of the following proportions.

22. $\dfrac{5}{7} = \dfrac{x}{3\frac{1}{2}}$

23. $\dfrac{0.5}{1.6} = \dfrac{0.1}{A}$

24. (a) If you drove your car 250 miles in 5.5 hours, what was your average speed (to the nearest tenth of a mile per hour)? (b) How far could you drive in 7 hours at this average speed?

25. In December, Mario was told he would receive a bonus of $4000. However, $\dfrac{1}{10}$ was withheld for state taxes, $\dfrac{1}{5}$ was withheld for federal taxes, and $200 was withheld for social security tax. How much cash did Mario receive after these deductions?

Percent (Calculators Recommended)

Mathematics at Work!

Percents are often used to measure and report the attitudes and characteristics of the general population.

The concept of percent is an important part of our daily lives. Income tax is based on a percent of income, property tax is based on a percent of the value of a home, a professional athlete's skills are measured as percents (batting percentage, free-throw percentage, pass completion percentage), profit is calculated as a percent of investment, commissions are calculated as percents of sales, and tips for service are figured as a percent of the bill.

Even reading a newspaper or magazine with understanding requires some knowledge of percents. Below are excerpts from *The Los Angeles Times* during one week in March, 1994.

In reference to a meeting of top business leaders:

> **. . . 35% admitted eating at a fast food restaurant during the previous week.**

In reference to the reformed KGB:

> **. . . a 30% cut in staff.**

In reference to the economy in India:

> **. . . about 40% of revenue comes from import fees.**

Discussing results of new state tests in California:

> **Only 7% of fourth graders showed a substantial grasp of mathematical concepts; . . . and 30% of the high school sophomores demonstrated only a "superficial understanding" of what they read.**

What to Expect in Chapter 7

Chapter 7 deals with ideas and applications related to percent, one of the most useful and most important mathematical concepts in our daily lives. Section 7.1 introduces percent by explaining that percent means hundredths (or per hundred) and therefore percents can be represented in fraction form with denominator 100. Percent of profit and the relationship between percents and decimals are also included in Section 7.1. Section 7.2 discusses the relationships among percents, fractions, and decimals. Sections 7.3 and 7.4 lay the groundwork for applications with percents in Sections 7.6 and 7.7. In Section 7.5 estimation with percents is presented to reinforce basic concepts and to promote a general understanding of percent. In Sections 7.6 and 7.7, we emphasize Polya's four-step process for solving problems:

1. Understanding the problem.
2. Devise a plan.
3. Carry out the plan.
4. Look back over the results.

7.1   Decimals and Percents

OBJECTIVES

1. Understand that **percent means hundredths.**
2. Relate percent to fractions with denominator 100.
3. Compare profit to investment and write the ratio as a percent.
4. Be able to change percents to decimals and decimals to percents.

Understanding Percent

The word **percent** comes from the Latin *per centum,* meaning **per hundred.** So, **percent means hundredths,** or the **ratio of a number to 100.** The symbol % is called the **percent sign.** As we shall see, this sign has the same meaning as the fraction $\frac{1}{100}$. For example,

$$\frac{35}{100} = 35\left(\frac{1}{100}\right) = 35\% \quad \text{and} \quad \frac{60}{100} = 60\left(\frac{1}{100}\right) = 60\%$$

In Figure 7.1 the large square is partitioned into 100 small squares, and each small square represents $\frac{1}{100}$, or 1%, of the large square. Thus, in Figure 7.1, the portion of the large square that is shaded is $\frac{27}{100} = 27\left(\frac{1}{100}\right) = 27\%$.

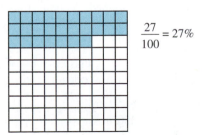

$$\frac{27}{100} = 27\%$$

Figure 7.1

If a fraction has a denominator of 100, then (with no change in the numerator) the numerator can be read as a percent by dropping the denominator and adding the % sign.

EXAMPLE 1

Each fraction is changed to a percent.

(a) $\dfrac{20}{100} = 20 \cdot \dfrac{1}{100} = 20\%$ Remember that percent means hundredths, and the % sign indicates hundredths, or $\dfrac{1}{100}$.

(b) $\dfrac{85}{100} = 85\%$

(c) $\dfrac{6.4}{100} = 6.4\%$ Note that the decimal point is not moved. That is, the numerator is unchanged.

(d) $\dfrac{3\frac{1}{2}}{100} = 3\frac{1}{2}\%$ Note that since the denominator is 100, the numerator is unchanged even though it contains a fraction.

(e) $\dfrac{240}{100} = 240\%$ If the numerator is more than 100, then the number is greater than 1 and the percent is more than 100%.

Percent of Profit

Percent of profit is the ratio of money made to money invested. Generally, two investments do not involve the same amount of money and therefore, the comparative success of each investment cannot be based on the **amount** of profit. In comparing investments, the investment with the **greater percent** of profit is considered the better investment.

The use of percent gives an effective method of comparison because each ratio of money made (profit) to money invested has the same denominator (100).

EXAMPLE 2

Calculate the percent of profit for both (a) and (b) and tell which is the better investment.

(a) $150 made as profit by investing $300

(b) $200 made as profit by investing $500

Solution

In each case, find the ratio of dollars profit to dollars invested and reduce the ratio so that it has a denominator of 100. Do not reduce to lowest terms.

(a) $\dfrac{\$150 \text{ profit}}{\$300 \text{ invested}} = \dfrac{3 \cdot 50}{3 \cdot 100} = \dfrac{50}{100} = 50\%$

(b) $\dfrac{\$200 \text{ profit}}{\$500 \text{ invested}} = \dfrac{5 \cdot 40}{5 \cdot 100} = \dfrac{40}{100} = 40\%$

Investment (a) is better than investment (b) because 50% is greater than 40%. Obviously, $200 profit is more than $150 profit, but the $500 risked as an investment in (b) is considerably more than the $300 risked in (a).

EXAMPLE 3

Of (a) and (b), which is the better investment?

(a) $40 made as a profit by investing $200.

(b) $75 made as a profit by investing $300.

Solution

Write each ratio as hundredths and compare the percents.

(a) $\dfrac{\$40 \text{ profit}}{\$200 \text{ invested}} = \dfrac{2 \cdot 20}{2 \cdot 100} = \dfrac{20}{100} = 20\%$

(b) $\dfrac{\$75 \text{ profit}}{\$300 \text{ invested}} = \dfrac{3 \cdot 25}{3 \cdot 100} = \dfrac{25}{100} = 25\%$

Investment (b) is better because 25% is more than 20%.

Decimals and Percent

One way to change a decimal to a percent is to first change the decimal to fraction form with denominator 100 and then change the fraction to percent form. For example,

Decimal Form		Fraction Form		Percent Form
0.47	=	$\dfrac{47}{100}$	=	47%
0.93	=	$\dfrac{93}{100}$	=	93%

In both of these illustrations the decimals are in hundredths and so changing to fraction form is relatively easy. If the decimal is not in hundredths, we can multiply by 1 in the form of $\dfrac{100}{100}$ as follows:

$$0.325 = 0.325 \cdot \dfrac{100}{100} = 0.325 \,(100) \cdot \dfrac{1}{100} = 32.5 \cdot \dfrac{1}{100} = 32.5\%$$

Now, looking at the fact that $0.325 = 32.5\%$, we can see that the decimal point in 0.325 was moved two places to the right (giving 32.5) and the % sign was written (giving the result 32.5%). This leads to the following general rule.

To Change a Decimal to a Percent:

Step 1: Move the decimal point two places to the right.

Step 2: Write the % sign.

(These two steps have the effect of multiplying by 100 then dividing by 100.)

The following relationships between decimals and percents provide helpful guidelines.

A decimal number that is

(a) less than 0.01 is less than 1%.

(b) between 0.01 and 0.10 is between 1% and 10%.

(c) between 0.10 and 1.00 is between 10% and 100%.

(d) more than 1 is more than 100%.

EXAMPLE 4

Change each decimal to an equivalent percent.

(a) 0.253   (b) 0.905   (c) 2.65   (d) 0.7   (e) 0.002

Solution

(a) 0.253 = 25.3%

       decimal point moved two places to the right   % sign added

(b) 0.905 = 90.5%

(c) 2.65 = 265%   Note that this is more than 100%.

(d) 0.7 = 70%   The decimal point is not written here. We could write 70.% or 70.0%.

(e) 0.002 = 0.2%   Note that this is less than 1%.

       decimal point moved two places to the right   % sign added

To change percents to decimals, we reverse the procedure for changing decimals to percents. For example,

$$39\% = 39\left(\frac{1}{100}\right) = \frac{39}{100} = 0.39$$ Remember, the decimal point is moved two places to the left when you divide by 100.

To Change a Percent to a Decimal:

Step 1: Move the decimal point two places to the left.

Step 2: Delete the % sign.

EXAMPLE 5 Change each percent to an equivalent decimal.

(a) 76%   (b) 18.5%   (c) 100%   (d) 1.3%   (e) 0.25%

Solution

(a) 76% = 0.76   ← % sign deleted

   ↑   ↖
   understood   decimal point moved
   decimal point   two places left

(b) 18.5% = 0.185

(c) 100% = 1.00 = 1

(d) 1.3% = 0.013

(e) 0.25% = 0.0025   Note that 0.25% is less than 1% and its decimal equivalent is less than 0.01.

CLASSROOM PRACTICE

Change from decimals to percents.

1. 0.34   **2.** 1.75

Change from percents to decimals.

3. 200%   **4.** 1.5%

ANSWERS: **1.** 34%   **2.** 175%   **3.** 2.00 (or 2)   **4.** 0.015

Exercises 7.1

What percent of each square is shaded?

1.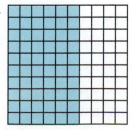

2.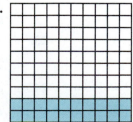

3.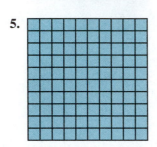

4.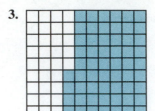

5.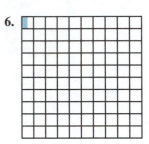

6.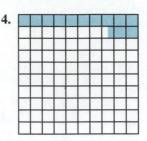

Change the following fractions to percents.

7. $\dfrac{20}{100}$ **8.** $\dfrac{9}{100}$ **9.** $\dfrac{15}{100}$

10. $\dfrac{62}{100}$ **11.** $\dfrac{53}{100}$ **12.** $\dfrac{68}{100}$

13. $\dfrac{125}{100}$ **14.** $\dfrac{200}{100}$ **15.** $\dfrac{336}{100}$

16. $\dfrac{13.4}{100}$ **17.** $\dfrac{0.48}{100}$ **18.** $\dfrac{0.5}{100}$

19. $\dfrac{2.14}{100}$ **20.** $\dfrac{1.62}{100}$

In Exercises 21–24, write the ratio of profit to investment as hundredths and then as percents. Compare the percents and tell which investment is better, (a) or (b).

21. (a) a profit of \$36 on a \$200 investment

    (b) a profit of \$51 on a \$300 investment

22. (a) a profit of \$40 on a \$400 investment

    (b) a profit of \$60 on a \$600 investment

23. (a) a profit of \$150 on a \$1500 investment

    (b) a profit of \$200 on a \$2000 investment

24. (a) a profit of \$300 on a \$2000 investment

    (b) a profit of \$360 on a \$3000 investment

Change the following decimals to percents.

25. 0.02	**26.** 0.09	**27.** 0.1
28. 0.7	**29.** 0.36	**30.** 0.52
31. 0.40	**32.** 0.65	**33.** 0.025
34. 0.035	**35.** 0.055	**36.** 0.004
37. 1.10	**38.** 1.75	**39.** 2
40. 2.3		

Change the following percents to decimals.

41. 2%	**42.** 7%	**43.** 18%
44. 20%	**45.** 30%	**46.** 80%
47. 0.26%	**48.** 0.52%	**49.** 125%
50. 120%	**51.** 232%	**52.** 215%
53. 17.3%	**54.** 10.1%	**55.** 13.2%
56. 6.5%		

57. Suppose that sales tax is figured at 7.25%. Change 7.25% to a decimal.

58. The interest rate on a loan is 6.4%. Change 6.4% to a decimal.

59. The sales commission for the clerk in a retail store is figured at 8.5%. Change 8.5% to a decimal.

60. In calculating his sales commission, Mr. Howard multiplies by the decimal 0.12. Change 0.12 to a percent.

61. To calculate what your maximum house payment should be, a banker multiplied your income by 0.28. Change 0.28 to a percent.

62. The discount you earn by paying cash is found by multiplying the amount of your purchase by 0.02. Change 0.02 to a percent.

63. The rate of profit based on sales price is 45%. Change 45% to a decimal.

64. The discount during a special sale on dresses is 30%. Change 30% to a decimal.

65. Suppose the state license fee is figured by multiplying the cost of your car by 0.065. Change 0.065 to a percent.

66. With 22.2% of its residents without health insurance coverage, New Mexico recently ranked as the state with the highest percentage of people that are health-care uninsured. Write 22.2% as a decimal.

67. Per capita property tax collections in California increased by 133% from 1980 to 1991. Write 133% as a decimal.

68. According to the Aluminum Association, the percentage of aluminum cans recycled in 1990 was 63.6% and this percentage dropped to 62.4% in 1991. Write 63.6% and 62.4% as decimals.

 The Recycle Bin (from Section 6.3)

Solve for the variable in each of the following proportions.

1. $\dfrac{R}{100} = \dfrac{4}{5}$

2. $\dfrac{R}{100} = \dfrac{2}{3}$

3. $\dfrac{23}{100} = \dfrac{A}{500}$

4. $\dfrac{75}{100} = \dfrac{A}{60}$

5. $\dfrac{90}{100} = \dfrac{36}{B}$

6. $\dfrac{10}{100} = \dfrac{7.4}{B}$

7.2 Fractions and Percents

OBJECTIVES

1. Be able to change fractions and mixed numbers to percents.
2. Be able to change percents to mixed numbers and fractions.

Changing Fractions and Mixed Numbers to Percents

As we discussed in Section 7.1, if a fraction has denominator 100, we can change it to a percent by writing the numerator and adding the % sign. If the denominator is a factor of 100 (2, 4, 5, 10, 20, 25, or 50), we can write the fraction in an equivalent form with denominator 100 and then change it to a percent. For example,

$$\frac{1}{4} = \frac{1}{4} \cdot \frac{25}{25} = \frac{25}{100} = 25\%$$

$$\frac{3}{5} = \frac{3}{5} \cdot \frac{20}{20} = \frac{60}{100} = 60\%$$

$$\frac{17}{50} = \frac{17}{50} \cdot \frac{2}{2} = \frac{34}{100} = 34\%$$

However, most fractions do not have denominators that are factors of 100. A more general approach (easily applied with calculators) is to change the fraction to decimal form by dividing the numerator by the denominator, and then change the decimal to a percent.

To Change a Fraction to a Percent:

Step 1: Change the fraction to a decimal.
(Divide the numerator by the denominator.)

Step 2: Change the decimal to a percent.

EXAMPLE 1   Change $\frac{5}{8}$ to a percent.

Solution

(a) Divide:

$$\begin{array}{r} .625 \\ 8\overline{)5.000} \\ \underline{4\,8} \\ 20 \\ \underline{16} \\ 40 \\ \underline{40} \\ 0 \end{array}$$

(b) Change .625 to a percent: $\frac{5}{8} = 0.625 = 62.5\%$

EXAMPLE 2   Change $\frac{18}{20}$ to a percent.

Solution

(a) Divide:

$$\begin{array}{r} .9 \\ 20\overline{)18.0} \\ \underline{18\,0} \\ 0 \end{array}$$

(b) $\frac{18}{20} = 0.9 = 90\%$

EXAMPLE 3   Change $2\frac{1}{4}$ to a percent.

Solution

(a) $2\frac{1}{4} = \frac{9}{4}$

$$
\begin{array}{r}
2.25 \\
4\,)\overline{9.00} \\
\underline{8} \\
1\,0 \\
\underline{8} \\
20 \\
\underline{20} \\
0
\end{array}
$$

(b) $2\frac{1}{4} = \frac{9}{4} = 2.25 = 225\%$

Agreement for Rounding Off Decimal Quotients

In this text:

1. Decimal quotients that are exact with four decimal places (or less) will be written with four decimal places (or less).

2. Decimal quotients that are not exact will be divided to the fourth place, and the quotient will be rounded off to the third place (thousandths).

Using a Calculator

Most calculators will give answers accurate to 8 or 9 decimal places. So, if you use a calculator to perform the long division when changing a fraction to a decimal, be sure to follow statement 2 in the agreement in the preceding box.

EXAMPLE 4   Using a calculator, $\frac{1}{3} = 0.3333333$.

(a) Rounding off the decimal quotient to the third decimal place:

$\frac{1}{3} = 0.333 = 33.3\%$ The answer is rounded off and is not exact.

(b) Without a calculator, we can divide and use fractions:

$$3\overline{)1.00}\;\;.33\tfrac{1}{3}$$
$$\underline{9}$$
$$10$$
$$\underline{9}$$
$$1$$

$$\frac{1}{3} = 0.33\tfrac{1}{3} = 33\tfrac{1}{3}\% \quad\text{or}\quad 33.\overline{3}\%$$

$33\tfrac{1}{3}\%$ and $33.\overline{3}\%$ are exact, and 33.3% is rounded off.

Any of these answers is acceptable, but be aware that 33.3% is a rounded-off answer.

EXAMPLE 5

During the years from 1921 to 1981, the New York Yankees baseball team played in 33 World Series Championships and won 22 of them. What percent of these championships did the Yankees win?

Solution

(a) The percent won can be found by using a calculator and changing the fraction $\dfrac{22}{33}$ to decimal form and then changing the decimal to a percent. Using a calculator,

$$\frac{22}{33} = 0.6666666 \approx 0.667 = 66.7\%$$

(b) Without a calculator, we can divide and use fractions:

$$3\overline{)2.00}\;\;.66\tfrac{2}{3}$$
$$\underline{18}$$
$$20$$
$$\underline{18}$$
$$2$$

$$\frac{22}{33} = \frac{11\cdot 2}{11\cdot 3} = \frac{2}{3} = 0.66\tfrac{2}{3} = 66\tfrac{2}{3}\% \quad\text{or}\quad 66.\overline{6}\%$$

The Yankees won $66\tfrac{2}{3}\%$ of these championships.

(**Note:** Any one of the answers $66\tfrac{2}{3}\%$, $66.\overline{6}\%$, or 66.7% is acceptable.)

EXAMPLE 6   Using a calculator, $\dfrac{1}{7} = 0.1428571$.

(a) Rounding off the decimal quotient to the third decimal place:

$$\frac{1}{7} = 0.143 = 14.3\%$$

(b) Using long division, we can write the answer with a fraction or continue to divide and round off the decimal answer. The choice is yours.

$$
\begin{array}{r}
0.14\frac{2}{7} = 14\frac{2}{7}\% \\
7\overline{)1.00} \\
\underline{7} \\
30 \\
\underline{28} \\
2
\end{array}
$$

or:
$$
\begin{array}{r}
0.1428 = 14.3\% \\
7\overline{)1.0000} \\
\underline{7} \\
30 \\
\underline{28} \\
20 \\
\underline{14} \\
60 \\
\underline{56} \\
4
\end{array}
$$
by rounding off the decimal quotient to the third decimal place

Changing Percents to Fractions and Mixed Numbers

To Change a Percent to a Fraction or Mixed Number:

Step 1: Write the percent as a fraction with denominator 100 and drop the % sign.

Step 2: Reduce the fraction.

EXAMPLE 7   Change each percent to an equivalent fraction or mixed number in reduced form.

(a) 60%   (b) $7\dfrac{1}{4}\%$ (c) 130%

Solution

(a) $60\% = \dfrac{60}{100} = \dfrac{3 \cdot 20}{5 \cdot 20} = \dfrac{3}{5}$

(b) $7\dfrac{1}{4}\% = \dfrac{7\dfrac{1}{4}}{100} = \dfrac{\dfrac{29}{4}}{100} = \dfrac{29}{4} \cdot \dfrac{1}{100} = \dfrac{29}{400}$

(c) $130\% = \dfrac{130}{100} = \dfrac{13 \cdot 10}{10 \cdot 10} = \dfrac{13}{10} = 1\dfrac{3}{10}$

A Common Misunderstanding

The fractions $\dfrac{1}{4}$ and $\dfrac{1}{2}$ are often confused with the percents $\dfrac{1}{4}\%$ and $\dfrac{1}{2}\%$. The differences can be clarified by using decimals.

Percent	Decimal	Fraction
$\dfrac{1}{4}\%$ (or 0.25%)	0.0025	$\dfrac{1}{400}$
$\dfrac{1}{2}\%$ (or 0.5%)	0.005	$\dfrac{1}{200}$
25%	0.25	$\dfrac{1}{4}$
50%	0.50	$\dfrac{1}{2}$

Thus,

$$\dfrac{1}{4} = 0.25 \qquad \text{and} \qquad \dfrac{1}{4}\% = 0.0025$$

$$0.25 \neq 0.0025$$

Similarly,

$$\dfrac{1}{2} = 0.50 \qquad \text{and} \qquad \dfrac{1}{2}\% = 0.005$$

$$0.50 \neq 0.005$$

You can think of $\dfrac{1}{4}$ as being one-fourth of a dollar (a quarter) and $\dfrac{1}{4}\%$ as being one-fourth of a penny. $\dfrac{1}{2}$ can be thought of as one-half of a dollar and $\dfrac{1}{2}\%$ as one-half of a penny.

Some percents are so common that their decimal and fraction equivalents should be memorized. Their fractional values are particularly easy to work with, and many times calculations involving these fractions can be done mentally.

Common Percent—Decimal—Fraction Equivalents

$1\% = 0.01 = \dfrac{1}{100}$ $33\dfrac{1}{3}\% = 0.33\dfrac{1}{3} = \dfrac{1}{3}$ $12\dfrac{1}{2}\% = 0.125 = \dfrac{1}{8}$

$25\% = 0.25 = \dfrac{1}{4}$ $66\dfrac{2}{3}\% = 0.66\dfrac{2}{3} = \dfrac{2}{3}$ $37\dfrac{1}{2}\% = 0.375 = \dfrac{3}{8}$

$50\% = 0.50 = \dfrac{1}{2}$ $62\dfrac{1}{2}\% = 0.625 = \dfrac{5}{8}$

$75\% = 0.75 = \dfrac{3}{4}$ $87\dfrac{1}{2}\% = 0.875 = \dfrac{7}{8}$

$100\% = 1.00 = 1$

CLASSROOM PRACTICE

Change to percents.

1. $\dfrac{3}{20}$ **2.** $\dfrac{3}{8}$ **3.** $\dfrac{2}{3}$

Change to fractions.

4. 35%   **5.** 40%   **6.** $87\dfrac{1}{2}\%$

ANSWERS: **1.** 15%   **2.** 37.5%   **3.** $66\dfrac{2}{3}\%$ **4.** $\dfrac{7}{20}$ **5.** $\dfrac{2}{5}$ **6.** $\dfrac{7}{8}$

Exercises 7.2

Change the following fractions and mixed numbers to percents.

1. $\dfrac{3}{100}$ **2.** $\dfrac{16}{100}$ **3.** $\dfrac{7}{100}$ **4.** $\dfrac{29}{100}$ **5.** $\dfrac{1}{2}$ **6.** $\dfrac{3}{4}$

7. $\dfrac{1}{4}$ **8.** $\dfrac{1}{20}$ **9.** $\dfrac{11}{20}$ **10.** $\dfrac{7}{10}$ **11.** $\dfrac{3}{10}$ **12.** $\dfrac{6}{8}$

13. $\frac{1}{5}$ 14. $\frac{2}{5}$ 15. $\frac{4}{5}$ 16. $\frac{1}{50}$ 17. $\frac{13}{50}$ 18. $\frac{1}{25}$

19. $\frac{12}{25}$ 20. $\frac{24}{25}$ 21. $\frac{1}{8}$ 22. $\frac{5}{8}$ 23. $\frac{7}{8}$ 24. $\frac{1}{9}$

25. $\frac{5}{9}$ 26. $\frac{2}{7}$ 27. $\frac{3}{7}$ 28. $\frac{5}{6}$ 29. $\frac{7}{11}$ 30. $\frac{5}{11}$

31. $1\frac{1}{14}$ 32. $1\frac{1}{6}$ 33. $1\frac{1}{20}$ 34. $1\frac{1}{4}$ 35. $1\frac{3}{4}$ 36. $1\frac{1}{5}$

37. $1\frac{3}{8}$ 38. $2\frac{1}{2}$ 39. $2\frac{1}{10}$ 40. $2\frac{1}{15}$

Change the following percents to fractions or mixed numbers.

41. 10%   42. 5%   43. 15%   44. 17%   45. 25%

46. 30%   47. 50%   48. $12\frac{1}{2}\%$ 49. $37\frac{1}{2}\%$ 50. $16\frac{2}{3}\%$

51. $33\frac{1}{3}\%$ 52. $66\frac{2}{3}\%$ 53. 33%   54. $\frac{1}{2}\%$ 55. $\frac{1}{4}\%$

56. 1%   57. 100%   58. 125%   59. 120%   60. 150%

61. 0.3%   62. 2.5%   63. 62.5%   64. 0.2%   65. 0.75%

Find the missing forms of each number.

	Fraction Form	Decimal Form	Percent Form
66.	$\frac{5}{8}$	(a)	(b)
67.	$\frac{11}{20}$	(a)	(b)
68.	(a)	0.09	(b)
69.	(a)	1.75	(b)
70.	(a)	(b)	36%
71.	(a)	(b)	10.5%

72. A department store offers a 30% discount during a special sale on men's suits. Change 30% to a fraction reduced to lowest terms.

73. In a sophomore class of 250 students, 10 represent the sophomore class on the student council. What percent of the class is on the student council?

74. Malcolm planned to drive 300 miles on a trip. After 3 miles, what percent of the trip had he driven?

75. (a) There are 12 inches in a foot. What percent of a foot is an inch?

(b) There are 4 quarts in a gallon. What percent of a gallon is a quart?

(c) There are 16 ounces in a pound. What percent of a pound is an ounce?

76. A desk once belonging to George Washington was recently sold at auction. It was expected to bring $80,000 but actually sold for $165,000. Write the actual selling price as a percent of the expected price.

77. According to an article in *The Boston Globe* (January 23, 1994), only about 250 of the nation's 175,000 dairy farmers still deliver milk door-to-door. What percent of dairy farmers still deliver milk?

Check Your Number Sense

78. For each pair of fractions given below, tell which fraction would be easier to change to a percent mentally. Explain your reasoning.

(a) $\dfrac{7}{20}$ or $\dfrac{2}{9}$?   (b) $\dfrac{3}{10}$ or $\dfrac{5}{16}$?   (c) $\dfrac{1}{12}$ or $\dfrac{1}{25}$?

♻ *The Recycle Bin*   (from Section 6.4)

Solve the following problems.

1. A salesman makes $9 for every $100 worth of merchandise he sells. What will he make if he sells $80,000 worth of merchandise?

2. An architect drew plans for a new building using a scale of $\dfrac{1}{2}$ inch to represent 36 feet. How many feet would 3 inches represent?

3. If gasoline costs $1.12 per gallon, how many gallons can be bought for $16.80?

4. A certain type of fertilizer is sold in bags containing 10 pounds with a recommended coverage of 3000 square feet. (a) If your lawn is in the shape of a rectangle 90 feet by 80 feet, how many pounds of this fertilizer do you need to cover your lawn? (b) How many bags do you need to buy?

7.3   Solving Percent Problems Using the Proportion $\dfrac{R}{100} = \dfrac{A}{B}$ (optional)

OBJECTIVES

1. Recognize three types of percent problems.
2. Solve percent problems using the proportion $\dfrac{R}{100} = \dfrac{A}{B}$

The Basic Proportion $\dfrac{R}{100} = \dfrac{A}{B}$

Many people have difficulty working with percent simply because they do not know whether to add, subtract, multiply, or divide. **There are only three basic types of problems using percent.** In this section, we will develop a technique for solving all three types of problems by using just one basic proportion of the form $\dfrac{R}{100} = \dfrac{A}{B}$. The terms represented in this basic proportion are explained in the following box.

$R\%$ = Rate (as a percent)

B = Base (number we are finding the percent of)

A = Amount or percentage (a part of the base)

The relationship among R, B, and A is given in the basic proportion

$$\frac{R}{100} = \frac{A}{B}$$

As an example of how to set up a basic proportion, consider the statement

25% of 60 is 15.

We can set this statement in the form of the basic proportion as follows:

25%   of   60   is   15.
  ↕   ↕   ↕
Rate   Base   Amount

The basic proportion

$$\frac{R}{100} = \frac{A}{B} \quad \text{becomes} \quad \frac{25}{100} = \frac{15}{60}.$$

Using the Basic Proportion

If any two of the values R, A, and B are known, then the third can be found by substituting the known values into the basic proportion and solving for the one unknown term.

The Three Basic Types of Percent Problems and the Proportion $\dfrac{R}{100} = \dfrac{A}{B}$

Type 1: Find the Amount, given the Base and the Percent.

$$\text{What is 65\% of 500?} \qquad \frac{65}{100} = \frac{A}{500}$$

R and B are known. The object is to find A.

Type 2: Find the Base, given the Percent and the Amount.

$$57\% \text{ of what number is 51.3?} \qquad \frac{57}{100} = \frac{51.3}{B}$$

R and A are known. The object is to find B.

Type 3: Find the Percent, given the Base and the Amount.

$$\text{What percent of 170 is 204?} \qquad \frac{R}{100} = \frac{204}{170}$$

A and B are known. The object is to find R.

The following examples illustrate how to substitute into the basic proportion and how to solve the resulting proportion for the unknown term by using the methods discussed in Chapter 6. **Remember that B is the number you are taking the percent of.** B is the number that follows the word **of.**

EXAMPLE 1

What is 65% of 500?

Solution

In this problem, $R\% = 65\%$ and $B = 500$. We want to find the value of A. Substitution in the basic proportion gives

$$\frac{65}{100} = \frac{A}{500}$$

$65 \cdot 500 = 100 \cdot A$ The product of the extremes must equal the product of the means.

$$\frac{32{,}500}{100} = \frac{\cancel{100} \cdot A}{\cancel{100}}$$

$$325 = A$$

So 65% of 500 is __325__.

EXAMPLE 2

57% of what number is 51.3?

Solution

In this problem, $R\% = 57\%$ and $A = 51.3$. We want to find the value of B. Substitution in the basic proportion gives

$$\frac{57}{100} = \frac{51.3}{B}$$

$57 \cdot B = 100 \cdot 51.3$ The product of the extremes must equal the product of the means.

$$\frac{\cancel{57} \cdot B}{\cancel{57}} = \frac{5130}{57}$$

$$B = 90$$

So 57% of __90__ is 51.3.

EXAMPLE 3

What percent of 170 is 204?

Solution

In this problem $B = 170$ and $A = 204$. We want to find the value of R. (Do you expect $R\%$ to be more than 100% or less than 100%?) Substitution in the basic proportion gives

$$\frac{R}{100} = \frac{204}{170}$$

$$170 \cdot R = 100 \cdot 204$$

$$\frac{\cancel{170} \cdot R}{\cancel{170}} = \frac{20{,}400}{170}$$

$$R = 120 \quad \text{(or } R\% = 120\%)$$

So __120%__ of 170 is 204.

CLASSROOM PRACTICE

Solve each problem for the unknown quantity.

1. 12% of 80 is _____.

2. _____% of 50 is 5.5.

3. 75% of _____ is 90.

4. Find 30% of 250.

5. Find 140% of 65.

ANSWERS: **1.** 9.6 **2.** 11% **3.** 120 **4.** 75 **5.** 91

Exercises 7.3

Use the basic proportion $\dfrac{R}{100} = \dfrac{A}{B}$ to solve each of the following problems for the unknown quantity.

1. 10% of 90 is _____.

2. 5% of 72 is _____.

3. 15% of 50 is _____.

4. 25% of 60 is _____.

5. 75% of 32 is _____.

6. 60% of 40 is _____.

7. 100% of 47 is _____.

8. 80% of 80 is _____.

9. 2% of _____ is 5.

10. 20% of _____ is 14.

11. 3% of _____ is 27.

12. 30% of _____ is 45.

13. 100% of _____ is 62.

14. 50% of _____ is 35.

15. 150% of _____ is 69.

16. 110% of _____ is 440.

17. _____% of 50 is 75.

18. _____% of 120 is 48.

19. _____% of 70 is 21.

20. _____% of 15 is 5.

21. _____% of 44 is 66.

22. _____% of 50 is 10.

23. _____% of 54 is 18.

24. _____% of 100 is 38.

Each of the following problems is one of the three types discussed, with slightly changed wording. Remember that B (the base) is the number you are finding the percent **of.**

25. _____ is 50% of 35.

26. _____ is 31% of 85.

27. 32 is 20% of _____.

28. 79 is 100% of _____.

29. 23 is _____% of 10.

30. 18 is _____% of 10.

31. 40 is $33\dfrac{1}{3}$% of _____.

32. 76.8 is 15% of _____.

33. 142.6 is 23% of _____.

34. 20.25 is 25% of _____.

35. 48 is _____% of 24.

36. 8 is _____% of 12.

37. _____ is 96% of 35.

38. _____ is 84% of 52.

39. _____ is 18% of 425.

40. _____ is 28% of 640.

41. Find 18% of 345.

42. Find 13.5% of 95.

43. Find 150% of 70.

44. What percent of 32 is 24?

45. 200 is 125% of what number?

46. 450 is 150% of what number?

47. What percent of 100 is 66.5?

48. Find 86.5% of 100.

49. What percent of 96 is 16?

50. 160 is 80% of what number?

51. According to U.S. Census information from 1980 to 1990, Mesa, Arizona had the largest population growth among the nation's 100 most populated cities. Between 1980 and 1990, the population of Mesa had increased by 89%, to 288,091. What was Mesa's population in 1980?

52. Because of the increasing cost of breakfast cereals, more and more people are buying private-label brands rather than national brands. In a recent year, the sale of private-label cereals rose from 170 million boxes to 180 million boxes. What was the percent increase in sales (to the nearest tenth of a percent)?

53. A Massachusetts highway department study done in the summer of 1993 noted that 68% of the fans attending Red Sox games at Fenway Park travel there by private automobile. The park seats 34,142 people at a sold-out game. About how many people (to the nearest ten) would you expect to arrive by private automobile at a sold-out game?

54. The same study noted in Exercise 53 found that 40% of the fans of Yankee Stadium travel there by car. Yankee Stadium seats 57,545 at sold-out games. About how many people (to the nearest ten) would you expect to arrive at a sold-out Yankee game by some means other than by automobile?

7.4 Solving Percent Problems Using the Equation $R \times B = A$

OBJECTIVES

1. Recognize three types of percent problems.
2. Solve percent problems using the equation $R \times B = A$.
3. Estimate percents and amounts related to percents.

The Basic Equation $R \times B = A$

In Section 7.3, we introduced the three types of percent problems and the use of the basic proportion $\dfrac{R}{100} = \dfrac{A}{B}$ to solve these problems. If both sides of this proportion are multiplied by B and R is changed to decimal form (or fraction form), we have the following result.

$$\frac{R}{100} = \frac{A}{B}$$

$$\frac{R}{100} \times B = \frac{A}{B} \times B$$

or $R \times B = A$ where R is the value of $R\%$ in decimal form.

We will call this last equation, $R \times B = A$, the **basic equation for solving percent problems.**

Now, consider the statement

25% of 60 is 15.

We can set this statement in the form of the basic equation as follows:

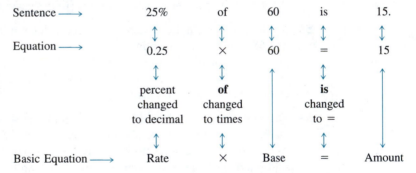

Sentence ⟶	25%	of	60	is	15.
Equation ⟶	0.25	×	60	=	15
	percent changed to decimal	**of** changed to times		**is** changed to =	
Basic Equation ⟶	Rate	×	Base	=	Amount

The terms represented in the basic equation are explained in the following box.

R = Rate or percent (as a decimal or fraction)

B = Base (number we are finding the percent of)

A = Amount or percentage (a part of the base)

of means to multiply. (The symbol × is used in the equation.)

is means equals (=).

The relationship among R, B, and A is given in the basic equation

$R \times B = A$

Using the Basic Equation

Even though there are just three basic types of problems that involve percent, many people have a difficult time differentiating among them and determining whether to multiply or divide in a particular problem. Most of these difficulties can be avoided by using the basic equation $R \times B = A$ and the equation-solving skills learned in Chapter 6. If the values of two of the three quantities in the equation $R \times B = A$ are known, then these values can be substituted into the equation, and the third value can be determined by solving the resulting equation.

The Three Basic Types of Percent Problems and the Equation $R \times B = A$

Type 1: Find the Amount, given the Base and the Percent.

What is 45% of 70?

$$R \times B = A$$
$$0.45 \times 70 = A$$

R and B are known. The object is to find A.

Type 2: Find the Base, given the Percent and the Amount.

30% of what number is 18.36?

$$R \times B = A$$
$$0.30 \times B = 18.36$$

R and A are known. The object is to find B.

Type 3: Find the Percent, given the Base and the Amount.

What percent of 84 is 16.8?

$$R \times B = A$$
$$R \times 824 = 16.8$$

A and B are known. The object is to find R.

Remember:

(a) **Of** means to multiply when used with decimals or fractions.

(b) **Is** means $=$.

(c) The percent is changed to decimal or fraction form.

Problem Type 1: Finding the Value of *A*

EXAMPLE 1 | What is 45% of 70?

Solution

Here $R = 45\% = 0.45$ and $B = 70$.

Substituting into the equation $R \times B = A$ gives

$$0.45 \times 70 = A$$
$$31.5 = A$$

$$\begin{array}{r} 70 \\ \times\ 0.45 \\ \hline 3\ 50 \\ 28\ 0 \\ \hline 31.50 \end{array}$$

So, 45% of 70 is __31.5__ .

(**Note:** The multiplication can be done by hand, as shown here, or with a calculator. In either case, the equations should be written so the = signs are aligned one below the other, as shown in the solution.)

EXAMPLE 2 | What is 18% of 200?

Solution

In this problem, $R = 18\% = 0.18$ and $B = 200$.
Substituting into the equation $R \times B = A$ gives

$$0.18 \times 200 = A$$
$$36 = A$$

So, 18% of 200 is __36__ .

Type 1 problems are the most common and are generally the easiest to solve. The rate (*R*) and base (*B*) are known. The unknown amount (*A*) is found by multiplying the rate times the base.

Problem Type 2: Finding the Value of *B*

EXAMPLE 3 | 30% of what number is 18.36?

Solution

In this case, $R = 30\% = 0.30$ and $A = 18.36$.
Substituting into the equation $R \times B = A$ gives

$0.30 \times B = 18.36$

$$\frac{\cancel{0.30} \times B}{\cancel{0.30}} = \frac{18.36}{0.30}$$ Divide both sides by 0.30, the coefficient of B.

$B = 61.2$

$$\begin{array}{r} 61.2 \\ .30\overline{)18.36\,0} \\ \underline{18\,0} \\ 36 \\ \underline{30} \\ 6\,0 \\ \underline{6\,0} \\ 0 \end{array}$$

So, 30% of __61.2__ is 18.36.

EXAMPLE 4   82% of _____ is 246?

Solution

Here $R = 82\% = 0.82$ and $A = 246$.
Substituting into the equation $R \times B = A$ gives

$0.82 \times B = 246$

$$\frac{\cancel{0.82} \times B}{\cancel{0.82}} = \frac{246}{0.82}$$ Divide both sides by 0.82, the coefficient of B. The division can be performed with a calculator.

$B = 300$

So, 82% or __300__ is 246.

Problem Type 3: Finding the Value of R

EXAMPLE 5   What percent of 84 is 16.8?

Solution

In this problem, $B = 84$ and $A = 16.8$.
Substituting into the equation $R \times B = A$ gives

$R \times 84 = 16.8$

$$\frac{R \times \cancel{84}}{\cancel{84}} = \frac{16.8}{84}$$ Divide both sides by 84, the coefficient of R.

$$\begin{array}{r} 0.2 \\ 84\overline{)16.8} \\ \underline{16\,8} \\ 0 \end{array}$$

$R = 0.2 = 20\%$

So, __20%__ of 84 is 16.8.

EXAMPLE 6

_____ % of 195 is 234?

Solution

In this case, $B = 195$ and $A = 234$.
Substituting into the equation $R \times B = A$ gives

$$R \times 195 = 234$$

$$\frac{R \times \cancel{195}}{\cancel{195}} = \frac{234}{195} \qquad \text{Divide both sides by 195, the coefficient of } R.$$

$$R = 1.2 = 120\% \qquad R \text{ is more than 100\% because } A \text{ is greater than } B.$$

So, __120%__ of 195 is 234.

When you work with applications, the wording of a problem is unlikely to be exactly as shown in Examples 1–6. The following examples show some alternative wording, but each problem is still one of the three basic types.

EXAMPLE 7

Find 65% of 42.

Solution

Here the Rate = 65% and the Base = 42. This is a Type 1 problem:

65% of 42 is _____.

$$.65 \times 42 = A \qquad\qquad .65$$
$$27.3 = A \qquad\qquad \underline{\times\quad 42}$$
$$\qquad\qquad\qquad 1.30$$
$$\qquad\qquad\qquad \underline{26.0}$$
$$\qquad\qquad\qquad 27.30$$

EXAMPLE 8

What percent of 125 gives a result of 31.25?

Solution

Here the Rate is unknown. The Base = 125 and Amount = 31.25. This is a Type 3 problem:

_____ % of 125 is 31.25.

$$R \times 125 = 31.25$$

$$\frac{R \times \cancel{125}}{\cancel{125}} = \frac{31.25}{125} \qquad\qquad\qquad \begin{array}{r} 0.25 \\ 125 \overline{)31.25} \\ \underline{25\,0} \\ 6\,25 \\ \underline{6\,25} \\ 0 \end{array}$$

$$R = 0.25 = 25\%$$

For easy reference, we repeat the list given in section 7.2 of common percent-decimal-fraction equivalents. This list serves to emphasize the relationships between fractions and percents. In Examples 9–12, we show the use of fractions in place of percents.

Common Percent—Decimal—Fraction Equivalents

$100\% = 1.00 = 1$

$25\% = .25 = \dfrac{1}{4}$

$50\% = .50 = \dfrac{1}{2}$

$75\% = .75 = \dfrac{3}{4}$

$33\dfrac{1}{3}\% = .33\dfrac{1}{3} = \dfrac{1}{3}$

$66\dfrac{2}{3}\% = .66\dfrac{2}{3} = \dfrac{2}{3}$

$12\dfrac{1}{2}\% = .125 = \dfrac{1}{8}$

$37\dfrac{1}{2}\% = .375 = \dfrac{3}{8}$

$62\dfrac{1}{2}\% = .625 = \dfrac{5}{8}$

$87\dfrac{1}{2}\% = .875 = \dfrac{7}{8}$

These examples show how using fraction equivalents for percents can simplify your work.

EXAMPLE 9

What is 25% of 24?

$$A = \dfrac{1}{\overset{}{\underset{1}{\cancel{4}}}} \times \overset{6}{\cancel{24}} = 6 \quad \text{Type 1 problem}$$

$$25\% = \dfrac{1}{4}$$

EXAMPLE 10

Find 75% of 36.

$$A = \dfrac{3}{\overset{}{\underset{\cancel{1}}{\cancel{4}}}} \times \overset{9}{\cancel{36}} = 27 \quad \text{Type 1 problem}$$

$$75\% = \dfrac{3}{4}$$

EXAMPLE 11

What is the value of $33\dfrac{1}{3}\%$ of 72?

$$A = \dfrac{1}{\overset{}{\underset{1}{\cancel{3}}}} \times \overset{24}{\cancel{72}} = 24 \quad \text{Type 1 problem}$$

$$33\dfrac{1}{3}\% = \dfrac{1}{3}$$

EXAMPLE 12 $37\frac{1}{2}\%$ of what number is 300?

$$\frac{3}{8} \times B = 300 \qquad \text{Type 2 problem}$$

$$\frac{\cancel{8}}{\cancel{3}} \times \frac{\cancel{3}}{\cancel{8}} \times B = \frac{8}{\cancel{3}} \times \overset{100}{\cancel{300}} \qquad \frac{8}{3} \text{ is the reciprocal of } \frac{3}{8}.$$

$$B = 800$$

Remember that a percent is just another form of a fraction and that when you find a percent of a number, the amount will be:

(a) smaller than the number if the percent is less than 100%.

(b) greater than the number if the percent is more than 100%.

CLASSROOM PRACTICE

Solve each problem for the unknown quantity.

1. 5% of 70 is _____.

2. _____% of 80 is 9.6.

3. 15% of _____ is 13.5.

4. Find 25% of 80.

5. Find 180% of 20.

ANSWERS: **1.** 3.5   **2.** 12%   **3.** 90   **4.** 20   **5.** 36

Exercises 7.4

Solve each problem for the unknown quantity. You may use a calculator to perform calculations; however, you should first write the basic equation $R \times B = A$ and other resulting equations one below the other.

1. 10% of 70 is _____.

2. 5% of 62 is _____.

3. 15% of 60 is _____.

4. 25% of 72 is _____.

5. 75% of 12 is _____.

6. 60% of 30 is _____.

7. 100% of 36 is _____.

8. 80% of 50 is _____.

9. 2% of _____ is 3.

10. 20% of _____ is 17.

11. 3% of _____ is 21.

12. 30% of _____ is 21.

13. 100% of _____ is 75.

14. 50% of _____ is 42.

15. 150% of _____ is 63.

16. 110% of _____ is 330.

17. _____ % of 60 is 90.

18. _____% of 150 is 60.

19. _____% of 75 is 15.

20. _____% of 12 is 4.

21. _____% of 34 is 17.

22. _____% of 30 is 6.

23. _____% of 48 is 16.

24. _____% of 100 is 35.

Each of the following problems is one of the three types discussed, with slightly changed wording. Remember that *B* is the number you are finding the percent of.

25. _____ is 50% of 25.

26. _____ is 31% of 76.

27. 22 is 20% of _____.

28. 86 is 100% of _____.

29. 13 is _____% of 10.

30. 15 is _____% of 10.

31. 24 is $33\frac{1}{3}$% of _____.

32. 92.1 is 15% of _____.

33. 119.6 is 23% of _____.

34. 9.5 is 25% of _____.

35. 36 is _____% of 18.

36. 60 is _____% of 40.

37. _____ is 96% of 17.

38. _____ is 84% of 32.

39. _____ is 18% of 325.

40. _____ is 28% of 460.

41. Find 18% of 244.

42. Find 15.2% of 75.

43. Find 120% of 60.

44. What percent of 32 gives a result of 8?

45. 100 is 125% of what number?

Use fractions to solve the following problems. Do the work mentally if you can after you have set up the related equation.

46. Find 50% of 32.

47. Find $66\frac{2}{3}$% of 60.

48. What is $12\frac{1}{2}$% of 80?

49. What is $62\frac{1}{2}$% of 16?

50. $33\frac{1}{3}$% of 75 is _____.

51. 25% of 150 is _____.

52. 75% of what number is 21?

53. 50% of what number is 35?

54. $37\frac{1}{2}$% of _____ is 61.2.

55. 100% of _____ is 76.3.

56. Magazine publishers will often give significant discounts to subscribers. The cover price of *Newsweek* is $79.65 for 6 months, but a 6-month subscription costs only $18.57! What is the percent **savings** (to the nearest tenth) for someone who subscribes to *Newsweek*?

57. Only 27% of American deaths in the Revolutionary War occurred in battle. All other mortalities were the result of exposure, disease, and starvation, for example. Of the estimated 25,300 deaths, how many men died in battle (to the nearest hundred)?

58. During his presidency from 1945 to 1953, Harry Truman vetoed 250 congressional bills, and 12 of those vetoes were overridden. What percent of Truman's vetoes were overridden?

59. William O. Douglas served as an associate justice on the Supreme Court for 36 years from 1939 to 1975, the longest tenure of any Supreme Court justice in history. He lived to be 82. What percent of his life (to the nearest one percent) did he spend on the Supreme Court?

60. The decade from 1960 to 1970 saw the largest 10-year increase in the number of male elementary and high school teachers in our nation's history—from 402,000 to 691,000. What was the percent increase from 1960 to 1970 (to the nearest one percent)?

$$\frac{289,000}{402,000}$$

7.5   Estimating with Percents

The purpose of this section is to reinforce your understanding of percent before you begin work with applications in Section 7.6 and Chapter 8. The approach is to provide you with several choices for an approximate answer to a percent problem and for you to choose, through mental calculations and reasoning only, the answer closest to the correct answer. You may then check your choice by actually performing the calculations with a calculator. The important idea is that you first try to estimate the answer mentally. That is, test your general understanding of percents before you perform any physical calculations.

For example, suppose that the problem is "Find 10% of 900," and the choices are (a) 9, (b) 90, (c) 900, (d) 9000. You might arrive at the correct choice, 90, by reasoning that 900 is 100% of 900 and 9000 is still a larger percent. Also, 9 is only 1% of 900. Thus, the only reasonable choice is (b).

Or, you might simply think that $10\% = \dfrac{1}{10}$ and $\dfrac{1}{10}$ of 900 is 90. Or, you might change 10% to the decimal 0.10 and multiply mentally, $0.10 \times 900 = 90$. In any case, the idea is for you to arrive at your choice by understanding the relationship of percent to the size of a number.

Exercises 7.5

In each of the following exercises, one of the choices is the correct answer to the problem. You are to work through the entire set of exercises by reasoning and calculating mentally. After you have checked your answers, you should work out the answers to any problems you missed and try to understand why your reasoning was incorrect.

1. Write 3% as a fraction.

(a) $\frac{3}{1}$ (b) $\frac{3}{10}$ (c) $\frac{3}{100}$ (d) $\frac{3}{1000}$

2. Write 7% as a fraction.

(a) $\frac{7}{1}$ (b) $\frac{7}{10}$ (c) $\frac{7}{100}$ (d) $\frac{7}{1000}$

3. Write $12\frac{1}{2}\%$ as a fraction, reduced.

(a) $\frac{1}{8}$ (b) $\frac{3}{8}$ (c) $\frac{5}{8}$ (d) $\frac{7}{8}$

4. Write $87\frac{1}{2}\%$ as a fraction, reduced.

(a) $\frac{1}{8}$ (b) $\frac{3}{8}$ (c) $\frac{5}{8}$ (d) $\frac{7}{8}$

5. Write $2\frac{1}{2}\%$ as a decimal.

(a) 0.025 (b) 0.25 (c) 2.5 (d) 25

6. Write $12\frac{1}{2}\%$ as a decimal.

(a) 0.0125 (b) 0.125 (c) 1.25 (d) 12.5

7. Find 10% of 530.

(a) 5.3 (b) 53 (c) 10.6 (d) 106

8. Find 20% of 530.

(a) 1.06 (b) 10.6 (c) 106 (d) 212

9. Find 30% of 530.

(a) 106 (b) 159 (c) 212 (d) 265

10. Find 40% of 530.

(a) 21.2 (b) 212 (c) 2120 (d) 21,200

11. What is 100% of 8.02?

(a) 0.802   (b) 8.02   (c) 80.2   (d) 802

12. What is 200% of 8.02?

(a) 1.604   (b) 16.04   (c) 160.4   (d) 1604

13. What is 110% of 300?

(a) 30   (b) 33   (c) 300   (d) 330

14. What is 120% of 300?

(a) 60   (b) 300   (c) 360   (d) 630

15. Write 0.25% as a decimal.

(a) 0.0025   (b) 0.025   (c) 0.25   (d) 2.5

16. Write 75% as a decimal.

(a) 0.0075   (b) 0.075   (c) 0.75   (d) 7.5

17. Write $\frac{1}{3}$ as a percent.

(a) $\frac{1}{3}\%$ (b) $33\frac{1}{3}\%$ (c) 66%   (d) $66\frac{2}{3}\%$

18. Write $\frac{2}{3}$ as a percent.

(a) $\frac{2}{3}\%$ (b) $33\frac{1}{3}\%$ (c) 66%   (d) $66\frac{2}{3}\%$

19. Write $\frac{5}{8}$ as a percent.

(a) $12\frac{1}{2}\%$ (b) 25%   (c) $37\frac{1}{2}\%$ (d) $62\frac{1}{2}\%$

20. Write $\frac{3}{8}$ as a percent.

(a) 12.5%   (b) 25%   (c) 37.5%   (d) 62.5%

21. 10% of what number is 74?

(a) 7.4   (b) 74   (c) 740   (d) 7400

22. 10% of what number is 46?

(a) 4.6   (b) 46   (c) 460   (d) 4600

23. 30% of what number is 15?

    (a) 45   (b) 50   (c) 450   (d) 500

24. 30% of what number is 21?

    (a) 63   (b) 70   (c) 630   (d) 700

25. What percent of 836 is 83.6?

    (a) 1%   (b) 10%   (c) 100%   (d) 1000%

26. What percent of 1500 is 15?

    (a) 1%   (b) 10%   (c) 100%   (d) 1000%

27. What percent of 25 is 250%

    (a) 1%   (b) 10%   (c) 100%   (d) 1000%

28. What percent of 93.7 is 937?

    (a) 1%   (b) 10%   (c) 100%   (d) 1000%

29. What is 150% of 50?

    (a) 25   (b) 50   (c) 75   (d) 100

30. What is 150% of 100?

    (a) 50   (b) 100   (c) 150   (d) 1500

31. What is 1% of 100?

    (a) 1   (b) 10   (c) 100   (d) 1000

32. What is 1% of 200?

    (a) 2   (b) 20   (c) 200   (d) 2000

33. What is 1% of 83.6?

    (a) 0.0836   (b) 0.836   (c) 8.36   (d) 83.6

34. What is 1% of 94.7?

    (a) 0.0947   (b) 0.947   (c) 9.47   (d) 94.7

35. What percent of 1000 is 250?

    (a) 2.5%   (b) 25%   (c) 250%   (d) 2500%

36. What percent of 1000 is 10?

    (a) 1%   (b) 10%   (c) 100%   (d) 1000%

37. 10% of what number is 10?

    (a) 1   (b) 10   (c) 100   (d) 1000

38. 5% of what number is 5?

    (a) 1   (b) 10   (c) 100   (d) 1000

39. 75% of what number is 36?

    (a) 27   (b) 36   (c) 48   (d) 75

40. 25% of what number is 40?

    (a) 10   (b) 20   (c) 50   (d) 160

41. Choose the best estimate for 30% of 92.17.

    (a) 9   (b) 18   (c) 27   (d) 36

42. Choose the best estimate for 80% of 327.

    (a) 25   (b) 33   (c) 65   (d) 250

43. Choose the best estimate for 19.3% of 516.

    (a) 10   (b) 20   (c) 100   (d) 200

44. Choose the best estimate for 8.7% of 3500.

    (a) 9   (b) 35   (c) 350   (d) 3500

45. Approximately what percent of 1000 is 1585?

    (a) 50%   (b) 66%   (c) 80%   (d) 150%

46. Approximately what percent of 2000 is 3120?

    (a) 20%   (b) 30%   (c) 150%   (d) 200%

47. 947 is approximately 50% of what number?

    (a) 500   (b) 1000   (c) 1800   (d) 4500

48. 345 is approximately 50% of what number?

    (a) 34   (b) 170   (c) 700   (d) 1500

49. 72 is approximately $33\frac{1}{3}\%$ of what number?

    (a) 21   (b) 24   (c) 100   (d) 200

50. 53 is approximately $66\frac{2}{3}\%$ of what number?

    (a) 15   (b) 75   (c) 150   (d) 200

7.6   Applications with Percents: Discount, Sales Tax, and Tipping

OBJECTIVES

1. Know Pólya's four-step process for solving problems.
2. Be familiar with the terms **discount** and **sales tax.**
3. Apply skills related to types of percent problems to solving applications.

Pólya's Four-Step Process for Solving Problems

George Pólya (1887–1985), a famous professor at Stanford University, studied the process of discovery learning. Among his many accomplishments, he developed the following four-step process as an approach to problem solving:

1. Understand the problem.

2. Devise a plan.

3. Carry out the plan.

4. Look back over the results.

There are a variety of types of applications discussed throughout this text and subsequent courses in mathematics, and you will find these four steps helpful as guidelines for understanding and solving all of them. Applying the necessary skills to solve exercises, such as adding fractions or solving equations, is not the same as accumulating the knowledge to solve problems. Problem solving can involve careful reading, reflection, and some original or independent thought.

Basic Steps for Problem Solving

1. Understand the problem. For example:

   (a) Read the problem.

   (b) Understand all the words.

2. Devise a plan using, for example, one or all of the following:

   (a) Guess or estimate or make a list of possibilities.

   (b) Draw a picture or diagram.

   (c) Use a variable and form an equation.

continued

3. Carry out the plan. For example:

(a) Try all the possibilities you have listed.

(b) Solve any equation that you may have set up.

4. Look back over the results. For example:

(a) Can you see an easier way to solve the problem?

(b) Does your solution actually work? Does it make sense?

(c) If there is an equation, check your answer in the equation.

In this section, we will be concerned with applications that involve percent and with the types of percent problems that have been discussed in the first four sections of Chapter 7. The use of a calculator is recommended, though you should keep in mind that it is a tool to enhance and not replace the necessary skills and abilities related to problem solving. Study the following examples carefully, as they illustrate some basic plans for problem solving with percents.

Discount

A **discount** is a reduction in the **original price** (or **marked price**) of an item. The new reduced price is called the **sale price,** and the discount is the difference between the original price and the sale price. The **rate of discount** (or **percent of discount**) is a percent that represents what part the discount is of the original price.

(**Note:** In this text, when solving money problems, rounded-off answers will be rounded to the next higher cent even if the next digit is less than 5.)

EXAMPLE 1 | A refrigerator that regularly sells for $975 is on sale at a 30% discount. (a) What is the amount of the discount? (b) What is the sale price?

Solution

First (a) find the amount of the discount, and then (b) find the sale price by subtracting the discount from the original price.

(a) Since the rate of discount is 30%, we find 30% of $975.
30% of $975 is _____ .

$$0.30 \times 975 = A$$

$$
\begin{array}{r r}
 & \$975 \quad \text{original price} \\
\times & .30 \quad \text{rate of discount} \\
\hline
292.50 = A & \$292.50 \quad \text{discount}
\end{array}
$$

The discount is $292.50.

(b) Find the sale price by subtracting the discount from the original price.

$975.00   original price
− 292.50   discount
$682.50   sale price

The sale price is $682.50.

Sales Tax

Sales tax is a tax charged on goods sold by retailers; it is assessed by states for income to operate state services. The rate of the sales tax varies from state to state (or even city to city in some cases). In fact, some states do not have a sales tax.

EXAMPLE 2   If the sales tax rate is 6%, what would be the final cost of the refrigerator in Example 1?

Solution

First (a) find the amount of sales tax, and then (b) add this tax to the sale price found in Example 1 to find the final cost.

(a) Find the amount of the sales tax.

6% of $682.50 is _____ .

$0.06 \times 682.50 = A$ $682.50   sale price
$ 40.95 = A$ $\times 0.06$ tax rate
                               $40.9500   sales tax

The sales tax is $40.95.

(b) Add the sales tax to the sale price to find the final cost.

$682.50   sale price
+ 40.95   sales tax
$723.45   final cost

The final cost of the refrigerator is $723.45.

EXAMPLE 3   An auto dealer paid $7566 for a large order of a special part. This was **not** the original price. He received a 3% discount off the original price because he paid cash. What was the original price?

Solution

Since $7566 is not the original price, do **not** take 3% of $7566. In fact, we are to find the original price. Therefore, we must reason that if a price is discounted by 3%, then 97% of the original price is what remains ($100\% - 3\% = 97\%$). Thus, $7566 represents 97% of the original price.

97% of _____ is $7566

$.97 \times B = 7566$

$$\frac{.97 \times B}{.97} = \frac{7566}{.97}$$

$B = 7800$

```
                78 00.
        .97.) 7566.00.
              679
              776
              776
               00
                0
                00
                0
                0
```

The original price was $7800.

(**Note:** To check this result, take 3% of $7800 to find the discount. Then subtract this discount from $7800. You will get the discounted price of $7566.)

Tipping

Tipping (or **leaving a tip**) is the custom of leaving a percent of a bill (usually at a restaurant) as a payment to the waiter or waitress for providing good service. The amount of the tip is usually 10 to 20% of the total amount charged (including tax), depending on the quality of the service. The "usual" tip is 15%, but in the case of particularly bad service, no tip is required.

Most people do not carry a calculator with them when dining out, so the calculation of a 15% tip can be an interesting exercise in percents. The following rule of thumb uses the basic facts that 5% is half of 10% and that 10% of a decimal number can be found by moving the decimal point 1 place to the left.

Rule of Thumb for Calculating a 15% Tip

1. For ease of calculation, round off the amount of the bill to the nearest whole dollar.

2. Find 10% of the rounded-off amount by moving the decimal point 1 place to the left.

3. Divide the answer in Step 1 by 2. (This represents 5% of the rounded-off amount, or one-half of 10%.)

4. Add the two amounts found in Steps 2 and 3. This sum is the amount of the tip.

Note: There are other rules of thumb that people use to calculate the amount of a tip. For example, you might always round up instead of rounding off to the nearest whole dollar, or you might round off to the nearest ten dollars, or you might leave a tip that makes the total of your expenses a whole-dollar amount. Consider two bills of $7.95 and $7.20. Some might round off each bill to $10.00 and leave a tip of 15% as $1.00 + $0.50 = $1.50 in each case. However, according to the rule in the text, we will calculate the tips as follows:

Round off $7.95 to $8.00 and calculate 15% as $0.80 + $0.40 = $1.20.

Round off $7.20 to $7.00 and calculate 15% as $0.70 + $0.35 = $1.05.

EXAMPLE 4

You and a friend dine out and the bill comes to $28.40, including tax. If you plan to leave a 15% tip, what should be the amount of the tip?

Solution

For ease of computation, round off $28.40 to $28.00.

Find 10% of $28.00 by moving the decimal point one place to the left:

10% of 28.00 = 2.80

Now find 5% of $28.00 by dividing $2.80 by 2:

2.80 ÷ 2 = 1.40

Adding gives

$2.80
+ 1.40
———
$4.20   amount of tip

We will say that the tip should be $4.20 as a textbook answer. However, practically speaking, you might leave either $4.25 or $4.50.

Exercises 7.6

The following problems may involve several calculations, and calculators are recommended. Follow Polya's four-step process for problem solving. After you have read the problem so that you understand it, write down the known information and your plan for solving the problem. Then follow through with your plan.

1. A store owner received a 3% discount from the manufacturer when she bought $15,500 worth of dresses. (a) What was the amount of the discount? (b) What did she pay for the dresses?

2. If the sales tax in a certain state is figured at 6%: (a) How much tax is there on a purchase of $30.20? (b) What is the total amount paid for the purchase?

3. If sales tax was figured at 6%: (a) How much tax was paid on the purchase of three textbooks priced at $55.00, $25.50, and $34.95? (b) What would be the total cost of all three books?

a) 82.50
b) 1925⁰⁰

4. A new briefcase was priced at $275. If it was to be marked down 30%: (a) What would be the discount? (b) What would be the new price?

5. The discount on a fur coat was $150. This was a 20% discount. (a) What was the original selling price of the coat? (b) What was the sale price? (c) What was the total amount paid for the coat if a 6% sales tax was added to the sales price?

6. The discount on men's suits was $50, and they were on sale for $200. (a) What was the original selling price? (b) What was the rate of discount?

7. In order to get more subscribers, a book club offered three books for a total price of $7.02. The total selling price was originally $17.55 for all three books. (a) What was the amount of the discount? (b) Based on the original selling price, what was the rate of the discount on these three books? *10.53/7.55*

$ 656.40

8. An auto supply store received a shipment of auto parts and a bill for $845.30. Some of the parts were not as ordered, and they were returned immediately. The value of the parts returned was $175.50. The terms of the billing provided the store with a 2% discount if it paid cash (for the parts it kept) within two weeks. What did the store pay for the parts it kept if it paid cash within two weeks?

9. Towels were on sale at a discount of 30%. If the sale price was $3.01, what was the original price?

10. Computer disks were on sale for $5.24 per box. What was the original price per box if the sale price represents a discount of 20%.

In Exercises 11–16, use the rule-of-thumb stated in the text to calculate a 15% tip.

11. You invited your friend to lunch at the local coffeehouse and the bill totaled $12.60. Your friend offered to leave the tip. What amount did he leave as a 15% tip? *15% of 13*

$24⁰⁰

12. A lawyer took four clients to dinner. The bill for the meals and drinks was $150.00 plus a 6% sales tax. What amount did she leave as a tip (rounding up to the nearest dollar) if she left 15%? *15% of 159*

13. A bill for a family dining at a restaurant was $38.40. What would a 15% tip have been? What total amount should they have left on the table if they wanted to go before the waiter came to pick up the money? *15% of 38*

14. On a recent business trip, Ken and Joe had breakfast at the restaurant next to the motel where they were staying. Ken's breakfast bill was $6.95 and Joe's was $8.75. How much should each man have left as a tip if each tipped 15%?

15. Mrs. Chung had two large pizzas delivered to her home for $26.00. If she tipped the delivery person 15%, what total amount did she pay the driver?

$6.30
$48.30

16. Juan took his date out for dinner before the senior prom. The total bill, including tax, was $42.00. How much did he tip the waitress if he left 15%? What were his total expenses for the meal?

17. Linda is enrolled in freshman calculus. She has the choice of buying the text in hardback form for $60.00 or in paperback form for $46.50. Tax is figured at 5% of the selling price. The bookstore buys back hardback books for 40% of the selling price and paperback books for 30% of the selling price. (a) Which book is the more economical buy for Linda if she sells her book back to the bookstore at the end of the semester? (b) How much does she save?

18. A student missed 3 problems on a mathematics test and received a grade of 85%. If all the problems were of equal value, how many problems were on the test?

19. Suppose you sell your home for $100,000 and you owe the Savings and Loan $60,000 on the first trust deed. You pay a real estate agent a commission of 6% of the selling price and other fees and taxes totaling $1200. How much cash do you have from the sale?

$16.88
$5.63

20. Sheets are marked $22.50 and pillowcases $7.50. What is the sale price of each item if each item is discounted 25% off the marked price?

21. In the roofing business, shingles are sold by the "square," which is enough material to cover a 10-foot by 10-foot square (or 100 square feet). A roofing supplier has a closeout on shingles at a 30% discount. (a) If the original price was $230 per square, what is the sale price per square? (b) How much would a roofer pay for 34 squares?

Writing and Thinking about Mathematics

22. Make up your own rule of thumb for leaving a 20% tip.

23. A man weighed 220 pounds. He lost 22 pounds in three months. Then he gained back 22 pounds over the next four months. (a) What percent of his weight did he lose in the first three months? (b) What percent of his weight did he gain back? The loss and gain are the same amount, but the two percents are different. Explain why.

Collaborative Learning Exercise

24. With the class separated into teams of two to four students, each team is to analyze the following discussion on discount and decide how to answer the related questions. Then each team leader is to present the team's answers and related ideas to the class for general discussion.

> In Example 1 on page 354, we calculated a 30% discount to find the sale price of a refrigerator. Then, in Example 2, we calculated the sales tax at 6% of the sale price. The final cost of the refrigerator was then found by adding the sales tax to the sale price.

Could most of these calculations have been avoided by simply discounting the original price by 24% to find the final cost (30% − 6% = 24%)? Explain your reasoning.

7.7 Applications with Percents: Commission, Profit, and Others

OBJECTIVES

1. Be familiar with and understand the term **commission**.
2. Understand that percent of profit is a ratio and can be based on cost or on selling price.
3. Know how to calculate percent of profit.

Commission

A **commission** is a fee paid to an agent or salesperson for a service. Commissions are usually a percent of a negotiated contract or a percent of sales. The amount that is multiplied by the percent is called the **base of the commission.**

EXAMPLE 1   A saleswoman earns a fixed salary of $900 a month plus a commission of 8% on whatever amount she sells over $6500 in merchandise. What did she earn the month that she sold $11,800 in merchandise?

Solution

First find the base of her commission by subtracting $6500 from the total amount she sold. Then find 8% of this base and add her monthly salary.

(a) Subtract $6500 from $11,800.

$$
\begin{array}{ll}
\$11,800 & \text{total amount she sold} \\
-\ \ 6,500 & \text{amount on which she does not earn a commission} \\
\hline
\$\ \ 5,300 & \text{base of commission}
\end{array}
$$

(b) Find 8% of this base.

8% of $5300 is _____.

$$
\begin{array}{llll}
0.08 \times 5300 = A & & \$5300 & \text{base of commission} \\
424 = A & & \times\ \ \ .08 & \text{rate of commission} \\
& & \hline
& & \$\ \ 424 & \text{commission}
\end{array}
$$

(c) Her monthly pay is the sum of her salary and her commission.

$\begin{array}{ll} \$\ \ 900.00 & \text{fixed salary} \\ +\ \ 424.00 & \text{commission} \\ \hline \$1324.00 & \text{total pay for the month} \end{array}$

She earned $1324 for the month.

Percent of Profit

In business, a company's **profit** is the difference between income (or revenue) and costs, such as wages, materials, and rent. However, manufacturers and retailers are also concerned with the profit on each item produced or sold, which is simply the difference between its selling price to the customer and the cost to the company. This is the type of profit that will be discussed in this section.

Terms Related to Profit

Profit: The difference between selling price and cost:

$$(\text{Profit} = \text{Selling price} - \text{Cost})$$

Percent of profit: There are two types; both are ratios with profit in the numerator.

1. Percent of profit **based on cost** is the ratio of profit to cost:

$$\frac{\text{Profit}}{\text{Cost}} = \%\ \text{of profit based on cost}$$

2. Percent of profit **based on selling price** is the ratio of profit to selling price:

$$\frac{\text{Profit}}{\text{Selling price}} = \%\ \text{of profit based on selling price}$$

EXAMPLE 2

A company manufactures and sells plastic boxes that cost $21 each to produce and are sold for $28 each. (a) What is the profit on each box? (b) What is the percent of profit based on cost? (c) What is the percent of profit based on selling price?

Solution

(a) To find the profit, subtract the cost from the selling price.

$\begin{array}{ll} \$28 & \text{selling price} \\ -\ 21 & \text{cost} \\ \hline \$\ 7 & \text{profit} \end{array}$

(b) To find the percent of profit based on cost, divide the profit by the cost.

$$\frac{\$7 \text{ profit}}{\$21 \text{ cost}} = \frac{1}{3} = 0.33\frac{1}{3} = 33\frac{1}{3}\% \text{ profit based on cost}$$

(c) To find the percent of profit based on selling price, divide the profit by the selling price.

$$\frac{\$7 \text{ profit}}{\$28 \text{ selling profit}} = \frac{1}{4} = 0.25 = 25\% \text{ profit based on selling price}$$

Note that in parts (b) and (c) the profit is $7 and this does not change in either case. What changes, and gives different percents, is the denominator.

Percent of profit **based on cost** is higher than percent of profit **based on selling price.** The business community reports whichever percent serves its purposes better. Your responsibility as an investor or consumer is to know which percent is reported and what it means to you.

EXAMPLE 3   Women's coats were on sale for $250. (a) If the coats cost the store owner $200, what was his percent of profit based on cost? (b) What was his percent of profit based on selling price?

Solution

(a) First find the profit.

$$\begin{array}{ll} \$250 & \text{selling price} \\ -\ \$200 & \text{cost} \\ \hline \$\ 50 & \text{profit} \end{array}$$

(b) Find the percent of profit based on cost.

$$\frac{\$50 \text{ profit}}{\$200 \text{ cost}} = 25\% \text{ profit based on cost}$$

(c) Find the percent of profit based on selling price.

$$\frac{\$50 \text{ profit}}{\$250 \text{ selling price}} = 20\% \text{ profit based on selling price}$$

Exercises 7.7

1. A realtor works on a 6% commission. What is his commission on a house he sold for $95,000?

2. A realtor selling commercial property works on a 4% commission. What is her commission on a building she sold for $875,000?

3. A sales clerk receives a monthly salary of $500 plus a commission on 6% on all sales over $2500. What did the clerk earn the month she sold $16,000 in merchandise?

$675

4. A shoe saleswoman worked on a fixed salary of $300 per month plus a 5% commission. How much did she make during the month in which she sold $7500 worth of shoes?

5. If a salesman works on a 10% commission only (no monthly salary), how much merchandise will he have to sell to earn $2800 in one month?

6. A basketball player made 120 of 300 shots she attempted. (a) What percent of her shots did she make? (b) What percent did she miss?

7. In one season, a basketball player missed 15% of his free throws. How many free throws did he make if he attempted 180 free throws?

8. A set of golf clubs cost the golf pro $400 and he sold them in the pro shop for $550. (a) What was his profit? (b) What was his percent of profit based on cost? (c) What was his percent of profit based on selling price?

9. Men's suits were on sale for $300 and they cost the store owner $250. (a) What was the profit for the store? (b) What was the store's percent of profit based on cost? (c) What was the store's percent of profit based on selling price?

10. The cost of a television set to a store owner was $350, and he sold the set for $490. (a) What was his profit? (b) What was his percent of profit based on cost? (c) What was his percent of profit based on selling price?

11. A car dealer bought a used car for $1500. He marked up the price so he would make a profit of 25% based on his cost. (a) What was the selling price? (b) If the customer paid 8% of the selling price in taxes and fees, what did the customer pay for the car?

.6% **12.** The property taxes on a house were $750. What was the tax rate if the house was valued at $125,000?

13. A computer programmer was told he would be given a bonus of 5% of any money his programs could save the company. How much would he have to save the company to earn a bonus of $6000?

14. You want to purchase a new home for $98,000. The bank will lend you 80% of the purchase price, but you must pay a loan fee of 2% of the amount of the loan and other fees totaling $850. How much cash do you need to purchase the home?

15. The author of a book was told she would have to cut the number of pages by 12% in order for the book to sell at a competitive price and still show a profit for the publisher. (a) What percent of the pages were in the final form of the book? (b) If the book contained 220 pages in its final form, how many pages did it originally contain? (c) How many pages were cut?

16. A pigeon egg will incubate in 18 days, while a duck egg requires 30 days to hatch. Expressed as a percent, how much longer does it take for the duck egg to hatch?

17. In 1991, Ford Motor Company produced 81,594 Mustangs, which was about 10.6% of its total production. About how many cars did Ford produce in 1991 (to the nearest thousand)?

18. Along with cattle, sheep, and poultry, Alaskan farmers also raise reindeer as livestock. In fact, in a recent year 37,000 reindeer accounted for 71% of Alaska's livestock population. To the nearest thousand, what was the total livestock population in Alaska that year?

19. In 1980, there were 11.3 billion shares of stock traded on the New York Stock Exchange. That volume increased steadily through 1987 when 47.8 billion shares were traded. Write the number of shares traded in 1987 as a percent of the number traded in 1980.

20. In 1980, the U.S. government's public debt per capita was $3,985 and by 1990 that debt had increased to $13,000 per capita. Write the per capita debt in 1990 as a percent of the per capita debt in 1980.

Writing and Thinking about Mathematics

21. Joel worked in a men's clothing store on a straight 8% commission. His friend, who worked at the same store, earned a monthly salary of $500 plus a 4% commission. (a) How much did each make during the month in which each sold $18,500 worth of clothing? (b) What percent more did Joel make than his friend? Explain why there can be more than one answer to part (b).

22. (a) What topic (or topics) discussed in Chapter 7 have you found to be the most interesting? Briefly, explain why. (b) What topic (or topics) have you found to be the most difficult to learn? Briefly, explain why.

23. (a) What topic (or topics) discussed in the text to this point have you found to be the most interesting? Briefly, explain why. (b) What topic (or topics) have you found to be the most difficult to learn? Briefly, explain why.

Collaborative Learning Exercise

24. With the class separated into teams of two to four students, each team is to analyze the following discussion of atomic half-life and answer the related questions as best they can. A general classroom discussion should follow with the class coming to an understanding of the concepts involved.

> The half-life of a radioactive chemical element is the time that it takes for 50% of the atoms in the element sample to decay. Scientists can use methods such as carbon-14 dating to determine the ages of archaeological artifacts by comparing the amount of carbon-14 remaining in an artifact with the level of carbon-14 that would have been present originally. The half-life of carbon-14 is 5730 years. What percent of 10 grams would be left after 5730 years? How many grams? What percent would be left after 11,460 years? How many grams? What percent would be left after 17,190 years? How many grams? Describe a pattern in your calculations over these periods of time. What percent would be left after six such periods of time?

Summary: Chapter 7

Key Terms and Ideas

Percent means hundredths, or the **ratio of a number to 100.**
Percent of profit is the ratio of money made to money invested.
Problems involving **discount, commission, sales tax, profit,** and **tipping** all relate to percent.

Terms Related to Profit

Profit: The difference between selling price and cost:

$$(\text{Profit} = \text{Selling price} - \text{Cost})$$

Percent of profit: There are two types; both are ratios with profit in the numerator.

1. Percent of profit **based on cost** is the ratio of profit to cost:

$$\frac{\text{Profit}}{\text{Cost}} = \% \text{ of profit based on cost}$$

2. Percent of profit **based on selling price** is the ratio of profit to selling price:

$$\frac{\text{Profit}}{\text{Selling price}} = \% \text{ of profit based on selling price}$$

Basic Relationships for Percent Problems

R = Rate or percent (as a decimal or fraction)

B = Base (number we are finding the percent of)

A = Amount or percentage (a part of the base)

Of means to multiply. (The symbol $\times$ is used in the equation.)

Is means equals ($=$).

The relationship among R, B, and A is given in the basic equation $R \times B = A$.

This relationship is also given in the basic proportion $\dfrac{R}{100} = \dfrac{A}{B}$.

Procedures

To Change a Decimal to a Percent:

Step 1: Move the decimal point two places to the right.

Step 2: Write the % sign.

(These two steps have the effect of multiplying by 100 then dividing by 100.)

To Change a Percent to a Decimal:

Step 1: Move the decimal point two places to the left.

Step 2: Delete the % sign.

To Change a Fraction to a Percent:

Step 1: Change the fraction to a decimal.
(Divide the numerator by the denominator.)

Step 2: Change the decimal to a percent.

**To Change a Percent to a Fraction
or Mixed Number:**

Step 1: Write the percent as a fraction with denominator 100 and drop
the % sign.

Step 2: Reduce the fraction.

Use the following four-step process as an approach to problem solving:

1. Understand the problem.

2. Devise a plan.

3. Carry out the plan.

4. Look back over the results.

Rule of Thumb for Calculating a 15% Tip

1. For ease of calculation, round off the amount of the bill to the nearest whole dollar.

2. Find 10% of the rounded-off amount by moving the decimal point 1 place to the left.

3. Divide the answer in Step 1 by 2. (This represents 5% of the rounded-off amount, or one-half of 10%.)

4. Add the two amounts found in Steps 2 and 3. This sum is the amount of the tip.

Review Questions: Chapter 7

1. Percent means _____.

Change the following fractions to percents.

2. $\dfrac{85}{100}$

3. $\dfrac{18}{100}$

4. $\dfrac{37}{100}$

5. $\dfrac{16\frac{1}{2}}{100}$

6. $\dfrac{15.2}{100}$

7. $\dfrac{115}{100}$

Change the following decimals to percents.

8. 0.06

9. 0.3

10. 0.67

11. 0.027

12. 3

13. 1.2

Change the following percents to decimals.

14. 35%

15. 4%

16. 0.25%

17. $\dfrac{1}{4}\%$

18. 7.1%

19. 132%

Change the following numbers to percents.

20. $\dfrac{6}{10}$

21. $\dfrac{3}{20}$

22. $\dfrac{4}{25}$

23. $\dfrac{3}{8}$

24. $\dfrac{5}{12}$

25. $1\dfrac{4}{15}$

Change the following percents to fractions or mixed numbers.

26. 14% **27.** 40% **28.** 66%

29. $12\frac{1}{2}\%$ **30.** 400% **31.** $33\frac{1}{2}\%$

Solve each problem for the unknown quantity.

32. 30% of 52 is _____. **33.** 15% of 17 is _____.

34. 3% of _____ is 7. **35.** 42% of _____ is 18.

36. _____% of 36 is 7.2. **37.** _____% if 48 is 16.

38. 75 is _____% of 300. **39.** _____ is 6% of 18.25.

40. 5 is 10% of _____. **41.** 14 is $5\frac{1}{2}\%$ of _____.

42. _____ is $6\frac{1}{2}\%$ of 15. **43.** 62 is _____% of 31.

Choose the correct answer by reasoning and calculating mentally.

44. Write $37\frac{1}{2}\%$ as a fraction, reduced.

(a) $\frac{1}{4}$ (b) $\frac{1}{3}$ (c) $\frac{3}{8}$ (d) $\frac{5}{8}$

45. Find 10% of 72.6.

(a) 0.726 (b) 7.26 (c) 72.6 (d) 726

46. Write $5\frac{1}{2}\%$ as a decimal.

(a) 0.055 (b) 0.55 (c) 5.5 (d) 550

47. What is 200% of 18.3?

(a) 3.66 (b) 36.6 (c) 366 (d) 3660

48. Write $\frac{3}{4}$ as a percent.

(a) 0.75% (b) 7.5% (c) 75% (d) 750%

49. A shirt was marked 25% off. What would you pay for the shirt if the original price was $15 and you had to pay 6% sales tax?

50. Men's topcoats were on sale for $180. This was a discount of $30 from the original price. (a) If the store owner paid $120 for the coats, what was his percent of profit based on his cost? (b) What was his percent of profit based on the selling price?

51. A student received a grade of 75% on a statistics test. If there were 32 problems on the test, all of equal value, how many problems did the student miss?

52. A salesman works on a 9% commission on his sales over $10,000 each month, plus a base salary of $600 per month. How much did he make the month he sold $25,000 in merchandise?

53. The property taxes on a house were $1800. What was the tax rate if the house was valued at $150,000?

54. Mary's allowance each week was $15. She saved for 6 weeks, then she spent $5 on a movie, $35 on clothes, and $20 on a gift for her parents' anniversary. What percent of her savings did she spend on each item?

55. The discount on a new car was $1500, including a rebate from the company. (a) What was the original price of the car if the discount was 15% of the original price? (b) What would be paid for the car if taxes and license fees totaled $650?

Test: Chapter 7

In Exercises 1–5, change each number to a percent.

1. $\dfrac{101}{100}$ **2.** 0.003 **3.** 0.173

4. $\dfrac{8}{25}$ **5.** $2\dfrac{3}{8}$

In Exercises 6–8, change each percent to a decimal.

6. 70% **7.** 180% **8.** 9.3%

In Exercises 9–11, change each percent to a fraction or mixed number with the fraction part reduced.

9. 130% **10.** $35\dfrac{1}{2}\%$ **11.** 8.6%

In Exercises 12–17, find the unknown quantity.

12. 10% of 56 is _____. **13.** 12 is _____% of 36.

14. 33 is 110% of _____. **15.** _____ is 62% of 475.

16. 6.765 is 33% of _____. **17.** 16.55 is _____% of 50.

Choose the correct answer by reasoning and calculating mentally.

18. Write 11% as a fraction.

(a) $\dfrac{11}{1}$ (b) $\dfrac{11}{10}$ (c) $\dfrac{11}{100}$ (d) $\dfrac{11}{1000}$

19. Write $\frac{2}{3}$ as a percent.

(a) 66%   (b) 67%   (c) $33\frac{1}{3}\%$ (d) $66\frac{2}{3}\%$

20. Find 25% of 100.

(a) 2.5   (b) 25   (c) 250   (d) 2500

21. 10% of what number is 62?

(a) 6.2   (b) 62   (c) 620   (d) 6200

22. Using the rule of thumb stated in the text, calculate the 15% tip for each of the following amounts.

(a) $6.72   (b) $25.35   (c) $17.95

23. A hardware store has light fixtures on sale for $50.00. This represents a 20% discount. (a) What was the original selling price? (b) If the fixtures cost the store $40, what was the store's percent of profit based on cost? (c) What was the store's percent of profit based on selling price?

24. A customer received a 2% discount on the purchase of a new dining room set because she paid in cash. (a) If she paid $1176, what was the original selling price? (b) What was the amount of the discount?

25. A salesman earns $900 each month plus a commission of 8% of his sales over $20,000. How much did he earn for the month in which his sales were $50,000?

1. The number 0 is called the additive _____.

2. Name the property of addition that is illustrated.

   $17 + 15 = 15 + 17$

3. Round off as indicated:

   (a) 12,943 (nearest thousand)   (b) 3.09672 (nearest thousandth)

First estimate the result, then perform the indicated operations to find the actual result.

4. $\begin{array}{r} 8695 \\ 457 \\ + 1206 \\ \hline \end{array}$ 5. $\begin{array}{r} 8500 \\ - 4675 \\ \hline \end{array}$ 6. $72\overline{)6696}$ 7. $\begin{array}{r} 182 \\ \times\ 36 \\ \hline \end{array}$

8. Evaluate the expression $3(5 + 2^2) - 6 - 2 \cdot 3^2$.

9. List the squares of the prime numbers less than 30.

10. (a) Find the LCM, and (b) state the number of times each number divides into the LCM for the following set of numbers: 56, 60, 75.

Evaluate each of the following expressions. Reduce all fractions.

11. $\dfrac{15}{28} \cdot \dfrac{7}{25}$

12. $\dfrac{3}{4} + \dfrac{7}{8} \div \dfrac{1}{2}$

13. $5.6 + 7\dfrac{1}{2} - 3\dfrac{5}{8}$

14. $\dfrac{\dfrac{1}{5} + \dfrac{7}{20}}{2\dfrac{1}{2} - 1\dfrac{3}{5}}$

15. Solve the proportion: $\dfrac{7}{x} = \dfrac{3\dfrac{1}{2}}{11}$

16. Find 75% of 104.   17. 82% of _____ is 328.

18. Write each of the following numbers in scientific notation.

    (a) 563,000,000   (b) 0.0000942

19. Find the equivalent measures in the metric system.

    (a) 345 cm = _____ mm   (b) 1.6 m = _____ cm

    (c) 830 mm = _____ cm   (d) 5200 m = _____ km

20. A clothing store has dresses on sale for $80.00. This represents a 20% discount. (a) What was the original selling price? (b) If the dresses cost the store $60,

what was the store's percent of profit based on cost? (c) What was the store's percent of profit based on selling price?

21. You ordered two pizzas to be delivered to your home. One was for $11.75 and the other was for $15.80. (a) If you wanted to give the driver a 15% tip, what would be the amount of the tip? (b) How much would you pay the driver?

22. You paid $4500 as a down payment on a new car and made 48 monthly payments of $350 each. (a) Approximately how much did you pay for the car? (b) Exactly how much did you pay for the car? (c) How much more did you pay for the car if the car could have been bought originally for $17,040 in cash? (d) What percent more did you pay by making monthly payments?

Consumer Applications

Mathematics at Work!

The interest rate that you pay on a loan will depend on the bank from which you borrow and the purpose of the loan.

At some time in your life, you will probably borrow money (this is the same as using a credit card), lend money, or have a savings account. In any case, you will need to understand the concept of **interest,** money paid for the use of money. If you have a loan (or a savings account) with a bank, you should know what type of interest you are paying (or getting), **simple interest** or **compound interest,** and what the rates are. You should know that the terms for loans and investments are negotiable; so it may pay for you to shop around for the best terms.

The formula for calculating **simple interest** (involving only one payment at the end of the term of the loan) is

$$I = P \times r \times t$$

where
I = Interest (earned or paid) P = Principal (the original amount)
r = rate of interest t = time (in years)

Interest paid on interest earned is called **compound interest.** When interest is compounded, the total amount A accumulated (including principal and interest) is given by the formula

$$A = P\left(1 + \frac{r}{n}\right)^{nt}$$

where n is the number of compounding periods in one year. For example, if interest is compounded quarterly, then $n = 4$. If interest is compounded daily, then $n = 360$. (**Note:** When calculating interest in this text, we will use 30 days per month and 360 days per year, a common practice in business.)

Suppose that $5000 is invested at 8% for 4 years and interest is compounded. Do you think there will be a difference in the accumulated amount if (a) interest is compounded quarterly and (b) interest is compounded daily? Do you think that this difference will be (a) about $5, (b) about $20, (c) about $50, or (d) over $100? (See Exercise 13 in Section 8.2.)

What to Expect in Chapter 8

Interest is money paid for the use of money, and the related concepts of simple interest (Section 8.1) and compound interest (Section 8.2) provide important applications of percent and interest that every adult in our society should understand.

Some of the many expenses in becoming a homeowner and other expenses in maintaining a home you own are discussed in Section 8.3. This discussion and related problems are designed to be informative regardless of whether you own a home, or rent an apartment, or anticipate owning a home. Section 8.4 presents a similar discussion of the expenses involved in buying a car and in owning a car.

Note that all the consumer applications in this chapter involve percent in some way.

8.1   Simple Interest

OBJECTIVES

1. Understand the concept of **simple interest.**
2. Find simple interest using the formula $I = P \times r \times t$.

Understanding Simple Interest

Interest is money paid for the use of money. The money that is invested or borrowed is called the **principal.** The **rate** is the **percent of interest** and is almost always stated as an **annual (yearly) rate.**

Interest is either paid or earned, depending on whether you are the borrower or the lender. In either case, the calculations involved are the same. Although interest rates can vary from year to year (or even daily, as in the case of home loans) and from one part of the world to another, the concept of interest is the same everywhere.

Some loans (called **notes**) are based on **simple interest** and involve only one payment (including interest and principal) at the end of the term of the loan. Such loans are usually made for a period of one year or less. (**Compound interest,** involving interest paid on interest, will be discussed in Section 8.2.)

The following formula is used to calculate simple interest.

Formula for Calculating Simple Interest

$$I = P \times r \times t$$

where

I = Interest (earned or paid)

P = Principal (the amount invested or borrowed)

r = rate of interest (stated as an annual or yearly rate and used in decimal or fraction form)

t = time (in years or fraction of a year)

Note: For the purpose of calculations, we will use 360 days as the number of days in one year (30 days in a month). This approximation is a common practice in business and banking, although with the advent of computers, many lending institutions now base their calculations on 365 days per year and pay or charge interest on a daily basis.

EXAMPLE 1

You want to borrow $2000 at 12% for one year. How much interest would you pay?

Solution

Use the formula for simple interest: $I = P \times r \times t$ with

$P = \$2000, r = 12\% = 0.12,$ and $t = 1$ year

$I = \$2000 \times 0.12 \times 1 = \$240$

You would pay $240 interest.

EXAMPLE 2

You decided you might need the $2000 for only 90 days. How much interest would you pay? (The interest rate would still be stated at 12%, the annual rate.)

Solution

Use the formula for simple interest: $I = P \times r \times t$ with

$P = \$2000, r = 12\% = 0.12,$ and $t = 90$ days $= \dfrac{90}{360}$ year $= \dfrac{1}{4}$ year

$I = \$2000 \times 0.12 \times \dfrac{1}{4} = \dfrac{240}{4} = \$60$

or

$I = \$2000 \times 0.12 \times 0.25 = \$60$

You would pay $60 interest if you borrowed the money for 90 days.

EXAMPLE 3

Sylvia borrowed $2400 at 10% interest for 30 days. How much interest did she have to pay?

Solution

Use $P = \$2400, r = 10\% = 0.10, t = 30$ days $= \dfrac{1}{12}$ year.

$I = 2400 \times 0.10 \times \dfrac{1}{12} = \$20$

She would have to pay $20 in interest.

If you know the values of any three of the variables in the formula $I = P \times r \times t$, you can find the value of the fourth variable by substituting into the formula and solving the resulting equation for the unknown variable. This procedure is illustrated in Examples 4 and 5.

EXAMPLE 4

How much (what principal) would you need to invest if your investment returned 9% interest and you wanted to make $100 in interest in 30 days?

Solution

Here the principal is unknown, while the interest (I = \$100), the rate of interest (r = 9% = 0.09), and the time (t = 30 days) are all known. However, before substituting into the formula, we must change t to a fraction of a year:

$$t = \frac{30}{360} = \frac{1}{12} \text{ year.}$$

$$I = P \times r \times t$$

$$100 = P \times 0.09 \times \frac{1}{12}$$

$$100 = P \times \frac{9}{100} \times \frac{1}{12}$$

$$100 = P \times \frac{3}{400}$$

$$\frac{400}{3} \times \frac{100}{1} = P \times \frac{3}{400} \times \frac{400}{3} \quad \text{Multiply both sides by } \frac{400}{3}.$$

$$\frac{40,000}{3} = P$$

$$P = \$13,333.34$$

or, using $\frac{1}{12}$ = 0.083 as a rounded-off decimal,

$$100 = P \times 0.09 \times 0.083$$

$$100 = P \times 0.00747$$

$$\frac{100}{0.00747} = \frac{P \times 0.00747}{0.00747}$$

$$P = \$13,386.89$$

Using the rounded-off decimal 0.083 leads to an error in the result because any rounded-off value that is used in a calculation produces some error. The correct result can be found by using 0.083333333, which is the display on your calculator for 1 ÷ 12.

EXAMPLE 5

Stuart wants to borrow \$1500 at 10% and is willing to pay \$250 in simple interest. How long can he keep the money?

Solution

Here time is unknown and the principal is \$1500, interest is \$250, and the rate is 10% = 0.10.

Substituting into the formula $I = P \times r \times t$ gives

$$250 = 1500 \times 0.10 \times t$$

$$250 = 150 \times t$$

$$\frac{250}{150} = \frac{\cancel{150} \times t}{\cancel{150}}$$

$$\frac{5}{3} = t$$

or

$$t = 1\frac{2}{3} \text{ years or 1 year 8 months}$$

Stuart can borrow the money for $1\frac{2}{3}$ years.

CLASSROOM PRACTICE

1. Jack borrowed $3000 at 6% for one year. How much interest did he pay?

2. Marcie lent her brother $1500 at 10% interest for 9 months. How much interest did she earn?

3. What interest rate would you be paying if you borrowed $1000 for 6 months and paid $60 in interest?

ANSWERS: **1.** $180   **2.** $112.50   **3.** 12%

Exercises 8.1

1. What is the simple interest paid on $500 at 6% for one year?

2. What is the simple interest paid on $2000 at 8% for one year?

3. How much interest would be paid on a loan of $1000 at 18% for 6 months? (**Note:** This interest rate may seem high, but the interest rates on some credit cards are even higher.)

4. How much interest would be paid on a loan of $3000 at 12% for 9 months?

5. You invested $2000 at 8% for 60 days. How much interest did your money earn?

6. Stacey lent her brother $1500 for 8 months at 10% interest. How much interest did she earn?

7. What principal will earn $50 in interest if it is invested at 8% for 90 days?

8. What principal will earn $75 in interest if it is invested for 60 days at 9%?

9. How long will it take for $1000 invested at 5% to earn $50 in simple interest?

10. What length of time will it take to earn $70 in simple interest if $2000 is invested at 7%?

11. What will be the interest earned in one year on a savings account of $800 if the bank pays 4% interest?

12. If interest is paid at 6% for one year, what will a principal of $1800 earn?

13. If a principal of $900 is invested at a rate of 9% for 90 days, what will be the interest earned?

14. A loan of $5000 is made at 8% for a period of 6 months. How much interest is paid?

15. If you borrow $750 for 30 days at 18%, how much interest will you pay?

16. How much interest is paid on a 60-day loan of $500 at 12%?

17. Find the simple interest paid on a savings account of $2800 for 120 days at 3.5%.

18. A savings account of $5300 is left for 90 days drawing interest at a rate of 5%. (a) How much interest is earned? (b) What is the amount in the account at the end of 90 days?

19. Every 6 months a stock pays 10% in dividends (interest on investment). What will be the earnings of $14,600 invested for 6 months? (Remember that rates of interest are given as annual rates.)

20. If you charge $1000 worth of merchandise at a local department store at 18% interest, how much will you owe at the end of 60 days?

21. You buy an oven on sale from $500 to $450, but you don't make a payment for 60 days and are charged interest at a rate of 18%. (a) How much do you pay for the oven by waiting 60 days to pay? (b) How much did you save by buying the oven on sale? (Sales tax is not included here.)

22. A friend borrows $500 from you for a period of 8 months and pays you interest at 6%. (a) How much interest are you paid? (b) If you had asked for 8%, how much more interest would you have earned?

23. What principal would have to be invested at 8% for 60 days to earn interest of $500?

24. What rate of interest is charged if a loan of $2500 for 90 days is paid off with $2562.50?

25. How many days must you leave $1000 in a savings account at 5.5% to earn $11.00?

26. Determine the missing item in each row.

Principal	Rate	Time	Interest
$ 400	16%	90 days	$ (a)
$ (b)	15%	120 days	$ 5.00
$ 560	12%	(c)	$ 5.60
$2700	(d)	40 days	$25.50

27. Determine the missing item in each row.

Principal	Rate	Time	Interest
$ 500	18%	30 days	$ (a)
$ 500	18%	(b)	$15.00
$ 500	(c)	90 days	$22.50
$ (d)	18%	30 days	$ 1.50

28. (a) If Carlos has a savings account of $25,000 drawing interest at 8%, how much interest will he earn in 6 months? (b) How long must he leave the money in the account to earn $1500?

29. Ms. Lee has accumulated $240,000 and she wants to live on the interest each year. If she needs $2000 a month to live on, what interest rate must she earn on her money?

30. Mr. Smith has a savings account of $2500 that draws 4.5% interest. How many days will it take for him to earn $56.25?

31. A bank decides to lend $5 million to a contractor to build new homes. How much interest will the bank earn in one year if the interest rate is 9.2%?

32. A credit card company has $120 million loaned to its customers at 18.9%. How much interest will it earn in one month?

33. A small airline company borrowed $7.5 million to buy some new airplanes. The loan rate was 7.5% and the airline paid $562,500 in interest. What was the length of time of the loan?

34. A department store keeps $15 million in merchandise in stock. If the store pays interest at 9% on a bank loan for this stock, how much interest will the store pay in 3 months' time?

35. Determine the missing item in each row.

Principal	Rate	Time	Interest
$1000	$10\frac{1}{2}\%$	60 days	$ _(a)_
$ 800	$13\frac{1}{2}\%$	_(b)_	$ 18.00
$2000	_(c)_	9 months	$172.50
_(d)_	$7\frac{1}{2}\%$	1 year	$ 85.00

8.2   Compound Interest

OBJECTIVES

1. Understand the concept of **compound interest.**
2. Find interest compounded periodically with repeated calculations.
3. Find compound interest using the formula $A = P\left(1 + \dfrac{r}{n}\right)^{nt}$.

Understanding Compound Interest

Interest paid on interest is called **compound interest.** To calculate compound interest, we can calculate the simple interest for each period of time that interest is now compounded, **using a new principal for each calculation.** This new principal is the **previous principal plus the earned interest.** The calculations can be performed in a step-by-step manner, as indicated in the following outline.

To Calculate Compound Interest:

1. Using the formula for simple interest, $I = P \times r \times t$, calculate the simple interest, where $t = \dfrac{1}{n}$ and n is the number of periods per year for compounding. For example:

For compounding annually, $n = 1$ and $t = \dfrac{1}{1} = 1$.

For compounding semiannually, $n = 2$ and $t = \dfrac{1}{2}$.

For compounding quarterly, $n = 4$ and $t = \dfrac{1}{4}$.

For compounding monthly, $n = 12$ and $t = \dfrac{1}{12}$.

For compounding daily, $n = 360$ and $t = \dfrac{1}{360}$.

2. Add this interest to the principal to create a new value for the principal.

3. Repeat Steps 1 and 2 however many times that interest is to be compounded.

In Examples 1 and 3 we show how compound interest can be calculated in the step-by-step manner just outlined. This process will help you develop a basic understanding of the concept of compound interest. (Remember that if calculations with dollars and cents involve three or more decimal places, round answers up to the next higher cent regardless of the digit in the thousandths place.)

After Example 3 we will discuss another formula and show how to calculate compound interest with a calculator and key marked $\boxed{x^y}$ (or $\boxed{y^x}$).

EXAMPLE 1

You deposit $1000 in an account that pays 8% interest compounded quarterly (every 3 months). How much interest would you earn in one year?

Solution

Use $t = \dfrac{1}{n} = \dfrac{1}{4}$ and the formula $I = P \times r \times t$ and calculate the interest four times.

(a) First period: $P = \$1000$.

$$I = 1000 \times \overset{0.02}{\cancel{0.08}} \times \frac{1}{\cancel{4}} = \$20 \text{ interest}$$

(b) Second period: $P = \$1000 + \$20 = \$1020$.

$$I = 1020 \times \underset{0.02}{0.08} \times \frac{1}{4} = \$20.40 \text{ interest}$$

(c) Third period: $P = \$1020 + \$20.40 = \$1040.40$.

$$I = 1040.40 \times \underset{0.02}{0.08} \times \frac{1}{4} = \$20.81 \text{ interest}$$

(d) Fourth period: $P = \$1040.40 + \$20.81 = \$1061.21$.

$$I = 1061.21 \times \underset{0.02}{0.08} \times \frac{1}{4} = \$21.23 \text{ interest}$$

$$
\begin{array}{r}
\$20.00 \\
20.40 \\
20.81 \\
+\ 21.23 \\
\hline
\$82.44
\end{array}
\quad \text{total interest earned in 1 year}
$$

The balance in the account will be $\$1000 + \$82.44 = \$1082.44$.

EXAMPLE 2

In Example 1, how much more interest will you earn by having the interest compounded quarterly for one year rather than calculated just as simple interest for the year?

Solution

Simple interest for one year would be $I = \$1000 \times 0.08 \times 1 = \$80.00$. The difference is

$$
\begin{array}{rl}
\$82.44 & \text{compound interest} \\
-\ 80.00 & \text{simple interest} \\
\hline
\$\ 2.44 & \text{more by compounding quarterly}
\end{array}
$$

EXAMPLE 3

If an account is compounded monthly at 10%, how much interest will $6000 earn in three months?

Solution

Use $t = \dfrac{1}{12}$ and calculate the interest three times.

(a) First period: $P = \$6000$.

$$I = \$6000 \times 0.10 \times \frac{1}{12} = \$50 \text{ interest}$$

(b) Second period: $P = \$6000 + \$50 = \$6050$.

$$I = \$6050 \times 0.10 \times \frac{1}{12} = \$50.42 \text{ interest}$$

(c) Third period: $P = \$6050 + \$50.42 = \$6100.42$.

$$I = \$6100.42 \times 0.10 \times \frac{1}{12} = \$50.84 \text{ interest}$$

The total interest earned will be

$$
\begin{array}{r}
\$\ 50.00 \\
50.42 \\
+\quad 50.84 \\
\hline
\$151.26
\end{array}
$$

Finding Compound Interest Using the Formula $A = P\left(1 + \dfrac{r}{n}\right)^{nt}$

The steps outlined in Examples 1 and 3 illustrate how the principal is adjusted for each time period of the compounding process and show the interest earned over each time period. The following compound interest formula can be used to find the total **amount** accumulated (also called the **future value** of the principal). To work with this formula, use a calculator and follow the process outlined in Example 4. Note that the steps follow the rules for order of operations.

Compound Interest Formula

When interest is compounded, the total, **amount A,** accumulated (including principal and interest) is given by the formula

$$A = P\left(1 + \frac{r}{n}\right)^{nt}$$

where

P = principal

r = annual interest rate (in decimal or fraction form)

t = length of time in years

n = number of compounding periods in one year

EXAMPLE 4   Mike invested \$4000 at 6% interest to be compounded monthly. What will be the amount in his account in 5 years?

Solution

Use $P = \$4000$, $r = 6\% = 0.06$ $n = 12$ times per year, $t = 5$ years.
Substituting into the formula gives

$$A = 4000\left(1 + \frac{0.06}{12}\right)^{12 \cdot 5}$$

$$= 4000(1 + 0.005)^{60}$$

$$= 4000(1.005)^{60}$$

$$= \$5395.41 \text{ total amount accumulated over 5 years}$$

The expression in the formula can be evaluated with a calculator by following the rules for order of operations.

STEP 1:   Enter 0.06.   The display should show 0.06.

STEP 2:   Press the key $\div$.

STEP 3:   Enter 12.   The display should show 12.

STEP 4:   Press the key $=$.   The display should show 0.005.

STEP 5:   Press the key $+$.

STEP 6:   Enter 1.

STEP 7:   Press the key $=$.   The display should show 1.005.
                                           This is the base for the exponent.

STEP 8:   Press the key x^y.   (This key is marked y^x on some calculators.)

STEP 9:   Enter 60.   This is the exponent.

STEP 10:   Press the key $=$.

You should now have 1.348850153 on your calculator display. This is the future value of $1. Since $4000 was invested, we need to multiply by 4000 to find the total amount A.

STEP 11:   Press the key $\times$.

STEP 12:   Enter 4000.   This is the principal.

STEP 13:   Press the key $=$.

The display should read 5395.40061.

The amount in Mike's account after 5 years would be $5395.41.

The sequence of steps used in Example 4 can be diagrammed as follows:

rate of interest		number of compounding periods		

(Enter) $\overbrace{0.06}$ (press) $\div$ (enter) $\overbrace{12}$ (press) $=$ (press) $+$

exponent

(Enter)   1   (press) $\boxed{=}$ (press) $\boxed{x^y}$ (enter) $\overbrace{60}$ (press) $\boxed{=}$

You now have the future value of $1 on the display.

principal

(Press) $\boxed{\times}$ (enter) $\overbrace{4000}$ (press) $\boxed{=}$

You now have the future value of $4000 on the display.

If we want to find the interest earned on an investment after using the compound interest formula, we subtract the original principal from the final amount:

$$I = A - P$$

EXAMPLE 5

How much interest did Mike earn in the investment described in Example 4?

Solution

$$\begin{aligned} I &= A - P \\ &= 5395.41 - 4000.00 \\ &= 1395.41 \end{aligned}$$

Mike earned $1395.41 in interest.

EXAMPLE 6

(a) Use the compound interest formula to find the value of $10,000 invested for 3 years if it is compounded daily at 12%. (b) Find the amount of interest earned.

Solution

(a) Follow the steps outlined in Example 4 with

$$P = \$10{,}000, \quad r = 12\% = 0.12, \quad n = 360, \quad t = 3.$$

Substituting into the formula gives

$$A = 10{,}000\left(1 + \frac{0.12}{360}\right)^{360(3)}$$

$$= 10{,}000(1 + 0.000333333)^{1080}$$

$$= 10{,}000(1.000333333)^{1080}$$

$$= 10{,}000(1.433243431)$$

$$= 14{,}332.44$$

The value (or Amount) will be $14,332.44.

(b) The interest earned will be

$$I = \$14{,}332.44 - \$10{,}000.00 = \$4332.44.$$

Exercises 8.2

In Exercises 1–8, use the formula for simple interest, $I = P \times r \times t$, repeatedly, as shown in Examples 1 and 3 in the text. Show the calculations for each period of compounding.

1. (a) If a bank compounds interest quarterly at 4% on a certificate of deposit, what will an investment of $13,000 be worth in 6 months? (b) in one year?

2. If a $9000 deposit in a savings account earns 5% compounded monthly, what will be the balance in the account in 6 months?

3. (a) If an account is compounded quarterly at a rate of 6%, how much interest will be earned on $5000 in one year? (b) What will be the total amount in the account? (c) How much more interest will be earned in the first year because the compounding is done quarterly rather than annually?

4. (a) How much interest will be earned in 2 years on a loan of $4000 compounded semiannually at 6%? (b) How much will be owed at the end of the 2-year period?

5. (a) How much interest will be earned on a savings account of $3000 in two years if interest is compounded annually at 5.5%? (b) If interest is compounded semiannually?

6. If interest is calculated at 10% compounded quarterly, what will be the value of $15,000 in 9 months?

7. Calculate the interest earned in six months on $10,000 compounded monthly at 8%.

8. Calculate the interest you will pay on a loan of $6500 for one year if the interest is compounded every 3 months at 18% and you make no monthly payments.

Note that in Exercises 9 and 10, payments are made so that the principal actually decreases.

9. Suppose that you borrow $4000 and agree to make four equal payments of $1000 each plus interest over the next four years. Interest is to be calculated at a rate of 6% based only on what you owe. How much interest will you pay?

10. Linda agreed to loan her son $400 under the following terms: He is to make payments of $100 plus interest every 60 days, and the interest rate is to be 6.6%. (a) How long will it take him to repay the loan? (b) How much interest will he pay?

In Exercises 11–20, use the formula for compound interest, $A = P\left(1 + \dfrac{r}{n}\right)^{nt}$, and the formula for finding interest, $I = A - P$, whenever each applies.

11. (a) Calculate the interest in one year on $5000 compounded monthly at 12%. (b) Suppose the interest is compounded semiannually. Is the accumulated value the same? (c) If not, explain why not in your own words.

12. (a) What will be the interest on $10,000 compounded daily at 10% for one year? (b) What is the difference between this and simple interest at 10% for one year?

13. (a) Find the value of $5000 compounded quarterly at 8% for 4 years. (b) What do you think the difference in interest would be if the money was compounded daily: about $5, $20, $50, or over $100? (c) Find the exact difference in interest.

14. (a) What would be the value of a $20,000 savings account at the end of 5 years if interest is calculated at 7% compounded annually? (b) How much more would be earned if the interest was compounded daily?

15. (a) Suppose that $3000 is invested at 5% and compounded monthly for one year. Find the accumulated value. (b) Is the accumulated amount the same if the original principal of $3000 is compounded annually for 12 years? If not, what is the difference?

16. (a) Find the value of $25,000 compounded daily at 5% for 20 years. (b) Do you think that the amount will be doubled or more than doubled if the rate is doubled to 10%? (c) Find the amount if the rate is 10%.

In Exercises 17–20, find the amount A and the interest earned I for the given information.

Compounding Period	Principal	Annual Rate	Time	A	$I = A - P$
17. Quarterly	$1000	10%	5 yr	(a)	(b)
18. Monthly	$1000	10%	5 yr	(a)	(b)
19. Daily	$2000	5%	10 yr	(a)	(b)
20. Daily	$7500	8%	20 yr	(a)	(b)

21. Use your calculator and choose the values of t (in years) to use in the formula for compound interest until you find how many years of daily compounding at 6% are needed for an investment of $6000 to approximately double in value. Write down the values you chose for t, why you chose those particular values, and the corresponding accumulated values of money. Explain why you agree or disagree with the idea that $10,000 would double in a shorter time period.

22. Use your calculator and choose the values of t (in years) to use in the formula for compound interest until you find how many years of daily compounding at 6% are needed for an investment of $10,000 to approximately triple in value. Write down the values you chose for t, why you chose those particular values, and the corresponding accumulated values of money. Explain why you agree or disagree with the idea that $30,000 would triple in a shorter time period.

8.3   Balancing a Checking Account

OBJECTIVES

1. Become familiar with the information in a bank statement.
2. Learn how to keep a checkbook register current.
3. Know how to balance a checking account.

Many people do not balance their checking accounts simply because they were never told how. Others trust the bank's calculations over their own. This is a poor practice because, for a variety of reasons, "computers do make errors." Also, because the bank's statement comes only once a month, you should know your current balance so your account will not be overdrawn. Overdrawn accounts pay a penalty and can lead to a bad credit rating.

Like many applications with mathematics, balancing a checking account follows a pattern of steps. The procedure and related ideas are given in the following lists.

The bank or the savings and loan company sends you a statement of your checking account each month. This statement contains a record of:

1. The beginning balance;

2. A list of all checks paid by the bank;

3. A list of deposits you made;

4. Any interest paid to you (some checking accounts do pay interest);

5. Any service charge by the bank; and

6. The closing balance.

Your checkbook register contains a record of:

1. All the checks you have written;

2. All the deposits you have made; and

3. The current balance.

Your current balance will not agree with the closing balance on the bank statement because:

1. You have not recorded the interest.

2. You have not recorded the service charge.

3. The bank does not have a record of all the checks you have written. (Several checks will be *outstanding* because they have not been received yet for payment by the bank by the date given on the bank statement.)

To Balance Your Checking Account
(sometimes called **reconciling** the bank
statement with your checkbook register):

1. Go through your checkbook register and the bank statement and put a check mark (✔) by each check paid and deposit recorded on the bank statement.

2. In your checkbook register:

   (a) Add any interest paid to your current balance.

   (b) Subtract any service charge from this balance.

   This number is your **true balance.**

3. Find the total of all the outstanding checks (no ✔ mark) in your checkbook register (checks not yet received by the bank).

4. On a reconciliation sheet:

   (a) Enter the balance from the bank statement.

   (b) Add any deposits you have made that are not recorded (no ✔ mark) on the bank statement.

   (c) Subtract the total of your outstanding checks (found in Step 3). This number should agree with your **true balance.** Now your checking account is balanced.

**If Your Checking Account
Does Not Balance:**

1. Go over your arithmetic in the balancing procedure.

2. Go over your arithmetic check by check in your checkbook register.

3. Make an appointment with bank personnel to find the reasons for any other errors.

EXAMPLE 1

YOUR CHECKBOOK REGISTER

Check No.	Date	Transaction Description	Payment (−)	(√)	Deposit (+)	Balance
		Balance brought forward				312.12
102	2-3	Painted Lady (cosmetics)	14.80	√		− 14.80 / 297.32
103	2-4	Nu – Grocers (groceries)	76.51			− 76.51 / 220.81
104	2-9	Wood Lumber (plywood)	53.70	√		− 53.70 / 167.11
	2-16	Deposit ( )		√	100.00	+ 100.00 / 267.11
105	2-21	Metal Box (nails)	10.14	√		− 10.14 / 256.97
	2-26	Deposit ( )			250.00	+ 250.00 / 506.97
	2-28	Bank Interest ( )		√		—— / 506.97
	2-28	Service Charge ( )	− 3.00	√		− 3.00 / 503.97
		( )				——
		( )				——
		True Balance				503.97

BANK STATEMENT
Checking Account Activity

Transaction Description	Amount	(√)	Running Balance	Date
Beginning Balance		√	312.12	2-1
Check #102	14.80	√	297.32	2-4
Check #104	53.70	√	243.62	2-9
Deposit	100.00	√	343.62	2-16
Check #105	10.14	√	333.48	2-24
Service Charge	3.00		330.48	2-28
Ending Balance			330.48	

RECONCILIATION SHEET

A. First mark √ beside each check and deposit listed in both your checkbook register and on the bank statement.

B. Second, in your checkbook register, add any interest paid and subtract any service charge listed on the bank statement.

C. Third, find the total of all outstanding checks.

Outstanding Checks	
No.	Amount
103	76.51
Total	76.51

Statement Balance	330.48
Add deposits not credited	+ 250.00
Total	580.48
Subtract total amount of checks outstanding	− 76.51
True Balance	503.97

EXAMPLE 2

YOUR CHECKBOOK REGISTER

Check No.	Date	Transaction Description	Payment (−)	(√)	Deposit (+)	Balance
		Balance brought forward				505.21
152	1-3	U – Auto Lease (car lease)	198.60	√		− 198.60 / 306.61
153	1-10	ABC Power Co. (electric bill)	75.00			− 75.00 / 231.61
154	1-10	Fly-By-Nite Airway (plane ticket)	112.00	√		− 112.00 / 119.61
	1-15	Deposit ( )		√	500.00	+ 500.00 / 619.61
155	1-15	Safe – 7 Drugs (medicine)	5.60	√		− 5.60 / 614.01
156	1-17	Milady Togs (dress)	49.80	√		− 49.80 / 564.21
157	2-03	U – Auto Lease (car lease)	198.60			− 198.60 / 365.61
	1-31	Bank Interest ( )		√		—— / 365.61
	1-31	Service Charge ( )	4.00	√		− 4.00 / 361.61
		( )				——
		True Balance				361.61

BANK STATEMENT
Checking Account Activity

Transaction Description	Amount	(√)	Running Balance	Date
Beginning Balance		√	505.21	1–01
Check #152	198.60	√	306.61	1–06
Check #154	112.00	√	194.61	1–14
Deposit	500.00	√	694.61	1–15
Check #155	5.60	√	689.01	1–18
Check #156	49.80	√	639.21	1–20
Service Charge	4.00	√	639.21	1–31
Ending Balance			635.21	

RECONCILIATION SHEET

A. First mark √ beside each check and deposit listed in both your checkbook register and on the bank statement.

B. Second, in your checkbook register, add any interest paid and subtract any service charge listed on the bank statement.

C. Third, find the total of all outstanding checks.

Outstanding Checks	
No.	Amount
153	75.00
157	198.60
Total	273.60

Statement Balance	635.21
Add deposits not credited	+ ——
Total	635.21
Subtract total amount of checks outstanding	− 273.60
True Balance	361.61

Exercises 8.3

For each of the following problems, you are given a copy of a checkbook register, the corresponding bank statement, and a reconciliation sheet. You are to find the true balance of the account on both the checkbook register and on the reconciliation sheet as shown in Example 1. Follow the directions on the reconciliation sheet.

1.

YOUR CHECKBOOK REGISTER

Check No.	Date	Transaction Description	Payment (−)	(√)	Deposit (+)	Balance
		Balance brought forward				0
	7-15	Deposit ( )		√	700.00	+700.00 / 700.00
1	7-15	Quiet Town Apt. (rent/deposit)	520.00			− 520.00 / 180.00
2	7-15	Pa Bell Telephone (phone installation)	32.16	√		− 32.16 / 147.84
3	8-15	XYZ Power Co. (gas/elect. hook up)	46.49	√		− 46.49 / 101.35
4	7-16	Foodway Stores (groceries)	51.90	√		− 51.90 / 49.45
	7-20	Deposit ( )			350.00	+ 350.00 / 399.45
5	7-23	Comfy Furniture (sofa, chair)	300.50	√		− 300.50 / 98.95
	8-1	Deposit ( )		√	350.00	+ 350.00 / 448.95
	7-31	Interest ( )			2.50	+ 2.50
	7-31	Service ( )	2.00			− 2.00
		True Balance				

BANK STATEMENT
Checking Account Activity

Transaction Description	Amount	(√)	Running Balance	Date
Beginning Balance	0.00	√	0.00	7-1
Deposit	700.00		700.00	7-15
Check #1	520.00		180.00	7-16
Check #4	51.90	√	128.10	7-17
Check #2	32.16	√	95.94	7-18
Check #3	46.49	√	49.45	7-18
Deposit	350.00	√	399.45	7-20
Interest	2.50		401.95	7-31
Service Charge	2.00	√	399.95	7-31

Ending Balance		399.95

RECONCILIATION SHEET

A. First mark √ beside each check and deposit listed in both your checkbook register and on the bank statement.

B. Second, in your checkbook register, add any interest paid and subtract any service charge listed on the bank statement.

C. Third, find the total of all outstanding checks.

Outstanding Checks		Statement Balance _____
No.	Amount	
		Add deposits not credited _____
		Total _____
		Subtract total amount of checks outstanding _____
_____		
Total _____		True Balance _____

2.

		YOUR CHECKBOOK REGISTER				
Check No.	Date	Transaction Description	Payment (−)	(√)	Deposit (+)	Balance
		Balance brought forward				1610.39
1234	12-7	Pearl City (pearl ring)	524.00	√		− 524.00 / 1086.39
1235	12-7	Comp-U-Tate (home computer)	801.60	√		− 801.60 / 284.79
1236	12-8	Sportz Hutz (skis)	206.25	√		− 206.25 / 78.54
1237	12-8	Guild Card Shop (Christmas cards)	25.50	√		− 25.50 / 53.04
	12-10	Deposit ( )			1000.00	+ 1000 / 1053.04
1238	12-14	Toys-R-We (stuffed panda)	80.41	√		− 80.41
1239	12-24	Meat Markette (turkey)	18.39	√		− 18.39
1240	12-24	Poodle Shoppe (pedigreed puppy)	300.00	√		− 300.00
1241	12-31	Homey Sav.&Loan (mortgage payment)	600.00			− 600.00
	12-31	Service Charge ( )	0			———
		True Balance				

	BANK STATEMENT		
	Checking Account Activity		
Transaction Description	Amount	(√)	Running Balance	Date
Beginning Balance		√	1610.39	12-01
Check #1234	524.00	√	1086.39	12-08
Check #1236	206.25	√	880.14	12-09
Check #1237	25.50	√	854.64	12-09
Deposit	1000.00		1854.64	12-10
Check #1235	801.60	√	1053.04	12-11
Check #1238	80.41	√	972.63	12-15
Check #1239	18.39	√	954.24	12-27
Check #1240	300.00	√	654.24	12-28
Ending Balance			654.24	

RECONCILIATION SHEET

A. First mark √ beside each check and deposit listed in both your checkbook register and on the bank statement.
B. Second, in your checkbook register, add any interest paid and subtract any service charge listed on the bank statement.
C. Third, find the total of all outstanding checks.

Outstanding Checks			
No.	Amount		Statement Balance _____
			Add deposits not credited + _____
			Total _____
			Subtract total amount of checks outstanding − _____
	_____		
Total _____			True Balance ======

3.

		YOUR CHECKBOOK REGISTER				
Check No.	Date	Transaction Description	Payment (−)	(√)	Deposit (+)	Balance
		Balance brought forward				756.14
271	6-15	Parts, Parts, Parts (spark plugs)	12.72			− 12.72 / 743.42
272	6-24	Firerock Tire Co. (2 tires)	121.40			———
273	6-30	Gus' Gas Station (tune-up)	75.68			———
	7-1	Deposit ( )			250.00	———
274	7-1	Prudent Ins Co. (car insurance)	300.00			———
	6-30	Service Charge ( )				———
		True Balance				

	BANK STATEMENT		
	Checking Account Activity		
Transaction Description	Amount	(√)	Running Balance	Date
Beginning Balance			756.14	6-01
Check #271	12.72		743.42	6-16
Check #272	121.40		622.02	6-26
Service Charge	1.00		621.02	6-30
Ending Balance			621.02	

RECONCILIATION SHEET

A. First mark √ beside each check and deposit listed in both your checkbook register and on the bank statement.
B. Second, in your checkbook register, add any interest paid and subtract any service charge listed on the bank statement.
C. Third, find the total of all outstanding checks.

Outstanding Checks			
No.	Amount		Statement Balance _____
			Add deposits not credited + _____
			Total ======
			Subtract total amount of checks outstanding − _____
Total _____			True Balance ======

4.

YOUR CHECKBOOK REGISTER

Check No.	Date	Transaction Description	Payment (−)	(√)	Deposit (+)	Balance
		Balance brought forward				12.14
419	1–2	Postmaster (stamps)	10.00			− 10.00 / 2.14
	1–3	Deposit ( )			525.50	——
420	1–17	E. Namel, DDS (dental check-up)	63.50			——
421	1–26	Cash ( )	100.00			——
422	1–31	Up –Top Apartments (rent)	350.00			——
	1–31	Service Charge ( )				——
		( )				——
		( )				——
		( )				——
		( )				——
		True Balance				

BANK STATEMENT
Checking Account Activity

Transaction Description	Amount	(√)	Running Balance	Date
Beginning Balance			12.21	1–01
Deposit	525.50		537.64	1–03
Check #419	10.00		527.64	1–03
Check #420	63.50		464.14	1–20
Check #421	100.00		364.14	1–26
Service Charge	2.00		362.14	1–31
Ending Balance			362.14	

RECONCILIATION SHEET

A. First mark √ beside each check and deposit listed in both your checkbook register and on the bank statement.

B. Second, in your checkbook register, add any interest paid and subtract any service charge listed on the bank statement.

C. Third, find the total of all outstanding checks.

Outstanding Checks			
No.	Amount		
		Statement Balance	———
		Add deposits not credited	+ ———
		Total	———
		Subtract total amount of checks outstanding	———
Total ———		True Balance	———

5.

YOUR CHECKBOOK REGISTER

Check No.	Date	Transaction Description	Payment (−)	(√)	Deposit (+)	Balance
		Balance brought forward				967.22
772	4–13	C.P. Hay (accountant)	85.00			——
	4–14	Deposit ( )			1200.00	——
773	4–14	E.Z. Pharmacy (aspirin)	4.71			——
774	4–15	I.R.S (income tax)	2000.00			——
775	4–30	Heavy Finance Co. (loan payment)	52.50			——
	5–1	Deposit ( )			600.00	——
	4–30	Interest ( )				——
	4–30	Service Charge ( )				——
		( )				——
		( )				——
		True Balance				

BANK STATEMENT
Checking Account Activity

Transaction Description	Amount	(√)	Running Balance	Date
Beginning Balance			967.22	4–01
Deposit	1200.00		2167.22	4–14
Check #772	85.00		2082.22	4–15
Check #773	4.71		2077.51	4–15
Interest	2.82		2080.33	4–30
Service Charge	4.00		2076.33	4–30
Ending Balance			2076.33	

RECONCILIATION SHEET

A. First mark √ beside each check and deposit listed in both your checkbook register and on the bank statement.

B. Second, in your checkbook register, add any interest paid and subtract any service charge listed on the bank statement.

C. Third, find the total of all outstanding checks.

Outstanding Checks			
No.	Amount		
		Statement Balance	———
		Add deposits not credited	+ ———
		Total	———
		Subtract total amount of checks outstanding	———
Total ———		True Balance	———

6.

YOUR CHECKBOOK REGISTER

Check No.	Date	Transaction Description	Payment (−)	(√)	Deposit (+)	Balance
			Balance brought forward			1403.49
86	9-1	Now Stationers (school supplies)	17.12			————
87	9-2	Young-At-Heart (clothes)	192.50			————
88	9-4	H.S.U. Bookstore (books)	56.28			————
89	9-7	Regent's Office (tuition)	380.00			————
90	9-7	Off-Campus Apts. (rent)	240.00			————
91	9-27	State Telephone Co. (phone bill)	24.62			————
92	9-30	Up-N-Up Foods (groceries)	47.80			————
	9-30	Service Charge ( )				————
		( )				————
		( )				————
					True Balance	

BANK STATEMENT
Checking Account Activity

Transaction Description	Amount	(√)	Running Balance	Date
Beginning Balance			1403.49	9-01
Check #86	17.12		1386.37	9-02
Check #87	192.50		1193.87	9-05
Check #88	56.28		1137.59	9-05
Check #90	240.00		897.59	9-10
Service Charge	4.00		893.59	9-30
Ending Balance			893.59	

RECONCILIATION SHEET

A. First mark √ beside each check and deposit listed in both your checkbook register and on the bank statement.
B. Second, in your checkbook register, add any interest paid and subtract any service charge listed on the bank statement.
C. Third, find the total of all outstanding checks.

Outstanding Checks	
No.	Amount

Statement Balance ————

Add deposits not credited + ————

Total ————

Subtract total amount of checks outstanding ————

Total ———— True Balance ————

7.

YOUR CHECKBOOK REGISTER

Check No.	Date	Transaction Description	Payment (−)	(√)	Deposit (+)	Balance
			Balance brought forward			602.82
14	6-20	Aisle Bridal (flowers)	402.40			————
	6-22	Deposit ( )			1000.00	————
15	6-24	Tuxedo Junction (tux)	155.65			————
16	6-28	D. Lohengrin (organist)	55.00			————
17	6-28	D-Lux Limo (limo rental)	125.00			————
18	6-30	C.C. Catering (food caterer)	700.00			————
19	7-1	Luv-Lee Stationers (thank-you cards)	35.20			————
	6-30	Service Charge ( )				————
		( )				————
		( )				————
					True Balance	

BANK STATEMENT
Checking Account Activity

Transaction Description	Amount	(√)	Running Balance	Date
Beginning Balance			602.82	6-01
Deposit	1000.00		1602.82	6-22
Check #14	402.40		1200.42	6-22
Check #15	155.65		1014.77	6-26
Service Charge	1.00		1013.77	6-30
Ending Balance			1013.77	

RECONCILIATION SHEET

A. First mark √ beside each check and deposit listed in both your checkbook register and on the bank statement.
B. Second, in your checkbook register, add any interest paid and subtract any service charge listed on the bank statement.
C. Third, find the total of all outstanding checks.

Outstanding Checks	
No.	Amount

Statement Balance ————

Add deposits not credited + ————

Total ————

Subtract total amount of checks outstanding ————

Total ———— True Balance ————

8

YOUR CHECKBOOK REGISTER

Check No.	Date	Transaction Description	Payment (−)	(√)	Deposit (+)	Balance
		Balance brought forward				278.32
326	8-12	J.J. Jones (birthday check)	40.00			_____
	8-15	Deposit ( )			500.00	_____
327	8-15	N.O. Payne, M.D. (physical exam)	260.00			_____
328	8-15	Local Waterworks (water/trash)	27.40			_____
	8-22	Deposit ( )			400.12	_____
329	8-27	Time-O-Life (magazine subscrip.)	12.50			_____
330	8-29	Bigger Mortgage (house payment)	750.00			_____
	8-31	Interest ( )				_____
	8-31	Service Charge ( )				_____
		( )				_____
					True Balance	

BANK STATEMENT
Checking Account Activity

Transaction Description	Amount	(√)	Running Balance	Date
Beginning Balance			278.32	8-01
Deposit	500.00		778.32	8-15
Check #326	40.00		738.32	8-15
Check #327	260.00		478.32	8-17
Deposit	400.12		878.44	8-22
Check #328	27.40		851.04	8-22
Interest	1.82		852.86	8-31
Service Charge	4.00		848.86	8-31
Ending Balance			848.86	

RECONCILIATION SHEET

A. First mark √ beside each check and deposit listed in both your checkbook register and on the bank statement.

B. Second, in your checkbook register, add any interest paid and subtract any service charge listed on the bank statement.

C. Third, find the total of all outstanding checks.

Outstanding Checks			
No.	Amount		Statement Balance _____
			Add deposits not credited + _____
			Total _____
			Subtract total amount of checks outstanding _____
	Total _____		True Balance _____

9

YOUR CHECKBOOK REGISTER

Check No.	Date	Transaction Description	Payment (−)	(√)	Deposit (+)	Balance
		Balance brought forward				147.02
203	2-3	Food Stoppe (groceries)	26.90			_____
204	2-8	Ekkon Oil (gasoline bill)	71.45			_____
205	2-14	Rose's Roses (flowers)	25.00			_____
206	2-14	I.M.R.U (alumni dues)	20.00			_____
	2-15	Deposit ( )			600.00	_____
207	2-26	SRO (theater tickets)	52.50			_____
208	2-28	MPG Mtg. (house payment)	500.00			_____
	2-28	Service Charge ( )				_____
		( )				_____
		( )				_____
					True Balance	

BANK STATEMENT
Checking Account Activity

Transaction Description	Amount	(√)	Running Balance	Date
Beginning Balance			147.02	2-01
Check #203	26.90		120.12	2-04
Check #204	71.45		48.67	2-14
Deposit	600.00		648.67	2-15
Check #205	25.00		623.67	2-15
Service Charge	3.00		620.67	2-28
Ending Balance			620.67	

RECONCILIATION SHEET

A. First mark √ beside each check and deposit listed in both your checkbook register and on the bank statement.

B. Second, in your checkbook register, add any interest paid and subtract any service charge listed on the bank statement.

C. Third, find the total of all outstanding checks.

Outstanding Checks			
No.	Amount		Statement Balance _____
			Add deposits not credited + _____
			Total _____
			Subtract total amount of checks outstanding _____
	Total _____		True Balance _____

10.

YOUR CHECKBOOK REGISTER

Check No.	Date	Transaction Description	Payment (−)	(√)	Deposit (+)	Balance
			Balance brought forward			*4071.82*
996	10-1	Red-E Credit (loan payment)	200.75		____	
997	10-10	United Ways (charity donation)	25.00		____	
998	10-21	MacIntosh Farms (barrel of apples)	42.20		____	
999	10-26	Fun Haus (costume rental)	35.00		____	
1000	10-28	Yum Yum Shoppe (Halloween candy)	12.14		____	
1001	10-29	B-Sharp, Inc. (piano tuners)	20.00		____	
1002	10-30	Food-2-Go! (party platter)	78.50		____	
1003	10-31	Cash ( )	300.00		____	
1004	10-31	Principal S&L (house payment)	1250.60		____	
	10-31	Interest ( )			____	
					True Balance	

BANK STATEMENT
Checking Account Activity

Transaction Description	Amount	(√)	Running Balance	Date
Beginning Balance			4071.82	10-01
Check #996	200.75		3871.07	10-05
Check #998	42.40		3828.87	10-23
Check #999	35.00		3793.87	10-30
Check #1003	300.00		3493.87	10-31
Interest	16.29		3510.16	10-31
Ending Balance			3510.16	

RECONCILIATION SHEET

A. First mark √ beside each check and deposit listed in both your checkbook register and on the bank statement.

B. Second, in your checkbook register, add any interest paid and subtract any service charge listed on the bank statement.

C. Third, find the total of all outstanding checks.

Outstanding Checks			Statement Balance	_____
No.	Amount			
			Add deposits not credited	+ _____
			Total	_____
			Subtract total amount of checks outstanding	− _____
	Total _____		True Balance	_____

8.4 Buying and Owning a Car

OBJECTIVES

1. Become aware of and learn how to calculate the expenses involved in buying a car.
2. Know how to calculate the percent of your income that is spent on your car.

Buying a Car

Buying a car is not as expensive or complicated as buying a home. However, more people buy cars than homes and, in many cases, buying a car is the most expensive purchase of a person's life. As with finances in general, paying cash for a car is cheaper than financing the car with a bank or savings and loan. If you are going to finance the purchase of a car, at least be aware of the expenses involved and study all the papers so that you know the total amount that you are paying for the car.

Expenses in Buying a Car

Purchase price: the selling price agreed on by the seller and the buyer

Sales tax: a fixed percent that varies from state to state

License fee: fixed by the state, often based on the type of car and its value

---

EXAMPLE 1

You are going to buy a new car for $18,500. The bank will lend you 70% of all the related expenses, including taxes and fees. If sales tax is figured at 8% and there is a license fee of $250, how much cash do you need in order to buy the car?

Solution

(a) First find the total of all the related expenses.

$18,500	selling price		$18,500	selling price	
× 0.08	tax rate		1,480	sales tax	
$ 1,480	sales tax		+ 250	license fee	
			$20,230	total expenses	

(b) Find 30% of all the expenses. (Since the bank will loan you 70%, you must provide 100% − 70% = 30% of the total expenses in cash.)

$20,230   total expenses
× 0.30
$6,069.00   cash

Owning a Car

Actually, the bank owns your car until all payments are made on the loan. The car is their security for the loan. However, in addition to the monthly payments, you must pay for insurance, any necessary repairs, and general maintenance costs.

> **Expenses in Owning a Car**
>
> **Monthly payments:** payments made if you borrowed money to buy the car
>
> **Auto insurance:** covers a variety of situations (liability, collision, towing, theft, and so on)
>
> **Operating costs:** basic items such as gasoline, oil, tires, tune-ups
>
> **Repairs:** replacing worn or damaged parts

EXAMPLE 2

In one month, Bonnie's car expenses were as follows: loan payment, $350; insurance, $80; gasoline, $100; oil and filter, $22; new headlight, $32. (a) What were her total car expenses for that month? (b) What percent of her car expenses was for the loan payment?

Solution

(a) Find her total expenses.

$350	loan payment
80	insurance
100	gasoline
22	oil and filter
+ 32	headlight
$584	total expenses

(b) Find the percent of the total spent on the loan payment.

$$\frac{350}{584} \approx 0.5993 \approx 60\%$$

Bonnie spent $584 on her car and the loan payment was about 60% of her expenses.

Exercises 8.4

1. To buy a used car for $6800, you must pay a 6% sales tax and a license fee of $120. If the bank will lend you 85% of your expenses, how much cash do you need to buy the car?

2. John wants to buy a new convertible for $25,000. His credit union will lend him 80% of his expenses. What amount of cash does he need to buy the car if the sales taxes are at 8.5% and the license fee is $250?

3. How much cash would you need to buy a car for $18,000 if the sales tax is calculated at 6%, the license fee is $200, and the loan company will let you borrow 75% of your expenses?

$1117.50

4. A used car is priced at $4500. Your old car is worth $800 on a trade-in. The sales tax is figured at 7.5% of the selling price and the license fee is $80. If the savings and loan will lend you $3000, how much cash do you need to buy the car?

subtract
1500
after finding
20% of total

5. Your old car is worth $1500 if you trade it in for a new car priced at $11,200. Sales tax is 6% and the license fee is $225. If the bank will lend you 80% of your expenses, how much cash do you need to buy the new car?

6. (a) If your car expenses for one month were $325 for the loan payment, $35 for insurance, $120 for gasoline, and $145 for two new tires, what were your total car expenses for the month? (b) If your income was $2000, what percent of your income was used for car expenses?

7. Nancy decided her old car needed painting. The paint job was priced at $1650, including repairing some dents, and she figured that driving her car cost an average of 24¢ per mile, including gas, oil, and insurance. What were her car expenses that month if she drove 1200 miles?

8. Art owns his car, but it needs a new transmission for $1200 (installed). (a) What were his car expenses the month that he had the new transmission installed and spent $65 for insurance, $75 for gas, $15 for oil and filter, and $300 for a tune-up? (b) If he took $1000 from his savings account to help pay for the transmission, what percent of his income of $2600 was used for the remainder of his car expenses?

9. Danielle decided she would like to have a new car, but she could not afford the car if expenses averaged more than 20% of her monthly income. If she figured the expenses would average $300 for a loan payment, $70 for insurance, $65 for gas, $8 for oil, $10 for tire wear, and $40 for general repairs, could she afford the car with a monthly income of $2100?

10. Suppose that you owned a car, your monthly income was $1800, and you figured you could spend 15% of this to operate a car. (a) What would you be able to spend on gas if insurance cost $85, oil and filter $20, and you estimated $60 per month for other expenses? (b) How many gallons of gas could you buy if gas cost $1.25 per gallon? (c) How many miles could you drive if your car averaged 19 miles per gallon?

8.5   Reading Graphs

OBJECTIVES

1. Learn how to read data and calculate information from the following types of graphs: bar graphs, circle graphs, line graphs, pictographs, and histograms.

Reading Graphs

Graphs are "pictures" of numerical information. Graphs appear almost daily in newspapers and magazines and frequently in textbooks and corporate reports. Communication with well-drawn graphs can be accurate, effective, and fast. Most computers can be programmed to draw graphs, and anyone whose work involves a computer in any way will probably be expected to understand graphs.

Five types of graphs and their main uses are:

1. Bar Graphs: to emphasize comparative amounts

2. Circle Graphs: to help in understanding percents or parts of a whole

3. Line Graphs: to indicate tendencies or trends over a period of time

4. Pictographs: to emphasize the topic being related as well as quantities

5. Histograms: to indicate data in classes

All graphs should:

1. Be clearly labeled   **2.** Be easy to read.   **3.** Have titles.

The following examples illustrate the five types of graphs listed. There are several questions related to each graph to test your understanding. Some questions can be answered directly from the graph while others involve some calculations.

These examples are given in a form with blanks for the answers to the questions. While the blanks are filled in, the student should double-check each answer by referring to the graph and by making any related calculations.

EXAMPLE 1

Bar Graph

Figure 8.1 shows a bar graph. Note that the scale on the left and the data on the bottom (months) are clearly labeled and the graph itself has a title. The questions following can be easily answered by looking at the graph.

(a) What were the sales in January? $100,000

(b) During what month were sales lowest? March

(c) During what month were sales highest? February and June

(d) What were the sales during each of the highest sales months? $150,000

(e) What were the sales in April? $75,000

The following questions will take some calculations after reading the graph.

(f) What was the amount of decrease in sales between February and March?

$150,000   February sales
−   50,000   March sales
$100,000   decrease in sales

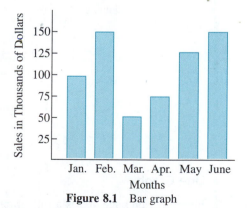

First Six Months Sales – 1994
Better Book Co.

Figure 8.1   Bar graph

(g) What was the percent of decrease?

$$\frac{100,000 \ \text{decrease}}{150,000 \ \text{February sales}} = \frac{2}{3} = 0.666 = 67\% \ \text{decrease}$$

EXAMPLE 2

Circle Graph

Figure 8.2 shows percents budgeted for various items in a home for one year. Suppose a family has an annual income of $15,000. Using Figure 8.2, calculate how much will be allocated to each item indicated in the graph.

Solution

Item		Amount
Housing	$0.25 \times \$15{,}000 =$	$\underline{\$3750.00}$
Food	$0.20 \times \$15{,}000 =$	$\underline{3000.00}$
Taxes	$0.05 \times \$15{,}000 =$	$\underline{750.00}$
Clothing	$0.07 \times \$15{,}000 =$	$\underline{1050.00}$
Savings	$0.10 \times \$15{,}000 =$	$\underline{1500.00}$
Education	$0.15 \times \$15{,}000 =$	$\underline{2250.00}$
Entertainment	$0.05 \times \$15{,}000 =$	$\underline{750.00}$
Transportation & Maintenance	$0.13 \times \$15{,}000 =$	$\underline{1950.00}$

What is the total of all amounts? $\underline{\$15{,}000.00}$

Home Budget for One Year

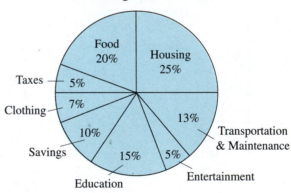

Figure 8.2 Circle Graph

EXAMPLE 3

Line Graph

Figure 8.3 on page 405 is a line graph that shows the relationships between daily high and low temperatures. You can see that temperatures tended to rise during the week but fell sharply on Saturday.

What was the lowest high temperature? _____66°_____

On what day did this occur? _____Sunday_____

What was the highest low temperature?   <u>   70°   </u>
On what day did this occur?   <u>   Friday   </u>

Find the average difference between the daily high and low temperatures for the week shown.

Solution

First find the differences; then average these differences.

Sunday $66 - 60 =$ <u>   6°   </u>

Monday $70 - 62 =$ <u>   8°   </u>   9.1°   average difference

Tuesday $76 - 66 =$ <u>   10°   </u> $7\overline{)64.0}$

Wednesday $72 - 66 =$ <u>   6°   </u> $\underline{63}$

Thursday $80 - 68 =$ <u>   12°   </u>   1 0

Friday $80 - 70 =$ <u>   10°   </u> $\underline{7}$

Saturday $74 - 62 =$ <u>   12°   </u>   3

<u>   64°   </u>   total of
                     differences

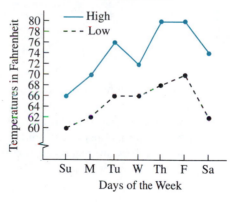

High & Low Temperatures
for One Week

Figure 8.3   Line graph

EXAMPLE 4

Pictograph

Figure 8.4 on page 406 shows a pictograph of new home construction in five counties in 1991. Answer the following questions.

(a) Which county had the most number of new homes built?

<u>   County C   </u>

How many?   <u>   2000 homes   </u>

(b) Which county had the least number of new homes built?

___County E___

How many? ___400 homes___

(c) What was the difference between new home construction in

Counties A and D? ___1550 − 800 = 750 homes difference___

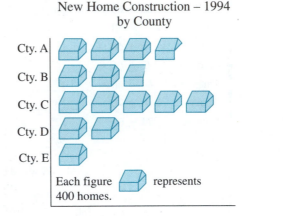

Figure 8.4   Pictograph

For bar graphs (see Example 1), we label the base line for each bar with individual names for people, months, days of the week, or other categories. Now we introduce a type of bar graph called a **histogram.** In a histogram, the base line is labeled with numbers that indicate the boundaries of a range of numbers called a **class.** The bars are placed next to each other with no space between them.

Terms Related to Histograms

Class: a range (or interval) of numbers that contains data items

Lower class limit: the smallest number that belongs to a class

Upper class limit: the largest number that belongs to a class

Class boundaries: numbers that are halfway between the upper limit of one class and the lower limit of the next class

Class width: the difference between the class boundaries of a class (the width of each bar)

Frequency: the number of data items in a class

EXAMPLE 5   Histogram

Figure 8.5 shows a histogram that summarizes the scores of 50 students on an English placement test. Answer the following questions by referring to the graph.

(a) How many classes are represented?   6

(b) What are the class limits of the first class?   201 and 250

(c) What are the class boundaries of the second class?

   250.5 and 300.5

(d) What is the width of each class?   50

(e) Which class has the greatest frequency?   second class

(f) What is this frequency?   16

(g) What percent of the scores are between 200.5 and 250.5? $\dfrac{2}{50} = 4\%$

(h) What percent of the scores are above 400? $\dfrac{12}{50} = 24\%$

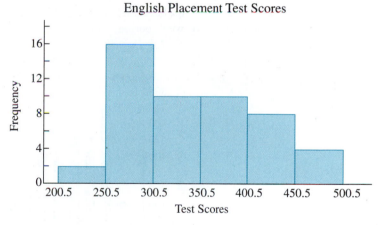

Figure 8.5   Histogram

Exercises 8.5

Answer the questions related to each of the graphs. Some questions can be answered directly from the graphs; others may require some calculations.

1. The following bar graph shows the numbers of students in five fields of study at a university.

Declared College Majors at Downstate University

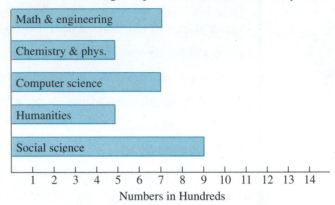

Numbers in Hundreds

(a) Which field of study has the largest number of declared majors?

(b) Which field of study has the smallest number of declared majors?

(c) How many declared majors are indicated in the entire graph?

(d) What percent are computer science majors?

2. The bar graph shows the numbers of vehicles that crossed one intersection during a two-week period.

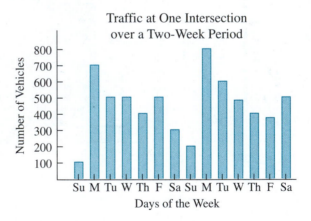

(a) On which day did the highest number of vehicles cross the intersection? How many crossed that day?

(b) What was the average number of vehicles that crossed the intersection on the two Sundays?

(c) What was the total number of vehicles that crossed the intersection during the two weeks?

(d) What percent of the total traffic was counted on Saturdays?

3. In comparing the following two graphs, assume that all five students graduated with comparable grades from the same high school.

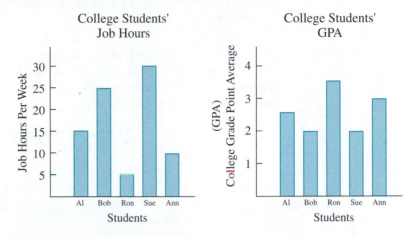

College Students'
Job Hours

College Students'
GPA

(a) Who worked the most hours per week?

(b) Who had the lowest GPA?

(c) If Ron spent 30 hours per week studying for his classes, what percent of his total work week (part-time work plus study time) did he spend studying?

(d) Which two students worked the most hours? Which two students had the lowest GPAs? Do you think that this is typical?

(e) Do you think that the two graphs shown here could be set as one graph? If so, show how you might do this.

4. The circle graph shown is based on a total budget of $34,500,000.

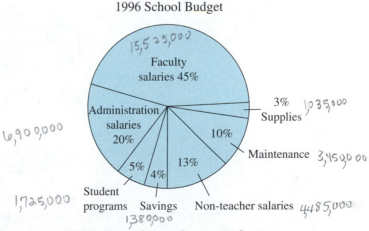

1996 School Budget

15,525,000

Faculty
salaries 45%

Administration
salaries
20%

6,900,000

3% 1,035,000
Supplies

10%

Maintenance 3,450,000

5% 4% 13%

1,725,000

Student
programs   Savings   Non-teacher salaries 4,485,000
1,380,000

(a) What amount will be allocated to each category?

(b) What percent will be for expenditures other than salaries?

22%

(c) How much will be spent on maintenance and supplies? $4,485,000$

(d) How much more will be spent on teachers' salaries than on administration salaries? $8,625,000$

5. Television station KCBA is off the air from 2 A.M. to 6 A.M., so there are only 20 hours of daily programming. Sports are not shown in the graph below because they are considered special events. In the 20-hour period shown, how much time (in minutes) is devoted daily to each category?

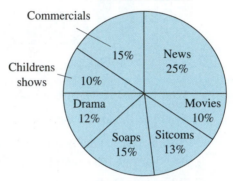

20-Hour TV Programming
at Station KCBA

6. Mike just graduated from college and decided that he should try to live within a budget. The circle graph shows the categories he chose and the percents he allowed. His beginning salary is $24,000.

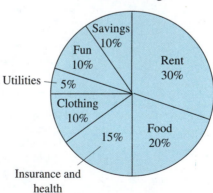

Mike's Budget

(a) How much did he budget for each category?

(b) What category was smallest in his budget?

(c) What total amount did he budget for food, clothing, and rent?

7. The pictograph shows the profits earned by four different local banks during the last year.

Profits Earned by
Four Local Banks

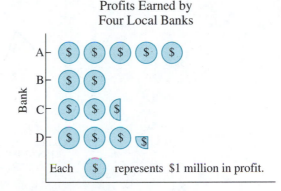

(a) Which bank earned the most money last year? How much?

(b) Which bank earned the least amount of money last year? How much?

(c) What percent of the total amount earned by all the banks was made by Bank B?

8. The pictograph shows the annual sales of boats by one manufacturer for the years 1988–1992.

Sales of Boats by
One Manufacturer, 1988–1992

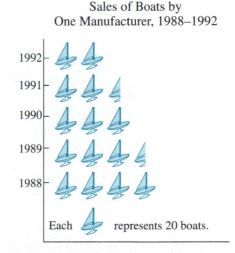

(a) During which year did the company sell the most boats? How many?

(b) What was the difference between the sales during the best year and the sales during the worst year?

(c) What was the average number of sales per year over the five-year period?

9.

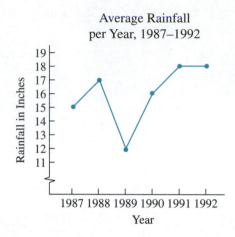

**Average Rainfall
per Year, 1987–1992**

(a) What year had the least rainfall?

(b) What was the most rainfall in a year?

(c) What year had the most rainfall?

(d) What was the average rainfall over the six-year period?

10.

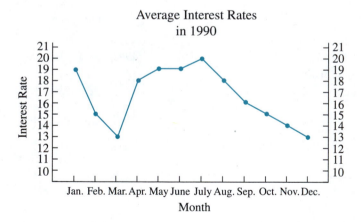

**Average Interest Rates
in 1990**

(a) During what month or months in 1990 were interest rates highest?

(b) Lowest?

(c) What was the average of the interest rates over the entire year?

11.

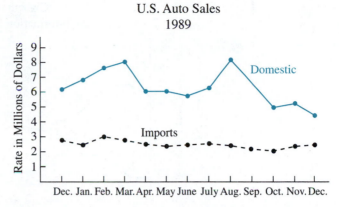

U.S. Auto Sales
1989

(a) During which month were domestic sales highest?

(b) How much higher were they than for the lowest month?

(c) What was the difference in import sales for the same two months?

(d) What was the difference between domestic and import sales in March?

(e) What percent of sales were imports in December 1988?

(f) In December 1989? (Answers will be approximate.)

12.

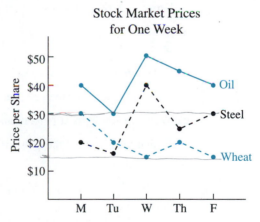

Stock Market Prices
for One Week

wheat (a) If on Monday morning you had 100 shares of each of the three stocks shown (oil, steel, wheat), and you held the stock all week, on which stock would you have lost money?

$1500 (b) How much would you have lost?   30-15 =15

steel (c) On which stock would you have gained money?

$1000 (d) How much would you have gained?   (30 - 20) = 10

oil (e) On which stock could you have made the most money if you had sold at the best time?

$5000 (f) How much could you have made?

13.

Growth of
the Information Economy

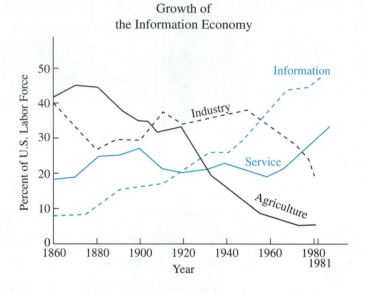

(a) What percent of workers were in each of the four areas in 1860?

(b) In 1980?

(c) Which area of work seems to have had the most stable percent of workers between 1860 and 1980?

(d) What is the difference between the highest and lowest percents for this area?

(e) Which area has had the most growth?

(f) What was its lowest percent and when?

(g) What was its highest percent and when?

(h) Which area has had the most decline?

14.

Tread Life for New Tires

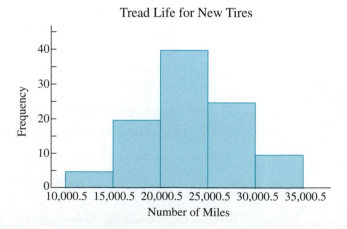

(a) How many classes are represented?

(b) What is the width of each class?

(c) Which class has the highest frequency?

(d) What is this frequency?

(e) What are the class boundaries of the second class?

(f) How many tires were tested?

(g) What percent of the tires were in the first class?

(h) What percent of the tires lasted more than 25,000 miles?

15.

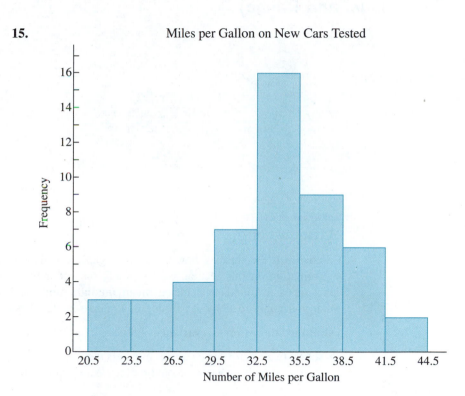

Miles per Gallon on New Cars Tested

(a) How many classes are represented?

(b) What is the class width?

(c) Which class has the smallest frequency?

(d) What is this frequency?

(e) What are the class limits for the third class?

(f) How many cars were tested?

(g) How many cars tested below 30 miles per gallon?

(h) What percent of the cars tested above 38 miles per gallon?

8.6 Statistics (Mean, Median, Mode, and Range)

OBJECTIVES

1. Develop an understanding of the terms **data** and **statistics.**
2. Learn how to calculate the **mean** and **range** of a set of data.
3. Learn how to locate the **median** and **mode** of a set of data.

Introduction to Statistics

Statistics is the study of how to gather, organize, analyze, and interpret numerical information. In this section, we will study only four measures (or four statistics) that are easily found or calculated: **mean, median, mode,** and **range.** Other measures that you might read about or study in a course in statistics are:

standard deviation and **variance** (both are measures of how "spread out" data are)

z-score (a measure that compares numbers that are expressed in different units—for example, your score on a biology exam and your score on a mathematics exam)

correlation coefficient (a measure of how two different types of data might be related—for example, to determine if there is a relationship between the amount of schooling a person has and the amount of that person's lifetime earnings)

You will need at least one semester of algebra to be able to study and understand these topics, so keep working hard.

The following terms and their definitions are necessary for understanding the topics and related problems in this section. They are listed here in one place for easy reference.

Terms Used in the Study of Statistics

Data: Value(s) measuring some characteristic of interest. (We will consider only numerical data.)

Mean

~~Statistic:~~ The arithmetic average of the data. (Add all the data and divide by the number of data items.)

Median: The middle data item. (Arrange the data in order and pick out the middle item.)

Mode: The single data item that appears the most number of times. (Some data may have more than one mode. We will leave the discussion of such a situation to a course in statistics.) In this text, if the data have a mode, there will be only one mode.)

Range: The difference between the largest and smallest data items.

In Examples 1–3, answer the questions about statistics using the data from Group A and Group B.

Group A: Annual Income for 8 Families

$18,000; $12,000; $15,000; $17,000; $35,000; $70,000; $15,000; $20,000

Group B: Grade Point Averages (GPA) for 11 Students

2.0; 2.0; 1.9; 3.1; 3.5; 2.9; 2.5; 3.6; 2.0; 2.4; 3.4

EXAMPLE 1

Find the mean income for the families in Group A.

Solution

Find the sum of the 8 incomes and divide by 8.

$$
\begin{array}{r}
\$\ 18,000 \\
12,000 \\
15,000 \\
17,000 \\
35,000 \\
70,000 \\
15,000 \\
+\ \ 20,000 \\
\hline
\$202,000
\end{array}
\qquad
\begin{array}{r}
\$25,250 \\
8\,)\overline{\,202,000} \\
\underline{16} \\
42 \\
\underline{40} \\
2\,0 \\
\underline{1\,6} \\
40 \\
\underline{40} \\
00 \\
\underline{0} \\
0
\end{array}
$$

You may want to use a calculator to do this arithmetic. The mean annual income is $25,250.

> **To Find the Median:**
>
> **1.** Arrange the data in order.
>
> **2.** If there is an **even** number of items, the median is the average of the two middle items.
>
> **3.** If there is an **odd** number of items, the median is the middle item.

EXAMPLE 2 Find the median income for Group A and the median GPA for Group B.

Solution

Arrange both sets of data in order.

Group A

$12,000; $15,000; $15,000; $17,000; $18,000; $20,000; $35,000; $70,000

Group B

1.9; 2.0; 2.0; 2.0; 2.4; 2.5; 2.9; 3.1; 3.4; 3.5; 3.6

For Group A, the median is the average of the 4th and 5th items because there is an even number of items (8 items).

$$\text{Median} = \frac{\$17,000 + \$18,000}{2} = \frac{35,000}{2} = \$17,500$$

For Group B, the median is the 6th item because, with an **odd** number of 11 items, the 6th item is the middle item.

$$\text{Median} = 2.5$$

EXAMPLE 3 Find the mode and the range for both Group A and Group B.

Solution

The mode is the most frequent item. From the arranged data in Example 2, we can see that:

for Group A, the mode is 15,000.

for Group B, the mode is 2.0.

The range is the difference between the largest and smallest items:

Group A range = $70,000 − $12,000 = $58,000

Group B range = 3.6 − 1.9 = 1.7

Commentary: Of the four statistics mentioned in this section, the mean and median are most commonly used. Many people feel that the mean (or arithmetic average) is relied on too much in reporting central tendencies for data such as income, housing costs, and taxes where a few very high items can *distort* the picture of a central tendency. As you can see in Group A data, the median of $17,500 is probably more representative of the data than the mean of $25,250 because the one high income of $70,000 raises the mean considerably.

When you read an article in a magazine or newspaper that reports means or medians, you should now have a better understanding of the implications.

Exercises 8.6

For each of the following problems, find (a) the mean, (b) the median, (c) the mode, and (d) the range of the given data.

1. Ten math students had the following scores on a final exam:

75, 83, 93, 65, 85,
85, 88, 90, 55, 71

2. Joe did the following number of sit-ups each morning for a week:

25, 52, 48, 42, 38, 58, 52

3. Fifteen college students reported the following hours of sleep the night before an exam:

4, 6, 6, 7, 6.5, 6.5, 7.5, 8.5
5, 6, 4.5, 5.5, 9, 3, 8

4. The local high school basketball team scored the following points per game during their 20-game season:

85, 60, 62, 70, 75, 52, 88, 50, 80, 72,
90, 85, 85, 93, 70, 75, 68, 73, 65, 82

5. Stacey went to six different repair shops to get the following estimates to repair her car. (The accident was not her fault; her car was parked at the time.)

$425, $525, $325, $300, $500, $325

6. Mike kept track of his golf scores for twelve rounds of eighteen holes each. His scores were:

85, 90, 82, 85, 87, 80,
78, 82, 88, 82, 86, 81

7. The local weather station recorded the following daily high temperatures for one month:

75, 76, 76, 78, 85, 82, 85, 88, 90, 90,
88, 95, 96, 92, 88, 88, 80, 80, 78, 80,
78, 76, 77, 75, 75, 74, 70, 70, 72, 73

8. The Big City fire department reported the following mileage for tires used on their nine fire trucks:

   14,000; 14,000; 11,000; 15,000;
   9,000; 14,000; 12,000; 10,000; 9,000

9. The city planning department issued the following numbers of building permits over a three-week period:

   17, 19, 18, 35, 30, 29, 23, 14,
   18, 16, 20, 18, 18, 25, 30

10. Police radar measured the following speeds of 35 cars on one street:

    28, 24, 22, 38, 40, 25, 24, 35, 25,
    23, 22, 50, 31, 37, 45, 28, 30, 30,
    30, 25, 35, 32, 45, 52, 24, 26, 18,
    20, 30, 32, 33, 48, 58, 30, 25

11. On a one-day fishing trip, Mr. and Mrs. Johnson recorded the following lengths of fish they caught (measured in inches):

    14.3; 13.6; 10.5; 15.5; 20.1;
    10.9; 12.4; 25.0; 30.2; 32.5

12. A machine puts out parts measured in thickness to the nearest hundredth of an inch. One hundred parts were measured and the results are tallied in the following chart:

Thickness Measured	0.80	0.83	0.84	0.85	0.87
Number of Parts	22	41	14	20	3

Summary: Chapter 8

Key Terms and Ideas

Formula for simple interest:

$$I = P \times r \times t \quad \text{where} \quad \begin{aligned} I &= \text{Interest} \\ P &= \text{Principal} \\ r &= \text{annual interest rate} \\ t &= \text{time (in years)} \end{aligned}$$

Formula for compound interest:

$$A = P\left(1 + \frac{r}{n}\right)^{nt}$$ where A = Amount accumulated

P = Principal

r = annual interest rate

t = time (in years)

n = number of corresponding periods per year

Some expenses in buying a car are:

purchase price, sales tax, license fee.

Some expenses in owning a car are:

monthly payments, insurance, operating costs, repairs.

Five types of graphs and their main uses are:

1. **Bar Graphs:** to emphasize comparative amounts

2. **Circle Graphs:** to help in understanding percents or parts of a whole

3. **Line Graphs:** to indicate tendencies or trends over a period of time

4. **Pictographs:** to emphasize the topic being related as well as quantities

5. **Histograms:** to indicate data in classes

All graphs should:

1. Be clearly labeled.   2. Be easy to read.   3. Have titles.

Terms related to histograms:

class, lower class limit, upper class limit, class boundaries, class width, frequency

Terms used in the study of statistics:

data, statistic, mean, median, mode, range

Procedures

> ### To Balance Your Checking Account:
>
> (sometimes called **reconciling** the bank statement with your checkbook register):
>
> **1.** Go through your checkbook register and the bank statement and put a check mark (✔) by each check paid and deposit recorded on the bank statement.
>
> **2.** In your checkbook register:
>
>   (a) Add any interest paid to your current balance.
>
>   (b) Subtract any service charge from this balance.
>
>   This number is your **true balance.**
>
> **3.** Find the total of all the outstanding checks (no ✔ mark) in your checkbook register (checks not yet received by the bank).
>
> **4.** On a reconciliation sheet:
>
>   (a) Enter the balance from the bank statement.
>
>   (b) Add any deposits you have made that are not recorded (no ✔ mark) on the bank statement.
>
>   (c) Subtract the total of your outstanding checks (found in Step 3). This number should agree with your **true balance.** Now your checking account is balanced.

To find the mean: Add all the data and divide by the number of data items.

> ### To Find the Median:
>
> **1.** Arrange the data in order.
>
> **2.** If there is an **even** number of items, the median is the average of the two middle items.
>
> **3.** If there is an **odd** number of items, the median is the middle item.

To find the range: Find the difference between the largest and the smallest data items.

Review Questions: Chapter 8

1. What will be the interest earned in one year on a savings account of $1500 if the bank pays $6\frac{1}{2}\%$ simple interest?

2. If a stock pays 12% dividends every 6 months, what will be the dividend paid on an investment of $13,450 in 6 months?

3. If a principal of $1000 is invested at a rate of 9% for 30 days, what will be the interest earned?

4. You made $50 on an investment at 10% simple interest for 3 months. What principal did you invest?

5. Determine the missing items in each row if the interest is simple interest.

Principal	Rate	Time	Interest
$ 200	12%	180 days	(a)
$ 300	18%	(b)	$ 81
$1000	(c)	1 year	$ 85
(d)	9%	18 months	$270

6. How much interest will be earned on a savings account of $6500 in 5 years if interest is compounded at 6% (a) annually? (b) quarterly? (c) monthly? (d) daily?

7. Complete the following table of values if interest is compounded annually at 8%.

	Principal	Amount in Account	Interest Earned
1st year	$10,000	(a)	(b)
2nd year	(c)	(d)	(e)
3rd year	(f)	(g)	(h)

8. The price of a new pickup truck that you wish to buy is $14,200. Tax is 6.5% of the purchase price, and the license fee is $350. If the bank will loan you 75% of the total expenses, how much cash will you need to buy the car?

Refer to the circle graph for Exercises 9–12.

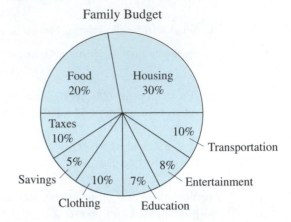

Family Budget

9. The circle graph shows a home budget. What amount will be spent on each category if the family income is $35,000?

10. How much more will the family spend for food than for clothing if their income is increased to $40,000 (to the nearest dollar)?

11. What fractional part of the family income is spent for food, housing, and taxes combined?

12. How much will the family spend for food, housing, and transportation combined if their income is reduced to $30,000 (to the nearest dollar)?

Refer to the following test scores for Exercises 13–16.

94, 86, 92, 70, 88, 91,
88, 70, 89, 88, 96, 92

13. Find the mean score.

14. Find the mode (if any).

15. Find the median score.

16. Find the range of the scores.

Test: Chapter 8

1. How much simple interest will be earned in one year on a savings account of $1500 if the bank pays 4.5% interest?

2. You made $42 in interest on an investment at 7% for 9 months. What principal did you invest?

3. If Ms. King had a savings account of $3000 earning 5.5% simple interest, how long would it take for her to earn $82.50 in interest?

4. If an investment of $7500 earns simple interest of $131.25 in 45 days, what is the rate of interest?

5. A certificate of deposit is earning 6% compounded quarterly. (a) If $1000 is deposited in the account, what will be the balance in the account at the end of one year? (b) at the end of five years?

6. How much more would be in the balance of the account in five years in Exercise 5 if the money was compounded daily?

7. Fred is buying a car for $17,500. Sales tax is 6% and his license and transfer fees total $513. (a) If he receives a $4000 trade-in allowance and a $1200 factory rebate, how much cash will he need to buy the car? (b) If he finances 70% of his expenses before the trade-in and the rebate, how much cash will he need?

Refer to the circle graph for Exercises 8–10.

Budget for Apartment Complex

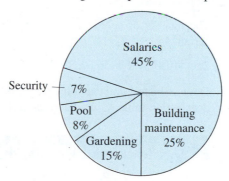

8. The budget for an apartment complex is shown in the graph. What fractional part of the budget is spent for salaries and security combined?

9. How much will be spent for building maintenance in one year if the budget is $250,000 per month?

10. How much will be spent for salaries and security combined in 6 months if the monthly budget is $150,000?

Refer to the bar graph for Exercises 11–13.

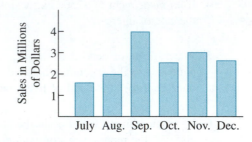

Ace Mfg. Company Sales

11. What were the total sales for the 6-month period?

12. What percent of the total sales for the 6 months were the July sales (to the nearest percent)?

13. What was the growth percent between August and September?

Find (a) the mean, (b) the median, (c) the mode (if any), and (d) the range for the groups of data in Exercises 14 and 15.

14. The number of hours of television viewing per day for a certain group:

3, 2, 2, 0, 1
4, 1, 2, 3, 2

15. The number of centimeters of precipitation during one six-month period in a certain area:

2.54, 10.16, 7.62, 20.32, 12.70, 7.62

16. Draw a histogram that illustrates the data given in the following frequency distribution table. Give the graph a title and label the scales appropriately.

	A Frequency Distribution of Mathematics Exam Scores	
Class Number	**Class Limits (Range of Exam Scores)**	**Frequency (Number of Students in Each Class)**
1	50–59	1
2	60–69	4
3	70–79	8
4	80–89	5
5	90–99	2
		20 students

All fractions should be reduced to lowest terms.

1. Write the following in standard notation: two hundred thousand, sixteen.

2. Write the following in decimal notation: three hundred and four thousandths.

3. Round off 16.996 to the nearest hundredth.

4. Find the decimal equivalent to $\dfrac{14}{35}$.

5. Find the decimal equivalent to $\dfrac{21}{40}$.

6. Write $\dfrac{9}{5}$ as a percent.

7. Write $1\dfrac{1}{2}\%$ as a decimal.

Perform the indicated operations for Exercises 8–19.

8. $\dfrac{2}{15} + \dfrac{11}{15} + \dfrac{7}{15}$

9. $4 - \dfrac{3}{11}$

10. $\dfrac{2}{5} \cdot \dfrac{1}{3} \cdot \dfrac{4}{7}$

11. $2\dfrac{4}{15} + 3\dfrac{1}{6} + 4\dfrac{7}{10}$

12. $70\dfrac{1}{4} - 23\dfrac{5}{6}$

13. $4\dfrac{5}{7} \cdot 2\dfrac{6}{11}$

14. $6 \div 3\dfrac{1}{3}$

15. $(700)(8000)$

16. $403 - 4.012$

17. $71 + 0.354 + 4.39$

18. $(0.27)(0.043)$

19. $27.404 \div 0.34$

20. Evaluate $(36 \div 3^2 \cdot 2) + 12 \div 4 - 2^2$.

21. Use the tests for divisibility to determine if 732 can be divided exactly by 2, 3, 4, 5, 9, and 10.

22. Find the prime factorization of 396.

23. Find the LCM of 14, 21, and 30.

Complete each of the following (Exercises 24–28).

24. Division by _____ is undefined.

25. In the proportion $\frac{7}{8} = \frac{140}{160}$, the extremes are _____ and _____.

26. 15% of _____ is 7.5

27. $9\frac{1}{4}$% of 200 is _____.

28. 65 is _____ % of 26.

29. Solve for x: $\dfrac{1\frac{2}{3}}{x} = \dfrac{10}{2\frac{1}{4}}$

30. In a certain company, three out of every five employees are male. How many female employees are there out of the 490 people working for this company?

31. The sum of two numbers is 521. If one of the numbers is 196, what is the other number?

32. Milt has the following scores on his first three math tests: 75, 87, and 79. To earn a B grade for the course, he must have at least an 80 average. What is the least score that he can get on his fourth test to maintain a B average?

33. An investment pays $6\frac{1}{4}$% simple interest. What is the interest on $4800 invested for 8 months?

34. How much is put into a savings account that pays $5\frac{1}{2}$% simple interest if after 100 days, the interest earned is $69.30?

35. Find the simple interest rate if an investment of $6200 earns $1395 in $2\frac{1}{2}$ years.

36. Five thousand dollars is invested in a savings account. (a) What is in the balance of the account after 9 years if interest is earned at 8% compounded quarterly? (b) If interest is earned at 8% compounded daily?

37. Ten thousand dollars is placed in an account paying 4% interest, compounded monthly. How much is in the account at the end of 20 years (to the nearest cent)?

Refer to the following test scores for Exercises 38–40:

   27, 36, 45, 72, 63, 36, 27, 18, 36, 90

38. Find the mean score.

39. Find the median score.

40. Find the range of the scores.

Introduction to Algebra: Signed Numbers

Mathematics at Work!

A car rental counter
is a common sight at
most airports.

Suppose that you are a travelling salesperson and your company has allowed you an expense of $50 per day for car rental. You have rented a car for $30 per day plus 5¢ per mile. How many miles can you drive for the $50 allowed?

Plan: Set up an equation relating the amounts of money spent and the money allowed in your budget, and solve the equation.

Solution: Let x = number of miles driven.

Money spent = Money budgeted

$$30 + 0.05x = 50$$

$$30 + 0.05x - 30 = 50 - 30$$

$$0.05x = 20$$

$$\frac{0.05x}{0.05} = \frac{20}{0.05}$$

$$x = 400 \text{ miles}$$

You would like to get the greatest number of miles possible for your money and the car rental agency down the street charges only $25 per day and 8¢ per mile. How many miles would you get for your $50 with this price structure? Is this a "better deal" for you? (See Exercise 43, Section 9.6.)

What to Expect in Chapter 9

Chapter 9 provides an introduction to **signed numbers** and several related topics from algebra. The concept of a **negative number** is introduced along with the rules for operating with signed numbers. In Section 9.1, number lines are used to provide a "picture" of signed numbers and their relationships. Signed numbers are **graphed** on number lines, and number lines are used to give an intuitive understanding of the very important idea of the magnitude (**absolute value**) of a signed number. This idea is the foundation for the rules of addition and subtraction with signed numbers developed in Sections 9.2 and 9.3.

Multiplication and division with signed numbers is discussed in Section 9.4; the fact that division by 0 is undefined is reinforced, this time in terms of positive and negative numbers. Section 9.5 provides an introduction to solving equations of the form $ax + b = c$ in which only one or two steps are required to find the solution.

The chapter closes with a section on solving problems by setting up and solving equations that have signed numbers as solutions.

9.1   Number Lines and Absolute Value

OBJECTIVES

1. Understand signed numbers and what integers are.
2. Know how to find the opposite of a signed number.
3. Be able to graph sets of signed numbers on number lines.
4. Know how to use the symbols of order: <, >, ≤, and ≥.
5. Learn how to find the absolute value of a signed number.

Number Lines

Draw a horizontal line, choose any point on the line, and label it with the number 0. (See Figure 9.1.)

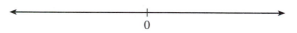

Figure 9.1

Now choose another point on the same line to the right of 0 and label it with the number 1 (Figure 9.2).

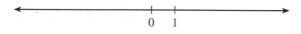

Figure 9.2

The line in Figure 9.2 is called a **number line.** The points corresponding to the whole numbers are determined by the distance between 0 and 1. The point corresponding to 2 is the same distance to the right of 1 as 1 is from 0, 3 is the same distance to the right of 2, and so on (Figure 9.3).

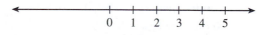

Figure 9.3

The **graph** of a number is indicated with a shaded dot at the corresponding point on the number line. The point or dot is the graph of the number, and the number is the **coordinate** of the point. We will use the terms **number** and **point** interchangeably. Thus, "number 3" and "point 3" are considered to have the same meaning.

The numbers 0, 1, and 5 form a set. Sets are indicated with braces and are named with capital letters. Thus, we can say $A = \{0, 1, 5\}$. The graph of A is shown in Figure 9.4.

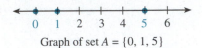

Graph of set $A = \{0, 1, 5\}$

Figure 9.4

On a horizontal number line, the point one unit to the left of 0 is the **opposite of 1.** It is also called **negative 1** and is symbolized -1. (The $-$ sign is called a negative sign.) The point two units to the left of 0 is the **opposite of 2** or **negative 2,** symbolized -2; and so on. (See Figure 9.5.)

The opposite of 1 is -1;   The opposite of -1 is $-(-1) = +1$;

the opposite of 2 is -2;   the opposite of -2 is $-(-2) = +2$;

the opposite of 3 is -3;   the opposite of -3 is $-(-3) = +3$;

and so on.   and so on.

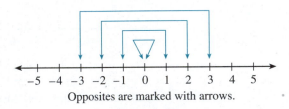

Opposites are marked with arrows.

Figure 9.5

The set of numbers consisting of the whole numbers and their opposites is called the set of **integers.** The counting numbers are called **positive integers.** (With positive integers the plus sign, $+$, is optional. Thus, $+3$ and 3 both represent positive 3.) The opposites of the positive integers are called **negative integers. The number 0 is neither positive nor negative, and 0 is its own opposite.** $(0 = -0)$ (See Figure 9.6.) Note that the opposite of a positive integer is a negative integer, and the opposite of a negative integer is a positive integer.

Integers $\ldots, -3, -2, -1, 0, 1, 2, 3, \ldots$

Positive Integers $1, 2, 3, 4, 5, \ldots$

Negative Integers $\ldots, -4, -3, -2, -1$

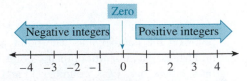

Figure 9.6

EXAMPLE 1   Find the opposite of 8.

Solution

-8

EXAMPLE 2   Find the opposite of -2.

Solution

$-(-2)$ read "the opposite of negative 2"

or $+2$ read "positive 2"

or 2   A number is understood to be positive if there is no sign in front of it.

EXAMPLE 3   Graph the set $B = \{-2, -1, 3\}$.

Solution

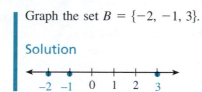

EXAMPLE 4   Graph the set $\{-4, -2, 0, 2, 4, \ldots\}$.

Solution

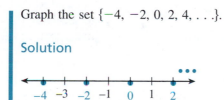

The integers are not the only numbers that can be represented on a number line. Fractions and decimal numbers such as $\frac{1}{2}, \frac{2}{3}, -\frac{5}{3}, -2.2$, and 1.7 can also be represented. (See Figure 9.7.)

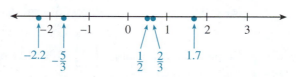

Figure 9.7

All the numbers that can be represented on a number line are called **signed numbers.** Signed numbers include all the integers, fractions, and decimals, as well as a type of number called **irrational numbers.** These numbers will be discussed later in the Additional Topics from Algebra chapter, which may or may not be included in your version of this textbook.

Order

On a horizontal number line, **smaller numbers are always to the left of larger numbers.** Each number is smaller than any number to its right and larger than any number to its left. Two symbols used to indicate order are

$<$ read "is less than"

and

$>$ read "is greater than"

Using <		OR	**Using >**	
$\frac{1}{4} < \frac{3}{4}$	$\frac{1}{4}$ is less than $\frac{3}{4}$		$\frac{3}{4} > \frac{1}{4}$	$\frac{3}{4}$ is greater than $\frac{1}{4}$
$0 < 5$	0 is less than 5		$5 > 0$	5 is greater than 0
$-3 < 1$	-3 is less than 1		$1 > -3$	1 is greater than -3
$-6.1 < -2.4$	-6.1 is less than -2.4		$-2.4 > -6.1$	-2.4 is greater than -6.1

Symbols for Order

$=$	is equal to	$\neq$	is not equal to
$<$	is less than	$>$	is greater than
$\leq$	is less than or equal to	$\geq$	is greater than or equal to

---

EXAMPLE 5

Determine whether each of the following statements is true or false.

(a) $6 < 10$ True, since 6 is less than 10.

(b) $2 > -1$ True, since 2 is greater than -1.

(c) $5 \geq 5$ True, since 5 is equal to 5.

(d) $-2.3 \leq 0$ True, since -2.3 is less than 0.

(e) $-3 < -10$ False, since -3 is greater than -10.

EXAMPLE 6

Graph the set of numbers $\left\{-\frac{1}{2}, 0, 1.5, 2\right\}$ on a number line.

Solution

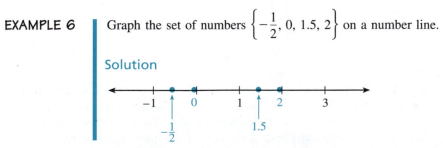

Absolute Value

On a number line, any number and its opposite lie the same number of units from 0. For example, both $+5$ and -5 are five units from 0. (See Figure 9.8.) The $+$ and $-$ signs indicate direction and the 5 indicates distance.

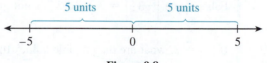

Figure 9.8

The distance a number is from 0 on a number line is called its **absolute value** and is symbolized by two vertical bars, $|\ \ |$. Thus, $|+5| = 5$ and $|-5| = 5$. Figure 9.9 illustrates this concept.

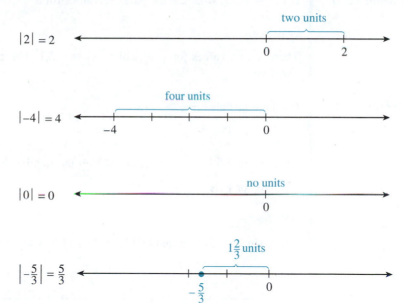

Figure 9.9

Remember: The absolute value of a number is never negative.

EXAMPLE 7   (a) $|5.6| = 5.6$

(b) $|-3.2| = 3.2$

EXAMPLE 8   True or False: $|-5| \leq 5$

Solution

True, since $|-5| = 5$ and $5 \leq 5$.

EXAMPLE 9   True or False: $\left|-3\frac{1}{2}\right| > 3\frac{1}{2}$.

Solution

False, since $\left|-3\frac{1}{2}\right| = 3\frac{1}{2}$ and $3\frac{1}{2}$ is not greater than $3\frac{1}{2}$.

EXAMPLE 10   If $|x| = 3$, what are the possible values for x?

Solution

$x = 3$ or $x = -3$ since $|3| = 3$ and $|-3| = 3$

EXAMPLE 11   If $|x| = -2.1$, what are the possible values for x?

Solution

There are no values for x for which $|x| = -2.1$. The absolute value can never be negative.

EXAMPLE 12   (a) True or False: $-6 > -13$.

Solution

True, since -6 is to the right of -13 on the number line.

(b) True or False: $|-6| > |-13|$.

Solution

False, since $|-6| = 6$ and $|-13| = 13$ and 6 is less than 13.

Exercises 9.1

Find the opposite of each number.

1. 14   **2.** 12   **3.** -10 **4.** -9 **5.** $1\frac{1}{3}$

6. $2\frac{1}{4}$ **7.** -5.3 **8.** -7.1 **9.** 30   **10.** 40

In each of the following exercises, draw a number line and graph the given set of numbers.

11. $\{0, 1, 2\}$ **12.** $\{0, 2, 4\}$

13. $\{-3, -1, 1\}$ **14.** $\{-3, -2, 0\}$

15. $\{1, 2, 3, \ldots, 10\}$ **16.** $\{-2, -1, 0, \ldots, 5\}$

17. $\{|-5|, |-2|, 0\}$

18. $\{|-3|, 0, 1\}$

19. $\left\{-2\frac{1}{2}, -1.3, -\frac{1}{8}, 0.75\right\}$

20. $\left\{-1\frac{3}{4}, -0.25, 0.125, 1\right\}$

Fill in the blank with the appropriate symbol, $<$, $>$, or $=$.

21. 5 _____ 3

22. -4 _____ 1

23. -3 _____ $-(-6)$

24. 7 _____ $-(-7)$

25. $\frac{2}{3}$ _____ $\frac{1}{2}$

26. $\frac{3}{4}$ _____ $\frac{1}{3}$

27. $-\frac{2}{3}$ _____ $-\frac{1}{2}$

28. $-\frac{3}{4}$ _____ $-\frac{1}{3}$

29. $|-2.3|$ _____ 2.3

30. $|4.6|$ _____ -4.6

Determine whether each statement is true or false.

31. $0 = -0$

32. $-10 < -8$

33. $-8 > -7.4$

34. $-6 \leq -6$

35. $-5.2 \leq -5.2$

36. $4 \leq -13$

37. $\left|3\frac{1}{2}\right| = -3\frac{1}{2}$

38. $|-4| \leq 4$

39. $-2.6 \leq |-2.6|$

40. $|-7| \geq 7$

41. $-7.1 < -8.1$

42. $-10.3 \leq -12$

Find the possible values for x in each of the following equations.

43. $|x| = 6$

44. $|x| = 9$

45. $|x| = 4.3$

46. $|x| = \frac{3}{2}$

47. $|x| = -2$

48. $|x| = -5$

49. $|x| = \frac{2}{3}$

50. $|x| = \frac{1}{4}$

9.2   Addition with Signed Numbers

OBJECTIVES

1. Develop an intuitive understanding of adding signed numbers along a number line.
2. Learn how to add signed numbers.

Addition with Signed Numbers

The sum of two signed numbers can be indicated with a plus (+) sign between the two numbers. For example,

(+2)　　+　　(+7)

  ↑　　　↑　　　↑

positive 2　plus　positive 7

The sum in this example is obviously +9 since we already know how to add 2 and 7:

$$(+2) + (+7) = +9$$

But the sums

$$(+2) + (-7) = ?$$

$$(-2) + (+7) = ?$$

and

$$(-2) + (-7) = ?$$

are not so obvious.

By using a number line, we can develop an intuitive idea of how to add signed numbers. Start at the first number to be added, then:

1. Move right if the second number is positive.

or

2. Move left if the second number is negative.

The distance moved is the absolute value of the second number.

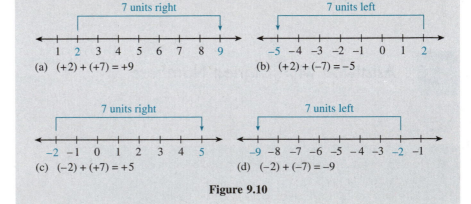

Figure 9.10

Find the following sums, using a number line as illustrated in Figure 9.10.

EXAMPLE 1 $(-5) + (+4) = ?$

Solution

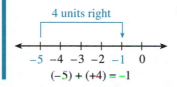

$(-5) + (+4) = -1$

EXAMPLE 2 $(-3) + (-8) = ?$

Solution

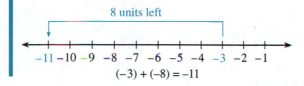

$(-3) + (-8) = -11$

EXAMPLE 3 $(+6) + (-10) = ?$

Solution

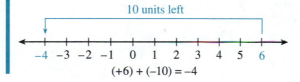

$(+6) + (-10) = -4$

EXAMPLE 4 $(-7) + (+7) = ?$

Solution

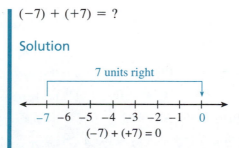

$(-7) + (+7) = 0$

Note: The sum of two opposites will always be 0.

EXAMPLE 5

$(-8.6) + (+5.2) = ?$

Solution

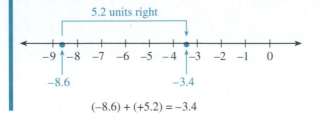

$(-8.6) + (+5.2) = -3.4$

Rules for Adding Signed Numbers

Signed numbers can be added mentally, without the use of number lines, once you know the following rules.

> **Rules for Adding Signed Numbers**
>
> 1. **To add two signed numbers with like signs,** add their absolute values and use the common sign.
>
> $(+6) + (+5) = +[|+6| + |+5|] = +[6 + 5] = +11$
>
> $(-2) + (-8) = -[|-2| + |-8|] = -[2 + 8] = -10$
>
> 2. **To add two signed numbers with unlike signs,** subtract their absolute values (smaller from larger) and use the sign of the number with the larger absolute value.
>
> $(+10) + (-15) = -[|-15| - |+10|] = -[15 - 10] = -5$
>
> $(+12) + (-9) = +[|+12| - |-9|] = +[12 - 9] = +3$

EXAMPLE 6

(a) $(+4) + (-1) = +(4 - 1) = +3$ (unlike signs)

(b) $(-5) + (-3) = -(5 + 3) = -8$ (like signs)

(c) $(+7) + (+10) = +(7 + 10) = +17$ (like signs)

(d) $(+6) + (-19) = -(19 - 6) = -13$ (unlike signs)

(e) $(+30) + (-30) = (30 - 30) = 0$ (opposites)

(f) $(+5.8) + (-7.2) = -(7.2 - 5.8) = -1.4$ (unlike signs)

(g) $\left(-3\frac{1}{2}\right) + \left(-4\frac{1}{4}\right) = -\left(3\frac{1}{2} + 4\frac{1}{4}\right) = -7\frac{3}{4}$ (like signs)

If more than two numbers are to be added, add any two, then add their sum to one of the other numbers until all numbers have been added.

In algebra, equations are written horizontally, so the ability to work with sums written horizontally is very important. However, there are some situations (as in long division) where numbers are written vertically. The rules for adding are the same whether the numbers are written horizontally or vertically.

EXAMPLE 7

$$(+9) + (+2) + (-6) = + (9 + 2) + (-6)$$
$$= + (11) + (-6)$$
$$= +5$$

EXAMPLE 8

$$\begin{array}{r} -12 \\ +8 \\ \hline -4 \end{array}$$

EXAMPLE 9

$$\begin{array}{r} -5 \\ 6 \\ -14 \\ \hline -13 \end{array}$$

EXAMPLE 10

$$\begin{array}{r} -1.4 \\ -2.3 \\ -4.9 \\ \hline -8.6 \end{array}$$

CLASSROOM PRACTICE

Find each sum.

1. $(-10) + (+3) =$

2. $(-5) + (+9) =$

3. $(-2) + (-3) =$

4. $(+1.6) + (-6.3) =$

ANSWERS: **1.** -7 **2.** $+4$ **3.** -5 **4.** -4.7

Exercises 9.2

Find each sum.

1. $(+6) + (-4)$

2. $(+8) + (-7)$

3. $(4) + (+6)$

4. $(5) + (-8)$

5. $(16) + (+3)$

6. $(-8) + (-2)$

7. $(-3) + (-6)$ **8.** $(-2) + (+2)$ **9.** $(+4) + (-4)$

10. $(13) + (12)$ **11.** $(6) + (-10)$ **12.** $(14) + (-17)$

13. $(+5) + (-3)$ **14.** $(+15) + (-18)$ **15.** $(-4) + (-12)$

16. $(-8) + (+8)$ **17.** $(+2) + (-6)$ **18.** $(-9) + (+5)$

19. $(-16) + (+3) + (+13)$ **20.** $(-5) + (+5) + (14)$

21. $(-1) + (-2) + (+7)$ **22.** $(+3) + (-4) + (-5)$

23. $(+6) + (+3) + (+5)$ **24.** $(-18) + (-5) + (-7)$

25. $(-1) + (+2) + (-4) + (+2)$

26. $\begin{array}{r} -4 \\ +8 \\ \hline \end{array}$	**27.** $\begin{array}{r} -5 \\ -10 \\ \hline \end{array}$	**28.** $\begin{array}{r} -13 \\ -6 \\ \hline \end{array}$	**29.** $\begin{array}{r} +16 \\ +25 \\ \hline \end{array}$	**30.** $\begin{array}{r} +14 \\ -8 \\ \hline \end{array}$
31. $\begin{array}{r} +20 \\ -7 \\ \hline \end{array}$	**32.** $\begin{array}{r} +2 \\ -5 \\ -3 \\ \hline \end{array}$	**33.** $\begin{array}{r} +8 \\ +3 \\ -1 \\ \hline \end{array}$	**34.** $\begin{array}{r} +10 \\ -4 \\ +2 \\ \hline \end{array}$	**35.** $\begin{array}{r} -16 \\ -8 \\ +12 \\ \hline \end{array}$
36. $\begin{array}{r} -15 \\ -20 \\ -6 \\ \hline \end{array}$	**37.** $\begin{array}{r} -4 \\ -17 \\ +11 \\ \hline \end{array}$	**38.** $\begin{array}{r} +13 \\ -5 \\ +17 \\ -25 \\ \hline \end{array}$	**39.** $\begin{array}{r} +14 \\ -14 \\ +37 \\ -37 \\ \hline \end{array}$	**40.** $\begin{array}{r} -8 \\ -5 \\ -13 \\ -22 \\ \hline \end{array}$
41. $\begin{array}{r} -100 \\ -50 \\ -85 \\ \hline \end{array}$	**42.** $\begin{array}{r} -96 \\ +14 \\ -83 \\ \hline \end{array}$	**43.** $\begin{array}{r} +750 \\ -632 \\ -198 \\ +200 \\ \hline \end{array}$	**44.** $\begin{array}{r} -300 \\ +450 \\ +325 \\ +500 \\ \hline \end{array}$	**45.** $\begin{array}{r} -25 \\ -95 \\ -48 \\ -20 \\ +67 \\ \hline \end{array}$

46. $\left(-3\frac{1}{6}\right) + \left(-2\frac{1}{3}\right)$ **47.** $\left(-1\frac{3}{5}\right) + \left(-2\frac{1}{5}\right)$

48. $\left(-10\frac{3}{4}\right) + \left(-9\frac{3}{8}\right)$ **49.** $\left(-4\frac{1}{5}\right) + \left(+6\frac{3}{10}\right)$

50. $\left(+2\frac{1}{2}\right) + \left(-7\frac{3}{4}\right)$ **51.** $(-10.3) + (-8.51)$

52. $(-25.6) + (+30)$ **53.** $(-18.4) + (+20)$

54. $\begin{array}{r} 10.4 \\ -16.5 \\ \hline \end{array}$	**55.** $\begin{array}{r} -33.82 \\ -35.46 \\ \hline \end{array}$	**56.** $\begin{array}{r} -16.9 \\ 14.3 \\ \hline \end{array}$	**57.** $\begin{array}{r} 25.4 \\ -19.5 \\ \hline \end{array}$
58. $\begin{array}{r} -10.1 \\ -3.4 \\ 5.5 \\ \hline \end{array}$	**59.** $\begin{array}{r} 8.6 \\ -2.5 \\ -7.2 \\ \hline \end{array}$	**60.** $\begin{array}{r} 24.3 \\ -18.5 \\ 16.1 \\ \hline \end{array}$	

9.3 Subtraction with Signed Numbers

OBJECTIVES

1. Understand the concept of the **opposite** of a signed number.
2. Learn how to subtract signed numbers.

Subtraction with Signed Numbers

In the intuitive discussion of **addition** using number lines, we used these rules:

1. Move *right* if the second number is positive.

or

2. Move *left* if the second number is negative.

With **subtraction,** we simply reverse the moves:

1. Move *left* if the second number is positive.

or

2. Move *right* if the second number is negative.

In other words, in subtraction, move in the opposite direction from the direction indicated by the number being subtracted. (See Figure 9.11.)

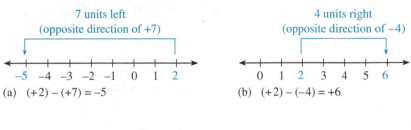

7 units left
(opposite direction of +7)

(a) $(+2) - (+7) = -5$

4 units right
(opposite direction of -4)

(b) $(+2) - (-4) = +6$

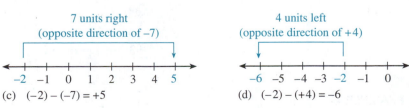

7 units right
(opposite direction of -7)

(c) $(-2) - (-7) = +5$

4 units left
(opposite direction of +4)

(d) $(-2) - (+4) = -6$

Figure 9.11

We can restate the previous discussion and the moves outlined in Figure 9.10 in terms of addition. **In subtraction, add the opposite of the number being subtracted.** For example,

$$(+4) \quad - \quad (+6) \quad = \quad (+4) \quad + \quad (-6) \quad = -2$$

positive 4   minus   positive 6   positive 4   plus   opposite of +6

$$(+5) \quad - \quad (-8) \quad = \quad (+5) \quad + \quad (+8) \quad = +13$$

positive 5   minus   negative 8   positive 5   plus   opposite of −8

Definition

The **difference** of two integers a and b is the sum of a and the opposite of b. Symbolically, $(a) - (b) = (a) + (-b)$.

EXAMPLE 1

$$(+2) - (-6) = (+2) + (+6) = +8$$

minus   plus   opposite of −6

EXAMPLE 2

$$(-3) - (-7) = (-3) + (+7) = +4$$

add opposite of − 7

EXAMPLE 3

$$(-5) - (+2) = (-5) + (-2) = -7$$

add opposite of +2

EXAMPLE 4

$$(-9) - (-9) = (-9) + (+9) = 0$$

add opposite of −9

The numbers may also be written vertically, one underneath the other. In this case, change the sign of the number being subtracted (the bottom number), then add.

EXAMPLE 5

Subtract	(Add)	
-10	-10	
-3	$+3$	← sign changed
	-7	

EXAMPLE 6

$+14$	$+14$	
$+9$	-9	← sign changed
	5	

EXAMPLE 7

$$
\begin{array}{r}
-8.6 \\
+4.3 \\
\hline
\end{array}
$$

$$
\begin{array}{r}
-8.6 \\
-4.3 \quad \leftarrow \text{sign changed} \\
\hline
-12.9
\end{array}
$$

A New Notation

We have been using a plus (+) sign for addition, a minus (−) sign for subtraction, and parentheses around each number being added or subtracted. In another notation more commonly used in algebra, the parentheses and the plus (+) and minus (−) signs are dropped. **The numbers are written horizontally, and the problem is considered as adding positive and negative numbers.** All positive and negative signs between numbers must be included.

If the first number to the left is positive, the + sign may be omitted with the understanding that the number is positive. For example,

$9 - 12$ is the same as $(+9) + (-12)$

So,

$9 - 12 = (+9) + (-12) = -3$

After some practice, you will be able to do the second step mentally and will simply write

$9 - 12 = -3$

EXAMPLE 8

$7 - 3 = (+7) + (-3) = 4$

or

$7 - 3 = 4$

EXAMPLE 9

$-13 + 6 = (-13) + (+6) = -7$

or

$-13 + 6 = -7$

EXAMPLE 10

$-4 - 8 = (-4) + (-8) = -12$

or

$-4 - 8 = -12$

EXAMPLE 11

(a) $6 - 9 = -3$

(b) $-20 + 12 = -8$

(c) $-3 - 14 = -17$

(d) $16.3 - 4.1 - 5.7 = 12.2 - 5.7 = 6.5$

EXAMPLE 12

$$-14.3 - 20 + 5 + 1.8 = -34.3 + 5 + 1.8$$
$$= -29.3 + 1.8$$
$$= -27.5$$

CLASSROOM PRACTICE

Evaluate each of the following expressions.

1. $(+8) - (-3)$

2. $(-4.1) - (-5.7)$

3. $-10 + 4$

4. $13 - 5 + 6.1$

ANSWERS: **1.** 11   **2.** 1.6   **3.** -6 **4.** 14.1

Exercises 9.3

Find each difference.

1. $(+5) - (+2)$ **2.** $(+16) - (+3)$ **3.** $(+8) - (-3)$

4. $(+12) - (-4)$ **5.** $(-5) - (+2)$ **6.** $(-10) - (+3)$

7. $(-10) - (-1)$ **8.** $(-15) - (-1)$ **9.** $(-3) - (-7)$

10. $(-2) - (-12)$ **11.** $(-4) - (+6)$ **12.** $(-9) - (+13)$

13. $(-13) - (-14)$ **14.** $(-12) - (-15)$ **15.** $(+9) - (-9)$

16. $(+11) - (-11)$ **17.** $(+15) - (-2)$ **18.** $(+20) - (-3)$

19. $(-17) - (+14)$ **20.** $(-16) - (+10)$ **21.** $(+3) - (+8)$

22. $(+1) - (+5)$ **23.** $(-5) - (-5)$ **24.** $(-7) - (-7)$

25. $(+7) - (+12)$ **26.** $(+8.4) - (+10.5)$ **27.** $(-7.31) - (5.2)$

28. $(-6.3) - (-6.3)$

29. $\left(-11\frac{7}{10}\right) - \left(-11\frac{7}{10}\right)$ **30.** $\left(4\frac{1}{2}\right) - \left(-2\frac{3}{4}\right)$

Subtract the bottom number from the top number.

31. $\begin{array}{r} 18 \\ -12 \\ \hline \end{array}$ **32.** $\begin{array}{r} 24 \\ 16 \\ \hline \end{array}$ **33.** $\begin{array}{r} -8 \\ -12 \\ \hline \end{array}$ **34.** $\begin{array}{r} -13 \\ -18 \\ \hline \end{array}$ **35.** $\begin{array}{r} -4 \\ +5 \\ \hline \end{array}$

36. $\begin{array}{r} 32 \\ -48 \\ \hline \end{array}$ **37.** $\begin{array}{r} -6.2 \\ -30.1 \\ \hline \end{array}$ **38.** $\begin{array}{r} -25.6 \\ -13.7 \\ \hline \end{array}$ **39.** $\begin{array}{r} -45.18 \\ -16.39 \\ \hline \end{array}$ **40.** $\begin{array}{r} -28.76 \\ -15.81 \\ \hline \end{array}$

Evaluate each of the following expressions.

41. $6 + 2$ **42.** $4 + 8$ **43.** $7 - 1$ **44.** $9 - 4$

45. $4 + 6$ **46.** $8 + 9$ **47.** $-3 - 1$ **48.** $-2 - 6$

49. $12 - 6$ **50.** $9 - 3$ **51.** $-13 + 4$ **52.** $-20 + 14$

53. $-10 + 9$ **54.** $-18 + 3$ **55.** $24 - 32$ **56.** $14 - 17$

57. $-12 - 6$ **58.** $-2 - 8$ **59.** $-15 + 18$ **60.** $-25 + 30$

61. $-20 + 21$ **62.** $-30 + 32$ **63.** $-7 + 7$ **64.** $-6 + 6$

65. $18 - 3$ **66.** $-4 + 16 - 8$ **67.** $-5 + 12 - 3$

68. $-20 - 2 + 6$ **69.** $14 - 5 - 12$ **70.** $13.1 + 15.2 - 6$

71. $-6.4 - 8.5 - 13.7$ **72.** $-4.1 - 7.3$ **73.** $30\frac{1}{2} + 12 - 18\frac{1}{4}$

74. $16\frac{1}{5} + 4\frac{3}{5} - 20\frac{3}{10}$ **75.** $15.63 + 6.14 - 21.77$

76. $13 - 4 + 6 - 5$ **77.** $16 - 3 - 7 - 1$

78. $19 - 5 - 8 - 6$ **79.** $-4 + 10 - 12 + 1$

80. $-8 + 14 - 10 + 3$

9.4 Multiplication, Division, and Order of Operations with Signed Numbers

OBJECTIVES

1. Learn how to multiply signed numbers.
2. Learn how to divide signed numbers.
3. Be able to simplify an expression with signed numbers by following the rules for order of operations.

Multiplication with Signed Numbers

Multiplication with whole numbers is shorthand for repeated addition. For example,

$$7 + 7 + 7 + 7 = 4 \cdot 7 = 28$$

$$3 + 3 + 3 + 3 + 3 + 3 = 6 \cdot 3 = 18$$

Multiplication with signed numbers can also be considered shorthand for repeated addition.

$$(-7) + (-7) + (-7) = 3(-7) = -21$$

$$(-1.5) + (-1.5) + (-1.5) + (-1.5) = 4(-1.5) = -6.0$$

Repeated addition of a negative number results in the product of a positive number and a negative number. Since the sum of negative numbers is negative, the product of a positive number with a negative number will be negative. In fact, **the product of any positive number with a negative number will be negative.**

EXAMPLE 1

(a) $4(-3) = (-3) + (-3) + (-3) + (-3) = -12$

(b) $5(-2) = -10$

(c) $3.1(-5) = -15.5$

(d) $-\dfrac{1}{3}\left(\dfrac{1}{2}\right) = -\dfrac{1}{6}$

The product of two negative numbers is not explained in terms of repeated addition. The following pattern leads, intuitively, to the rule for multiplying two negative numbers. Try to supply the missing products.

$$(+3)(-5) = -15$$
$$(+2)(-5) = -10$$
$$(+1)(-5) = -5$$
$$(0)(-5) = 0$$
$$(-1)(-5) = ?$$
$$(-2)(-5) = ?$$
$$(-3)(-5) = ?$$

Did you notice that the products are increasing 5 at a time? If you did, your intuition should lead to

$$(-1)(-5) = +5$$
$$(-2)(-5) = +10$$
$$(-3)(-5) = +15$$

Although one example does not prove a rule, this time intuition is correct. **The product of two negative numbers is positive.**

EXAMPLE 2

(a) $(-6)(-4) = +24$

(b) $-7(-9) = +63$

(c) $-2.3(-5.1) = +11.73$

(d) $\left(-\dfrac{2}{3}\right)\left(-\dfrac{1}{5}\right) = +\dfrac{2}{15}$

EXAMPLE 3   (a) $(0)(-12) = 0$ and $(0)(+12) = 0$

(b) $(-2.7)(0) = 0, (+2.7)(0) = 0,$ and $(0)(0) = 0$

As Examples 3(a) and (b) illustrate, **the product of 0 with any number is 0.**

Rules for Multiplying Signed Numbers

1. The product of two positive numbers is positive.

2. The product of two negative numbers is positive.

3. The product of a positive number and a negative number is negative.

4. The product of 0 with any number is 0.

The same rules can be stated using symbols.

If a and b are positive numbers,

1. $a \cdot b = ab$

2. $(-a)(-b) = ab$

3. $a(-b) = -ab$

4. $a \cdot 0 = 0$ and $(-a) \cdot 0 = 0$

Division with Signed Numbers

Now, consider a division problem, such as $42 \div 6$. We know that

$$\frac{42}{6} = 7 \quad \text{because} \quad 42 = 6 \cdot 7$$

That is, division is defined in terms of multiplication.

Definition

For any numbers a and b where $b \neq 0$,

$$\frac{a}{b} = x \quad \text{means} \quad a = b \cdot x$$

If a is any number, then $\dfrac{a}{0}$ is undefined. (**We cannot divide by 0.**) (See Section 1.7.)

Using the rules for multiplication with signed numbers, we can develop the rules for division with signed numbers.

EXAMPLE 4 $\dfrac{+28}{+4} = +7$ because $+28 = +4(+7).$

EXAMPLE 5 $\dfrac{-28}{+4} = -7$ because $-28 = (+4)(-7).$

EXAMPLE 6 $\dfrac{+28}{-4} = -7$ because $+28 = (-4)(-7).$

EXAMPLE 7 $\dfrac{-28}{-4} = +7$ because $-28 = (-4)(+7).$

The rules for division with signed numbers can be stated as follows.

Rules for Dividing Signed Numbers

1. The quotient of two positive numbers is positive.

2. The quotient of two negative numbers is positive.

3. The quotient of a positive number and a negative number is negative.

The rules can be stated using symbols.

If a and b are positive numbers,

1. $\dfrac{a}{b} = \dfrac{a}{b}$

2. $\dfrac{-a}{-b} = \dfrac{a}{b}$

3. $\dfrac{-a}{b} = -\dfrac{a}{b}$ *and* $\dfrac{a}{-b} = -\dfrac{a}{b}$

EXAMPLE 8 $\dfrac{21.7}{-7} = -3.1$

EXAMPLE 9 $\dfrac{-36.28}{-2} = +18.14$

EXAMPLE 10 $\dfrac{-\frac{3}{4}}{+\frac{1}{8}} = -\dfrac{3}{\cancel{4}}\left(\dfrac{\overset{2}{\cancel{8}}}{1}\right) = -\dfrac{6}{1} = -6$

If a product involves more than two signed numbers, multiply any two, then continue to multiply until all the numbers have been multiplied.

EXAMPLE 11 $\begin{aligned}(-3)(4)(-2) &= [(-3)(4)](-2)\\ &= [-12](-2)\\ &= +24\end{aligned}$

EXAMPLE 12 $\begin{aligned}(-5)(-3)(-10) &= +15(-10)\\ &= -150\end{aligned}$

EXAMPLE 13 $\begin{aligned}(-2)^3(-6.1) &= (-2)(-2)(-2)(-6.1)\\ &= +4(-2)(-6.1)\\ &= -8(-6.1)\\ &= +48.8\end{aligned}$

CLASSROOM PRACTICE

Find the value of each expression.

1. $(-5)(5)$

2. $(-3)(-3)(0)$

3. $\dfrac{-4}{0}$

4. $\dfrac{-15.5}{-5}$

5. $(-4)^3$

ANSWERS: **1.** -25 **2.** 0   **3.** undefined   **4.** 3.1   **5.** -64

Order of Operations with Signed Numbers

The rules for order of operations are the same for expressions with signed numbers as they were stated in Section 2.2 for expressions with whole numbers. The rules are restated here for easy reference.

Rules for Order of Operations

1. First, simplify within grouping symbols, such as parentheses ( ), brackets [ ], or braces { }. Start with the innermost grouping.

2. Second, find any powers indicated by exponents.

3. Third, moving from **left to right,** perform any multiplications or divisions in the order they appear.

4. Fourth, moving from **left to right,** perform any additions or subtractions in the order they appear.

Evaluate the following expressions using the rules for order of operations and your knowledge of positive and negative numbers.

EXAMPLE 14 $14.6 + 3(-2.5)$

Solution

$$14.6 + 3(-2.5) = 14.6 + (-7.5) \quad \text{Multiply first.}$$
$$= 7.1 \quad \text{Add.}$$

EXAMPLE 15 $-3(5 - 6) - 4 \cdot 2$

Solution

$$-3(5 - 6) - 4.2 = -3(-1) - 4 \cdot 2 \quad \text{Simplify within parentheses.}$$
$$= +3 - 8 \quad \text{Multiply, from left to right.}$$
$$= -5 \quad \text{Add (or subtract).}$$

EXAMPLE 16 $(-8)^2 \cdot 5 + 3.1(-2) \div \dfrac{1}{2}$

Solution

$$(-8)^2 \cdot 5 + 3.1(-2) \div \frac{1}{2} = 64 \cdot 5 + 3.1(-2) \div \frac{1}{2} \quad \text{Find the power indicated by the exponent.}$$

$$= 320 + (-6.2) \div \frac{1}{2} \quad \text{Multiply, from left to right.}$$

$$= 320 + (-6.2) \cdot \frac{2}{1} \quad \text{Divide.}$$

$$= 320 - 12.4$$

$$= 307.6 \quad \text{Subtract.}$$

EXAMPLE 17 | $[5 + 3(-2 - 4)] \div 13$

Solution

$$\begin{aligned}
[5 + 3(-2 - 4)] \div 13 &= [5 + 3(-6)] \div 13 \\
&= [5 - 18] \div 13 \\
&= [-13] \div 13 \\
&= -1
\end{aligned}$$

Exercises 9.4

Find the following products.

1. $5(-3)$	**2.** $4(-6)$	**3.** $-6(-4)$
4. $-2(-7)$	**5.** $-5(4)$	**6.** $-8(3)$
7. $14(2)$	**8.** $13(3)$	**9.** $-10(5)$
10. $-11(3)$	**11.** $(-7)3$	**12.** $(-2)9$
13. $6(-8)$	**14.** $9(-4)$	**15.** $-7(-9)$
16. $-8(-9)$	**17.** $0(-6)$	**18.** $0(-4)$
19. $(-6)(-5)(3)$	**20.** $(-2)(-1)(7)$	**21.** $4(-2)(-3)$
22. $5(-6)(-1)$	**23.** $(-7)(-2)(-3)$	**24.** $(-2)(-2)(-8)$
25. $(-5)(0)(-6)$	**26.** $(-6)(0)(-2)$	**27.** $(-1)^3$
28. $(-3)^3$	**29.** $(-2)^4$	**30.** $(-4)^3$

Find the following quotients.

31. $\dfrac{-12}{4}$	**32.** $\dfrac{-18}{2}$	**33.** $\dfrac{-14}{7}$
34. $\dfrac{-28}{7}$	**35.** $\dfrac{-20}{-5}$	**36.** $\dfrac{-30}{-3}$
37. $\dfrac{-50}{-10}$	**38.** $\dfrac{40}{-8}$	**39.** $\dfrac{30}{-6}$
40. $\dfrac{-25}{-25}$	**41.** $\dfrac{-8}{-8}$	**42.** $\dfrac{22}{11}$
43. $\dfrac{36}{9}$	**44.** $\dfrac{24}{8}$	**45.** $\dfrac{26}{-13}$
46. $\dfrac{-31}{0}$	**47.** $\dfrac{17}{0}$	**48.** $\dfrac{0}{-16}$
49. $\dfrac{0}{25}$	**50.** $\dfrac{0}{-20}$	

Evaluate each of the following expressions by using the rules for order of operations.

51. $4 \cdot 3 - 5 \cdot 7$ **52.** $5 \cdot 3 - 4 \cdot 8$ **53.** $18 \cdot 2 + 6 \div (-2)$

54. $7 \cdot 5 + 12 \div (-3)$ **55.** $16 \div (-4) \cdot 2$ **56.** $24 \div (-8) \div 3$

57. $-18 \cdot 2 \div 3$ **58.** $-20 \cdot 15 \div 3$ **59.** $\dfrac{15}{-3} + 4(-8)$

60. $\dfrac{16}{-2} + 3(-5)$ **61.** $(+12)(-6) \div 3 \cdot 2$ **62.** $(+14)(-2) \div 7 \cdot 5$

63. $(-5)^2 + (-5)^3$ **64.** $(-4)^2 + (-4)^3$

65. $(16 - 25)(32 - 21)$ **66.** $(8 - 10)(15 - 12)$

67. $(8.3 - 4.1)(-3.2)$ **68.** $(6.4 - 5.2)(-1.6)$

69. $[4 + 3(-1 - 5)] \div 2$ **70.** $[10 + 2(-3 - 8)] \div 3$

71. $(6 \cdot 10 - 5) \div 11 \cdot 5 - 3^2$ **72.** $(4 \cdot 11 - 4) \div 10 \cdot 4 - 5^2$

73. $6^2 - 4 \cdot 2 + 5(7 + 3^2)$ **74.** $5^2 - 6 \cdot 3 + 2(5 + 2^3)$

75. $(1.3)^2 - (1.4)^2$ **76.** $(1.1)^2 - (1.2)^2$

77. $(-8.4) \div \left(-\dfrac{1}{2}\right) \cdot \dfrac{3}{4}$ **78.** $(-7.2) \div \left(-\dfrac{1}{3}\right) \cdot \dfrac{5}{8}$

79. $\left(-\dfrac{2}{3}\right) \div \left(-\dfrac{7}{8}\right) \div (-2)$ **80.** $\left(-\dfrac{3}{4}\right) \div \left(-\dfrac{5}{16}\right) \div (-3)$

Fill in each blank with the term (**positive, negative, zero, undefined**) that correctly completes the statement.

81. The product of two negative numbers is _____.

82. The quotient of two negative numbers is _____.

83. The quotient of two positive numbers is _____.

84. The product of two positive numbers is _____.

85. The quotient of a positive number and a negative number is _____.

86. The product of three negative numbers is _____.

87. If x is any signed number, then $0 \cdot x$ is _____.

88. If x is any nonzero number, then $\dfrac{0}{x}$ is _____.

89. If x is any signed number, then $\dfrac{x}{0}$ is _____.

90. If x is a negative signed number, then x^2 is _____.

9.5   Introduction to Solving Equations

OBJECTIVES

1. Be able to translate English phrases into algebraic expressions.
2. Be able to translate algebraic expressions into English phrases.
3. Know how to solve linear equations of the form $ax + b = c$.

Translating English Phrases

Many word problems can be solved if you know how to:

1. Translate English phrases into algebraic expressions.

2. Set up an equation using the translated expressions.

3. Solve the resulting equation.

In this section, we will learn how to translate English phrases by finding key words and then develop techniques for solving equations. In Section 9.6, we will see how these skills can be used to set up equations and solve problems.

The following list contains some of the key words that appear in word problems. These words indicate what operations are to be performed either with known numbers (called **constants**) or with variables and constants.

Key Words (that indicate operations)

Addition	Subtraction	Multiplication	Division
add	subtract (from)	multiply	divide
sum	difference	product	quotient
plus	minus	times	
more than	less than	twice, double	
increased by	decreased by	of (with fractions)	

The following examples illustrate how English phrases can be translated into algebraic expressions. Different letters are used as variables to represent the unknown numbers and the key words are in boldface print.

English Phrase	Algebraic Expression
1. A number **plus** 7 The **sum** of a number and 7 7 **more than** a number	$x + 7$
2. 8 **times** a number The **product** of a number and 8 A number **multiplied by** 8	$8n$
3. The **quotient** of a number and 6 A number **divided by** 6	$\dfrac{n}{6}$
4. **Twice** a number **decreased** by 5.1 5.1 **less than twice** a number 5.1 **subtracted from two times** a number	$2y - 5.1$

The words **difference** and **quotient** deserve special mention because their use implies that the operations are with the numbers in the order given. For example:

"The **difference** between x and 10" means $x - 10$, and **not** $10 - x$.

"The **quotient** of y and 20" means $\dfrac{y}{20}$, and **not** $\dfrac{20}{y}$.

A skill closely related to translating English phrases into algebraic expressions is that of reversing the process; that is, looking at an algebraic expression and being able to create some phrase that has the same meaning as the expression. Generally in such cases, there may be more than one correct phrase.

EXAMPLE 1   Write an English phrase that indicates the meaning of each expression.

(a) $14x$ (b) $n + 3.5$ (c) $2x - 5$ (d) $\dfrac{y}{3}$

Solution

Algebraic Expression	Possible English Phrase
(a) $14x$	The product of 14 and a number
(b) $n + 3.5$	3.5 more than a number
(c) $2x - 5$	Twice a number decreased by 5
(d) $\dfrac{y}{3}$	A number divided by 3

Solving Equations ($ax + b = c$)

If an equation contains a variable, such as $5x + 1 = 16$, then any number may be substituted for x. For example:

Substituting $x = 8$ gives $5 \cdot 8 + 1 = 16$ (a false statement).

Substituting $x = 3$ gives $5 \cdot 3 + 1 = 16$ (a true statement).

We can substitute as many values for x as we choose. However, the object is not just to substitute values for the variable x, but to find the value for x that results in a true statement when this value is substituted for x in the equation. This procedure is called **solving the equation.** The value found is called a **solution** of the equation.

We just found that substituting $x = 3$ in the equation $5x + 1 = 16$ gives a true statement, and thus, $x = 3$ is a solution of the equation. In fact, $x = 3$ is the only solution.

Definition

A **first-degree equation** in x (or **linear equation** in x) is any equation that can be written in the form

$$ax + b = c$$

where a, b, and c are constants and $a \neq 0$.

(Note: A variable other than x may be used.)

EXAMPLE 2

Given the equation $3n + 15 = 27$, (a) show that 4 is a solution, and (b) show that 2 is not a solution.

Solution

(a) Since $3 \cdot 4 + 15 = 27$ is a true statement, 4 is a solution.

(b) Since $3 \cdot 2 + 15 = 27$ is a false statement, 2 is not a solution.

A fundamental fact of algebra, stated here without proof, is that **every first-degree equation has exactly one solution.**

Sometimes, with simple equations, the solution can be found by observation or by trial and error. However, by using the following principles, we can guarantee finding the solution in an efficient, organized manner that works regardless of how complicated the equation is. The basic idea is that **the same operation is to be performed on both sides of the equal sign** (that is, on both sides of the equation).

For example, if you add 17 to one side of an equation, you must add 17 to the other side as well. You cannot add 17 to one side and subtract 17 from the other side. As illustrated in Figure 9.12, you can think of the equal sign as the fulcrum (or balance point) on a balance scale.

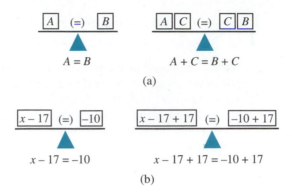

Figure 9.12

Principles Used in Solving a First-Degree Equation

In the four basic principles stated here, A and B represent algebraic expressions or constants.

1. The Addition Principle:

If $A = B$ is true, then $A + C = B + C$ is also true for any number C. (The same number may be added to both sides of an equation.)

2. The Subtraction Principle:

If $A = B$ is true, then $A - C = B - C$ is also true for any number C. (The same number may be subtracted from both sides of an equation.)

3. The Multiplication Principle:

If $A = B$ is true, then $C \cdot A = C \cdot B$ is also true for any number C. (Both sides of an equation may be multiplied by the same number.)

4. The Division Principle:

If $A = B$ is true, then $\dfrac{A}{C} = \dfrac{B}{C}$ is also true for any nonzero number C. (Both sides of an equation may be divided by the same nonzero number.)

Note that in the Division Principle, division by C is the same as multiplication by the reciprocal of C, namely $\dfrac{1}{C}$. Thus, we could list the Division and Multiplication Principles as one principle. The use of any of these principles with an equation gives a new equation with the same solution. Such equations are said to be **equivalent.** In the process of solving equations, we apply the principles just listed to find equivalent equations until the solution is obvious, as in the equations $x = 3$, $y = 15$, or $0.07 = R$.

EXAMPLE 3   Solve the equation $x + 10.3 = 25$.

Solution

$$x + 10.3 = 25 \qquad \text{Write the equation.}$$

$$x + 10.3 - 10.3 = 25 - 10.3 \qquad \text{Use the Subtraction Principle and subtract}$$
$$\text{10.3 from both sides.}$$

$$x + 0 = 14.7 \qquad \text{Simplify both sides.}$$

$$x = 14.7$$

EXAMPLE 4   Solve the equation $5y = 180$.

Solution

$$5y = 180 \qquad \text{Write the equation.}$$

$$\frac{5y}{5} = \frac{180}{5} \qquad \begin{array}{l}\text{Use the Division Principle and divide both sides by 5.}\\ \text{(Or use the Multiplication Principle and multiply both sides}\\ \text{by } \frac{1}{5}.) \text{ Simplify both sides.}\end{array}$$

$$y = 36$$

EXAMPLE 5   Solve the equation $\dfrac{n}{6} = 3.1$.

Solution

$$\frac{n}{6} = 3.1 \qquad \text{Write the equation.}$$

$$\frac{6}{1} \cdot \frac{n}{6} = \frac{3.1}{1} \cdot \frac{6}{1} \qquad \text{Use the Multiplication Principle and multiply both sides by } \frac{6}{1}.$$

$$n = 18.6 \qquad \text{Simplify both sides.}$$

More difficult problems may involve several steps. Keep in mind the following general ideas:

1. The goal is to isolate the variable on one side of the equation (either the right side or the left side).

2. First use the Addition and Subtraction Principles whenever numbers are subtracted or added to the variable.

3. Use the Multiplication and Division Principles to make 1 the coefficient of the variable.

EXAMPLE 6

Solve the equation $4x - 16 = 64$.

Solution

$$4x - 16 = 64 \qquad \text{Write the equation.}$$

$$4x - 16 + 16 = 64 + 16 \qquad \text{Use the Addition Principle and add 16 to both sides.}$$

$$4x + 0 = 80 \qquad \text{Simplify both sides.}$$

$$4x = 80$$

$$\frac{1}{4} \cdot 4x = \frac{1}{4} \cdot 80 \qquad \text{Use the Multiplication Principle and multiply both sides by } \frac{1}{4}. \text{ (Or, use the Division Principle and divide both sides by 4.)}$$

$$1 \cdot x = 20 \qquad \text{Simplify both sides.}$$

$$x = 20$$

Checking Solutions to Equations

Checking can be done by substituting the solution found back into the original equation to see if the resulting statement is true.

Check

$$4x - 16 = 64$$

$$4 \cdot 20 - 16 \overset{?}{=} 64$$

$$80 - 16 \overset{?}{=} 64$$

$$64 = 64 \qquad \text{A true statement}$$

CLASSROOM PRACTICE

Solve each of the following equations.

1. $x + 17 = 25$

2. $\frac{x}{2} = -25$

3. $3n = -21$

4. $13 = 5n - 2$

ANSWERS: **1.** $x = 8$ **2.** $x = -50$ **3.** $n = -7$ **4.** $n = 3$

Exercises 9.5

Write an algebraic expression described by each of the following English phrases. (Any variable may be used to represent the unknown number.)

1. a number plus 5

2. 8 more than a number

3. the sum of a number and 9

4. the product of a number and 9

5. the quotient of a number and 9

6. the difference of a number and 9

7. 0.13 less than a number

8. 0.55 decreased by a number

9. 0.13 decreased by a number

10. 16 plus a number

11. 4 more than twice a number

12. 1.5 increased by twice a number

13. 8 times a number plus 6

14. 6 more than 5 times a number

15. 1.8 less than the quotient of a number and 2

16. 2.5 more than the product of a number and 8

17. the difference of 3 and twice a number

18. the sum of 1 and three times a number

19. the product of 6 and a number increased by 4 times the number

20. 7 times a number decreased by the product of the number and 5

Write an English phrase that indicates the meaning of each expression. (There may be more than one correct phrase.)

21. $5x$

22. $n + 15$

23. $x - 6$

24. $y - 3.2$

25. $y + 4.91$

26. $7y$

27. $\dfrac{n}{17}$

28. $\dfrac{x}{10}$

29. $2x + 1$

30. $3x - 1$

31. $\dfrac{y}{3} + 20$

32. $\dfrac{a}{5} + 12$

33. $15x - 15$

34. $\dfrac{6}{x}$

35. $13 - x$

In Exercises 36–40 a set of numbers is given following the equation. Determine which number in the set is the solution by substituting the numbers into the equation.

36. $x - 17 = 10$; $\{27, 30, 33\}$

37. $2n + 3 = 13$; $\{0, 2, 5, 7\}$

38. $5n + 8 = 23$; $\{0, 1, 2, 3\}$

39. $\dfrac{n}{3} + 6 = 8$; $\{3, 6, 9, 12\}$

40. $\dfrac{x}{2} - 4 = 0$; $\{2, 4, 6, 8, 10\}$

Give an explanation (or a reason) for each step in the solution process.

41. $3x + 10 = 22$ _____

$3x + 10 - 10 = 22 - 10$ _____

$3x = 12$ _____

$\dfrac{3x}{3} = \dfrac{12}{3}$ _____

$x = 4$ _____

42. $2x - 1.5 = 3.5$ _____

$2x - 1.5 + 1.5 = 3.5 + 1.5$ _____

$2x = 5.0$ _____

$\dfrac{2x}{2} = \dfrac{5.0}{2}$ _____

$x = 2.5$ _____

Solve each of the following equations.

43. $x - 12 = 6$ **44.** $x - 10 = 9$ **45.** $n + 3 = 13$

46. $n + 7 = 15$ **47.** $7y = 42$ **48.** $9y = 72$

49. $\dfrac{n}{7} = 3.2$ **50.** $\dfrac{x}{3} = 10$ **51.** $2x + 1 = 14$

52. $3x - 1 = 14$ **53.** $18 = 4y - 6$ **54.** $35 = 5n + 7$

55. $\dfrac{n}{2} + 3 = 15$ **56.** $\dfrac{x}{5} - 9 = 6$ **57.** $\dfrac{x}{3} + 1.2 = 2.0$

58. $\dfrac{x}{4} - 3.5 = 1.5$ **59.** $70 = 2x - 6$ **60.** $82 = 3x + 4$

Writing and Thinking about Mathematics

61. We have indicated that, in an equation, the variable may appear on the right side of the equation or on the left side of the equation. The solution will not be affected. For example, the solution of the equation $3x + 1 = 25$ is the same as the solution of the equation $25 = 3x + 1$. Does this seem reasonable to you? Explain briefly.

9.6 Problem Solving with Equations

OBJECTIVES

1. Be able to solve number problems by using equations.
2. Be able to solve problems involving geometric concepts by using equations.

Solving Number Problems

The word problems discussed here can be solved by using the skills of translating English phrases into algebraic expressions that we developed in Section 9.5. The problem-solving approach is to follow the four steps outlined below:

> **Basic Steps for Solving Number Problems**
>
> 1. Read the problem carefully at least twice. Look for key words and phrases that can be translated into algebraic expressions.
>
> 2. Assign a variable as the unknown quantity and form an equation using the expressions you translated.
>
> 3. Solve the equation.
>
> 4. Look back over the problem and check that the answer is reasonable.

EXAMPLE 1   Four less than twice a number is equal to 20. Find the number.

Solution

STEP 1:   Let n = the unknown number.

STEP 2:   Translate "four less than twice a number" to $2n - 4$.

STEP 3:   Form the equation: $2n - 4 = 20$.

STEP 4: Solve the equation:

$$2n - 4 = 20$$

$$2n - 4 + 4 = 20 + 4$$

$$2n + 0 = 24$$

$$2n = 24$$

$$\frac{1}{2} \cdot 2n = \frac{1}{2} \cdot 24$$

$$1 \cdot n = 12$$

$$n = 12$$

Check

$$2(12) - 4 \overset{?}{=} 20$$

$$24 - 4 \overset{?}{=} 20$$

$$20 = 20$$

The number is 12.

EXAMPLE 2

A number is increased by 4.5 and the result is 30. What is the number?

Solution

STEP 1: Let x = the unknown number.

STEP 2: Translate "a number is increased by 4.5" to $x + 4.5$.

STEP 3: Form the equation: $x + 4.5 = 30$.

STEP 4: Solve the equation:

$$x + 4.5 = 30$$

$$x + 4.5 - 4.5 = 30 - 4.5$$

$$x + 0 = 25.5$$

$$x = 25.5$$

Check

$$25.5 + 4.5 \overset{?}{=} 30$$

$$30 = 30$$

So 25.5 is the correct number.

EXAMPLE 3

A span of a suspension bridge is the distance between its supports. The longest span on the Tacoma Narrows bridge in Washington State is 2800 feet. This is 40 feet more than twice the longest span of the Triboro bridge in New York City. What is the longest span of the Triboro bridge?

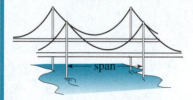

Solution

STEP 1:   Let x = the length of the longest span of the Triboro bridge.

STEP 2:   Translate "40 feet more than twice the longest span of the Triboro bridge" to $2x + 40$.

STEP 3:   Form the equation: $2x + 40 = 2800$.

STEP 4:   Solve the equation:

$$2x + 40 = 2800$$

$$2x + 40 - 40 = 2800 - 40$$

$$2x + 0 = 2760$$

$$2x = 2760$$

$$\frac{2x}{2} = \frac{2760}{2}$$

$$x = 1380$$

The longest span of the Triboro bridge is 1380 feet.
(The student should check this result in the original equation.)

Solving Geometry Problems

The concepts of perimeter, area, and volume of the geometric figures are described by various formulas that are in the form of equations. In this section, the necessary formula will be provided in each problem and all the values except one will be known. By substituting the known values in the formula, you can find the unknown value by using the techniques we have learned for solving equations.

For easy reference, Table 9.1 contains the formulas for perimeter (P) and area (A) for six geometric figures.

Table 9.1   Formulas for Finding Perimeter and Area of Six Geometric Figures

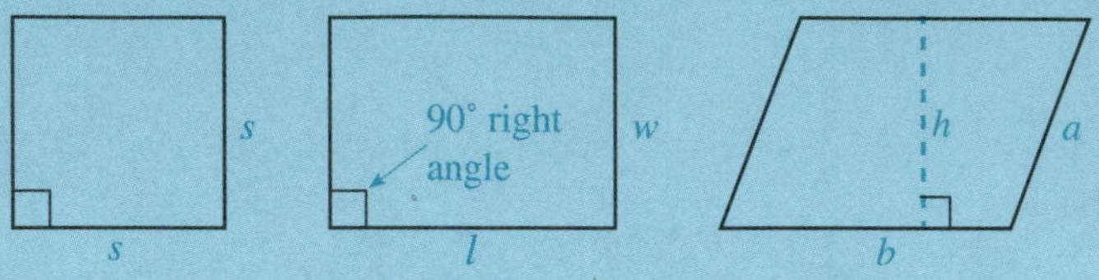

Square
$P = 4s$
$A = s^2$

Rectangle
$P = 2l + 2w$
$A = lw$

Parallelogram
$P = 2b + 2a$
$A = bh$

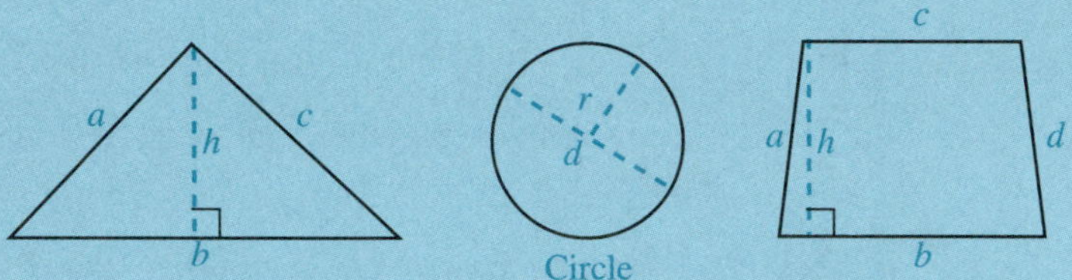

Triangle
$P = a + b + c$
$A = \frac{1}{2}bh$

Circle
$C = 2\pi r$
$C = \pi d$
$A = \pi r^2$

Trapezoid
$P = a + b + c + d$
$A = \frac{1}{2}h(b + c)$

EXAMPLE 4

If the length of a rectangle is 20 centimeters, and the perimeter is 70 centimeters, what is its width?

Solution

The formula for the perimeter of a rectangle is $P = 2l + 2w$. In this problem we know that $P = 70$ cm and that $l = 20$ cm. Substitution gives the equation

$$70 = 2 \cdot 20 + 2 \cdot w$$

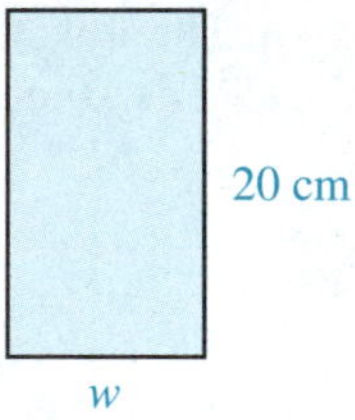

Now, we want to solve this equation for w.

$$70 = 40 + 2w$$

$$70 - 40 = 40 - 40 + 2w$$

$$30 = 2w$$

$$\frac{30}{2} = \frac{2w}{2}$$

$$15 = w$$

The width of the rectangle is 15 cm.
(Checking shows that $70 = 2 \cdot 20 + 2 \cdot 15$.)

EXAMPLE 5

Suppose that the area of a triangle is 45 square feet. If the base is 10 feet long, what is the height of the triangle?

Solution

The formula for the area of a triangle is $A = \dfrac{1}{2} \cdot b \cdot h$. In this problem, we know that $A = 45$ and $b = 10$. Substitution gives the equation

$$45 = \frac{1}{2} \cdot 10 \cdot h$$

Now, we want to solve this equation for h.

$$45 = \frac{1}{2} \cdot 10 \cdot h$$

$$45 = 5 \cdot h$$

$$\frac{45}{5} = \frac{\cancel{5} \cdot h}{\cancel{5}}$$

$$9 = h$$

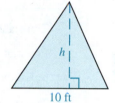
10 ft

The height of the triangle is 9 feet.

Exercises 9.6

Translate each of the following equations into a sentence in English. Do not solve the equation.

1. $n + 5 = 16$ **2.** $x - 7 = 8$ **3.** $4x - 8 = 20$

4. $\dfrac{y}{2} = 6.1$ **5.** $\dfrac{n}{3} + 1 = 13$ **6.** $15 = 25 - 2x$

Follow the steps outlined in this section to solve the following problems.

7. The $\underset{\text{(a)}}{\underline{\text{difference between a number and twelve}}}$ is equal to 16. What is the number?

Let $x =$ the unknown number.

Translation of (a): _____

Equation: _____

Solve the equation.

8. The sum of a number and fifteen is equal to 35. What is the number?

     (a)

Let n = the unknown number.

Translation of (a): _____

Equation: _____

Solve the equation.

9. Four less than three times a number is equal to 2.6. What is the number?

     (a)

Let y = the unknown number.

Translation of (a): _____

Equation: _____

Solve the equation.

10. Twenty is equal to one more than twice a number. What is the number?

     (a)

Let x = the unknown number.

Translation of (a): _____

Equation: _____

Solve the equation.

Before solving the following problems, make a guess as to what you think the answer is. (For example, do you think the number is greater than 10 or less than 10? Do you think the number is a fraction between 0 and 1? Do not be afraid to make a "wrong" guess.) Then, translate the phrases, form an equation, and solve the equation.

11. The sum of a number and 45 is equal to 56. What is the number?

12. The difference between a number and 1.4 is 2.7. What is the number?

13. What is the number whose product with 3 is equal to 51?

14. If the quotient of a number and 6 is 24, what is the number?

15. If the product of a number and four is increased by 3, the result is 23. Find the number.

16. If the product of a number and 7 is decreased by 5, the result is 44. Find the number.

17. Six less than twice a number is equal to 30. What is the number?

18. If 8 more than the quotient of a number and 3 is equal to 24, what is the number?

19. Eleven is the result if 9 is subtracted from the quotient of a number and 5. What is the number?

20. Ten and eight-tenths is equal to four less than twice a number. What is the number?

Use your knowledge of geometric figures to set up an equation to solve each of the following problems.

21. The perimeter of a rectangle is 40 meters. If the length is 15 meters, what is the width of the rectangle? ($P = 2l + 2w$)

22. The perimeter of a triangle is 35 inches. If two of the sides are each 12 inches long, what is the length of the third side? ($P = a + b + c$)

23. If the base of a triangle is 18 inches long and its area is 54 square inches, what is the height of the triangle? $\left(A = \frac{1}{2} \cdot b \cdot h \right)$

24. The area of a rectangle is 63 square centimeters. If the width of the rectangle is 7 centimeters, what is its length? ($A = lw$)

25. A parallelogram is a four-sided figure in which opposite sides are parallel. We can also show (in geometry) that these opposite sides have the same length. Suppose that the perimeter of a parallelogram is 180 yards. If one of two equal sides has a length of 50 yards, what is the length of each of the other two equal sides? ($P = 2a + 2b$)

26. If a triangle has a base length of 20 millimeters and a height of 10 millimeters, what is its area? $\left(A = \frac{1}{2} \cdot b \cdot h \right)$

27. The perimeter of a circle is called its circumference. If the circumference of a circle is 18.84 inches, what is its radius? ($C = 2\pi r$) (Use $\pi = 3.14$.)

28. The circumference of a circle is 69.08 centimeters. Find its diameter. ($C = \pi d$) (Use $\pi = 3.14$.)

29. In any triangle, the sum of the measures of the angles is 180°. If two of the angles each measure 45°, what is the measure of the third angle? ($\alpha + \beta + \gamma = 180$)

30. In any triangle, the sum of the measures of the angles is 180°. If two of the angles have measures of 30° and 50°, what is the measure of the third angle? ($\alpha + \beta + \gamma = 180$)

31. A rectangular solid is in the shape of a box. Consider a rectangular solid with a volume of 60 cubic inches. If it is 3 inches wide and 4 inches long, what is the height of the rectangular solid? ($V = lwh$)

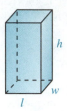

32. A rectangular solid is in the shape of a box. If it has a volume of 37.2 cubic feet and is 1.5 feet wide and 6.2 feet high, what is its length? ($V = lwh$)

33. The perimeter of a triangle is 36 centimeters. If two sides are equal in length and the third side is 8 centimeters long, what is the length of each of the other two sides? ($P = a + b + c$)

34. A length of wire is bent to form a triangle with two sides equal. If the wire is 25 inches long and the two equal sides are each 9 inches long, what is the length of the third side? ($P = a + b + c$)

35. A real estate agent says that the current value of a home is $40,000 more than twice its value when it was new. If the current value is $160,000, what was the value of the home when it was new?

36. A classic car is now selling for $1500 more than three times its original price. If the selling price is now $16,500, what was the car's original price?

37. The BART Trans-Bay Tubes, the underwater rapid transit tunnels below San Francisco Bay, are 19,008 feet long and are the longest underwater tunnels in North America. They are 2058 feet longer than three times the length of Sumner Tunnel under Boston Harbor. How long is the Sumner Tunnel?

38. According to the U.S. Department of Agriculture, the projected average annual expenditure by a middle-income family to raise a child, born in 1990, to age seventeen is $12,357. What is the projected total expenditure per child?

39. According to the National Association of Realtors, in April 1992 the median price of a single family home in Honolulu, Hawaii was $342,000. This was $32,000 more than four times the median price of a home in Detroit at that time. What was the median price of a home in Detroit?

40. In the National Park System, the smallest National Seashore is Fire Island in New York, which is 19,579 acres. It is 3024 acres less than one-sixth of the number of acres in the Gulf Islands in Florida and Mississippi (islands that make up the largest National Seashore Park). How many acres are there in the Gulf Islands?

41. Coffee is now often sold in a vacuum-packed brick. An 11.5-ounce brick of Hills Brothers coffee is a rectangular solid with a volume of 43.875 cubic inches. It is 6 inches long and 3.25 inches high. What is its width? ($V = lwh$)

42. A fine leaded window with colored glass is in the shape of a rectangle with a triangle on top. One of the longer sides of the rectangle is 2 feet longer than a shorter side and the sides of the triangle are all the same length as a shorter side of the rectangle. Find the lengths of the sides of the rectangle and the triangle if the perimeter of the figure is 34 feet. (**Hint:** Sketch the figure.)

43. Suppose that you are a traveling salesperson and your company has allowed you an expense of $50 per day for car rental. You have rented a car for $30 per day plus 5¢ per mile. How many miles can you drive for the $50 allowed?

> **Plan:**   Set up an equation relating the amounts of money spent and the money allowed in your budget, and solve the equation.
> **Solution:** Let x = number of miles driven.
>
> Money spent = Money budgeted
>
> $$30 + 0.05x = 50$$
> $$30 + 0.05x - 30 = 50 - 30$$
> $$0.05x = 20$$
> $$\frac{0.05x}{0.05} = \frac{20}{0.05}$$
> $$x = 400 \text{ miles}$$

You would like to get the greatest number of miles possible for your money and the car rental agency down the street charges only $25 per day and 8¢ per mile. How many miles would you get for your $50 with this price structure? Is this a "better deal" for you?

44. Under which of the following programs for a truck rental would you get the greatest number of miles for your $200 budget if you want the truck for two days?

(a) a fee of $40 per day plus 15¢ per mile

(b) a fee of $60 per day plus 10¢ per mile

Writing and Thinking about Mathematics

45. If you planned to travel (a) less than 800 miles or (b) more than 800 miles with a truck under the rental conditions discussed in Exercise 44, which of the two programs would you use? Explain why.

Collaborative Learning Exercise

46. As a result of his studies in discovery learning, George Pólya (1887–1985), a famous professor at Stanford University, developed the following four-step process as an approach to problem solving:

1. Understand the problem.

2. Devise a plan.

3. Carry out the plan.

4. Look back over the results.

With the class separated into teams of three to four students, each team is to discuss some nonmathematical problem that a member of the team solved today. (For example: what to have for breakfast, where to study, what route to take to school.) That is, can this person's thoughts and actions be described in terms of Pólya's four steps? As the team looks back over the solution, would there have been a better or more efficient way to solve this person's problem? The team leader is to report the results of this discussion to the class.

Summary: Chapter 9

Key Terms and Ideas

A shaded dot on a number line is the **graph** of the number corresponding to that point, and the number is the **coordinate** of the point.

The whole numbers and their opposites form the set of **integers.**

Integers	$\ldots, -3, -2, -1, 0, 1, 2, 3, \ldots$
Positive integers	$1, 2, 3, 4, 5, \ldots$
Negative integers	$\ldots, -3, -2, -1$

The number 0 is neither positive nor negative. (See Figure 9.13.)

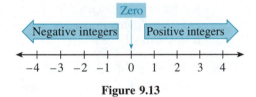

Figure 9.13

Signed numbers are numbers that can be represented on a number line.

Symbols for Order

$=$	is equal to		$\neq$	is not equal to
$<$	is less than		$>$	is greater than
$\leq$	is less than or equal to		$\geq$	is greater than or equal to

The **absolute value** of a number is its distance from 0 on a number line. **The absolute value of a number is never negative.**

The **difference** of two integers a and b is the sum of a and the opposite of b. Symbolically,

$$(a) - (b) = (a) + (-b).$$

For any numbers a and b where $b \neq 0$,

$$\frac{a}{b} = x \qquad \text{means} \qquad a = b \cdot x.$$

A **first-degree equation** in x (or **linear equation** in x) is any equation that can be written in the form

ax + b = c

where a, b, and c are constants and a $\neq$ 0.

Rules for Adding Signed Numbers

1. **To add two signed numbers with like signs,** add their absolute values and use the common sign.

   $(+6) + (+5) = +[|+6| + |+5|] = +[6 + 5] = +11$

   $(-2) + (-8) = -[|-2| + |-8|] = -[2 + 8] = -10$

2. **To add two signed numbers with unlike signs,** subtract their absolute values (smaller from larger) and use the sign of the number with the larger absolute value.

   $(+10) + (-15) = -[|-15| - |+10|] = -[15 - 10] = -5$

   $(+12) + (-9) = +[|+12| - |-9|] = +[12 - 9] = +3$

If a and b are positive numbers,

1. $a \cdot b = ab$

2. $(-a)(-b) = ab$

3. $a(-b) = -ab$

4. $a \cdot 0 = 0$ and $(-a) \cdot 0 = 0$

If a and b are positive numbers,

1. $\dfrac{a}{b} = \dfrac{a}{b}$

2. $\dfrac{-a}{-b} = \dfrac{a}{b}$

3. $\dfrac{-a}{b} = -\dfrac{a}{b}$ and $\dfrac{a}{-b} = -\dfrac{a}{b}$

Principles Used in Solving a First-Degree Equation

In the four basic principles stated here, A and B represent algebraic expressions or constants.

1. **The Addition Principle:**
   If $A = B$ is true, then $A + C = B + C$ is also true for any number C. (The same number may be added to both sides of an equation.)

2. **The Subtraction Principle:**
   If $A = B$ is true, then $A - C = B - C$ is also true for any number C. (The same number may be subtracted from both sides of an equation.)

3. **The Multiplication Principle:**
   If $A = B$ is true, then $C \cdot A = C \cdot B$ is also true for any number C. (Both sides of an equation may be multiplied by the same number.)

4. **The Division Principle:**
   If $A = B$ is true, then $\dfrac{A}{C} = \dfrac{B}{C}$ is also true for any nonzero number C. (Both sides of an equation may be divided by the same nonzero number.)

Review Questions: Chapter 9

Find the opposite of each number.

1. -4

2. $\dfrac{3}{8}$

3. 0

For each of the following, draw a number line and graph the given set of numbers.

4. $\{-1, 0, |-1|\}$

5. $\left\{-1\dfrac{1}{2}, -0.3, \dfrac{3}{4}, 1.8\right\}$

6. $\{-4, -3, \ldots, 2\}$

Fill in the blank with the appropriate symbol, $<$, $>$, or $=$.

7. -4 _____ $|-4|$ **8.** $\dfrac{2}{3}$ _____ $\dfrac{5}{8}$ **9.** $-(-6)$_____$|-6|$

Determine whether each statement is true or false.

10. $-\left(-\dfrac{1}{2}\right) \le \dfrac{1}{2}$ **11.** $-18 < -12$ **12.** $3.2 > |-3.2|$

Name the possible values for x.

13. $|x| = 0.4$ **14.** $|x| = -7$ **15.** $|x| = \dfrac{9}{10}$

Find each sum.

16. $(-14) + (+18)$ **17.** $(-17) + (+12)$

18. $(-15) + (+4) + (-9)$ **19.** $(+14) + (-23) + (+9)$

20. $\left(-2\dfrac{3}{5}\right) + \left(-1\dfrac{7}{10}\right)$ **21.** $(-17.3) + (+14)$

22. $\begin{array}{r} -312 \\ -422 \\ +610 \\ \hline \end{array}$ **23.** $\begin{array}{r} -7.63 \\ 11.01 \\ 2.24 \\ \hline \end{array}$

Find each difference.

24. $(+12) - (+7)$ **25.** $(+8) - (-8)$ **26.** $(-16) - (+5)$

27. $(-9) - (-9)$

Subtract the bottom number from the top number.

28. $\begin{array}{r} 16 \\ -10 \\ \hline \end{array}$ **29.** $\begin{array}{r} 16 \\ 10 \\ \hline \end{array}$ **30.** $\begin{array}{r} -62 \\ -12 \\ \hline \end{array}$ **31.** $\begin{array}{r} -62 \\ 12 \\ \hline \end{array}$

Evaluate each of the following expressions.

32. $-12.4 + 5.11 - 3$ **33.** $5\dfrac{3}{8} - 2\dfrac{1}{4} - 1\dfrac{2}{3}$

34. $-8 - 14 + 29 - 7$

Find the following products.

35. $12(-6)$ **36.** $-4(-16)$ **37.** $\left(-\dfrac{2}{3}\right)\left(-\dfrac{6}{7}\right)\left(\dfrac{7}{8}\right)$

38. $(-4.732)(0.75104)(0)(-5.2)$ **39.** $(3)(0.4)(-0.02)$

40. $(-5)^3$

Find the following quotients.

41. $\dfrac{-51}{3}$ **42.** $\dfrac{-24}{-72}$ **43.** $\dfrac{0}{-73}$ **44.** $\dfrac{-52}{0}$

45. $(-14) \div \left(-\dfrac{7}{8}\right)$ **46.** $\dfrac{435}{-1.5}$

Use the rules for order of operations to evaluate the following expressions.

47. $3(-4) + 12 \div 2$ **48.** $(-4)^2 \div (8 - 2 \cdot 3)$

49. $\dfrac{1}{4} + \dfrac{2}{3} \div \dfrac{5}{9} + \dfrac{3}{10}$ **50.** $[4 + 3(1 - 2^2)] \div 4$

Fill in the blanks with the correct term (**positive, negative, zero, undefined**).

51. The quotient when any signed number is divided by zero is _____.

52. The product of a negative number and zero is _____.

Solve each of the following equations.

53. $x + 10 = 2.6$ **54.** $x - 13 = 12$

55. $2n = -60$ **56.** $3.5n = 70$

57. $2n - 5 = 15$ **58.** $-34 = 6 + 4x$

59. Three times a number plus 15.2 is equal to 17. What is the number?

60. If the product of a number and 5 is decreased by 24, the result is -59. Find the number.

Test: Chapter 9

1. Find the opposite of 0.34.

2. Draw a number line and graph the following.

$$\left\{-3, -0.5, \dfrac{3}{4}, |-1|, 2.4\right\}$$

Complete the following with the appropriate symbol, $<$, $>$, or $=$.

3. -12 _____ -15 **4.** $|-12|$ _____ $-(-15)$

5. Name the possible values for x if $|x| = 17$.

Determine whether each statement is true or false.

6. $-\dfrac{3}{4} < \dfrac{1}{10}$ **7.** $-8.5 \geq 8.0$ **8.** $|-73| = 73$

Find each sum.

9. $(1.4) + (-6)$

10. $(-3) + (+7) + (-4) + (-9)$

11. $(-16) + (+3) + (-10)$

12. $(-18.3) + (15.7)$

Find each difference.

13. $(-2.5) - (1.5)$

14. $\left(-5\frac{3}{4}\right) - \left(-3\frac{1}{2}\right)$

Evaluate each of the following expressions.

15. $(-0.4)(-0.07)$

16. $(-3)(-5)(2)(-3)$

17. $\left(-\frac{3}{8}\right)\left(\frac{4}{5}\right)\left(-\frac{5}{6}\right)$

18. $(-2.1)^2$

19. $\frac{-5.6}{-7}$

20. $\frac{-3}{0}$

21. $\frac{65}{-13}$

22. $3\frac{1}{4} \div \left(-1\frac{1}{2}\right)$

Use the rules for order of operations to evaluate the following expressions.

23. $\left(-6\frac{1}{2}\right) \div \left(1\frac{1}{6}\right) - 3\frac{1}{14}$

24. $(-7)^2 - 6 \cdot 8 \div 2 + 3(6 - 5^2)$

Solve each of the following equations.

25. $x - 16 = -10$

26. $n + 1.4 = 3.57$

27. $17x = 5.1$

28. $3x + 1.9 = 7.9$

29. Four times a number is increased by 5 to give a result of 69. What is the number?

30. A rectangle has a perimeter of 230 feet. If the rectangle is 65 feet in length, what is its width? $(P = 2l + 2w)$

Choose the correct answer by performing the related operations mentally.

1. Write $\dfrac{11}{5}$ as a percent.

    (a) 200%   (b) 210%   (c) 220%   (d) 240%

2. Write 150% as a mixed number, reduced.

    (a) $1\dfrac{1}{2}$ (b) $1\dfrac{5}{8}$ (c) $1\dfrac{3}{4}$ (d) $2\dfrac{1}{2}$

3. Write $2\dfrac{1}{3}$ as a percent.

    (a) 213%   (b) 230%   (c) 233%   (d) $233\dfrac{1}{3}$%

4. Find 100% of 75.

    (a) 7.5   (b) 75   (c) 750   (d) 7500

5. Write the number 423.85 in its English word form.

6. Round off 166.075 to the nearest hundred.

7. Find the average (mean) of the numbers 5000, 6250, 9475, and 8672 (to the nearest unit).

Evaluate the following expressions. Reduce all fractions.

8.   75.63
       81.45
   +146.98

9.   8000.0
   −6476.9

10.   43.8
    × 2.7

11. $\dfrac{1}{2} + \dfrac{2}{3} \cdot \left(\dfrac{1}{2}\right)^2 - \dfrac{9}{10}$

12. $(-500)(7000)$

13. $\dfrac{16}{0}$

14. $\dfrac{0}{-3}$

15. $(+15) + (-10) + (-5)$

16. $(-2.4)(-6.1)$

17. $(-7)^2 \cdot 5 + 2.1(-6) \div \dfrac{1}{2}$

18. $(15 - 17)(21 - 32)$

19. What is the rate of interest if the simple interest earned on $5000 for 90 days is $156.25?

20. Find the value of $10,000 invested at 6% compounded daily for 5 years.

21. If a car averages 23.6 miles per gallon, how many miles will it go on 15 gallons of gas?

22. The cost of a sofa to a furniture store was $750 and was sold for $1250. (a) What was the store's profit? (b) What was the percent of profit based on cost? (c) What was the percent of profit based on selling price?

23. The scale on a map indicates that $1\frac{3}{4}$ inches represents 50 miles. How many miles apart are two cities marked 2 inches apart on the map?

Solve each of the following equations.

24. $6x + 1.4 = 20$ **25.** $3.2n - 20 = 50$

26. If one is added to twice a number, the result is equal to 19. What is the number?

Measurement

Mathematics at Work!

Prospective home-buyers will often look at many different properties before making a purchase.

When you are buying a home, the real estate agent will have a sheet of paper with a picture of the home, the important features of the home (such as how many bedrooms, how many bathrooms, dining room, family room, living room, size and type of garage, whether or not there is a swimming pool), and the size of each room. Another fact of interest to many homebuyers is the size of the lot (or the acreage). In residential property the lot size is generally stated in square feet; but, historically, people think of purchasing land by the acre, and many homeowners would prefer to know what part of an acre is in the lot (or, for larger parcels, how many acres are in the lot). In Section M.3, we state the fact that 43,560 ft^2 = 1 acre. Using this fact, answer the following question.

Suppose that the home you are considering buying sits on a rectangular-shaped lot that is 90 feet by 121 feet. What fraction of an acre is the size of this lot?

(See also Exercise 91 in Section M.3.)

What to Expect in this Chapter

This chapter discusses the interrelationships among the units of measure in two systems: the metric system and the U.S. customary system. The emphasis in Sections M.1 and M.2 is on understanding the metric system and on changing from one unit of measure to an equivalent unit of measure within the metric system. To aid your understanding and to illustrate the ease of working within the metric system, special charts (with units of measure as headings) are provided, which clearly relate changes in units to placement of the decimal point. Section M.1 deals with units of length and units of area, and Section M.2 covers units of weight and units of volume.

Section M.3 discusses measures in the U.S. customary system, gives tables of equivalent measures in the metric system and the U.S. customary system, and shows how to change from a unit in one system to an equivalent unit in the other. Section M.4 discusses addition and subtraction with denominate numbers (numbers that are labeled with units, such as 3 hours 15 minutes or 6 feet 4 inches).

M.1   # Metric System: Length and Area

OBJECTIVES

1. Learn the basic prefixes in the metric system.
2. Know that the **meter** is the basic unit of length.
3. Be able to change metric measures of length without a chart.
4. Know that area is measured in **square units.**
5. Be able to change metric measures of area without a chart.
6. Know that the **are** and **hectare** are the basic units of land area.

The Metric System

About 90% of the people in the world use the metric system of measurement. The United States is the only major industrialized country still committed to the U.S. customary system (formerly called the English system). Even in the United States, the metric system has been used for years in such fields as medicine, science, and in the military. Auto mechanics who work on imported cars must be familiar with the metric system because the measures of the auto parts are in metric units.

Length

The **meter** is the basic unit of length in the metric system. Smaller and larger units are named by putting a prefix in front of the basic unit: for example, **centi**meter and **kilo**meter. The prefixes we will use are listed here from smallest to largest unit size:

milli-	(thousandths)	**deka-**	(tens)
centi-	(hundredths)	**hecto-**	(hundreds)
deci-	(tenths)	**kilo-**	(thousands)

Other prefixes that indicate extremely small units are **micro-, nano-, pico-, femto-,** and **atto-.** Prefixes that indicate extremely large units are **mega-, giga-,** and **tera-.** Figure M.1 illustrates the relative sizes of millimeters and centimeters.

one millimeter (mm): About the width of the wire in a paper clip

1 mm

one centimeter (cm): About the width of a paper clip

1 cm

one meter (m): Just over 39 inches, or slightly longer than a yard
one kilometer (km): About 0.62 of a mile

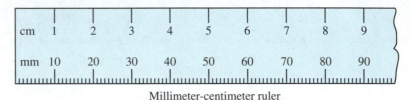

| cm | 1 | 2 | 3 | 4 | 5 | 6 | 7 | 8 | 9 |
| mm | 10 | 20 | 30 | 40 | 50 | 60 | 70 | 80 | 90 |

Millimeter-centimeter ruler

Figure M.1

Table M.1 contains the metric prefixes and their values. **These prefixes must be memorized, and memorized in order.** Table M.2 lists the measures of length, their respective relationships to the meter, and their abbreviations.

Table M.1   Metric Prefixes and Their Values

Prefix	Value	
milli	0.001	—thousandths
centi	0.01	—hundredths
deci	0.1	—tenths
basic unit	1	—ones
deka	10	—tens
hecto	100	—hundreds
kilo	1000	—thousands

Table M.2   Measures of Length

1 **milli**meter (mm)	= 0.001 meter
1 **centi**meter (cm)	= 0.01 meter
1 **deci**meter (dm)	= 0.1 meter
1 **meter** (m)	= 1.0 meter
1 **deka**meter (dam)	= 10 meters
1 **hecto**meter (hm)	= 100 meters
1 **kilo**meter (km)	= 1000 meters

As you can tell from studying Tables M.1 and M.2, the metric units of length are related to each other by powers of 10. That is, you simply multiply by 10 to get the equivalent measure in the next lower (or smaller) unit. For example,

1 m = 10 dm   and   1 cm = 10 mm

Conversely, you need only divide by 10 to get the equivalent measure in the next higher (or larger) unit. For example,

1 m = 0.1 dam   and   1 dam = 0.1 hm

Thus, you will have more of a smaller unit and less of a larger unit.

The following method of using a chart makes conversion from one metric unit to another quite easy.

A chart can be used to change from one unit of length to another.

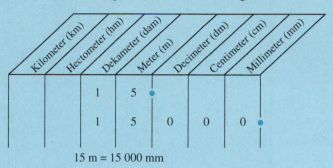

15 m = 15 000 mm

1. List each unit across the top. Memorize the unit prefixes in order.

2. Enter the given number so that each digit is in one column and the decimal point is on the given unit line.

3. Move the decimal point to the desired unit line.

4. Fill in the spaces with 0's using one digit per column.

Special Note: In the metric system,

1. A 0 is written to the left of the decimal point if there is no whole number part.   (0.25 m)

2. No commas are used in writing numbers. If a number has more than four digits (left or right of the decimal point), the digits are grouped in threes from the decimal point with a space between the groups.   (14 000 m)

The use of the chart on page M5 shows how the following equivalent measures can be found. Note that there are no periods in metric abbreviations.

EXAMPLES
1–5

1. 5 m = 0.005 km   (If there is no whole number part, 0 is written.)

2. 3 m = 300 cm

3. 16 cm = 0.16 m

4. 7.6 cm = 76 mm

5. 8 km = 800 dam

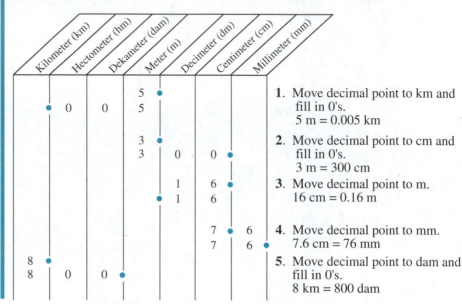

1. Move decimal point to km and fill in 0's.
5 m = 0.005 km

2. Move decimal point to cm and fill in 0's.
3 m = 300 cm

3. Move decimal point to m.
16 cm = 0.16 m

4. Move decimal point to mm.
7.6 cm = 76 mm

5. Move decimal point to dam and fill in 0's.
8 km = 800 dam

Make your own chart on a piece of paper and see if you agree with the results in Examples 6–10.

EXAMPLES
6–10

6. 4 m = 4000 mm

7. 3.1 cm = 31 mm

8. 50 cm = 0.5 m

9. 18 km = 18 000 m

10. 2.5 km = 2500 m

After some practice, you may not need a chart to help in changing units. The following technique can be used; however, the chart method is highly recommended.

Changing Metric Measures of Length Without a Chart

1. To change a measure that is

one unit smaller, multiply by 10.   3 cm = 30 mm

two units smaller, multiply by 100.   5 m = 500 cm

three units smaller, multiply by 1000.   14 m = 14 000 mm

And so on.

2. To change to a measure that is

one unit larger, divide by 10.   50 cm = 5 dm

two units larger, divide by 100.   50 cm = 0.5 m

three units larger, divide by 1000.   13 mm = 0.013 m

And so on.

EXAMPLES
11–14

Smaller Units → Larger Numbers

11. 42 m = 420 dm = 4200 cm = 42 000 mm

12. 17.3 m = 173 dm = 1730 cm = 17 300 mm

Larger Units → Smaller Numbers

13. 6 m = 0.6 dam = 0.06 hm = 0.006 km

14. 112 m = 11.2 dam = 1.12 hm = 0.112 km

Area

Area is a measure of the interior of (or surface enclosed by) a figure in a plane. For example, the two rectangles in Figure M.2 have different areas because they have different amounts of interior space, or different amounts of space are enclosed by the sides of the figures.

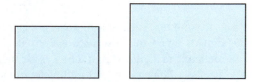

These two rectangles have different **areas**.

Figure M.2

Area is measured in **square units.** A square that is 1 centimeter long on each side is said to have an area of 1 square centimeter, or the area is 1 cm². A rectangle that is 7 cm on one side and 4 cm on the other side encloses 28 squares that have area 1 cm². So the rectangle is said to have an area of 28 square centimeters or 28 cm². (See Figure M.3.)

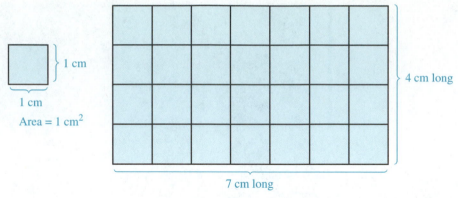

Area = 7 cm x 4 cm = 28 cm^2

There are 28 squares that are each 1 cm^2 in the large rectangle.

Figure M.3

Table M.3 shows metric area measures useful for relatively small areas. For example, the area of the floor of your classroom could be measured in square meters for carpeting, and the area of this page of paper could be measured in square centimeters. Other measures, listed in Table M.4 on page M11, are used for measuring land.

Table M.3   Measures of Small Area
$1 \text{ cm}^2 = 100 \text{ mm}^2$
$1 \text{ dm}^2 = 100 \text{ cm}^2 = 10\ 000 \text{ mm}^2$
$1 \text{ m}^2 = 100 \text{ dm}^2 = 10\ 000 \text{ cm}^2 = 1\ 000\ 000 \text{ mm}^2$

Note that each unit of area in the metric system is 100 times the next smaller unit of area—**not** just 10 times, as it is with length. For example, consider the rectangle in Figure M.3. If the sides are measured in millimeters, then the dimensions are 70 millimeters by 40 millimeters (each length is increased by a factor of 10) and the area is

7 cm $\times$ 4 cm = 28 cm^2

or

70 mm $\times$ 40 mm = 2800 mm^2

Thus, the measure of the area in square millimeters is 100 times the measure of the same area in square centimeters.

Examples 15 and 16 illustrate this idea in detail.

EXAMPLE 15   A square 1 centimeter on a side encloses 100 square millimeters.

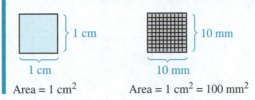

Area = 1 cm^2 Area = 1 cm^2 = 100 mm^2

EXAMPLE 16   A square 1 decimeter (10 cm) on a side encloses 1 square decimeter.
(1 dm^2 = 100 cm^2 = 10 000 mm^2)

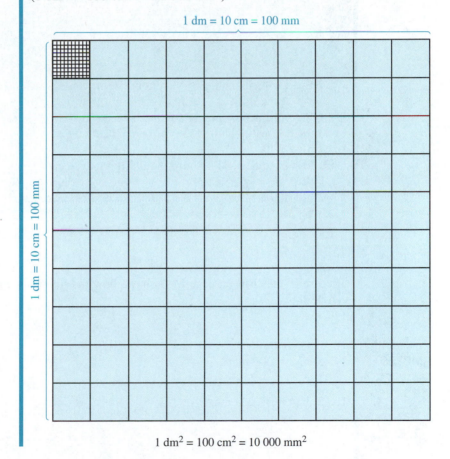

1 dm^2 = 100 cm^2 = 10 000 mm^2

A chart similar to that used on page M5 can be used to change measures of area. **The key difference is that there must be two digits in each column.** This corresponds to multiplication of the previous unit of area by 100.

A chart can be used to change from one unit of area to another.

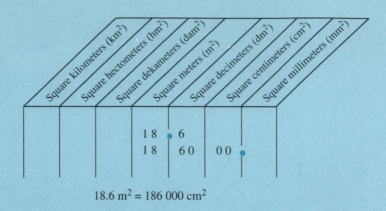

$18.6 \text{ m}^2 = 186\ 000 \text{ cm}^2$

1. List each area unit across the top. (Abbreviations will do.)

2. Enter the given number so that there are **two** digits in each column with the decimal point on the given unit line.

3. Move the decimal point to the desired unit line.

4. Fill in the spaces with 0's using two digits per column.

The use of the chart on page M11 shows how the following equivalent measures can be found.

EXAMPLES
17–21

17. $5 \text{ cm}^2 = 500 \text{ mm}^2$

18. $3 \text{ dm}^2 = 30\ 000 \text{ mm}^2$

19. $1.4 \text{ m}^2 = 14\ 000 \text{ cm}^2$

20. $1.4 \text{ m}^2 = 0.014 \text{ dam}^2$

21. $8 \text{ km}^2 = 8\ 000\ 000 \text{ m}^2$

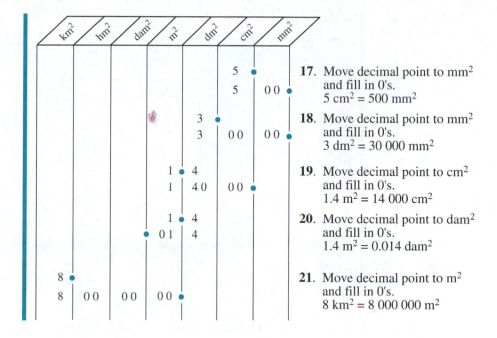

17. Move decimal point to mm² and fill in 0's.
5 cm² = 500 mm²

18. Move decimal point to mm² and fill in 0's.
3 dm² = 30 000 mm²

19. Move decimal point to cm² and fill in 0's.
1.4 m² = 14 000 cm²

20. Move decimal point to dam² and fill in 0's.
1.4 m² = 0.014 dam²

21. Move decimal point to m² and fill in 0's.
8 km² = 8 000 000 m²

Make your own chart on a piece of paper and see if you agree with the results below.

EXAMPLES
22–24

22. 8.52 m² = 852 dm² = 85 200 cm²

23. 147 cm² = 14 700 mm²

24. 3.8 cm² = 0.038 dm² = 0.000 38 m²

A square with each side 10 meters long encloses an area of 1 **are** (a). A **hectare** (ha) is 100 ares. The are and hectare are used in measuring land area. (See Table M.4.)

Table M.4   Measures of Land Area
1 a = 100 m²
1 ha = 100 a = 10 000 m²

EXAMPLE 25   A square 10 m on each side encloses 100 m^2 or 1 are.

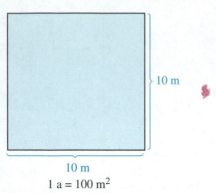

10 m

10 m

1 a = 100 m^2

EXAMPLE 26   Refer to Table M.4 on page M11 to confirm the following conversions.

(a) 3.2 a = 320 m^2

(b) 65 m^2 = 0.65 a

EXAMPLE 27   How many ares are in 1 km^2? (Note: One km is about 0.6 mile, so 1 km^2 is about $0.6 \times 0.6 = 0.36$ square mile.)

Remember that 1 km = 1000 m, so

$$1 \text{ km}^2 = (1000 \text{ m}) \times (1000 \text{ m})$$
$$= 1\,000\,000 \text{ m}^2$$
$$= 10\,000 \text{ a}$$

Divide m^2 by 100 to get ares because every 100 m^2 is equal to 1 are.

EXAMPLE 28   A farmer plants corn and beans as shown in the figure. How many ares and how many hectares are planted in corn? In beans? (From Example 27 we know 1 km^2 = 10 000 a.)

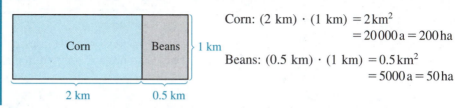

Corn

Beans

1 km

2 km   0.5 km

Corn: $(2 \text{ km}) \cdot (1 \text{ km}) = 2 \text{ km}^2$
$= 20\,000\,\text{a} = 200\,\text{ha}$

Beans: $(0.5 \text{ km}) \cdot (1 \text{ km}) = 0.5 \text{ km}^2$
$= 5000\,\text{a} = 50\,\text{ha}$

CLASSROOM PRACTICE

Change the following units of length as indicated.

1. 4 m = _____ mm

2. 3.1 cm = _____ mm

3. 50 cm = _____ m

4. 18 km = _____ m

Change the following units of area as indicated.

5. 22 cm^2 = _____ mm^2 **6.** 500 mm^2 = _____ cm^2

7. 3.7 dm^2 = _____ cm^2 = _____ mm^2

8. 3.6 a = _____ m^2

9. 0.73 ha = _____ a = _____ m^2

ANSWERS: **1.** 4000 mm   **2.** 31 mm   **3.** 0.5 m   **4.** 18 000 m   **5.** 2200 mm^2
6. 5 cm^2 **7.** 370 cm^2 = 37 000 mm^2 **8.** 360 m^2 **9.** 73 a = 7300 m^2

Exercises M.1

1. Write the six metric prefixes discussed in this section in order from largest to smallest.

Change the following units as indicated. Use the chart method until you are accustomed to the metric system.

2. 1 m = _____ cm   **3.** 5 m = _____ cm

4. 12 m = _____ cm   **5.** 6 m = _____ cm

6. 2 m = _____ mm   **7.** 0.3 m = _____ mm

8. 0.7 m = _____ mm   **9.** 1.4 m = _____ mm

10. 1.6 cm = _____ mm   **11.** 1.8 cm = _____ mm

12. 25 cm = _____ mm   **13.** 35 cm = _____ mm

14. 4 m = _____ dm   **15.** 16 m = _____ dm

16. 7 dm = _____ cm   **17.** 21 dm = _____ cm

18. 3 km = _____ m   **19.** 5 km = _____ m

20. 5.28 km = _____ m   **21.** 6.4 km = _____ m

22. 11 mm = _____ cm   **23.** 26 mm = _____ cm

24. 72 mm = _____ cm   **25.** 48 mm = _____ cm

26. Change 6 mm to decimeters.   **27.** Change 12 mm to centimeters.

28. Change 20 mm to meters.   **29.** Change 30 mm to meters.

30. Convert 145 mm to meters.   **31.** Convert 256 mm to meters.

32. Convert 25 cm to meters.   **33.** Convert 32 cm to meters.

34. How many meters are in 150 cm?

35. How many meters are in 170 cm?

36. How many kilometers are in 3000 m?

37. How many kilometers are in 2400 m?

38. Express 500 m in kilometers.   **39.** Express 400 m in kilometers.

40. Express 3.45 m in centimeters.   **41.** Express 4.62 m in centimeters.

42. 6.3 cm = _____ m   **43.** 5.2 cm = _____ m

44. 3.25 m = _____ mm   **45.** 6.41 m = _____ mm

46. How many centimeters are in 3 mm?

47. How many centimeters are in 5 mm?

48. Change 32 mm to meters.   **49.** Change 57 mm to meters.

50. What number of kilometers is equivalent to 20 000 m?

51. What number of kilometers is equivalent to 35 000 m?

52. Express 1.5 km in meters.   **53.** Express 2.3 km in meters.

54. How many kilometers are in 0.5 m?

55. How many kilometers are in 1.5 m?

Change the following units of area as indicated. Use the chart method until you become thoroughly familiar with the metric system.

56. 3 cm^2 = _____ mm^2 **57.** 5.6 cm^2 = _____ mm^2

58. 8.7 cm^2 = _____ mm^2 **59.** 3.61 cm^2 = _____ mm^2

60. Express 600 mm^2 in cm^2.   **61.** Express 28 mm^2 in cm^2.

62. How many cm^2 are in 1400 mm^2?

63. How many cm^2 are in $20\ 000 \text{ mm}^2$?

64. 4 dm^2 = _____ cm^2 = _____ mm^2

65. 7.3 dm^2 = _____ cm^2 = _____ mm^2

66. 57 dm^2 = _____ cm^2 = _____ mm^2

67. 0.6 dm^2 = _____ cm^2 = _____ mm^2

68. 17 m^2 = _____ dm^2 = _____ cm^2 = _____ mm^2

69. 2.9 m^2 = _____ dm^2 = _____ cm^2 = _____ mm^2

70. 0.03 m^2 = _____ dm^2 = _____ cm^2 = _____ mm^2

Change the following units of land area as indicated.

71. $7.8\ a = \underline{\hspace{1cm}}\ m^2$

72. $300\ a = \underline{\hspace{1cm}}\ m^2$

73. $0.04\ a = \underline{\hspace{1cm}}\ m^2$

74. $0.53\ a = \underline{\hspace{1cm}}\ m^2$

75. $8.69\ ha = \underline{\hspace{1cm}}\ a = \underline{\hspace{1cm}}\ m^2$

76. $7.81\ ha = \underline{\hspace{1cm}}\ a = \underline{\hspace{1cm}}\ m^2$

77. $0.16\ ha = \underline{\hspace{1cm}}\ a = \underline{\hspace{1cm}}\ m^2$

78. $0.02\ ha = \underline{\hspace{1cm}}\ a = \underline{\hspace{1cm}}\ m^2$

79. How many hectares are in 1 a?

80. How many hectares are in 15 a?

81. How many hectares are in $5\ km^2$?

82. Change $4.76\ km^2$ to hectares.

83. Change $0.3\ km^2$ to hectares.

84. Change 650 ha to ares.

M.2 Metric System: Weight and Volume

OBJECTIVES

1. Become familiar with the concepts of mass and weight.
2. Know that the **kilogram** is the basic unit of mass.
3. Know that in some fields the **gram** is a more convenient unit of mass.
4. Be able to change metric measures of mass.
5. Know that volume is measured in **cubic units.**
6. Be able to change metric measures of volume.
7. Know that the **liter** is the basic unit of liquid volume.

Mass (Weight)

Mass is the amount of material in an object. Regardless of where the object is in space, its mass remains the same. (See Figure M.4.) **Weight** is the force of the Earth's gravitational pull on an object. The farther an object is from Earth, the less the gravitational pull of the Earth. Thus, astronauts experience weightlessness in space, but their mass is unchanged.

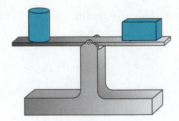

The two objects have the same **mass** and balance on an equal arm balance, regardless of their location in space.

Figure M.4

Because most of us do not stray far from the Earth's surface, in this text weight and mass will be used interchangeably. Thus, a **mass** of 20 kilograms will be said to **weigh** 20 kilograms.

The basic unit of mass in the metric system is the **kilogram,**[*] about 2.2 pounds. In some fields, such as medicine, the **gram** (about the mass of a paper clip) is more convenient as a basic unit than the kilogram.

Large masses, such as loaded trucks and railroad cars, are measured by the **metric ton** (1000 kilograms or about 2200 pounds.) (See Tables M.5 and M.6.)

Table M.5	Measures of Mass
1 **milli**gram (mg)	= 0.001 gram
1 **centi**gram (cg)	= 0.01 gram
1 **deci**gram (dg)	= 0.1 gram
1 gram (g)	= 1.0 gram
1 **deka**gram (dag)	= 10 grams
1 **hecto**gram (hg)	= 100 grams
1 **kilo**gram (kg)	= 1000 grams
1 metric ton (t)	= 1000 kilograms

Table M.6	Equivalent Measures of Mass	
1000 mg = 1 g		0.001 g = 1 mg
1000 g = 1 kg		0.001 kg = 1 g
1000 kg = 1 t		0.001 t = 1 kg
1 t = 1000 kg = 1 000 000 g = 1 000 000 000 mg		

The centigram, decigram, dekagram, and hectogram have little practical use and are not included in the exercises. For completeness, they are all included in the headings of the chart used to change units.

---

[*] Technically, a kilogram is the mass of a certain cylinder of platinum-iridium alloy kept by the International Bureau of Weights and Measures in Paris.

Originally, the basic unit was a gram, defined to be the mass of 1 cm^3 of distilled water at 4° Celsius. This mass is still considered accurate for many purposes, so that

1 cm^3 of water has a mass of 1 g.
1 dm^3 of water has a mass of 1 kg.
1 m^3 of water has a mass of 1000 kg, or 1 metric ton.

A chart can be used to change from one unit of mass to another.

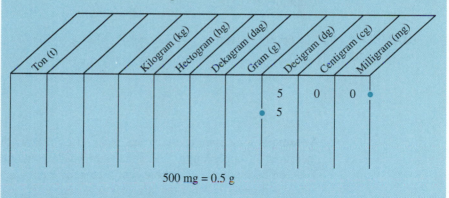

500 mg = 0.5 g

1. List each unit across the top.

2. Enter the given number so that there is **one** digit in each column with the decimal point on the given unit line.

3. Move the decimal point to the given unit line.

4. Fill in the spaces with 0's using one digit per column.

The use of the chart below shows how the following equivalent measures can be found.

EXAMPLES
1–5

1. 23 mg = 0.023 g   **2.** 6 g = 6000 mg   **3.** 49 kg = 49 000 g

4. 5 t = 5000 kg   **5.** 70 kg = 0.07 t

Ton			Kilogram	Hectogram	Dekagram	Gram	Decigram	Centigram	Milligram	
								2	3 •	**1.** Move decimal point to g and fill in 0's. 23 mg = 0.023 g
							• 0	2	3	
						6 •				**2.** Move decimal point to mg and fill in 0's. 6 g = 6000 mg
						6	0	0	0 •	
		4	9 •							**3.** Move decimal point to g and fill in 0's. 49 kg = 49 000 g
		4	9	0	0	0 •				
5 •										**4.** Move decimal point to kg and fill in 0's. 5 t = 5000 kg
5	0	0	0 •							
		7	0 •							**5.** Move decimal point to t and fill in 0's. 70 kg = 0.07 t
• 0	7	0								

Make your own chart on a piece of paper and see if you agree with the results in Examples 6–10.

EXAMPLES
6–10

6. 60 mg = 0.06 g   **7.** 135 mg = 0.135 g   **8.** 5700 kg = 5.7 t
9. 100 g = 0.1 kg   **10.** 78 g = 78 000 kg

Volume

Volume is a measure of the space enclosed by a three-dimensional figure and is measured in **cubic units.** The volume or space contained within a cube that is 1 cm on each edge is **one cubic centimeter,** or 1 cm³, as shown in Figure M.5. A cubic centimeter is about the size of a sugar cube.

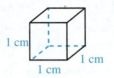

Volume = 1 cm³

Figure M.5

A rectangular solid that has edges of 3 cm and 2 cm and 5 cm has a volume of 3 cm × 2 cm × 5 cm = 30 cm³. We can think of the rectangular solid as being three layers of ten cubic centimeters, as shown in Figure M.6.

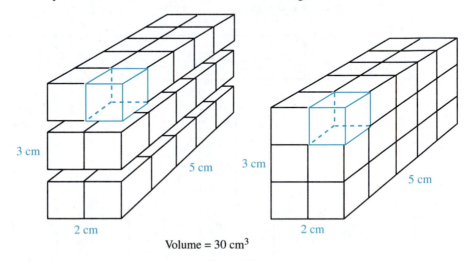

Volume = 30 cm³

Figure M.6

If a cube is 1 decimeter along each edge, then the volume of the cube is 1 cubic decimeter (or 1 dm³). In terms of centimeters, this same cube has volume

10 cm × 10 cm × 10 cm = 1000 cm³

That is, as shown in Figure M.7,

$$1 \text{ dm}^3 = 1000 \text{ cm}^3$$

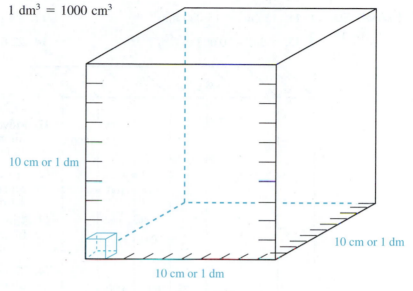

10 cm or 1 dm

10 cm or 1 dm

10 cm or 1 dm

Figure M.7

This relationship is true of cubic units in the metric system: equivalent cubic units can be found by multiplying the previous unit by 1000. Again, we can use a chart; however, this time **there must be three digits in each column.**

A chart can be used to change from one unit of volume to another.

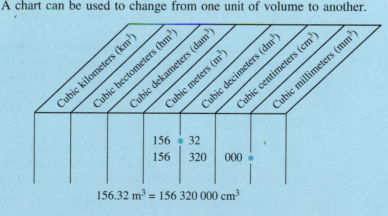

Cubic kilometers (km³)	Cubic hectometers (hm³)	Cubic dekameters (dam³)	Cubic meters (m³)	Cubic decimeters (dm³)	Cubic centimeters (cm³)	Cubic millimeters (mm³)
			156 • 32			
			156	320	000 •	

$$156.32 \text{ m}^3 = 156\,320\,000 \text{ cm}^3$$

1. List each volume unit across the top. (Abbreviations will do.)

2. Enter the given number so that there are **three** digits in each column with the decimal point on the given unit line.

3. Move the decimal point to the desired unit line.

4. Fill in the spaces with 0's using three digits per column.

The chart below shows how the following equivalent measures can be found.

EXAMPLES
11–14

11. $15 \text{ cm}^3 = 15\ 000 \text{ mm}^3$

12. $4.1 \text{ dm}^3 = 4100 \text{ cm}^3$

13. $8 \text{ dm}^3 = 0.008 \text{ m}^3$

14. $22.6 \text{ m}^3 = 22\ 600\ 000 \text{ cm}^3$

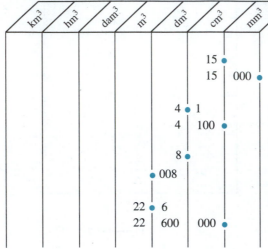

11. Move decimal point to mm^3 and fill in 0's.
$15 \text{ cm}^3 = 15\ 000 \text{ mm}^3$

12. Move decimal point to cm^3 and fill in 0's.
$4.1 \text{ dm}^3 = 4100 \text{ cm}^3$

13. Move decimal point to m^3 and fill in 0's.
$8 \text{ dm}^3 = 0.008 \text{ m}^3$

14. Move decimal point to cm^3 and fill in 0's.
$22.6 \text{ m}^3 = 22\ 600\ 000 \text{ cm}^3$

Mark your own chart on a piece of paper and see if you agree with the results in Examples 15–17.

EXAMPLES
15–17

15. $3.7 \text{ dm}^3 = 3700 \text{ cm}^3$

16. $0.8 \text{ m}^3 = 0.0008 \text{ dam}^3$

17. $4 \text{ m}^3 = 4000 \text{ dm}^3 = 4\ 000\ 000 \text{ cm}^3 = 4\ 000\ 000\ 000 \text{ mm}^3$

Volumes measured in cubic kilometers are so large that they are not used in everyday situations. Possibly, some scientists work with these large volumes. More practically, we are interested in m^3, dm^3, cm^3, and mm^3.

Liquid Volume

Liquid volume is measured in **liters** (abbreviated L). You are probably familiar with 1 L and 2 L bottles of soda on your grocer's shelf. A **liter** is the volume enclosed in a cube that is 10 cm on each edge. So, 1 liter is equal to

$$10 \text{ cm} \times 10 \text{ cm} \times 10 \text{ cm} = 1000 \text{ cm}^3 \qquad \text{or} \qquad 1 \text{ liter} = 1000 \text{ cm}^3$$

That is, the cubic box shown in Figure M.7 would hold 1 liter of liquid.

The prefixes kilo-, hecto-, deka-, deci-, centi-, and milli- all indicate the same parts of a liter as they do of the meter. The same type of chart used in Section M.1 **with one digit per column** will be helpful for changing units. The centiliter (cL), deciliter (dL), and dekaliter (daL) are not commonly used and are not included in the tables or exercises.

Table M.7 Measures of Liquid Volume	
1 **milli**liter (mL)	= 0.001 liter
1 **liter** (L)	= 1.0 liter
1 **hecto**liter (hL)	= 100 liters
1 **kilo**liter (kL)	= 1000 liters

Table M.8 Equivalent Measures of Volume			
1000 mL	= 1 L	1 mL	= 1 cm³
1000 L	= 1 kL	1 L	= 1 dm³
10 hL	= 1 kL	1 kL	= 1 m³

The use of the chart below shows how the following equivalent measures can be found.

EXAMPLES 18–21

18. $6 \text{ L} = 6000 \text{ mL} = 0.06 \text{ hL}$ **19.** $500 \text{ mL} = 0.5 \text{ L}$

20. $3 \text{ kL} = 3000 \text{ L}$ **21.** $72 \text{ hL} = 7.2 \text{ kL}$

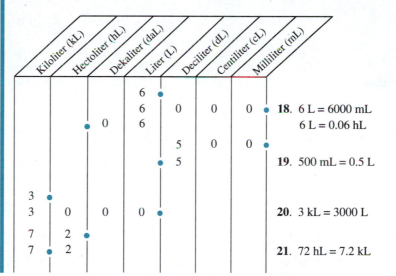

There is an interesting "crossover" relationship between liquid volume measures and cubic volume measures. Since

$$1 \text{ L} = 1000 \text{ mL} \quad \text{and} \quad 1 \text{ L} = 1000 \text{ cm}^3$$

we have

$1 \text{ ml} = 1 \text{ cm}^3$

Also,

$$1 \text{ kL} = 1000 \text{ L} = 1\,000\,000 \text{ cm}^3 \quad \text{and} \quad 1\,000\,000 \text{ cm}^3 = 1 \text{ m}^3$$

This gives

$1 \text{ kL} = 1000 \text{ L} = 1 \text{ m}^3$

Use Table M.8 on page M21 to confirm the following conversions.

EXAMPLES
22–27

22. 6000 mL = 6 L

23. 3.2 L = 3200 mL

24. 60 hL = 6 kL

25. 637 mL = 0.637 L

26. 70 mL = 70 cm^3

27. 3.8 kL = 3.8 m^3

CLASSROOM PRACTICE

Change the following units as indicated.

1. 500 mg = _____ g

2. 43 g = _____ mg

3. 62 g = _____ kg

4. 18 cm^3 = _____ mm^3

5. 7.9 dm^3 = _____ m^3

6. 2 mL = _____ L

7. 500 mL = _____ L

8. 16 mL = _____ cm^3

ANSWERS: **1.** 0.5 g   **2.** 43 000 mg   **3.** 0.062 kg   **4.** 18 000 mm^3 **5.** 0.0079 m^3
6. 0.002 L   **7.** 0.5 L   **8.** 16 cm^3

Exercises M.2

Change the following units of mass (weight) as indicated.

1. 2 g = _____ mg

2. 7 kg = _____ g

3. 3700 kg = _____ t

4. 34.5 mg = _____ g

5. 5600 g = _____ kg

6. 4000 kg = _____ t

7. 91 kg = _____ t

8. 73 kg = _____ mg

9. 0.7 g = _____ mg

10. 0.54 g = _____ mg

11. How many kilograms are there in 5 t?

12. How many kilograms are there in 17 t?

13. Change 2 t to kilograms.

14. Change 896 mg to grams.

15. Express 896 g in milligrams.

16. Express 342 kg in grams.

17. Convert 75 000 g to kilograms.

18. Convert 3000 mg to grams.

19. Convert 7 t to grams.

20. Convert 0.4 t to grams.

21. Change 0.34 g to kilograms.

22. Change 0.78 g to milligrams.

23. How many grams are in 16 mg?

24. How many milligrams are in 2.5 g?

25. 92.3 g = _____ kg

26. 3.94 g = _____ mg

27. 7.58 t = _____ kg

28. 5.6 t = _____ kg

29. 2963 kg = _____ t

30. 3547 kg = _____ t

Complete the following tables.

31. 1 cm^3 = _____ mm^3
1 dm^3 = _____ cm^3
1 m^3 = _____ dm^3
1 km^3 = _____ m^3

32. 1 dm = _____ cm
1 dm = _____ mm
1 dm^2 = _____ cm^2
1 dm^2 = _____ mm^2
1 dm^3 = _____ cm^3
1 dm^3 = _____ mm^3

33. 1 m = _____ dm
1 m = _____ cm
1 m^2 = _____ dm^2
1 m^2 = _____ cm^2
1 m^3 = _____ dm^3
1 m^3 = _____ cm^3

34. 1 km = _____ m
1 km^2 = _____ m^2
1 km^3 = _____ m^3
1 km = _____ dm
1 km^2 = _____ ha
1 km^3 = _____ kL

Change the following units of volume as indicated.

35. 73 m^3 = _____ dm^3

36. 0.9 m^3 = _____ dm^3

37. 525 cm^3 = _____ m^3

38. 400 m^3 = _____ cm^3

39. 8.7 m^3 = _____ cm^3

40. 63 dm^3 = _____ m^3

41. How many cm^3 are in 45 mm^3?

42. How many mm^3 are in 3.1 cm^3?

43. Change 19 mm^3 to dm^3.

44. Change 5 cm^3 to mm^3.

45. Convert 2 dm^3 to cm^3.

46. Convert 76.4 mL to liters.

47. Change 5.3 L to milliliters.

48. Change 30 cm^3 to milliliters.

49. Change 30 cm^3 to liters.

50. Change 5.3 mL to liters.

51. 48 kL = _____ L

52. 72 000 L = _____ kL

53. 290 L = _____ kL

54. 569 mL = _____ L

55. 80 L = _____ mL = _____ cm^3

56. 7.3 L = _____ mL = _____ cm^3

M.3   U.S. Customary Measurements and Metric Equivalents

OBJECTIVES

1. Become familiar with common units of measure in the U.S. customary system.
2. Be able to change measures within the U.S. customary system.
3. Know how to use formulas for converting between Fahrenheit and Celsius measures of temperature.
4. Learn how to use tables to convert units of measure between the U.S. customary system and the metric system.

U.S. Customary Equivalents

In the U.S. customary system (formerly the English system), the units are not systematically related as are the units in the metric system. Historically some of the units were associated with parts of the body, which would vary from person to person. For example, a foot was the length of a person's foot, and a yard was the distance from the tip of one's nose to the tip of one's fingers with arm outstretched. A king might dictate his own foot to be the official "foot," but, of course, the next king might have a different-sized foot.

There is considerably more stability now because the official weights and measures are monitored by the government.

In this section we will discuss the common units for length, area, liquid volume, weight, and time, and how to find equivalent measures. The basic relationships are listed in Table M.9. The measures of time are universal.

To change units, you must either have a table of equivalent values with you or memorize the basic equivalent values. Most people know some of these values but not all.

There are several methods used to change from one unit to another. One is to use proportions and solve these proportions; another is to substitute ratios (commonly used in science courses); and another is to substitute equivalent values for just one unit and multiply. We will illustrate the third technique because it is probably the simplest.

Table M.9   U.S. Customary Units of Measure

Length
1 foot (ft)   = 12 inches (in.)
1 yard (yd)   = 3 ft
1 mile (mi)   = 5280 ft

Area
1 ft^2 $= 144 \text{ in.}^2$
1 yd^2 $= 9 \text{ ft}^2$
1 acre $= 4840 \text{ yd}^2$
    $= 43,560 \text{ ft}^2$

Weight
1 pound (lb) = 16 ounces (oz)
1 ton (t)   = 2000 lb

Liquid Volume
1 pint (pt)   = 16 fluid ounces (fl oz)
1 quart (qt)   = 2 pt = 32 fl oz
1 gallon (gal)   = 4 qt

Time
1 minute (min) = 60 seconds (sec)
1 hour (hr)   = 60 min
1 day   = 24 hr

Convert the measures in Examples 1–9 as indicated.

EXAMPLE 1   4 ft = _____ in.

Solution

Think of 4 ft as 4(1 ft) and replace 1 ft with 12 in.

4 ft = 4(1 ft) = 4(12 in.) = 48 in.

EXAMPLE 2   6 ft = _____ in.

Solution

6 ft = 6(1 ft) = 6(12 in.) = 72 in.

EXAMPLE 3   36 in. = _____ ft

Solution

In this case, changing from a smaller to a larger unit, a fraction is necessary. We know

12 in. = 1 ft

so,

$$\frac{12 \text{ in.}}{12} = \frac{1 \text{ ft}}{12} \quad \text{or} \quad 1 \text{ in.} = \frac{1}{12} \text{ ft}$$

Now,

$$36 \text{ in.} = 36(1 \text{ in.}) = 36\left(\frac{1}{12} \text{ ft}\right) = 3 \text{ ft}$$

EXAMPLE 4

12 ft = _____ yd

Solution

Since 3 ft = 1 yd, we have 1 ft = $\frac{1}{3}$ yd.

So,

$$12 \text{ ft} = 12 \,(1 \text{ ft}) = 12\left(\frac{1}{3} \text{ yd}\right) = 4 \text{ yd}$$

EXAMPLE 5

How many acres are there in 87,120 ft²?

Solution

Since 43,560 ft² = 1 acre, we have 1 ft² = $\dfrac{1}{43,560}$ acres, so

$$87,120 \text{ ft}^2 = 87,120 \,(1 \text{ ft}^2) = 87,120 \left(\frac{1}{43,560} \text{ acres}\right) = 2 \text{ acres}$$

EXAMPLE 6

6 qt = _____ pt

Solution

$$6 \text{ qt} = 6(1 \text{ qt}) = 6(2 \text{ pt}) = 12 \text{ pt}$$

EXAMPLE 7

2.5 lb = _____ oz

Solution

$$2.5 \text{ lb} = 2.5(1 \text{ lb}) = 2.5(16 \text{ oz}) = 40 \text{ oz}$$

EXAMPLE 8

36 hr = _____ days

Solution

Since 24 hr = 1 day, we know 1 hr = $\dfrac{1}{24}$ day.

Substituting,

$$36 \text{ hr} = 36(1 \text{ hr}) = 36\left(\dfrac{1}{24} \text{ day}\right) = 1\dfrac{1}{2} \text{ days or } 1.5 \text{ days}$$

EXAMPLE 9

3 hr = _____ sec

Solution

In this case, two substitutions are made. First, change hours to minutes, then change minutes to seconds.

$$3 \text{ hr} = 3(1 \text{ hr}) = 3(60 \text{ min}) = 180 \text{ min}$$
$$180 \text{ min} = 180(1 \text{ min}) = 180(60 \text{ sec}) = 10,800 \text{ sec}$$

So,

$$3 \text{ hr} = 10,800 \text{ sec}$$

CLASSROOM PRACTICE

Convert the following measures as indicated.

1. 2 ft = _____ in.   **2.** 8 in. = _____ ft

3. 2 gal = _____ qt   **4.** 3 t = _____ lb

5. 45 min = _____ hr   **6.** 3 ft^2 = _____ in.^{2}

ANSWERS: **1.** 24 in.   **2.** $\dfrac{2}{3}$ ft   **3.** 8 qt   **4.** 6000 lb   **5.** $\dfrac{3}{4}$ hr   **6.** 432 in.^{2}

U.S. Customary and Metric Equivalents

We begin the following discussion of equivalent measures between the U.S. customary and metric systems with measures of temperature.

Temperature: U.S. customary measure is in **degrees Fahrenheit** (°F). Metric measure is in **degrees Celsius** (°C).

The two scales are shown here on thermometers. Approximate conversions can be found by reading along a ruler or the edge of a piece of paper held horizontally across the page.

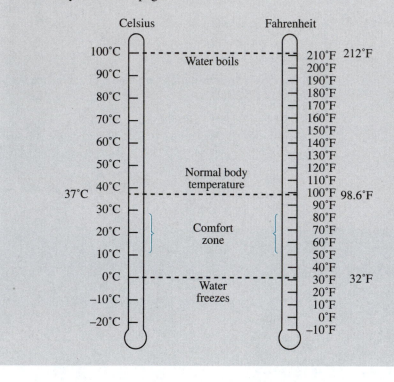

EXAMPLE 10

Hold a straight edge horizontally across the two thermometers and you will read:

100° C = 212° F   Water boils at sea level.
40° C = 104° F   hot day in the desert
20° C = 68° F   comfortable room temperature

Two formulas that give exact conversions are given here.

F = Fahrenheit temperatures and C = Celsius temperature.

$$C = \frac{5(F - 32)}{9} \qquad F = \frac{9 \cdot C}{5} + 32$$

A calculator will give answers accurate to eight digits. Answers that are not exact may be rounded off to whatever place of accuracy you choose.

EXAMPLE 11   Let $F = 86°$ and find the equivalent measure in Celsius.

Solution

$$C = \frac{5(86 - 32)}{9} = \frac{5(54)}{9} = 30 \qquad \text{Thus, } 86° \text{ F} = 30° \text{ C.}$$

EXAMPLE 12   Let $C = 40°$ and convert this to degrees Fahrenheit.

Solution

$$F = \frac{9 \cdot 40}{5} + 32 = 72 + 32 = 104 \qquad \text{Thus, } 40° \text{ C} = 104° \text{ F.}$$

In the tables of Length Equivalents, Area Equivalents, Volume Equivalents, and Mass Equivalents, (Tables M.10–M.13), the equivalent measures are rounded off. Any calculations with these measures (with or without a calculator) cannot be any more accurate than the measure in the table. Figure M.8 shows some length equivalents.

Table M.10	Length Equivalents	
U.S. to Metric	**Metric to U.S.**
1 in. = 2.54 cm (exact)	1 cm = 0.394 in.
1 ft = 0.305 m	1 m = 3.28 ft
1 yd = 0.914 m	1 m = 1.09 yd
1 mi = 1.61 km	1 km = 0.62 mi

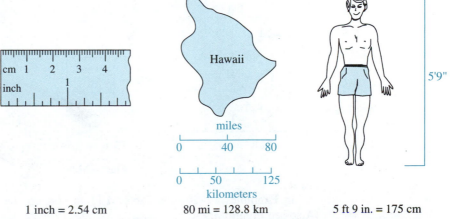

1 inch = 2.54 cm   80 mi = 128.8 km   5 ft 9 in. = 175 cm

Figure M.8

In Examples 13–16, use Table M.10 on page M29 to convert measurements as indicated. (Also see Figure M.8.)

EXAMPLE 13   6 ft = _____ cm

6 ft = 72 in. = 72(2.54 cm) = 183 cm (rounded off)

or

6 ft = 6(0.305 m) = 1.83 m = 183 cm

EXAMPLE 14   25 mi = _____ km

25 mi = 25(1.61 km) = 40.25 km

EXAMPLE 15   30 m = _____ ft

30 m = 30(3.28 ft) = 98.4 ft

EXAMPLE 16   10 km = _____ mi

10 km = 10(0.62 mi) = 6.2 mi

Table M.11	Area Equivalents	
U.S. to Metric	**Metric to U.S.**
1 in.^{2} = 6.45 cm^2	1 cm^2 = 0.155 in.^{2}
1 ft^2 = 0.093 m^2	1 m^2 = 10.764 ft^2
1 yd^2 = 0.836 m^2	1 m^2 = 1.196 yd^2
1 acre = 0.405 ha	1 ha = 2.47 acres

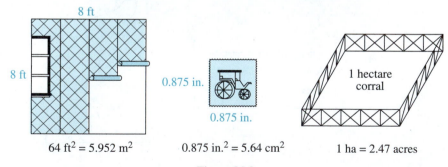

64 ft^2 = 5.952 m^2 0.875 in.^{2} = 5.64 cm^2 1 ha = 2.47 acres

Figure M.9

In Examples 17–20, use Table M.11 to convert the measures as indicated. (Also see Figure M.9.)

EXAMPLE 17

$40 \text{ yd}^2 = \underline{\hspace{1cm}} \text{ m}$

$40 \text{ yd}^2 = 40(0.863 \text{ m}^2) = 33.44 \text{ m}^2$

EXAMPLE 18

$5 \text{ acres} = \underline{\hspace{1cm}} \text{ ha}$

$5 \text{ acres} = 5(0.405 \text{ ha}) = 2.025 \text{ ha}$

EXAMPLE 19

$5 \text{ ha} = \underline{\hspace{1cm}} \text{ acres}$

$5 \text{ ha} = 5(2.47 \text{ acres}) = 12.35 \text{ acres}$

EXAMPLE 20

$100 \text{ cm}^2 = \underline{\hspace{1cm}} \text{ in.}^2$

$100 \text{ cm}^2 = 100(0.155 \text{ in.}^2) = 15.5 \text{ in.}^2$

Table M.12 Volume Equivalents	
U.S. to Metric	**Metric to U.S.**
$1 \text{ in.}^3 = 16.387 \text{ cm}^3$	$1 \text{ cm}^3 = 0.06 \text{ in.}^3$
$1 \text{ ft}^3 = 0.028 \text{ m}^3$	$1 \text{ m}^3 = 35.315 \text{ ft}^3$
$1 \text{ qt} = 0.946 \text{ L}$	$1 \text{ L} = 1.06 \text{ qt}$
$1 \text{ gal} = 3.785 \text{ L}$	$1 \text{ L} = 0.264 \text{ gal}$

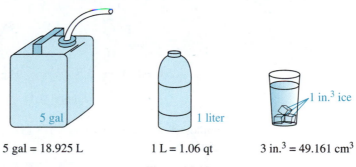

5 gal = 18.925 L	1 L = 1.06 qt	3 in.³ = 49.161 cm³

Figure M.10

In Examples 21–24, use Table M.12 to convert the measures as indicated. (Also see Figure M.10.)

EXAMPLE 21

$20 \text{ gal} = \underline{\hspace{1cm}} \text{ L}$

$20 \text{ gal} = 20(3.785 \text{ L}) = 75.7 \text{ L}$

EXAMPLE 22

42 L = _____ gal

42 L = 42(0.264 gal) = 11.088 gal

or

42 L = 11.1 gal (rounded off)

EXAMPLE 23

6 qt = _____ L

6 qt = 6(0.946 L) = 5.676 L

or

6 qt = 5.7 L (rounded off)

EXAMPLE 24

10 cm^3 = _____ in.^{3}

10 cm^3 = 10(0.06 in.^{3}) = 0.6 in.^{3}

Table M.13 Mass Equivalents	
U.S. to Metric	**Metric to U.S.**
1 oz = 28.35 g	1 g = 0.035 oz
1 lb = 0.454 kg	1 kg = 2.205 lb

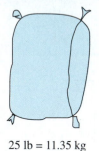 

25 lb = 11.35 kg   9 kg = 19.85 lb

Figure M.11

In Examples 25 and 26, use Table M.13 to convert the measures as indicated. (Also see Figure M.11.)

EXAMPLE 25

5 lb = _____ kg

5 lb = 5(0.454 kg) = 2.27 kg

EXAMPLE 26

15 kg = _____ lb

15 kg = 15(2.205 lb) = 33.075 lb

or

15 kg = 33.1 lb (rounded off)

Exercises M.3

Convert the following measures as indicated. Use Table M.9 as a reference.

1. 5 ft = _____ in.

2. 3 ft = _____ in.

3. 1.5 ft = _____ in.

4. 2.5 ft = _____ in.

5. 48 in. = _____ ft

6. 120 in. = _____ ft

7. 30 in. = _____ ft

8. 18 in. = _____ ft

9. 3 yd = _____ ft

10. 4 yd = _____ ft

11. $2\frac{1}{3}$ yd = _____ ft

12. $1\frac{2}{3}$ yd = _____ ft

13. 3 mi = _____ ft

14. 4 mi = _____ ft

15. 6 ft = _____ yd

16. 9 ft = _____ yd

17. 10,560 ft = _____ mi

18. 15,840 ft = _____ mi

19. Convert 2 pt to fluid ounces.

20. Convert 3 pt to fluid ounces.

21. Express 3 qt in pints.

22. Express 1.5 qt in pints.

23. Convert 5 gal to quarts.

24. Convert 3 gal to quarts.

25. 12 qt = _____ gal

26. 20 qt = _____ gal

27. 15 qt = _____ gal

28. 13 qt = _____ gal

29. 5 lb = _____ oz

30. 3 lb = _____ oz

31. 3.5 lb = _____ oz

32. 1.75 lb = _____ oz

33. $2\frac{1}{2}$ t = _____ lb

34. 3 t = _____ lb

35. 5 hr = _____ min

36. 4 hr = _____ min

37. How many minutes are in $1\frac{1}{2}$ hr?

38. How many minutes are in $3\frac{1}{2}$ hr?

39. Change 30 min to hours.

40. Change 15 min to hours.

41. Express 3 days in hours.

42. Express 4 days in hours.

43. Convert 0.5 hr to seconds.

44. Convert $\frac{2}{3}$ hr to seconds.

45. How many seconds are in 5 min?

46. How many seconds are in 3 min?

47. Change 240 sec to minutes.

48. Change 300 sec to minutes.

49. How many days are in 48 hr?

50. How many days are in 72 hr?

51. $25°C =$ _____ $° F$

52. $80° C =$ _____ $° F$

53. $10° C =$ _____ $° F$

54. $35° C =$ _____ $° F$

55. $50° F =$ _____ $° C$

56. $100° F =$ _____ $° C$

57. Change $32° F$ to degrees Celsius.

58. Change $41° F$ to degrees Celsius.

59. How many meters are in 3 yd?

60. How many meters are in 5 yd?

61. Change 60 mi to kilometers.

62. Change 100 mi to kilometers.

63. Convert 200 km to miles.

64. Convert 65 km to miles.

65. How many inches are in 50 cm?

66. How many inches are in 100 cm?

67. 3 in.$^2 =$ _____ cm^2

68. 16 in.$^2 =$ _____ cm^2

69. 600 ft$^2 =$ _____ m^2

70. 300 ft$^2 =$ _____ m^2

71. 100 yd$^2 =$ _____ m^2

72. 250 yd$^2 =$ _____ m^2

73. 1000 acres $=$ _____ ha

74. 250 acres $=$ _____ ha

75. How many acres are in 300 ha?

76. How many acres are in 400 ha?

77. Change 5 m^2 to square feet.

78. Change 10 m^2 to square feet.

79. Change 30 cm^2 to square inches.

80. Change 50 cm^2 to square inches.

81. 10 qt $=$ _____ L

82. 20 qt $=$ _____ L

83. 10 L $=$ _____ qt

84. 25 L $=$ _____ qt

85. 42 L $=$ _____ gal

86. 50 L $=$ _____ gal

87. 10 lb $=$ _____ kg

88. 500 kg $=$ _____ lb

89. 16 oz $=$ _____ g

90. 100 g $=$ _____ oz

91. Suppose that the home you are considering buying sits on a rectangular shaped lot that is 270 feet by 121 feet. What fraction of an acre is the size of this lot?

92. A new manufacturing building covers an area of 3 acres. How many square feet of ground does the new building cover?

93. A painting of a landscape is on a rectangular canvas that measures 3 feet by 4 feet. (a) How many square inches of wall space will the painting cover when it is hanging? (b) How many square yards?

94. Mr. Zimmer decided to build a fence around a pasture of 5 acres so that some of his cattle could graze there. (a) If he put 50 head of cattle in the field, how many square feet were there in the field for each head of cattle? (b) How many square yards?

Denominate Numbers

OBJECTIVES

1. Know the difference between abstract numbers and denominate numbers.
2. Be able to simplify mixed denominate numbers.
3. Know how to add like denominate numbers.
4. Know how to subtract like denominate numbers.

Denominate Numbers

Numbers with no units of measure attached are called **abstract numbers.** Numbers with units of measure attached are called **denominate numbers.** The numbers discussed in Sections M.1–M.3 were all denominate numbers. In this section, we will do more than just convert from one unit to another. Here we will discuss:

(a) mixed denominate numbers (denominate numbers with two or more units).

(b) adding denominate numbers.

(c) subtracting denominate numbers.

Examples of mixed denominate numbers that we commonly use are

5 ft 8 in.;
3 lb 4 oz;
and 1 hr 45 min.

With mixed numbers, we make sure that the fraction part is less than 1. Thus, we write $5\frac{1}{2}$, **not** $4\frac{3}{2}$. Similarly, in **simplified mixed denominate numbers,** the number of the smaller unit is less than 1 of the larger unit. The following examples illustrate the technique for simplifying mixed denominate numbers.

Examples 1–3 show how to simplify mixed denominate numbers so that the number of smaller units is less than 1 of the larger unit.

EXAMPLE 1

3 ft 14 in.

Solution

Since 14 in. is more than 1 ft, we write

$$
\begin{aligned}
3 \text{ ft } 14 \text{ in. } &= 3 \text{ ft } + 12 \text{ in. } + 2 \text{ in.} \\
&= 3 \text{ ft } + 1 \text{ ft } + 2 \text{ in.} \\
&= \quad\quad 4 \text{ ft } \quad\quad + 2 \text{ in.} \\
&= 4 \text{ ft } 2 \text{ in.}
\end{aligned}
$$

EXAMPLE 2

5 lb 30 oz

Solution

Since 30 oz is more than 1 lb, we write

$$
\begin{aligned}
5 \text{ lb } 30 \text{ oz } &= 5 \text{ lb } + 16 \text{ oz } + 14 \text{ oz} \\
&= 5 \text{ lb } + 1 \text{ lb } + 14 \text{ oz} \\
&= \quad\quad 6 \text{ lb } \quad\quad + 14 \text{ oz} \\
&= 6 \text{ lb } 14 \text{ oz}
\end{aligned}
$$

EXAMPLE 3

2 hr 70 min

Solution

Since 70 min is more than 1 hr, we write

$$
\begin{aligned}
2 \text{ hr } 70 \text{ min } &= 2 \text{ hr } + 60 \text{ min } + 10 \text{ min} \\
&= 2 \text{ hr } + 1 \text{ hr } + 10 \text{ min} \\
&= \quad\quad 3 \text{ hr } \quad\quad + 10 \text{ min} \\
&= 3 \text{ hr } 10 \text{ min}
\end{aligned}
$$

Adding and Subtracting Denominate Numbers

Understanding how to simplify mixed denominate numbers helps in both adding and subtracting such numbers. **Like denominate numbers** are denominate numbers with the same units. For example, 5 ft 10 in. and 2 ft 3 in. are like denominate numbers. Also, 3 hr 5 min and 4 hr 15 min are like denominate numbers.

To Add Like Denominate Numbers:

1. Write the numbers in column form so that like units are aligned.

2. Add the numbers in each column.

3. Simplify the resulting sum if necessary.

Add the like denominate numbers in Examples 4–6 and simplify the sum if necessary.

EXAMPLE 4

$$
\begin{array}{r}
3 \text{ ft } 2 \text{ in.} \\
2 \text{ ft } 8 \text{ in.} \\
+ \ 5 \text{ ft } 5 \text{ in.} \\
\hline
\end{array}
$$

10 ft 15 in. = 10 ft + 12 in. + 3 in.
          = 11 ft 3 in.

EXAMPLE 5

$$
\begin{array}{r}
2 \text{ hr } 15 \text{ min} \\
+4 \text{ hr } 50 \text{ min} \\
\hline
\end{array}
$$

6 hr 65 min = 6 hr + 60 min + 5 min
          = 7 hr 5 min

EXAMPLE 6

$$
\begin{array}{r}
3 \text{ gal } 2 \text{ qt} \\
1 \text{ gal } 3 \text{ qt} \\
+5 \text{ gal } 2 \text{ qt} \\
\hline
\end{array}
$$

9 gal 7 qt = 9 gal + 4 qt + 3 qt
         = 10 gal 3 qt

To Subtract Like Denominate Numbers:

1. Write the numbers in column form so that like units are aligned.

2. If necessary for subtraction, borrow 1 of the larger units and rewrite the top number.

3. Subtract the like units.

Subtract the like denominate numbers in Examples 7–9.

EXAMPLE 7

$\begin{array}{r} 8 \text{ lb } 14 \text{ oz} \\ - \ 3 \text{ lb } 10 \text{ oz} \\ \hline 5 \text{ lb } \ \ 4 \text{ oz} \end{array}$

EXAMPLE 8

$\begin{array}{r} 6 \text{ ft } 5 \text{ in.} \\ - \ 2 \text{ ft } 8 \text{ in.} \\ \hline \end{array}$ Here 5 in. is smaller than 8 in., and we cannot subtract. So, borrow 1 ft = 12 in. from 6 ft.

$\begin{array}{r} 6 \text{ ft } 5 \text{ in.} = 5 \text{ ft } 17 \text{ in.} \\ -2 \text{ ft } 8 \text{ in.} = 2 \text{ ft } \ \ 8 \text{ in.} \\ \hline 3 \text{ ft } \ \ 9 \text{ in.} \end{array}$ (12 in. + 5 in. = 17 in.)

EXAMPLE 9

$\begin{array}{r} 13 \text{ hr } 20 \text{ min} \\ -10 \text{ hr } 50 \text{ min} \\ \hline \end{array}$ Here 20 min is smaller than 50 min, and we cannot subtract. So, borrow 1 hr = 60 min from 13 hr.

$\begin{array}{r} 13 \text{ hr } 20 \text{ min} = 12 \text{ hr } 80 \text{ min} \\ -10 \text{ hr } 50 \text{ min} = 10 \text{ hr } 50 \text{ min} \\ \hline 2 \text{ hr } 30 \text{ min} \end{array}$ (60 min + 20 min = 80 min)

CLASSROOM PRACTICE

Perform the indicated operations and simplify.

1. $\begin{array}{r} 2 \text{ ft } 3 \text{ in.} \\ 6 \text{ ft } 8 \text{ in.} \\ +1 \text{ ft } 4 \text{ in.} \\ \hline \end{array}$

2. $\begin{array}{r} 5 \text{ hr } 30 \text{ min} \\ -1 \text{ hr } 45 \text{ min} \\ \hline \end{array}$

ANSWERS: **1.** 10 ft 3 in.   **2.** 3 hr 45 min

Exercises M.4

Simplify the following mixed denominate numbers.

1. 3 ft 20 in.   **2.** 4 ft 18 in.

3. 6 lb 20 oz   **4.** 3 lb 24 oz

5. 5 min 80 sec   **6.** 14 min 90 sec

7. 2 days 30 hr   **8.** 5 days 36 hr

9. 8 gal 5 qt   **10.** 2 gal 6 qt

11. 4 pt 20 fl oz   **12.** 3 pt 24 fl oz

Add and simplify if necessary.

13. $\begin{array}{r} 2 \text{ ft } 8 \text{ in.} \\ 5 \text{ ft } 4 \text{ in.} \\ +1 \text{ ft } 7 \text{ in.} \\ \hline \end{array}$

14. $\begin{array}{r} 3 \text{ ft } 5 \text{ in.} \\ 6 \text{ ft } 5 \text{ in.} \\ +2 \text{ ft } 3 \text{ in.} \\ \hline \end{array}$

15.   10 lb 10 oz
  + 7 lb 8 oz

16.   4 lb 5 oz
  +4 lb 7 oz

17.   8 min 35 sec
  +9 min 35 sec

18.   5 min 10 sec
  +14 min 35 sec

19.   2 hr 15 min 45 sec
  +1 hr 55 min 30 sec

20.   5 hr 20 min 30 sec
  +2 hr 35 min 40 sec

21.   2 days 20 hr 50 min
  +3 days 5 hr 45 min

22.   1 day 15 hr 40 min
  +2 days 10 hr 20 min

23.   5 gal 2 qt
   3 gal 2 qt
  +4 gal 3 qt

24.   4 gal 3 qt
   1 gal 2 qt
  +3 gal 1 qt

25.   4 gal 3 qt 10 fl oz
  +2 gal 3 qt 10 fl oz

26.   5 gal 3 qt 15 fl oz
  +1 gal 2 qt 8 fl oz

27.   5 yd 2 ft 8 in.
  +6 yd 2 ft 10 in.

28.   3 yd 1 ft 7 in.
  +2 yd 2 ft 3 in.

Subtract.

29.   5 yd 1 ft 7 in.
  −2 yd 2 ft 5 in.

30.   8 yd 2 ft 3 in.
  −7 yd 1 ft 8 in.

31.   9 gal 2 qt 4 fl oz
  −5 gal 3 qt 6 fl oz

32.   20 gal 1 qt 13 fl oz
  −14 gal 2 qt 10 fl oz

33.   15 hr 30 min
  −12 hr 45 min

34.   6 hr 20 min
  −4 hr 40 min

35.   15 min 20 sec
  −10 min 30 sec

36.   30 min 15 sec
  −20 min 25 sec

37.   8 lb 4 oz
  −3 lb 12 oz

38.   20 lb 10 oz
  −10 lb 14 oz

39.   6 ft 5 in.
  −2 ft 9 in.

40.   3 ft 8 in.
  −1 ft 10 in.

41. What is 1 ft 6 in. more than 3 ft 8 in.?

42. Find the sum of 5 hr 30 min, 7 hr 15 min, and 3 hr 45 min.

43. What is the difference between 6 hr and 3 hr 20 min?

44. What is 4 lb 12 oz less than 10 lb 10 oz?

45. What is 7 qt more than 3 gal 1 qt?

46. What is the sum of 5 gal 3 qt, 2 gal 3 qt, and 1 gal 3 qt?

Chapter Summary

Key Terms and Ideas

Metric Prefixes and Their Values

Prefix	Value	
milli-	0.001	—thousandths
centi-	0.01	—hundredths
deci-	0.1	—tenths
basic unit	1	—ones
deka-	10	—tens
hecto-	100	—hundreds
kilo-	1000	—thousands

The **meter (m)** is the basic unit of **length** in the metric system.

Area is a measure of the interior (or enclosure) of a surface. Area is measured in **square units.** The **are (a)** and **hectare (ha)** are measures of land area.

Mass is the amount of material in an object. **Weight** is the force of the Earth's gravitational pull on an object. The concepts of mass and weight are used interchangeably in this text. Two basic units of mass in the metric system are the **gram (g)** and the **kilogram (kg).**

Volume is a measure of the space enclosed by a three-dimensional figure. Volume is measured in **cubic units.**

Liquid volume is measured in **liters (L).**

Temperature conversions between **Fahrenheit (F)** and **Celsius (C)** are given by the following two formulas:

$$C = \frac{5(F - 32)}{9} \qquad F = \frac{9 \cdot C}{5} + 32$$

The text contains several tables of conversions and equivalents for the U.S. customary system and the metric system of measurement.

Denominate numbers are numbers with units of measure attached.

Abstract numbers are numbers with no related measurement units.

Procedures

Charts can be used for finding equivalent measures in the metric system.

1. Use one digit per column when changing units of length, mass (weight), and liquid volume.

2. Use two digits per column when changing units of area.

3. Use three digits per column when changing units of volume.

To Add Like Denominate Numbers:

1. Write the numbers in column form so that like units are aligned.

2. Add the numbers in each column.

3. Simplify the resulting sum if necessary.

To Subtract Like Denominate Numbers:

1. Write the numbers in column form so that like units are aligned.

2. If necessary for subtraction, borrow 1 of the larger units and rewrite the top number.

3. Subtract the like units.

Chapter Review Questions

Change the following metric units as indicated.

1. 15 m = _____ cm

2. 35 mm = _____ dm

3. 37 cm^2 = _____ mm^2

4. 17 mm^2 = _____ cm^2

5. 3 ha = _____ a

6. 3 ha = _____ m^2

7. 5 L = _____ cm^3

8. 36 L = _____ mL

9. 13 dm^3 = _____ cm^3

10. 68 cm^3 = _____ mm^3

11. 5 kg = _____ g

12. 3.4 g = _____ mg

13. 6.71 t = _____ kg

14. 19 mg = _____ g

15. 8 kg = _____ g

16. 4290 g = _____ kg

Convert the following measures as indicated.

17. 2.75 ft = _____ in.

18. $6\frac{1}{2}$ yd = _____ ft

19. 74 in. = _____ ft

20. 2 ft = _____ yd

21. $4\frac{3}{4}$ lb = _____ oz

22. 1.3 t = _____ lb

23. 54 oz = _____ lb

24. 800 lb = _____ t

25. 3.2 hr = _____ min

26. $2\frac{5}{12}$ days = _____ hr

27. 168 sec = _____ min

28. 84 hr = _____ days

29. 1.6 pt = _____ fl oz

30. $2\frac{3}{8}$ gal = _____ qt

31. $3\frac{1}{4}$ qt = _____ gal

32. 22 fl oz = _____ pt

Simplify the following mixed denominate numbers.

33. 5 ft 19 in.

34. 2 yds 5 ft 28 in.

35. 3 lb 22 oz

36. 2 t 3120 lb

37. 5 days 28 hr

38. 62 min 73 sec

39. 1 pt 40 fl oz

40. 2 gal 11 qt 3 pt

Add and simplify if necessary.

41.   2 lb 4 oz
        +3 lb 7 oz

42.   12 hr 23 min 12 sec
        + 6 hr 21 min 4 sec

43.   1 ft 5 in.
       2 ft 3 in.
    +1 ft 9 in.

44.   1 gal 2 qt 1 pt
       2 gal   1 pt
    +   3 qt 1 pt

Find each difference.

45.   5 lb 12 oz
    −3 lb   5 oz

46.   3 gal 3 qt 1 pt
    −1 gal 2 qt 1 pt

47.   6 hr 12 min
    −5 hr 23 min

48.   4 yd 2 ft 5 in.
    −1 yd 2 ft 9 in.

Use the appropriate formula to convert the degrees as indicated.

49. $72°$ F = _____ $°$ C

50. $3°$ C = _____ $°$ F

Use the appropriate table to convert the following measures as indicated.

51. 5 in. = _____ cm

52. 2 ft 3 in. = _____ cm
(Use in. to cm equivalents.)

53. 25.4 cm = _____ in.

54. 20 cm = _____ in.

55. $1\frac{1}{2}$ yd = _____ m

56. 2 ft 3 in = _____ m
(Use ft to m equivalents.)

57. 7 m = _____ ft

58. 3.2 m = _____ yd

59. 200 mi = _____ km

60. $12\frac{3}{4}$ mi = _____ km

61. 88 km = _____ mi

62. 0.4 km = _____ mi

63. 6 acres = _____ ha

64. 5 yd^2 = _____ m^2

65. 2.1 m^2 = _____ ft^2

66. $\frac{3}{5}$ ha = _____ acres

67. 40 lb = _____ kg

68. 1 lb 2 oz = _____ g

69. 454 g = _____ oz

70. 0.08 kg = _____ lb

Chapter Test

1. Which is longer, 20 mm or 20 cm? How much longer?

2. Which is heavier, 10 g or 10 kg? How much heavier?

3. Which has the greater volume, 15 mL or 15 cm^3? How much greater?

Change the following units as indicated.

4. 37 cm = _____ m

5. 23 m = _____ cm

6. 2 L = _____ cm^3

7. 1200 g = _____ kg

8. 5.6 t = _____ kg

9. 75 a = _____ m^2

10. 11 000 mm = _____ m

11. 4 cm^3 = _____ mm^3

12. 960 mm^2 = _____ cm^2

13. 83.5 mg = _____ g

Convert the following measures as indicated. Use the appropriate table where necessary.

14. $5\frac{3}{4}$ lb = _____ oz

15. 3 ft 8 in. = _____ yd

16. 1 day 4 hr = _____ min

17. 50 cm = _____ in.

18. 7.8 kg = _____ lb

19. 25 L = _____ gal

20. $16\frac{3}{5}$ acres = _____ ha

21. 200 km = _____ mi

22. 30 qt = _____ L

23. 50g = _____ oz

24. Use the formula $C = \dfrac{5(F - 32)}{9}$ to convert 68° F to degrees Celsius.

25. Use the formula $F = \dfrac{9 \cdot C}{5} + 32$ to convert 30° C to degrees Fahrenheit.

26. Simplify the following mixed denominate number.

   1 gal 4 qt 3 pt 18 fl oz

Add the following and simplify if necessary.

27.　　2 days 5 hr　7 min
   　 +5 days 7 hr 41 min

28.　　2 ft 9 in.
   　 +1 ft 4 in.

Find the following difference.

29.　　3 qt 1 pt 12 fl oz
   　 −1 qt 1 pt　5 fl oz

30.　　10 lb　4 oz
   　 − 8 lb 12 oz

1. The number 0 is called the additive identity. The number _____ is called the multiplicative identity.

2. Write the number six hundred thirty and fifty-five thousandths in decimal form.

3. Find the difference between the product of 63.5 and 19.2 and the sum of 284.6 and 173.9.

4. Find the quotient of 89.5 and 17.6 to the nearest tenth.

5. Find the LCM of the set of numbers 35, 49, and 105.

6. Evaluate the expression: $16 + (12 \cdot 3 + 2^3) \div 4 - 20$

7. Evaluate the expression: $(0.5)^2 + 2\frac{1}{5} \div \frac{11}{6} - 0.3$

8. Evaluate the expression: $(1.5)^2 - (1.2)^2$

9. Find (a) the mean, (b) the median, (c) the mode, and (d) the range of the following set of numbers:

    26, 51, 49, 41, 41, 61, 46

10. If a savings account earned $75 in interest at 5% for 9 months, what was the original amount in the account.

11. If an investment pays 6% interest compounded daily, what will be the value of $8000 in 5 years?

12. If a stock pays a 10% dividend every quarter, what will be the total of the dividends paid on a $6000 investment over 4 years? (Assume that the dividends are reinvested each quarter.)

13. 18 is _____ % of 90.

14. 70% of _____ is 91.

15. 37.5% of 103.4 is _____.

16. 86% of 14,000 is _____.

17. (a) What were your car expenses for the month in which you paid $320 for four new tires, $25 for oil and filter, $80 for gas, $80 for insurance, and $125 for a tune-up? (b) If your income is $2520 per month, what percent of your income did you spend on your car that month?

18. If your car averages 22.5 miles per gallon, how many miles will it go on 18 gallons of gas?

19. An air-conditioning fan makes 150 revolutions per minute. How many revolutions will the fan make if it runs for 3 hours?

20. Which is the better buy of frozen orange juice?

    (a) 16 fl oz for $1.79

    (b) 12 fl oz for $1.09

21. Calculators are on sale at a discount of 28%, and the original price was $62.50. (a) What is the approximate amount of the discount? (b) What is the sale price?

22. Add and simplify:

    12 hr 23 min 12 sec
 + 6 hr 39 min 48 sec

23. Find the difference:

    4 yd 2 ft 5 in.
 −1 yd 2 ft 9 in.

24. Change the units as indicated.

  (a) 35.6 cm = _____ mm

  (b) 18.72 cm = _____ m

  (c) 195 cm^2 = _____ mm^2

  (d) 0.54 L = _____ mL

  (e) 78 000 cm^3 = _____ m^3

25. The circle graph shows a family budget for one year. What amount will be spent in each category if the family income is $45,000?

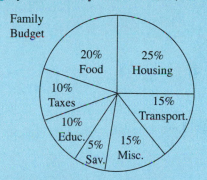

Family Budget

26. Write each of the following numbers in scientific notation.

  (a) 890,000,000

  (b) 0.0000632

27. Perform the indicated operations.

  (a) $(-5) + (-4) + (-12)$

  (b) $(16 - 18)(13 - 5)$

28. Solve each of the following equations.

  (a) $3x + 10 = -11$

  (b) $4y - 14 = 24.8$

29. Five more than six times a number is equal to 71. Find the number.

30. If three times a number is decreased by 13.5, the result is equal to -21.6. What is the number?

Geometry

Mathematics at Work!

View of Wrigley Field, home of the Chicago Cubs.

As illustrated in the figure shown here, the shape of a baseball infield is a square 90 feet on each side. (A square is a four-sided plane figure in which all four sides are the same length and the sides next to each other meet at 90° angles.) Do you think that the distance from home plate to second base is more than 180 feet or less than 180 feet? More than 90 feet or less than 90 feet? The distance from the pitcher's mound to home plate is $60\frac{1}{2}$ feet. Is the pitcher's mound exactly halfway between home plate and second base? Do the two diagonals of the square (as illustrated in the figure) intersect at the pitcher's mound? What is the distance from home plate to second base (to the nearest tenth of a foot)? (See Exercise 61 in Section G.6.)

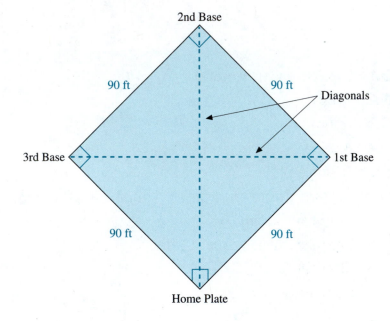

What to Expect in this Chapter

This chapter provides an introduction to some of the basic concepts of plane geometry. Many of these ideas are considered to be fundamental knowledge for future courses in mathematics.

In Section G.1, we introduce three undefined terms: **point, line,** and **plane.** These ideas form the basis for the study of plane geometry as organized and developed by Euclid (about 300 B.C.). The concept of **length** is developed through a discussion of **perimeter** (**circumference** for a circle) and the formulas for the perimeters of six types of geometric figures: square, rectangle, parallelogram, triangle, circle, and trapezoid. In Section G.2, the concept of **area** is discussed and the formulas for the areas of these same six types of geometric figures are presented.

The concept of **volume** and formulas for finding the volume of five three-dimensional figures are presented in Section G.3. The figures discussed are the rectangular solid, rectangular pyramid, right circular cylinder, right circular cone, and sphere.

Section G.4 begins with the definition of an **angle** and discusses how to use a protractor for measuring angles. We show that an angle is classified according to its measure as acute, right, obtuse, or straight. A detailed discussion of **triangles** appears in Section G.5. A triangle can be classified two ways: by the lengths of its sides and by the measures of its angles. Similar triangles have the following two properties: corresponding angles have the same measure and corresponding sides are proportional.

In Section G.6, we discuss square roots and the fact that most square roots are irrational numbers and can only be approximated in decimal form even with a calculator that gives 8- or 9-digit accuracy. This section provides valuable understanding and interesting applications of decimal numbers such as the Pythagorean Theorem.

G.1   Length and Perimeter

OBJECTIVES

1. Recognize the terms **point, line,** and **plane** and know that they are undefined terms.
2. Understand the concepts of length and perimeter.
3. Learn how to apply the appropriate formula to find the perimeter of a given geometric figure.
4. Know how to label answers with the correct units.

Introduction to Geometry

Plane geometry is the study of the properties of figures in a plane. Geometry is an important part of the development of mathematics because basic mathematical concepts are geometric as well as arithmetic and algebraic.

The three most basic ideas in plane geometry are **point, line,** and **plane.** These terms are considered so fundamental that any attempt to define their meanings would use terms more complicated and more difficult to understand than these terms. Thus, they are simply not defined and are called **undefined terms.** As we will see, these undefined terms provide the foundation for the study of geometry and the definitions of other geometric terms such as line segment, ray, angle, triangle, circle, and so on.

Undefined Term	Representation	Discussion
1. Point	A • point A	A dot represents a point. Points are labeled with capital letters.
2. Line	ℓ A B line ℓ or line $\overleftrightarrow{AB}$	·A line has no beginning or end. Lines are labeled with small letters or by two points on the line.
3. Plane	P plane P	Flat surfaces, such as a table top or wall, represent planes. Planes are labeled with capital letters.

In a more formal approach to plane geometry than we will use in this text, these undefined terms along with certain assumptions known as axioms, other statements called theorems, and a formal system of logic are studied in detail. This approach to the study of geometry is credited to Euclid (about 300 B.C.), which is why the plane geometry courses generally given in high school are known as courses in Euclidean geometry.

In this chapter, we will discuss formulas and properties related to several plane figures, including rectangles, triangles, and circles, and some three-dimensional figures such as rectangular solids and spheres. The formulas do not depend on any particular measurement system, and both the metric system and U.S. customary system of measure will be used in the exercises.

Be sure to label all answers with the correct unit of measure indicating length, area, or volume.

Length and Perimeter

A **formula** is a general statement (usually an equation) that relates two or more variables. The geometric figures and formulas discussed in this section illustrate applications of measures of length. From the metric system, some units of length are meter (m), decimeter (dm), centimeter (cm), millimeter (mm), and kilometer (km). From the U.S. customary system, some units of length are foot (ft), inch (in.), yard (yd), and mile (mi). Equivalent measures in both the metric system and the U.S. customary system were discussed in the Measurement chapter, which may or may not be included in this version of the text.

Important Terms

Perimeter: Total distance around a plane geometric figure

Circumference: Perimeter of a circle

Radius: The distance from the center of a circle to a point on the circle

Diameter: Distance from one point on a circle to another point on the circle, measured through the center

---

For the formulas related to circles, we designate the length of a diameter with the letter d and the length of a radius with the letter r. Note that a diameter of a circle is twice as long as a radius. That is, $d = 2r$.

Six geometric figures and the formulas for finding their perimeters:

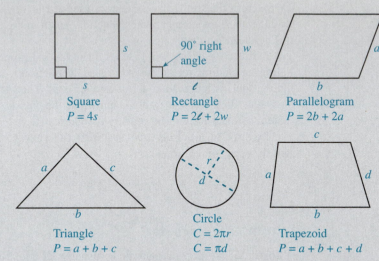

Square
$P = 4s$

Rectangle
$P = 2\ell + 2w$

Parallelogram
$P = 2b + 2a$

Triangle
$P = a + b + c$

Circle
$C = 2\pi r$
$C = \pi d$

Trapezoid
$P = a + b + c + d$

NOTE: π is the symbol used for the constant number 3.1415926535. . . . This constant is an infinite decimal with no pattern to its digits. For our purposes, we will use $\pi = 3.14$, but you should be aware that 3.14 is only an approximation to π.

EXAMPLE 1   Find the perimeter of a rectangle with length 20 in. and width 15 in.

Solution

Sketch the figure first.

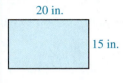

20 in.

15 in.

$P = 2\ell + 2w$
$P = 2 \cdot 20 + 2 \cdot 15$
$\quad = 40 + 30 = 70$ in.

The perimeter is 70 inches.

EXAMPLE 2   Find the circumference of a circle with diameter 3 m.

Solution

Sketch the figure first.

$$C = \pi d$$
$$C = 3.14(3) = 9.42 \text{ m}$$

The circumference is 9.42 m. [**NOTE:** Remember that 9.42 m is only an approximate answer because 3.14 is an approximation of π.]

EXAMPLE 3

Find the perimeter of a triangle with sides of 4 cm, 0.7 dm, and 80 mm. Write your answer in millimeters.

Solution

Sketch the figure and change all the units to millimeters.

4 cm = 40 mm   0.7 dm = 70 mm

80 mm

$$P = a + b + c$$
$$P = 40 + 70 + 80$$
$$= 190 \text{ mm}$$

The perimeter is 190 mm.

EXAMPLE 4

Find the circumference of a circle with radius 6 ft.

Solution

Sketch the figure first.

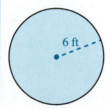

6 ft

$$C = 2\pi r$$
$$C = 2 \cdot 3.14 \cdot 6$$
$$= 37.68 \text{ ft}$$

The circumference is 37.68 ft.

EXAMPLE 5

Find the perimeter of a figure that is a semicircle (half of a circle) and its diameter with the diameter 20 centimeters long.

Solution

A sketch of the figure will help.

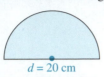

$d = 20$ cm

Now we find the perimeter of the figure by adding the length of the diameter to the length of the semicircle.

$$\text{Length of semicircle} = \frac{1}{2}\pi d = \frac{1}{2}(3.14) \cdot 20 = (1.57) \cdot 20 = 31.4$$

Perimeter of figure $= 31.4 + 20 = 51.4$ cm

CLASSROOM PRACTICE

1. Find the circumference of a circle with radius 4 in.

2. Find the perimeter of a square with sides of 3.6 cm.

ANSWERS: **1.** 25.12 in.   **2.** 14.4 cm

Exercises G.1

1. Match each formula for perimeter to its corresponding geometric figure.

_____ (a) square   A. $P = 2l + 2w$

_____ (b) parallelogram   B. $P = 4s$

_____ (c) circle   C. $P = 2b + 2a$

_____ (d) rectangle   D. $C = 2\pi r$

_____ (e) trapezoid   E. $P = a + b + c$

_____ (f) triangle   F. $P = a + b + c + d$

2. Find the perimeter of a triangle with sides of 4 cm, 8.3 cm, and 6.1 cm.

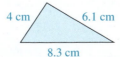

4 cm   6.1 cm

8.3 cm

3. Find the perimeter of a rectangle with length 35 mm and width 17 mm.

17 mm

35 mm

4. Find the perimeter of a square with sides of 13.3 m.

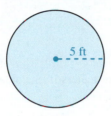

13.3 m

13.3 m

5. Find the circumference of a circle with radius 5 ft. (Use $\pi = 3.14$.)

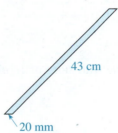

5 ft

6. Find the perimeter of a parallelogram with one side 43 cm and another side 20 mm. Write your answer in millimeters.

43 cm

20 mm

7. Find the circumference of a circle with diameter 6.2 yd. (Use $\pi = 3.14$.)

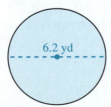

6.2 yd

8. Find the perimeter of a rectangle with length 50 m and width 50 dm. Write your answer in meters.

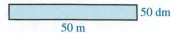

9. Find the perimeter of a triangle with sides of 5 cm, 55 mm, and 0.3 dm. Write your answers in centimeters.

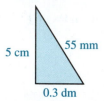

10. Find the circumference of a circle in meters if its radius is 70 cm. (Use $\pi = 3.14$.)

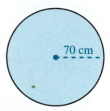

11. Find the perimeter of a square in meters if one side is 4 km long.

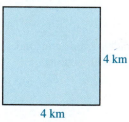

12. Find the perimeter of a triangle with sides of 1 yd, 4 ft, and 5 ft. Write your answer in feet.

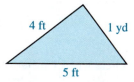

13. Find the perimeter of the trapezoid shown here.

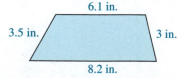

14. Find the perimeter of a parallelogram with sides of $3\frac{1}{2}$ ft and 14 in. Write your answer in both inches and feet.

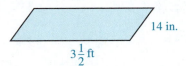

14 in.

$3\frac{1}{2}$ ft

15. Find the circumference of a circle with diameter 1.5 in. Write your answer in centimeters.

1.5 in.

Find the perimeter of each of the following figures with the indicated dimensions. (Use $\pi = 3.14$.)

16.

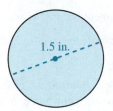

3 cm

17.

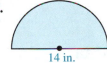

14 in.

18.

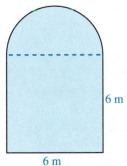

6 m

6 m

19.

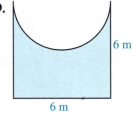

6 m

6 m

20.

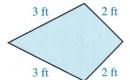

3 ft   2 ft

3 ft   2 ft

21.

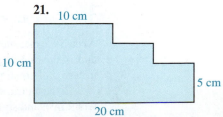

10 cm

10 cm

5 cm

20 cm

22.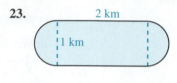
4 ft

3 ft

4 ft

23.
2 km

1 km

24. First draw a square with sides that are each 2 inches long. Next, draw a circle inside the square that just touches each side of the square. (This circle is said to be inscribed in the square.)

(a) What is the length of a diameter of this circle?

(b) Do you think that the circumference is less than 8 inches or more than 8 inches? Why?

25. With a piece of string and a ruler, measure the circumference and the diameter of each of several circular figures in your home. With a calculator, divide each circumference by the length of the corresponding diameter. What result do you observe? Discuss your results with other students in the class.

G.2   Area

OBJECTIVES

1. Understand the concept of **area.**
2. Learn how to apply the appropriate formula to find the area of a given geometric figure.
3. Know how to label answers with the correct units.

The Concept of Area

Area is a measure of the interior of (or surface enclosed by) a figure in a plane and is measured in **square units.** The concept of area is illustrated in Figure G.1 in terms of square inches.

In the metric system, some units of area are square meters (m^2), square centimeters (cm^2), and square millimeters (mm^2). In the U.S. customary system, some units of area are square feet (ft^2), square inches (in.^{2}), and square yards (yd^2).

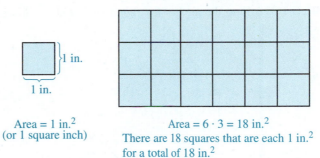

Area = 1 in.^{2}
(or 1 square inch)

Area = $6 \cdot 3 = 18$ in.^{2}
There are 18 squares that are each 1 in.^{2}
for a total of 18 in.^{2}

Figure G.1

Formulas for Area

The following formulas for the areas of the geometric figures shown are valid regardless of the system of measurement used. **Be sure to label your answers with the correct units.**

Note: In triangles and other figures, we have used the letter h to represent the **height** of the figure. The height is also called the **altitude.**

Six geometric figures and the formulas for finding their areas:

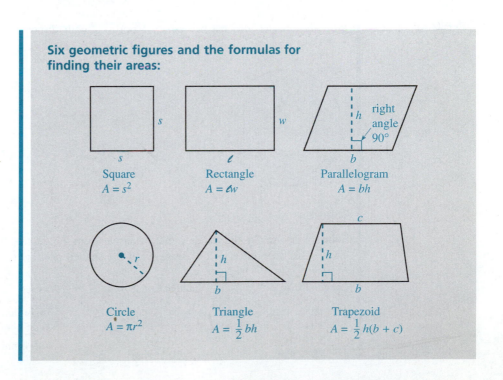

Square
$A = s^2$

Rectangle
$A = \ell w$

Parallelogram
$A = bh$

Circle
$A = \pi r^2$

Triangle
$A = \frac{1}{2}bh$

Trapezoid
$A = \frac{1}{2}h(b + c)$

EXAMPLE 1   Find the area of the figure shown here with the indicated dimensions.

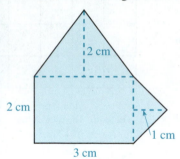

Solution

There are two triangles and one rectangle.

Rectangle	**Larger Triangle**	**Smaller Triangle**
$A = lw$	$A = \dfrac{1}{2}bh$	$A = \dfrac{1}{2}bh$
$A = 2 \cdot 3 = 6 \text{ cm}^2$	$A = \dfrac{1}{2} \cdot 3 \cdot 2 = 3 \text{ cm}^2$	$A = \dfrac{1}{2} \cdot 2 \cdot 1 = 1 \text{ cm}^2$

$$\text{Total area} = 6 \text{ cm}^2 + 3 \text{ cm}^2 + 1 \text{ cm}^2$$
$$= 10 \text{ cm}^2$$

EXAMPLE 2   Find the area of the washer (shaded portion) with dimensions as shown. (Use $\pi = 3.14$.)

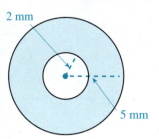

Solution

Subtract the area of the inside (smaller) circle from the area of the outside (larger) circle. This difference will be the area of the shaded portion.

Larger Circle

$A = \pi r^2$
$A = 3.14(5^2)$
$\quad = 3.14(25)$
$\quad = 78.50 \text{ mm}^2$

Smaller Circle

$A = \pi r^2$
$A = 3.14(2^2)$
$\quad = 3.14(4)$
$\quad = 12.56 \text{ mm}^2$

Washer

$\quad\;\; 78.50 \text{ mm}^2$
$\underline{-12.56 \text{ mm}^2}$
$\quad\;\; 65.94 \text{ mm}^2$ area of washer

EXAMPLE 3

Find the area of a trapezoid with altitude 6 in. and parallel sides of length 1 ft and 2 ft.

Solution

First draw a figure and label the lengths of the known parts. The units of length must be the same throughout, either all in inches or all in feet.

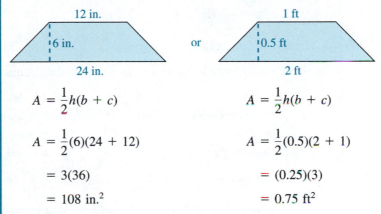

$$A = \frac{1}{2}h(b + c)$$

$$A = \frac{1}{2}(6)(24 + 12)$$

$$= 3(36)$$

$$= 108 \text{ in.}^2$$

$$A = \frac{1}{2}h(b + c)$$

$$A = \frac{1}{2}(0.5)(2 + 1)$$

$$= (0.25)(3)$$

$$= 0.75 \text{ ft}^2$$

The two correct areas are 108 in.^{2} and 0.75 ft^2. [**NOTE:** Since 1 ft^2 = (12 in.) (12 in.) = 144 in.^{2}, we have 0.75 ft^2 = 0.75(144) in.^{2} = 108 in.^{2} The two answers are equivalent. They are simply in different units of area.]

EXAMPLE 4

Find the area enclosed by a semicircle and a diameter if the diameter is 10 in.

Solution

First sketch the figure. (A semicircle is half of a circle.)

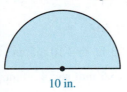

10 in.

For a circle, $A = \pi r^2$.

Thus, for a semicircle, $A = \frac{1}{2}\pi r^2$.

For this semicircle, $d = 10$, so $r = 5$.

$$A = \frac{1}{2}(3.14)(5)^2$$

$$= 1.57 \cdot 25$$

$$= 39.25 \text{ in.}^2$$

The area enclosed by the semicircle and a diameter is 39.25 in.^{2}

1. Find the area of a circle with diameter 6 in.

2. Find the area of a triangle with base 10 cm and height 5 cm.

ANSWERS: **1.** 28.26 in.^{2} **2.** 25 cm^2

Exercises G.2

1. Match each formula for area to its corresponding geometric figure.

———— (a) square   A. $A = lw$

———— (b) parallelogram   B. $A = bh$

———— (c) circle   C. $A = s^2$

———— (d) rectangle   D. $A = \pi r^2$

———— (e) trapezoid   E. $A = \dfrac{1}{2}bh$

———— (f) triangle   F. $A = \dfrac{1}{2}h(b + c)$

First sketch the figure and label the given parts; then find the area of each of the following figures.

2. A rectangle 35 in. long and 25 in. wide

3. A triangle with base 2 cm and altitude 6 cm

4. A triangle with base 5 mm and altitude 8 mm

5. A circle of radius 5 yd (Use $\pi = 3.14$.)

6. A circle of radius 1.5 ft (Use $\pi = 3.14$.)

7. A trapezoid with parallel sides of 3 in. and 10 in., and altitude of 35 in.

8. A trapezoid with parallel sides of 3.5 mm and 4.2 mm, and altitude of 1 cm

9. A parallelogram with altitude 10 cm to a base of 5 mm

Find the area of each of the following figures with the indicated dimensions. (Use $\pi = 3.14$.)

10.

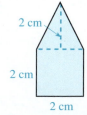

2 cm

2 cm

2 cm

11.

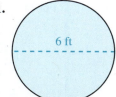

6 ft

12.

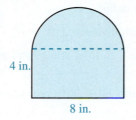

4 in.

8 in.

13.

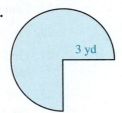

3 yd

14.

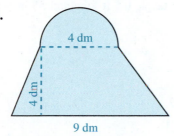

4 dm

4 dm

9 dm

15.

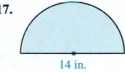

10 cm

3 cm

2 cm

10 cm

4 cm

16 cm

16.

3 cm

17.

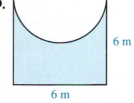

14 in.

18.

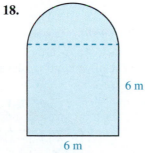

6 m

6 m

19.

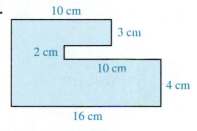

6 m

6 m

Find the area of the shaded portions in ares. (Use $\pi = 3.14$.)

20. 70 m

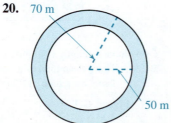

50 m

21.

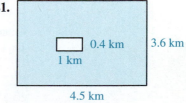

0.4 km

1 km

3.6 km

4.5 km

22. Find the area of a circle of radius 1 ft.

23. Find the area of a circle with diameter 1 ft.

24. Find the area of a rectangle 2 m long and 60 cm wide. Write your answer in both cm² and m².

25. Find the area of a square with sides of 50 cm. Write your answer in both cm² and m².

26. Find the area of a rectangle 0.5 ft long and 35 in. wide. Write your answer in both in.² and ft².

Volume

OBJECTIVES

1. Understand the concept of **volume.**
2. Learn how to apply the appropriate formula to find the volume of a given geometric figure.
3. Know how to label answers with the correct units.

The Concept of Volume

Volume is the measure of the space enclosed by a three-dimensional figure and is measured in **cubic units.** The concept of volume is illustrated in Figure G.2 in terms of cubic inches.

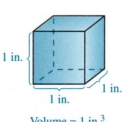

1 in.
1 in.
1 in.
Volume = 1 in.³
(or 1 cubic inch)

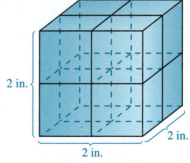

2 in.
2 in.
2 in.
Volume = length x height x width = 2 x 2 x 2 = 8 in.³
There are a total of 8 cubes that are each
1 in.³ for a total of 8 in.³

Figure G.2

In the metric system, some of the units of volume are cubic meters (m³), cubic decimeters (dm³), cubic centimeters (cm³), and cubic millimeters (mm³). In the U.S. customary system, some of the units of volume are cubic feet (ft³), cubic inches (in.³), and cubic yards (yd³).

Formulas for Volume

The following formulas for the volumes of the common geometric solids shown are valid regardless of the measurement system used. Always be sure to label your answers with the correct units.

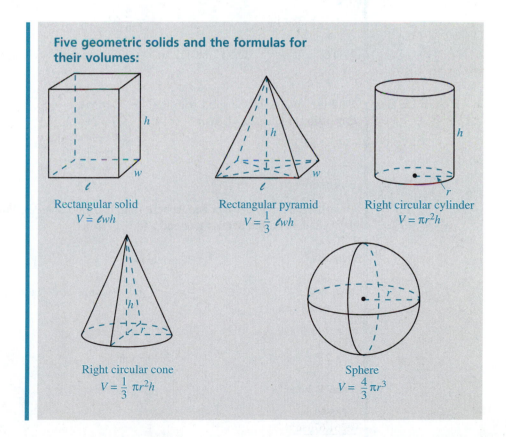

Five geometric solids and the formulas for their volumes:

Rectangular solid
$V = \ell wh$

Rectangular pyramid
$V = \dfrac{1}{3}\,\ell wh$

Right circular cylinder
$V = \pi r^2 h$

Right circular cone
$V = \dfrac{1}{3}\,\pi r^2 h$

Sphere
$V = \dfrac{4}{3}\,\pi r^3$

EXAMPLE 1

Find the volume of the rectangular solid with length 8 in., width 4 in., and height 1 ft. Write your answer in cubic inches and in cubic feet.

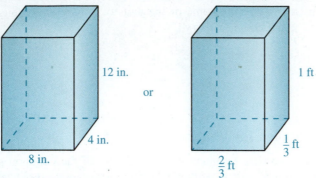

12 in.

or

1 ft

4 in.

$\frac{1}{3}$ ft

8 in.

$\frac{2}{3}$ ft

Solution

$$V = \ell wh$$

$$V = 8 \cdot 4 \cdot 12$$

$$= 384 \text{ in.}^3$$

$$V = \ell wh$$

$$V = \frac{2}{3} \cdot \frac{1}{3} \cdot 1$$

$$= \frac{2}{9} \text{ ft}^3$$

NOTE: $1 \text{ ft}^3 = (12 \text{ in.})(12 \text{ in.})(12 \text{ in.}) = 1728 \text{ in.}^3$ and $\frac{2}{9}(1728 \text{ in.}^3) = 384 \text{ in.}^3$

EXAMPLE 2

Find the volume of the solid with the dimensions indicated. (Use $\pi = 3.14$.)

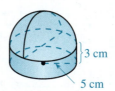

3 cm

5 cm

Solution

On top of the cylinder is a hemisphere (one-half of a sphere). Find the volume of the cylinder and the hemisphere and add the results.

Cylinder

$$V = \pi r^2 h$$

$$V = 3.14(5^2)(3)$$

$$= 235.5 \text{ cm}^3$$

Hemisphere

$$V = \frac{1}{2} \cdot \frac{4}{3}\pi r^3 \quad \text{(one-half volume of a sphere)}$$

$$V = \frac{2}{3}(3.14)(5^3)$$

$$= 261.67 \text{ cm}^3$$

Total Volume

235.50 cm³
+261.67 cm³
497.17 cm³ (or 497.17 mL) total volume

Exercises G.3

1. Match each formula for volume to its corresponding geometric figure.

_____ (a) rectangular solid

A. $V = \frac{4}{3}\pi r^3$

_____ (b) rectangular pyramid

B. $V = \frac{1}{3}\pi r^2 h$

_____ (c) right circular cylinder

C. $V = lwh$

_____ (d) right circular cone

D. $V = \pi r^2 h$

_____ (e) sphere

E. $V = \frac{1}{3}lwh$

Find the volume of each of the following solids in a convenient unit. (Use $\pi = 3.14$.)

2. A rectangular solid with length 5 in., width 2 in., and height 7 in.

3. A right circular cylinder 15 in. high and 1 ft in diameter.

4. A sphere with radius 4.5 cm.

5. A sphere with diameter 12 ft.

6. A right circular cone 3 dm high with a 2-dm radius.

7. A rectangular pyramid with length 8 cm, width 10 mm, and height 3 dm.

Find the volume of each of the following solids with the dimensions indicated.

8.

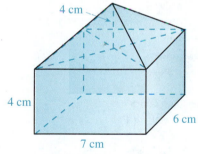

4 cm

4 cm

6 cm

7 cm

9.

12 dm

25 dm

10 dm

10.

8 cm

3 cm

11.

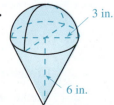

3 in.

6 in.

12.

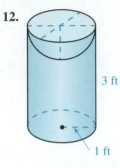

3 ft

1 ft

13.

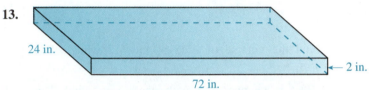

24 in.

2 in.

72 in.

G.4   Angles

OBJECTIVES

1. Know the definition of an **angle.**
2. Learn to use a protractor to measure angles.
3. Learn how to classify an angle by its measure as **acute, right, obtuse,** or **straight.**
4. Be able to determine when two angles are **complementary, supplementary,** or **equal** and know the meanings of these terms.
5. Know the meanings of the terms **vertical angles** and **adjacent angles.**

Measuring Angles

We begin the discussion of angles with two definitions using the undefined terms, **point** and **line.**

Term	Definition	Illustration
1. Ray	A **ray** consists of a point (called the **endpoint**) and all the points on a line on one side of that point.	ray $\overrightarrow{PQ}$ with endpoint P
2. Angle	An **angle** consists of two rays with a common endpoint (called a **vertex**).	$\angle AOB$ with vertex O

In an angle, the two rays are called the **sides** of the angle.

Every angle has a **measurement** or **measure** associated with it. Suppose that a circle is divided into 360 equal arcs. If two rays are drawn from the center of the circle through two consecutive points of division on the circle, then that angle is said to **measure one degree** (symbolized 1°). For example, in Figure G.3, a device called a protractor shows that the measure of $\angle AOB$ is 60 degrees. (We write m$\angle AOB = 60°$.)

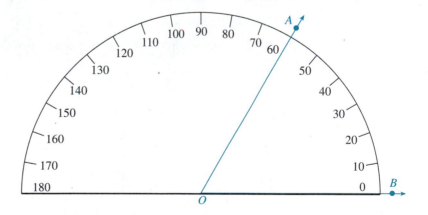

The protractor shows m $\angle AOB = 60°$

Figure G.3

To measure an angle with a protractor, lay the bottom edge of the protractor along one side of the angle with the vertex at the marked center point. Then read the measure from the protractor where the other side of the angle crosses it. (See Figure G.4.)

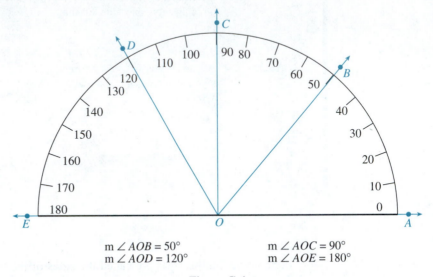

$$\text{m} \angle AOB = 50° \qquad\qquad \text{m} \angle AOC = 90°$$
$$\text{m} \angle AOD = 120° \qquad\qquad \text{m} \angle AOE = 180°$$

Figure G.4

Three common ways of labeling angles (Figure G.5) are:

1. Using three capital letters with the vertex as the middle letter.

2. Using single numbers such as 1, 2, 3.

3. Using the single capital letter at the vertex when the meaning is clear.

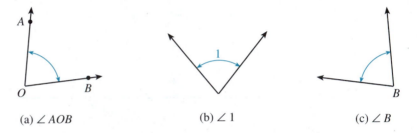

(a) $\angle AOB$ (b) $\angle 1$ (c) $\angle B$

Three ways of labeling angles

Figure G.5

Classifying Angles

Angles can be classified (or named) according to their measures.

> **Note:** We will use the two inequality symbols
>
> < (read "is less than")
>
> > (read "is greater than")
>
> to indicate the relative sizes of the measures of angles.

Types of Angles

Name	Measure	Illustration	
1. Acute	$0° < m\angle A < 90°$		$\angle A$ is an acute angle.
2. Right	$m\angle A = 90°$		$\angle A$ is a right angle.
3. Obtuse	$90° < m\angle A < 180°$		$\angle A$ is an obtuse angle.
4. Straight	$m\angle A = 180°$		The rays are in opposite directions. $\angle A$ is a straight angle.

The following figure is used for Examples 1 and 2.

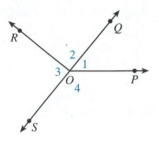

EXAMPLE 1 Use a protractor to check the measures of the following angles.

(a) m∠1 = 45°   (b) m∠2 = 90°   (c) m∠3 = 90°   (d) m∠4 = 135°

EXAMPLE 2 Tell whether each of the following angles is acute, right, obtuse, or straight.

(a) ∠1   (b) ∠2   (c) ∠POR

Solution

(a) ∠1 is acute since 0° < m∠1 < 90°.

(b) ∠2 is a right angle since m∠2 = 90°.

(c) ∠POR is obtuse since m∠POR = 45° + 90° = 135° > 90°.

Two angles are:

1. **complementary** if the sum of their measures is 90°.

2. **supplementary** if the sum of their measures is 180°.

3. **equal** if they have the same measure.

EXAMPLE 3 In the figure shown, (a) ∠1 and ∠2 are complementary since m∠1 + m∠2 = 90°; (b) ∠COD and ∠COA are supplementary since m∠COD + m∠COA = 70° + 110° = 180°; (c) ∠AOD is a straight angle since m∠AOD = 180°; (d) ∠BOA and ∠BOD are supplementary; and in this case m∠BOD = m∠BOD = 90°.

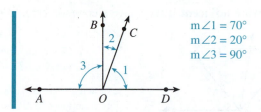

$$m\angle 1 = 70°$$
$$m\angle 2 = 20°$$
$$m\angle 3 = 90°$$

EXAMPLE 4

In the figure below, $\overleftrightarrow{PS}$ is a straight line and $m\angle QOP = 30°$. Find the measures of (a) $\angle QOS$ and (b) $\angle SOP$. (c) Are any pairs supplementary?

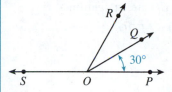

Solution

(a) $m\angle QOS = 150°$.

(b) $m\angle SOP = 180°$.

(c) Yes. $\angle QOP$ and $\angle QOS$ are supplementary and $\angle ROP$ and $\angle ROS$ are supplementary.

If two lines intersect, then two pairs of vertical angles are formed. **Vertical angles** are also called **opposite angles.** (See Figure G.6.)

$\angle 1$ and $\angle 3$ are vertical angles.
$\angle 2$ and $\angle 4$ are vertical angles.

Figure G.6

Vertical angles are equal. That is, **vertical angles have the same measure.** (In Figure G.6, $m\angle 1 = m\angle 3$ and $m\angle 2 = m\angle 4$.)

Two angles are **adjacent** if they have a common side. (See Figure G.7.)

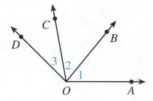

∠1 and ∠2 are adjacent. They have the common side $\overrightarrow{OB}$.

∠2 and ∠3 are adjacent. They have the common side $\overrightarrow{OC}$.

Figure G.7

EXAMPLE 5

In the figure below, $\overleftrightarrow{AC}$ and $\overleftrightarrow{BD}$ are straight lines.

(a) Name an angle adjacent to ∠EOD.

(b) What is m∠AOD?

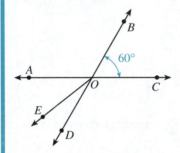

Solution

(a) There are three angles adjacent to ∠EOD:
∠AOE, ∠BOE, and ∠COD.

(b) Since ∠BOC and ∠AOD are vertical angles, they have the same measure. So, m∠AOD = 60°.

Exercises G.4

1. (a) If ∠1 and ∠2 are complementary, and m∠1 = 15°, what is m∠2? (b) If m∠1 = 3°, what is m∠2? (c) If m∠1 = 45°, what is m∠2? (d) If m∠1 = 75°, what is m∠2?

2. (a) If ∠3 and ∠4 are supplementary and m∠3 = 45°, what is m∠4? (b) If m∠3 = 90°, what is m∠4? (c) If m∠3 = 110°, what is m∠4? (d) If m∠3 = 135°, what is m∠4?

3. The supplement of an acute angle is an obtuse angle.

(a) What is the supplement of a right angle?

(b) What is the supplement of an obtuse angle?

(c) What is the complement of an acute angle?

4. In the figure shown, $\overleftrightarrow{DC}$ is a straight line and m∠$BOA = 90°$.

(a) What type of angle is ∠AOC?

(b) What type of angle is ∠BOC?

(c) What type of angle is ∠BOA?

(d) Name a pair of complementary angles.

(e) Name two pairs of supplementary angles.

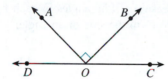

5. An **angle bisector** is a ray that divides an angle into two angles with equal measures. If $\overrightarrow{OX}$ bisects ∠COD and m∠$COD = 50°$, what is the measure of each of the equal angles formed?

Given m∠$AOB = 30°$ and m∠$BOC = 80°$ and $\overrightarrow{OX}$ and $\overrightarrow{OY}$ are angle bisectors, find the measures of the following angles.

6. ∠AOX **7.** ∠BOY **8.** ∠COX

9. ∠YOX **10.** ∠AOY **11.** ∠AOC

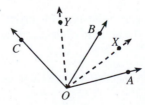

12. In the figure shown, (a) name all the pairs of supplementary angles. (b) Name all the pairs of complementary angles.

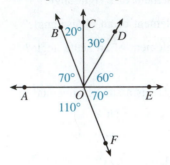

Use a protractor to measure all the angles in each figure. Each line segment may be extended as a ray to form the side of an angle.

13.

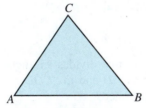

14.

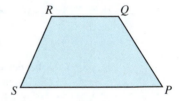

15.

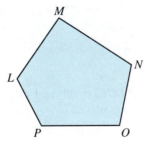

16.

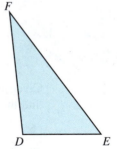

17. Name the type of angle formed by the hands on a clock (a) at six o'clock; (b) at three o'clock; (c) at one o'clock; (d) at five o'clock.

18. What is the measure of each angle formed by the hands of the clock in Exercise 17?

19. The figure below shows two intersecting lines.

   (a) If m∠1 = 30°, what is m∠2?

   (b) Is m∠3 = 30°? Give a reason for your answer other than the fact that ∠1 and ∠3 are vertical angles.

   (c) Name four pairs of adjacent angles.

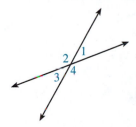

20. In the figure shown, $\overleftrightarrow{AB}$ is a straight line.

   (a) Name two pairs of adjacent angles.

   (b) Name two vertical angles if there are any.

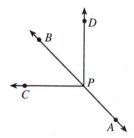

21. Given that m∠1 = 30° in the figure shown, find the measures of the other three angles.

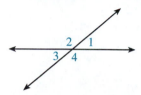

22. In the figure shown on the next page, ℓ, m, and n are straight lines with m∠1 = 20° and m∠6 = 90°.

   (a) Find the measures of the other four angles.

   (b) Which angle is supplementary to ∠6?

(c) Which angles are complementary to $\angle 1$?

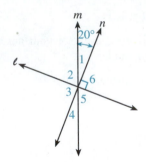

23. In the figure shown, $m\angle 2 = m\angle 3 = 40°$. Find all other pairs of angles that have equal measures.

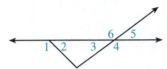

G.5   Triangles

OBJECTIVES

1. Learn how to classify a triangle by its sides: **scalene, isosceles,** or **equilateral.**
2. Learn how to classify a triangle by its angles: **acute, right,** or **obtuse.**
3. Know the following two important statements about any triangle:
   (a) The sum of the measures of the angles is 180°.
   (b) The sum of the lengths of any two sides must be greater than the length of the third side.
4. Know that in similar triangles,
   (a) corresponding angles have the same measure, and
   (b) corresponding sides are proportional.

Classifying Triangles

A **triangle** consists of the three line segments that join three points that do not lie on a straight line. The line segments are called the **sides** of the triangle, and the points are called the **vertices** of the triangle. If the points are labeled A, B, and C, the triangle is symbolized $\triangle ABC$. (See Figure G.8.)

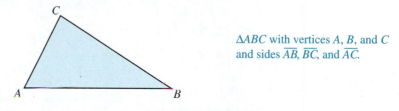

$\triangle ABC$ with vertices A, B, and C and sides $\overline{AB}$, $\overline{BC}$, and $\overline{AC}$.

Figure G.8

The sides of a triangle are said to determine three angles, and these angles are labeled by the vertices. Thus, the angles of $\triangle ABC$ are $\angle A$, $/B$, and $/C$. (Since the definition of angle involves rays, we can think of the sides of the triangle extended as rays to form these angles.)

Triangles are classified in two ways:

1. according to lengths of their sides, and

2. according to the measures of their angles.

The corresponding names and properties are listed in the following tables.

Note: We will indicate the length of a line segment, such as $\overline{AB}$, by writing only the letters, such as AB.

Triangles Classified by Sides

Name	Property	Illustration
1. Scalene	No two sides are equal.	$\triangle ABC$ is scalene since no two sides are equal.

continued

2. Isosceles   At least two sides are equal. $\triangle PQR$ is isosceles since $PR = QR$.

3. Equilateral   All three sides are equal. $\triangle XYZ$ is equilateral since $XY = XZ = YZ$.

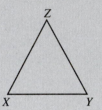

Triangles Classified by Angles

Name	Property	Illustration
1. Acute	All three angles are acute.	$\angle A$, $\angle B$, $\angle C$ are all acute, so $\triangle ABC$ is acute.

Name	Property	Illustration
2. Right	One angle is a right angle.	$m\angle P = 90°$ so $\triangle PQR$ is a right triangle.

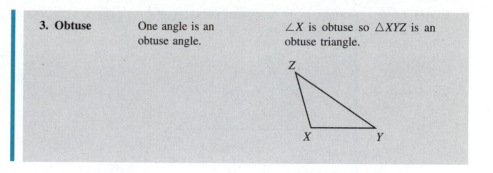

3. Obtuse	One angle is an obtuse angle.	$\angle X$ is obtuse so $\triangle XYZ$ is an obtuse triangle.

Every triangle is said to have six parts, namely three angles and three sides. Two sides of a triangle are said to **include** the angle at their common endpoint or vertex. The third side is said to be **opposite** this angle.

The sides in a right triangle have special names. The longest side, opposite the right angle, is called the **hypotenuse** and the other two sides are called **legs.** (See Figure G.9.)

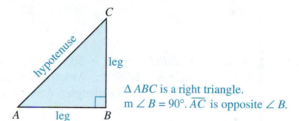

$\triangle ABC$ is a right triangle.
$m \angle B = 90°$. $\overline{AC}$ is opposite $\angle B$.

Figure G.9

Two important statements can be made about any triangle:

1. The sum of the measures of the angles is 180°.

2. The sum of the lengths of any two sides must be greater than the length of the third side.

EXAMPLE 1

In the figure of $\triangle ABC$, $AB = AC$. What kind of triangle is $\triangle ABC$?

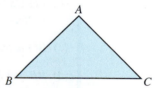

Solution

$\triangle ABC$ is isosceles because two sides are equal.

EXAMPLE 2

Suppose the lengths of the sides of △*PQR* are stated as shown in the figure below. Is this possible?

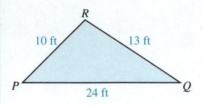

Solution

This is not possible because *PR* + *QR* = 10 ft + 13 ft = 23 ft and *PQ* = 24 ft, which is longer than the sum of the other two sides. In any triangle, the sum of the lengths of any two sides must be greater than the length of the third side.

EXAMPLE 3

In the figure △*BOR*, m∠*B* = 50° and m∠*O* = 70°. (a) What is m∠*R*? (b) What kind of triangle is △*BOR*? (c) Which side is opposite ∠*R*? (d) Which sides include ∠*R*? (e) Is △*BOR* a right triangle? Why or why not?

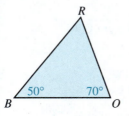

Solution

(a) The sum of the measures of the angles must be 180°. Since 50° + 70° = 120°,

   m∠*R* = 180° − 120° = 60°

(b) △*BOR* is an acute triangle since all the angles are acute. Also, △*BOR* is scalene because no two sides are equal.

(c) $\overline{BO}$ is opposite ∠*RO*.

(d) $\overline{RB}$ and $\overline{RO}$ include ∠*R*.

(e) △*BOR* is not a right triangle because none of the angles is a right angle.

Similar Triangles

On an intuitive level, two triangles are said to be **similar** if they have the same "shape." They may or may not have the same "size." More formally (and more usefully), similar triangles have the following two important properties.

Two triangles are similar if:

1. their **corresponding angles are equal.**

2. their **corresponding sides are proportional.** (See Figure G.10.)

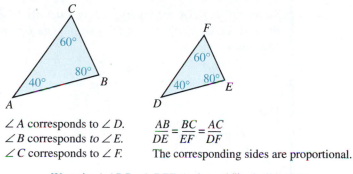

∠ A corresponds to ∠ D.
∠ B corresponds to ∠ E.
∠ C corresponds to ∠ F.

$$\frac{AB}{DE} = \frac{BC}{EF} = \frac{AC}{DF}$$

The corresponding sides are proportional.

We write $\triangle ABC \sim \triangle DEF$. (~ is read "is similar to.")

Figure G.10

EXAMPLE 4

Consider the two triangles $\triangle ABC$ and $\triangle AXY$ as shown with m∠ ABC = m∠ AXY = 90°.

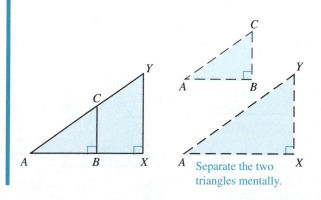

Separate the two triangles mentally.

Solution

We can show that the corresponding angles are equal as follows:

m∠ *CAB* = m∠ *YAX* because they are the same angle.

m∠ *CBA* = m∠ *YXA* because both are right angles (90°).

m∠ *BCA* = m∠ *XYA* because the sum of the measures of the angles in each triangle must be 180°.

Therefore, the corresponding angles are equal, and the triangles are similar.

EXAMPLE 5

Refer to the figure used in Example 4. If $AB = 4$ cm, $AX = 8$ cm, and $BC = 3$ cm, find XY.

Solution

Since $\overline{AB}$ and $\overline{AX}$ are corresponding sides (they are opposite equal angles) and $\overline{BC}$ and $\overline{XY}$ are corresponding sides (they are opposite equal angles), we have the proportion:

$$\frac{AB}{AX} = \frac{BC}{XY}$$

Thus,

$$\frac{4 \text{ cm}}{8 \text{ cm}} = \frac{3 \text{ cm}}{XY}$$

$$4 \cdot XY = 3 \cdot 8$$

$$\frac{\cancel{4} \cdot XY}{\cancel{4}} = \frac{24}{4}$$

$$XY = 6 \text{ cm}$$

Exercises G.5

Name each of the following triangles with the indicated measures of angles and lengths of sides.

1.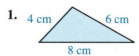
4 cm   6 cm   8 cm

2.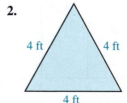
4 ft   4 ft   4 ft

3.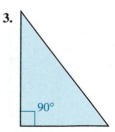
90°

4.

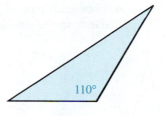

110°

5.

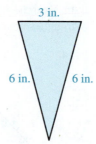

3 in.

6 in.   6 in.

6.

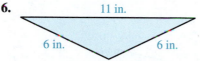

11 in.

6 in.   6 in.

7.

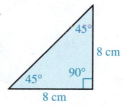

45°

8 cm

45°   90°

8 cm

8.

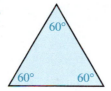

60°

60°   60°

9.

80°   40°

In Exercises 10–13, assume that the given triangles are similar and find the values for x and y.

10.

60° x

y 50°

70°

70°

11.

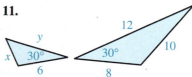

12

y

x 30°   30°   10

6   8

12.

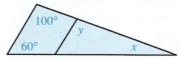

100°

y

60° x

13.

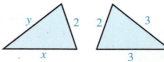

y 2   2   3

x 3

In Exercises 14–17, determine whether each pair of triangles is similar. If the triangles are similar, explain why and indicate the similarity by using the ~ symbol.

14.

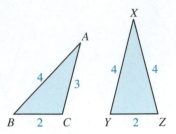

15.

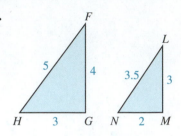

16.

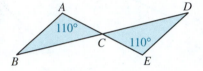

17.

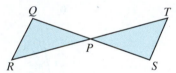

18. In the figure below, m∠Q = m∠S. Is △PQR ~ △PST? Explain your reasoning.

19. In △XYZ and △UVW below, m∠Z = 30° and m∠W = 30°.

(a) If both triangles are isosceles, what are the measures of the other four angles? (In an isosceles triangle, the angles opposite the equal sides must be equal.)

(b) Are the triangles similar? Explain.

20. The Pythagorean Theorem (see Section G.6) states that in a right triangle, the square of the hypotenuse is equal to the sum of the squares of the two legs.

(a) Are the triangles △ABC and △DEF right triangles? Explain.

(b) If so, which sides are the hypotenuses?

(c) Are the triangles similar? Explain.

(d) If so, which angles are equal?

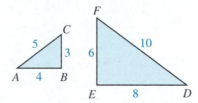

Writing and Thinking about Mathematics

21. (a) Is it possible to form a triangle, $\triangle STV$, if $ST = 12$ centimeters, $TV = 9$ centimeters, and $SV = 15$ centimeters? Explain your reasoning.

(b) If so, what kind of triangle is $\triangle STV$?

(c) Use a ruler to draw this triangle, if possible.

G.6 Square Roots and the Pythagorean Theorem

OBJECTIVES

1. Memorize the squares of the whole numbers from 1 to 20.
2. Memorize the perfect square whole numbers from 1 to 400.
3. Understand the terms **square root, radical sign, radicand,** and **radical.**
4. Know how to use a calculator to find square roots.
5. Be able to simplify radical expressions.
6. Understand the terms **right triangle, hypotenuse,** and **leg.**
7. Know and be able to use the Pythagorean Theorem.

Square Roots and Irrational Numbers

A number is **squared** when it is multiplied by itself. If a whole number is squared, the result is called a **perfect square.** For example, squaring 7 gives $7^2 = 49$, and 49 is a perfect square. Table G.1 shows the perfect square numbers found by squaring the whole numbers from 1 to 20. A more complete table (of powers, roots, and prime factorizations) is located in the back of the book.

Table G.1 Squares of Whole Numbers from 1 to 20			
$1^2 = 1$	$6^2 = 36$	$11^2 = 121$	$16^2 = 256$
$2^2 = 4$	$7^2 = 49$	$12^2 = 144$	$17^2 = 289$
$3^2 = 9$	$8^2 = 64$	$13^2 = 169$	$18^2 = 324$
$4^2 = 16$	$9^2 = 81$	$14^2 = 196$	$19^2 = 361$
$5^2 = 25$	$10^2 = 100$	$15^2 = 225$	$20^2 = 400$

Since $5^2 = 25$, the number 5 is called the **square root** of 25. We write $\sqrt{25} = 5$. Similarly,

$$\sqrt{49} = 7 \quad \text{since} \quad 7^2 = 49$$

$$\sqrt{1} = 1 \quad \text{since} \quad 1^2 = 1$$

$$\sqrt{0} = 0 \quad \text{since} \quad 0^2 = 0$$

The symbol $\sqrt{}$ is called a **radical sign.**

The number under the radical sign is called the **radicand.**

The complete expression, such as $\sqrt{25}$, is called a **radical.**

Table G.2 contains the square roots of the perfect square numbers from 1 to 400. (Notice that Table G.2 is just another way of looking at Table G.1.) Both tables should be memorized.

Table G.2 Square Roots of Perfect Squares from 1 to 400			
$\sqrt{1} = 1$	$\sqrt{36} = 6$	$\sqrt{121} = 11$	$\sqrt{256} = 16$
$\sqrt{4} = 2$	$\sqrt{49} = 7$	$\sqrt{144} = 12$	$\sqrt{289} = 17$
$\sqrt{9} = 3$	$\sqrt{64} = 8$	$\sqrt{169} = 13$	$\sqrt{324} = 18$
$\sqrt{16} = 4$	$\sqrt{81} = 9$	$\sqrt{196} = 14$	$\sqrt{361} = 19$
$\sqrt{25} = 5$	$\sqrt{100} = 10$	$\sqrt{225} = 15$	$\sqrt{400} = 20$

The square roots of some numbers are not found as easily as those in the tables. In fact, most square roots are irrational numbers (**nonrepeating infinite decimals**); that is, most square roots can only be approximated with decimals.

Decimal approximations of $\sqrt{2}$ are given here to emphasize this idea and to help you understand that there is no finite decimal number whose square is 2.

$$
\begin{array}{r}
1.4 \\
\times\ 1.4 \\
\hline
56 \\
1\ 4 \\
\hline
1.96
\end{array}
\qquad
\begin{array}{r}
1.414 \\
\times\quad 1.414 \\
\hline
5656 \\
1414 \\
5656 \\
1\ 414 \\
\hline
1.999396
\end{array}
\qquad
\begin{array}{r}
1.41421 \\
\times\qquad 1.41421 \\
\hline
141421 \\
282842 \\
565684 \\
141421 \\
565684 \\
1\ 41421 \\
\hline
1.9999899241
\end{array}
\qquad
\begin{array}{r}
1.41422 \\
\times\qquad 1.41422 \\
\hline
282844 \\
282844 \\
565688 \\
141422 \\
565688 \\
1\ 41422 \\
\hline
2.0000182084
\end{array}
$$

So, $\sqrt{2}$ is between 1.41421 and 1.41422.

To find (or approximate) a square root with a calculator, use the key labeled $\boxed{\sqrt{x}}$.

EXAMPLE 1   Use a calculator to find $\sqrt{2}$ accurate to nine decimal places.

Solution

STEP 1:   Enter 2.

STEP 2:   Press the key $\boxed{\sqrt{x}}$.

STEP 3:   The display will show 1.414213562.

Similarly, with a calculator,

   $\sqrt{3} = 1.732050808$ accurate to nine decimal places

and

   $\sqrt{5} = 2.236067977$ accurate to nine decimal places

Simplifying Radical Expressions

The square roots just discussed and many others are irrational numbers, and their decimal forms are infinite nonrepeating decimals. However, there are situations, particularly in algebra, in which we are not interested in the decimal representation of a square root, but in the simplified form of a radical expression. To simplify radicals in general, we need the following property of square roots.

> **Property of Square Roots**
>
> For nonnegative numbers a and b,
> $$\sqrt{ab} = \sqrt{a}\sqrt{b}$$

With this property, we can simplify a radical expression by finding a perfect square factor. For example,

$$\sqrt{18} = \sqrt{9 \cdot 2} = \sqrt{9} \cdot \sqrt{2} = 3\sqrt{2} \quad \text{(9 is a square factor.)}$$

The expression $3\sqrt{2}$ is simplified because the radicand, 2, has no perfect square factor. That is, **a radical is considered to be in simplest form when the radicand has no square number factor.**

EXAMPLE 2 Simplify $\sqrt{50}$.

Solution

$$\sqrt{50} = \sqrt{25 \cdot 2} = \sqrt{25} \cdot \sqrt{2} = 5\sqrt{2} \quad \text{(25 is a perfect square factor.)}$$

EXAMPLE 3 Simplify $\sqrt{75}$.

Solution

$$\sqrt{75} = \sqrt{25 \cdot 3} = \sqrt{25} \cdot \sqrt{3} = 5\sqrt{3}$$

EXAMPLE 4 Simplify $\sqrt{450}$.

Solution

$$\sqrt{450} = \sqrt{9 \cdot 50} = \sqrt{9 \cdot 25 \cdot 2} = \sqrt{9} \cdot \sqrt{25} \cdot \sqrt{2} = 3 \cdot 5\sqrt{2} = 15\sqrt{2}$$

or

$$\sqrt{450} = \sqrt{225 \cdot 2} = \sqrt{225} \cdot \sqrt{2} = 15\sqrt{2}$$

The Pythagorean Theorem

The following discussion involving right triangles serves as an application of squares and square roots and uses the terms defined below.

Right triangle: A triangle with a right angle (90°)

Hypotenuse: The longest side of a right triangle; the side opposite the right angle

Leg: Each of the other two sides of a right triangle

Pythagoras (c. 585–502 B.C.), a famous Greek mathematician, is given credit for discovering the following theorem (even though historians have found that the facts of the theorem were known before the time of Pythagoras).

The Pythagorean Theorem

In a right triangle, the square of the hypotenuse is equal to the sum of the squares of the two legs:

$$c^2 = a^2 + b^2$$

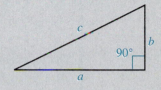

EXAMPLE 5

Show that a triangle with sides of 3 inches, 4 inches, and 5 inches must be a right triangle.

Solution

If the triangle is a right triangle, then the sides must satisfy the property stated in the Pythagorean Theorem: $c^2 = a^2 + b^2$. Or, in this case, $5^2 = 3^2 + 4^2$. Since $25 = 9 + 16$ is a true statement, the triangle is a right triangle.

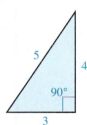

EXAMPLE 6

Find the length of the hypotenuse of a right triangle with legs of length 12 cm and 5 cm.

Solution

Let c = the length of the hypotenuse. Now, by the Pythagorean Theorem,

$$c^2 = 12^2 + 5^2$$
$$c^2 = 144 + 25$$
$$c^2 = 169$$
$$c = \sqrt{169} = 13$$

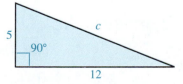

The length of the hypotenuse is 13 centimeters.

In most right triangles, all three sides do not have whole number values. Examples 7 and 8 illustrate right triangles with irrational numbers as the lengths of the hypotenuses.

EXAMPLE 7 | Find the length of the hypotenuse of a right triangle in which both legs have length 1 meter.

Solution

$$c^2 = 1^2 + 1^2$$
$$c^2 = 1 + 1 = 2$$
$$c = \sqrt{2}$$

The length of the hypotenuse is $\sqrt{2}$ meters (or about 1.41 meters).

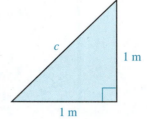

EXAMPLE 8 | A guy wire is attached to the top of a telephone pole and anchored in the ground 10 feet from the base of the pole. If the pole is 20 feet high, what is the length of the guy wire?

Solution

Let x = the length of the guy wire.
Then, by the Pythagorean Theorem,

$$x^2 = 10^2 + 20^2$$
$$x^2 = 100 + 400$$
$$x^2 = 500$$
$$x = \sqrt{500} = \sqrt{100 \cdot 5} = \sqrt{100} \cdot \sqrt{5} = 10\sqrt{5}$$

The guy wire is $10\sqrt{5}$ feet long (or about 22.4 feet long).

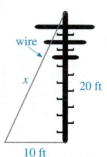

Exercises G.6

State whether or not each number is a perfect square in Exercises 1–10.

1. 144	**2.** 169	**3.** 81	**4.** 16	**5.** 400
6. 225	**7.** 242	**8.** 48	**9.** 45	**10.** 40

11. Show by squaring 1.732 and 1.733 that $\sqrt{3}$ is between these two numbers.

12. Show by squaring 2.236 and 2.237 that $\sqrt{5}$ is between these two numbers.

Simplify the radical expressions in Exercises 13–36.

13. $\sqrt{12}$ **14.** $\sqrt{28}$ **15.** $\sqrt{24}$

16. $\sqrt{32}$ **17.** $\sqrt{48}$ **18.** $\sqrt{288}$

19. $\sqrt{363}$ **20.** $\sqrt{242}$ **21.** $\sqrt{500}$

22. $\sqrt{300}$ **23.** $\sqrt{128}$ **24.** $\sqrt{125}$

25. $\sqrt{72}$ **26.** $\sqrt{98}$ **27.** $\sqrt{605}$

28. $\sqrt{150}$ **29.** $\sqrt{169}$ **30.** $\sqrt{196}$

31. $\sqrt{800}$ **32.** $\sqrt{80}$ **33.** $\sqrt{90}$

34. $\sqrt{40}$ **35.** $\sqrt{256}$ **36.** $\sqrt{361}$

Use the Pythagorean Theorem to determine whether or not each of the triangles in Exercises 37 and 38 is a right triangle.

37.
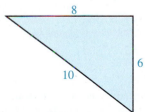

38.
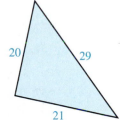

Find the length of the hypotenuse of each of the following right triangles.

39.
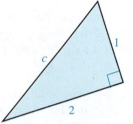

40.
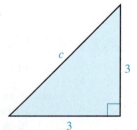

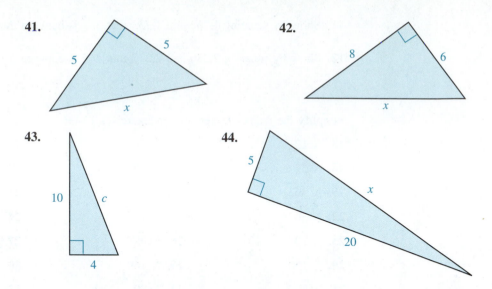

41.

42.

43.

44.

First mentally estimate the value of each of the following radicals and write down your estimate. Then use a calculator to find the value accurate to four decimal places.

45. $\sqrt{8}$ **46.** $\sqrt{34}$ **47.** $\sqrt{45}$

48. $\sqrt{10}$ **49.** $\sqrt{40}$ **50.** $\sqrt{20}$

51. $\sqrt{75}$ **52.** $\sqrt{200}$ **53.** $\sqrt{300}$

54. $\sqrt{80}$ **55.** $\sqrt{500}$ **56.** $\sqrt{95}$

57. The base of a fire-engine ladder is 20 feet from a building and reaches to a fourth floor window 60 feet above ground level. How long is the ladder extended (to the nearest tenth of a foot)?

58. A forestay that helps to support a ship's mast reaches from the top of a mast 20 meters high to a point on the deck 10 meters from the base of the mast. What is the length of the stay (to the nearest tenth of a meter)?

59. The Xerox Center building in Chicago is 500 feet tall and at a certain time of day it casts a shadow that is 150 feet long. At that time of day, what is the distance (to the nearest tenth of a foot) from the tip of the shadow to the top of the Xerox building?

60. If an airplane passes directly over your head at an altitude of 1 mile, how far is the airplane from your position (to the nearest tenth of a mile) after it has flown 4 miles farther at the same altitude?

61. The shape of a baseball infield is a square with each side 90 feet long. (a) Find the distance (to the nearest tenth of a foot) from home plate to second base. (b) The diagonals of the square intersect halfway from home plate to second base. If the pitcher's mound is $60\frac{1}{2}$ feet from home plate, is the pitcher's mound closer to home plate or to second base? (See the figure on page G2.)

62. A square is said to be inscribed in a circle if each corner of the square lies on the circle. (a) Find the circumference and area of a circle with diameter 20 mm. (b) Find the perimeter and area of a square inscribed in the circle.

63. Before painting a picture on canvas, an artist must stretch the canvas on a rectangular wooden frame. To be sure that the corners of the canvas are true right angles, the artist can measure the diagonals of the stretched canvas. What should be the diagonal measure, to the nearest tenth of an inch, of a canvas whose sides are 24 inches and 38 inches in length?

64. While installing windows in a new home, a builder measures the diagonals of rectangular window casements to verify that their corners are true right angles. What should be the diagonal measure, to the nearest tenth of an inch, of a window casement with dimensions 36 inches by 54 inches?

65. To create a square inside a square, a quilting pattern requires four triangular pieces like the one shaded in the figure shown here. If the square in the center measures 10 cm on a side, and the two legs of each triangle must be of equal length, how long are the legs of each triangle, to the nearest tenth of a centimeter?

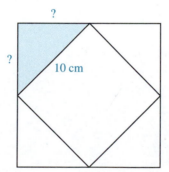

66. The shape of home plate in baseball can be created by cutting off two triangular corners from a square as in the figure shown here. If each of the triangles has a hypotenuse of 12 inches and legs of equal length, what is the length of one side of the original square, to the nearest tenth of an inch?

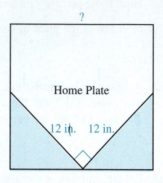

Collaborative
Learning Exercise

67. In algebra, the fraction $\frac{1}{2}$ is used as an exponent to indicate square root. That is, we can write $\sqrt{a} = a^{\frac{1}{2}}$. Similarly, a cube root is indicated as $\sqrt[3]{a} = a^{\frac{1}{3}}$ and a fourth root is indicated as $\sqrt[4]{a} = a^{\frac{1}{4}}$.

With the class separated into teams of two to four, each team is to write what they think is a good definition of the terms "cube root" and "fourth root." With a calculator the teams are to find both the cube root and fourth root (accurate to four decimal places) of each of the following numbers:

(a) 8   (b) 16   (c) 27   (d) 125

The team leader is to read the results, with classroom discussion to follow.

Chapter Summary

Key Terms and Ideas

Formulas

Figure	Perimeter	Area
Square	$P = 4s$	$A = s^2$
Rectangle	$P = 2l + 2w$	$A = lw$
Parallelogram	$P = 2b + 2a$	$A = bh$
Triangle	$P = a + b + c$	$A = \frac{1}{2}bh$
Trapezoid	$P = a + b + c + d$	$A = \frac{1}{2}h(b + c)$
Circle	$C = \pi d = 2\pi r$	$A = \pi r^2$

Formulas

Figure	Volume
Rectangular solid	$V = lwh$
Rectangular pyramid	$V = \frac{1}{3}lwh$
Right circular cylinder	$V = \pi r^2 h$
Right circular cone	$V = \frac{1}{3}\pi r^2 h$
Sphere	$V = \frac{4}{3}\pi r^3$

An **angle** consists of two rays with a common endpoint (called a vertex). **Adjacent** angles have a common side.

Classification of Angles

1. Acute $0° < m\angle A < 90°$

2. Right $m\angle A = 90°$

3. Obtuse $90° < m\angle A < 180°$

4. Straight $m\angle A = 180°$

Two angles are:

1. complementary if the sum of their measures is 90°.

2. supplementary if the sum of their measures is 180°.

3. equal if they have the same measure.

Vertical angles (or **opposite angles**) are equal.

Triangles can be classified by sides and by angles.

By Sides	By Angles
1. Scalene	1. Acute
2. Isosceles	2. Right
3. Equilateral	3. Obtuse

Rules and Properties

1. The sum of the measures of the angles of any triangle is 180°.

2. The sum of the lengths of any two sides of a triangle must be greater than the length of the third side.

Two triangles are similar if:

1. their corresponding angles are equal.

2. their corresponding sides are proportional.

Property of Square Roots

For nonnegative numbers a and b,

$$\sqrt{ab} = \sqrt{a}\sqrt{b}$$

Pythagorean Theorem

In a right triangle the square of the hypotenuse is equal to the sum of the squares of the two legs:

$$c^2 = a^2 + b^2$$

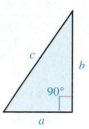

Chapter Review Questions

Find (a) the perimeter and (b) the area of each of the following figures. (Use $\pi = 3.14$.)

1.

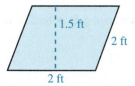

6 in.   4 in.   8 in.   12 in.

2.

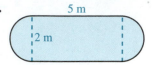

5 m   2 m

3.

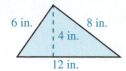

1.5 ft   2 ft   2 ft

4.

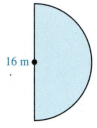

16 m

5. Find the area of a square with sides 3.5 mm.

6. Find the area of a circle with radius 14 cm. (Use $\pi = 3.14$.)

7. Find the area of a trapezoid with parallel sides of 7 in. and 15 in., and altitude of 1 ft.

8. What is the perimeter of a parallelogram with one side 3 feet long and another side 2.5 feet long?

9. What is the volume of a sphere with radius 10 cm? (Use $\pi = 3.14$.)

10. What is the volume of a right circular cone of height 10 cm and diameter 10 cm? (Use $\pi = 3.14$.)

Find the volume of each of the following solids with the dimensions indicated. (Use $\pi = 3.14$.)

11.

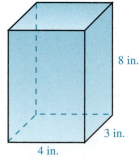

8 in.   3 in.   4 in.

12.

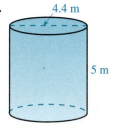

4.4 m   5 m

13. In the figure, $\overleftrightarrow{AC}$ and $\overleftrightarrow{BE}$ are straight lines.

(a) Name an angle complementary to $\angle AOB$.

(b) Name two vertical angles with measure 120°.

(c) Name an angle equal to $\angle COD$.

(d) Name an angle that is adjacent to $\angle DOE$.

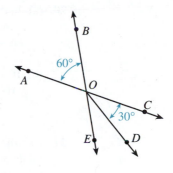

14. In the figure, $\overleftrightarrow{KN}$ is a straight line.

(a) What is the measure of $\angle KOM$?

(b) What type of angle is $\angle KOM$?

(c) What angle is complementary to $\angle MON$?

(d) Which angles are vertical angles?

(e) Which angles are right angles?

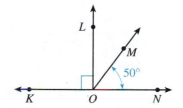

15. Name the type of angle formed by the hands on a clock (a) at two o'clock, (b) at nine o'clock, (c) at four o'clock, (d) at six o'clock.

16. Given that m$\angle 1$ = m$\angle 3$ and that m$\angle 1$ = 70°, find the measures of the other three angles in the figure shown.

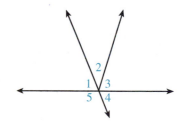

Name each of the following triangles with the indicated measures of angles and lengths of sides.

17.

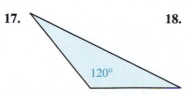

18.

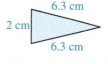

19.

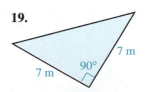

20. (a) In the figure to the right, is $\triangle ABC \sim$ $\triangle ADE$? Give reasons for your answer.

(b) If $AE = 10$ in., what is the length of AC?

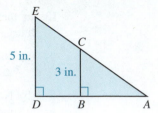

21. Use the Pythagorean Theorem to determine whether or not $\triangle PQR$ is a right triangle.

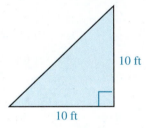

22. Find the length of the hypotenuse in the right triangle shown.

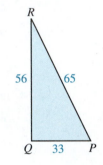

Simplify each of the following radical expressions.

23. $\sqrt{75}$ **24.** $\sqrt{121}$ **25.** $\sqrt{600}$ **26.** $\sqrt{5000}$

Chapter Test

1. In the figure shown below, $\overleftrightarrow{AD}$ and $\overleftrightarrow{BE}$ are straight lines.

   (a) What type of angle is $\angle BOC$?

   (b) What type of angle is $\angle AOE$?

   (c) Name a pair of vertical angles.

   (d) Name two pairs of supplementary angles.

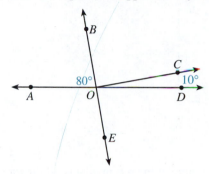

2. (a) If $\angle 1$ and $\angle 2$ are complementary and $m\angle 1 = 35°$, what is $m\angle 2$?

   (b) If $\angle 3$ and $\angle 4$ are supplementary and $m\angle 3 = 15°$, what is $m\angle 4$?

3. Find (a) the circumference and (b) the area of the following circle. (Use $\pi = 3.14$.)

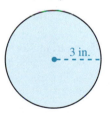

4. Find (a) the perimeter and (b) the area of the trapezoid with dimensions shown here.

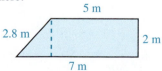

5. (a) Find the perimeter of a rectangle that is $8\frac{1}{2}$ inches wide and $11\frac{2}{3}$ inches long. (b) Find the area of the rectangle.

6. Find the area of a circle that has a diameter of 14 inches.

7. Find the area of a right triangle if its legs are 4 centimeters and 5 centimeters long.

8. Find the volume of the cylinder shown here with the given dimensions.

10 cm

5 cm

9. Find the volume of the rectangular solid shown here with the given dimensions.

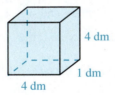

4 dm

1 dm

4 dm

10. Name the type of each of the following triangles based on the measures shown.

(a)

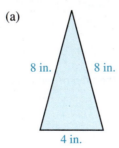

8 in.   8 in.

4 in.

(b)

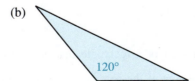

120°

(c)

3 cm   6 cm

7 cm

11. (a) Find the measure of ∠ x.

(b) What kind of triangle is △RST?

(c) Which side is opposite ∠ S?

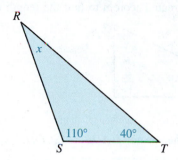

R

x

110°   40°

S   T

12. Find the value of x given that △ABC ~ △ADE.

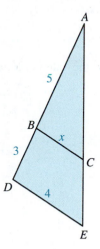

A

5

B

x

3

C

D   4

E

Simplify the following radicals.

13. $\sqrt{20}$ 14. $\sqrt{45}$

15. $\sqrt{68}$ 16. $\sqrt{700}$

17. $\sqrt{900}$ 18. $\sqrt{338}$

19. Use the Pythagorean Theorem to determine whether or not the triangle is a right triangle.

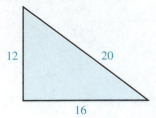

20. Use the Pythagorean Theorem to find the length of the hypotenuse.

21. Determine whether or not $\triangle AOB \sim \triangle COD$. Explain your reasoning.

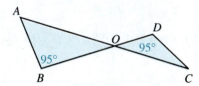

22. Given that $\triangle ABC \sim \triangle ADE$, find the value of x and the value of y.

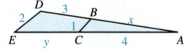

1. Find the prime factorization for each of the following numbers.

   (a) 150   (b) 263   (c) 104

2. Find the LCM of 15, 50, and 70.

3. Evaluate the expression $(18 + 3^2 \cdot 2) \div 6 - 4$.

4. Write $5\frac{1}{2}\%$ as a decimal.

5. Write 6.9 as a percent.

Perform the indicated operations for Exercises 6–13. Reduce all fractions to lowest terms.

6. $\dfrac{3}{4} + \dfrac{5}{8} + \dfrac{1}{10}$ 7. $5 - \dfrac{7}{10}$

8. $2\frac{1}{2} \cdot 1\frac{1}{3} \cdot 7\frac{1}{2}$ 9. $\dfrac{13}{14} \div \dfrac{39}{49}$

10. $800(8000)$ 11. $304 - 7.013$

12. $0.8336 \div 0.16$

13. $(3.05)(0.66)$

14. 75% of 96 is _____.

15. 75% of 960 is _____.

16. 75% of 9600 is _____.

17. 75% of 96,000 is _____.

18. 20% of _____ is 13.

19. _____% of 92 is 13.8.

20. Solve for x: $\dfrac{x}{1.7} = \dfrac{4.6}{7.82}$.

21. Solve for x: $\dfrac{4\frac{1}{2}}{x} = \dfrac{9}{10}$.

22. If interest is compounded daily at 7% for five years, what is the future value of $10,000?

23. (a) If your car expenses in one month were $375 for the loan payment, $60 for insurance, and $59 for gasoline, what were your total car expenses? (b) If your income was $1630, what percent of your income was used for car expenses?

Change the following metric units as indicated.

24. 23 g = _____ mg

25. 6 m = _____ mm

26. 5.6 km = _____ m

27. 3.8 L = _____ mL

28. 9.11 mm^2 = _____ cm^2

29. 4500 cm^3 = _____ m^3

30. 4.4 dm^2 = _____ cm^2 = _____ mm^2

31. Find the average (mean) of the numbers 5000, 6250, 9475, and 8672.

Perform the indicated operation and reduce all fractions.

32. $\dfrac{16}{0}$

33. $\dfrac{0}{-3}$

34. $(-500)(7000)$

35. $(-2.4)(-6.1)$

36. $(-27) + (-10) + (-5)$

37. $(15 - 17)(21 - 32)$

38. $(-7)^2 \cdot 5 + 2.1(-6) \div \dfrac{1}{2}$

39. $\dfrac{1}{2} + \dfrac{2}{3} \cdot \left(\dfrac{1}{2}\right)^2 - \dfrac{9}{10}$

40. Find (a) the circumference and (b) the area of a circle with diameter 14 cm. (Use $\pi = 3.14$.)

41. Find (a) the perimeter and (b) the area of a rectangle with length 2 m and width 35 cm.

42. Find the volume of a right circular cylinder 18 in. high and 1 ft in diameter.

43. If $\triangle ABC \sim \triangle DEC$ and $m\angle A = 65°$
and $m\angle B = 45°$, find the measures of
the other angles in the figure.

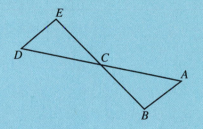

44. A right triangle has legs of 4.2 m and 0.6 m. What is the length of the
hypotenuse?

Solving

Equations

Mathematics at Work!

Scientists routinely make use of formulas in their work.

Formulas are mathematical tools used in almost all fields of endeavor. For example, there are formulas for interest in finance, standard deviation in statistics, plant growth in biology, dosages in medicine, perimeter, area, and volume in geometry, mixtures in chemistry, force in physics, electrical impulses in electronics and computer science, navigation on the oceans and in the air.

A formula is a general statement that relates two or more variables. While many formulas may already be familiar to you, there are situations where the recognized form that is usually used may not be the most convenient. In these cases, you should solve for one of the other variables in the formula.

For example, consider the formula for simple interest, $I = Prt$. This formula may be solved for I (interest); when P, r, and t have known values, we simply substitute these values into the formula and multiply. However, if someone wants to know how much time (in years) is needed to earn various amounts of interest, the formula may be more useful in the following form (solved for t): $t = \dfrac{I}{Pr}$.

Suppose \$5000 (P) is invested at 10% (r). Find the time needed to earn

(a) \$100 in interest: $t = \dfrac{100}{5000(0.10)} = \dfrac{100}{500} = 0.2$ yr $=$ 72 days

(b) \$150 in interest: $t = \dfrac{150}{5000(0.10)} = \dfrac{150}{500} = 0.3$ yr $= 108$ days

(c) \$300 in interest: $t = \dfrac{300}{5000(0.10)} = \dfrac{300}{500} = 0.6$ yr $= 216$ days

Solving formulas for other variables is discussed in Section E.4.

What to Expect in this Chapter

The important topic of solving equations was introduced in Chapter 9 with examples and exercises at an elementary level. This chapter expands on the basic techniques used in solving equations. Because the process of solving equations is discussed more completely, the equations and applications that are presented are at the level of a beginning algebra course.

In Section E.1, we show how to combine like terms in algebraic expressions by using the distributive property and how to evaluate expressions for various values of the variable. These ideas are used in Section E.2 in solving first-degree equations and in checking solutions.

Before solving word problems in Section E.3, we review changing English phrases into algebraic expressions. The word problems presented involve number problems, consecutive integers, and geometry. In Section E.4, we show that formulas are simply equations, and they can be solved for variables other than what is usually given in the standard form.

E.1 Simplifying and Evaluating Expressions

OBJECTIVES

1. Know the meanings of the terms **constant, variable, term,** and **coefficient.**
2. Recognize like terms and be able to combine like terms.
3. Understand the distributive property of multiplication over addition.
4. Be able to evaluate algebraic expressions for given values of the variables by using the rules for order of operations.

Combining Like Terms

A single number is called a **constant** and a symbol (or letter) used to indicate more than one number is called a **variable.** Any constant or variable or the indicated product of constants and powers of variables is called a **term.** Examples of terms are

$$18, \quad \frac{3}{4}, \quad 5xy, \quad -4x^2, \quad -10x, \quad \text{and} \quad \frac{5}{9}x^3y$$

A number written next to a variable (as in $-10x$) or a variable written next to another variable (as in xy) indicates multiplication. In the term $-4x^2$, the constant -4 is called the **numerical coefficient** of x^2 (or simply the **coefficient** of x^2). If a term has only one variable, then the exponent of the variable is called the **degree** of the term. Constants are said to be of zero degree.

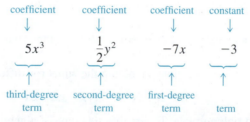

Like terms (or **similar terms**) are terms that contain the same variables raised to the same powers. That is, if a variable is raised to a certain power in one term, it is raised to the same power in the other like terms. Constants are like terms.

Like Terms

-7, 19, and $\dfrac{7}{8}$	are like terms because each term is a constant.
$-3a$, $14a$, and $5a$	are like terms because each term contains the same variable, a, raised to the same power, 1.
$4xy^2$ and $8xy^2$	are like terms because each term contains the same two variables, x and y, with x first-degree in both terms and y second-degree in both terms.

Unlike Terms

$9x$ and $3x^2$	are unlike terms (**not** like terms) because the variable x is not raised to the same power in both terms.
$15ab^2$ and $-4a^2b$	are unlike terms since the variables are **not** raised to the same power in both terms.

If no number is written next to a variable, then the coefficient is understood to be 1. If a minus sign $(-)$ is written next to a variable, then the coefficient is understood to be -1. For example,

$$x = 1 \cdot x, \qquad a^3 = 1 \cdot a^3, \qquad \text{and} \qquad y = 1 \cdot y$$

$$-x = -1 \cdot x, \quad -a^3 = -1 \cdot a^3, \quad \text{and} \quad -y = -1 \cdot y$$

EXAMPLE 1

From the following list of terms, pick out the like terms.

$$5, \quad 4y, \quad -12, \quad -y, \quad 3x^2z, \quad -\frac{1}{2}y, \quad -2x^2z, \quad 0$$

Solution

(a) 5, -12, and 0 are like terms. All are constants.

(b) $4y$, $-y$, and $-\dfrac{1}{2}y$ are like terms. All have the same variable factor, y.

(c) $3x^2z$ and $-2x^2z$ are like terms. All have the same variable factor, x^2z.

Algebraic expressions involving the sums (or differences) of terms, such as

$$8 + x, \quad 5y^2 - 11y, \quad \text{and} \quad 2x^2 - 5x - 2$$

are called **polynomials.** To simplify this type of algebraic expression, we **combine like terms** whenever possible. For example,

$$7x + 2x = 9x \quad \text{and} \quad 8n + 3n + 2n = 13n$$

The technique for combining like terms is based on the following **distributive property of multiplication over addition** (or simply the **distributive property**).

> **Distributive Property of Multiplication over Addition**
>
> For integers a, b, and c,
>
> $$a(b + c) = ab + ac$$

Another form of the distributive property is

$$ba + ca = (b + c)a$$

This form is particularly useful when b and c are numerical coefficients because it leads directly to the explanation of combining like terms. Thus,

$$ba + ca = (b + c)a$$
$$7x + 2x = (7 + 2)x = 9x \quad \text{by the distributive property}$$

Intuitively, we can think of x as a label or name. Thus, just as 7 oranges plus 2 oranges is 9 oranges, $7x + 2x = 9x$. Similarly,

$$6n - n + 3n = (6 - 1 + 3)n = 8n \quad \text{Note that } -1 \text{ is the coefficient of } -n.$$

EXAMPLE 2   Simplify each expression by combining like terms whenever possible.

(a) $6x + 10x + 3$

(b) $4x^2 + x^2 - 8x^2$

(c) $2(x + 3) + 4(x - 7)$

(d) $3a - 52b + 9$

Solution

(a) $6x + 10x + 3 = (6 + 10)x + 3$ Use the distributive property with $6x + 10x$. Note
$$= 16x + 3$$
that the constant 3 is not combined with $16x$ because they are not like terms.

(b) $4x^2 + x^2 - 8x^2 = (4 + 1 - 8)x^2$ Note that $+1$ is the coefficient of x^2.
$$= -3x^2$$

(c) $2(x + 3) + 4(x - 7) = 2x + 6 + 4x - 28$ First use the distributive property
$$= 2x + 4x + 6 - 28 \quad \text{directly.}$$
$$= (2 + 4)x - 22 \quad \text{Combine like terms.}$$
$$= 6x - 22$$

(d) $3a - 52b + 9$ This expression is already simplified since it has no like terms.

To help in understanding the distributive property, consider the following application.

APPLICATION

What will be the total price of 5 hot dogs and 5 soft drinks at the ball game if one hot dog costs $3 and one soft drink costs $1?

Solution

You can think of the total cost of one hot dog and one soft drink and calculate as follows:

cost of 1 dog   total cost of
and 1 drink   1 dog and 1 drink
↓   ↓

$$5(3 + 1) \quad = \quad 5(4) \quad = \quad 20$$

Or, you can think of the cost of all the hot dogs plus all the soft drinks and use the distributive property as follows:

total cost   total cost
of 5 dogs   of 5 drinks

$$5(3 + 1) \quad = \quad 5 \cdot 3 \quad + \quad 5 \cdot 1 \quad = \quad 15 \quad + \quad 5 \quad = \quad 20$$

The total price calculated is the same with both methods, $20.

Evaluating Algebraic Expressions

Algebraic expressions can be evaluated for given values of the variables by substituting one value for each variable into the expression and then following the rules for order of operations. When like terms are present, the process of evaluation can be made easier by first combining like terms.

> **To Evaluate an Algebraic Expression:**
>
> **1.** Combine like terms.
>
> **2.** Substitute the values given for any variables. (**Note:** To indicate multiplication, enclose the numbers substituted in parentheses.)
>
> **3.** Follow the rules for order of operations.

The rules for order of operations are restated here for easy reference.

> **Rules for Order of Operations**
>
> **1.** First, simplify within grouping symbols, such as parentheses (   ), brackets [   ], or braces {   }. Start with the innermost grouping.
>
> **2.** Second, find any powers indicated by exponents.

3. Third, moving from **left to right,** perform any multiplications or divisions in the order in which they appear.

4. Fourth, moving from **left to right,** perform any additions or subtractions in the order in which they appear.

EXAMPLE 3

Evaluate the expression $6ab + ab + 4a - a$ for $a = -2$ and $b = +3$.

Solution

Combining like terms gives

$$6ab + ab + 4a - a = (6 + 1)ab + (4 - 1)a$$
$$= 7ab + 3a$$

Substituting $a = -2$ and $b = +3$ into the simplified expression and following the rules for order of operations gives the value of the expression.

$$7ab + 3a$$
$$= 7(-2)(3) + 3(-2) \quad \text{Note that the substituted numbers must be in parentheses to indicate multiplication.}$$
$$= -42 - 6$$
$$= -48$$

EXAMPLE 4

Evaluate the expression $-2y - 6y - 1 + 10$ for $y = -5$.

Solution

Combining like terms gives

$$-2y - 6y - 1 + 10 = (-2 - 6)y + 9 = -8y + 9$$

Substituting $y = -5$ gives

$$-8y + 9$$
$$= -8(-5) + 9 \quad \text{Note that } -5 \text{ is in parentheses to indicate multiplication.}$$
$$= 40 + 9$$
$$= 49$$

EXAMPLE 5

Evaluate the expression $7x^2 - 2x^2 + 2x - x^2 - 14x$ for $x = 3$.

Solution

Combining like terms

$$7x^2 - 2x^2 + 2x - x^2 - 14x = (7 - 2 - 1)x^2 + (2 - 14)x$$
$$= 4x^2 - 12x$$

Substituting $x = 3$ and evaluating:

$$4x^2 - 12x = 4(3)^2 - 12(3)$$
$$= 36 - 36$$
$$= 0$$

CLASSROOM PRACTICE

Simplify each expression.

1. $5x - 6x$ **2.** $3y + 4y - 2y + 6 - 8$

3. $2x^2 + 3x^2 + ab - 4ab$

Evaluate each expression for $x = -3$.

4. $-7x - 14$ **5.** $2x^2 - x + 1$

ANSWERS: **1.** $-1x$ (or $-x$)   **2.** $5y - 2$ **3.** $5x^2 - 3ab$ **4.** 7 **5.** 22

Exercises E.1

Simplify the following expressions by combining like terms whenever possible.

1. $6x + 2x$ **2.** $4x - 3x$ **3.** $5x + x$

4. $7x - 3x$ **5.** $-10a + 3a$ **6.** $-11y + 4y$

7. $-18y + 6y$ **8.** $-2x - 5x$ **9.** $-5x - 4x$

10. $-x - 2x$ **11.** $-7x - x$ **12.** $2x - 2x$

13. $5x - 5x$ **14.** $16p - 17p$ **15.** $9c - 10c$

16. $3x - 5x + 12x$ **17.** $2a + 14a - 25a$

18. $6c - 13c + 5c$ **19.** $40p - 30p - 10p$

20. $16x - 15x - 3x$ **21.** $2x + 3x - 7$

22. $5x - 6x + 2$ **23.** $7x - 8x + 5$

24. $-5x - 7x - 4$ **25.** $-8a - 3a - 2$

26. $-4x + x + 1 - 3$ **27.** $-2x + 5x + 6 - 5$

28. $4x + 7 - 8 + 3x$ **29.** $-5x - 1 + 8 + 9x$

30. $10y + 3 - 4 - 6y$

31. $6y^2 - y^2$ **32.** $15x^2 - 5x^2$

33. $2x^2 + 3x + 1$ **34.** $5x^2 - 2x + 3$

35. $x + y - x - 2y$ **36.** $a + b - 2a - b$

37. $3ab + a^2 - ab + 2a^2$

38. $xy + y^2 + 4xy + y^2$

39. $3(x + 2y) + 2(x - y)$
   [HINT: Use the distributive property.]

40. $5(x + 4y) + 3(2x - y)$
   [HINT: Use the distributive property.]

Evaluate each of the expressions below for $x = -3$, $y = 2$, $z = 3$, $a = -1$, and $c = -2$.

41. $x - 2$

42. $y - 2$

43. $z - 3$

44. $2x + z$

45. $3y - x$

46. $x - 4z$

47. $20 - 2a$

48. $10 + 2c$

49. $3c - 5$

50. $2x + 3x - 7$

51. $7a - a + 3$

52. $-3y - 4y + 6 - 2$

53. $-2x - 3x + 1 - 4$

54. $5y - 2y - 3y + 4$

55. $2x - 3x + x - 8$

56. $5y^2 - y^2 + 2y$

57. $4x^2 + 3x - 1$

58. $a^2 + 2c - 3ac$

59. $2a^2 + 3c - 4ac$

60. $x^2 + y^2 - z^2$

E.2 Solving Equations ($ax + b = c$)

OBJECTIVE

Know how to solve first-degree (or linear) equations of the form $ax + b = c$.

In Section 9.5, we provided an introduction to solving first-degree equations. (You may want to review that section at this time.) In this section, we repeat some of the definitions and ideas basic to solving equations and expand their applications to deal with equations that involve a few more steps, including combining like terms and use of the distributive property.

Solving First-Degree Equations

If an equation contains a variable, such as $2x + 4 = 10$, we want to find the value (or values) for the variable that will give a true statement when substituted for the variable. This procedure is called **solving the equation,** and the value (or values) found is called the **solution** (or **solutions**) of the equation. **To solve an equation means to find all the solutions.**

EXAMPLE 1 $2x = 14$

Show that 7 is a solution and that 5 is **not** a solution.

Solution

Since $2 \cdot 7 = 14$ is true, 7 is a solution.
But, $2 \cdot 5 = 14$ is false, so 5 is **not** a solution.

EXAMPLE 2 $y + 11 = 3$

Show that -8 is a solution and that 8 is **not** a solution.

Solution

Since $-8 + 11 = 3$ is true, -8 is a solution.
But, $8 + 11 = 3$ is false, so 8 is **not** a solution.

Definition

A **first-degree equation in x** is any equation that can be written in the form

$ax + b = c$ where a, b, and c are constants and $a \neq 0$.

(The variable may be a letter or symbol other than x.)

**Principles Used in Solving
a First-Degree Equation**

In the four basic principles stated here, A and B represent algebraic expressions or constants.

1. **The Addition Principle:**
   If $A = B$ is true, then $A + C = B + C$ is also true for any number C.
   (The same number may be added to both sides of an equation.)

2. **The Subtraction Principle:**
   If $A = B$ is true, then $A - C = B - C$ is also true for any number C.
   (The same number may be subtracted from both sides of an equation.)

3. **The Multiplication Principle:**
   If $A = B$ is true, then $C \cdot A = C \cdot B$ is also true for any number C.
   (Both sides of an equation may be multiplied by the same number.)

4. **The Division Principle:**
   If $A = B$ is true, then $\dfrac{A}{C} = \dfrac{B}{C}$ is also true for any nonzero number C.
   (Both sides of an equation may be divided by the same nonzero number.)

Note that in the Division Principle, division by C is the same as multiplication by the reciprocal of C, namely $\frac{1}{C}$. Thus, we could list the Division and Multiplication Principles as one principle. The use of any of these principles with an equation gives a new equation with the same solution. Such equations are said to be **equivalent.** To solve equations, apply the principles just listed. Find equivalent equations until the solution is obvious, as in the equations

$$x = -7, \quad y = 60, \quad \text{or} \quad R = 0.05.$$

EXAMPLE 3

$$x + 14 = 10$$
$$x + 14 - 14 = 10 - 14 \quad \text{Using the \textbf{Addition Principle}, add } - 14 \text{ to both sides.}$$
$$x + 0 = -4 \quad \text{Simplify.}$$
$$x = -4 \quad \text{Simplify again.}$$

EXAMPLE 4

$$8y = 72$$

$$\frac{8y}{8} = \frac{72}{8} \quad \text{Using the \textbf{Multiplication Principle}, divide both sides by 8. This is the same as multiplying by } \frac{1}{8}.$$

$$y = 9 \quad \text{Simplify to find the solution.}$$

Using the Basic Principles for Solving Equations

More difficult problems may involve several steps. Keep in mind the following general guidelines.

General Guidelines for Solving Equations

1. The goal is to isolate the variable on one side of the equation (either the right side or the left side).

2. First use the Addition and Subtraction Principles whenever numbers are subtracted or added to the variable.

3. Use the Multiplication and Division Principles to make 1 (or $+1$) the coefficient of the variable.

EXAMPLE 5

$$4x + 3 = 11 \quad \text{Write the equation.}$$
$$4x + 3 - 3 = 11 - 3 \quad \text{Add } -3 \text{ to both sides.}$$
$$4x = 8 \quad \text{Simplify.}$$

$$\frac{4x}{4} = \frac{8}{4} \quad \text{Divide both sides by 4, the coefficient of } x.$$

$$x = 2 \quad \text{Simplify.}$$

EXAMPLE 6

$$2x - 4 + 3x = 26 \qquad \text{Write the equation.}$$

$$5x - 4 = 26 \qquad \text{Combine like terms on the left side.}$$

$$5x - 4 + 4 = 26 + 4 \quad \text{Add } +4 \text{ to both sides.}$$

$$5x = 30 \qquad \text{Simplify.}$$

$$\frac{5x}{5} = \frac{30}{5} \qquad \text{Divide both sides by 5.}$$

$$x = 6 \qquad \text{Simplify.}$$

CHECK:

$$2x - 4 + 3x = 26$$
$$2(6) - 4 + 3(6) \stackrel{?}{=} 26$$
$$12 - 4 + 18 \stackrel{?}{=} 26$$
$$26 = 26$$

EXAMPLE 7

$$5x + 2 = 3x - 8 \qquad \text{Write the equation.}$$

$$5x + 2 - 2 = 3x - 8 - 2 \qquad \text{Add } -2 \text{ to both sides.}$$

$$5x = 3x - 10 \qquad \text{Simplify.}$$

$$5x - 3x = 3x - 10 - 3x \qquad \text{Add } -3x \text{ to both sides. Simplify;}$$
$$2x = -10 \qquad \begin{array}{l}\text{now one side has the term with the} \\ \text{variable, and the other side has} \\ \text{the constant term.}\end{array}$$

$$\frac{2x}{2} = \frac{-10}{2} \qquad \text{Divide both sides by 2.}$$

$$x = -5 \qquad \text{Simplify.}$$

CHECK:

$$5x + 2 = 3x - 8$$
$$5(-5) + 2 \stackrel{?}{=} 3(-5) - 8$$
$$-25 + 2 \stackrel{?}{=} -15 - 8$$
$$-23 = -23$$

EXAMPLE 8

$$3(x + 2) = 18 - x \qquad \text{Write the equation.}$$

$$3x + 6 = 18 - x \qquad \text{Use the distributive property.}$$

$$3x + 6 - 6 = 18 - x - 6 \qquad \text{Add } -6 \text{ to both sides.}$$

$$3x = 12 - x \qquad \text{Simplify.}$$

$$3x + x = 12 - x + x \quad \text{Add } -x \text{ to both sides.}$$

$$4x = 12 \qquad\qquad \text{Simplify.}$$

$$\frac{4x}{4} = \frac{12}{4} \qquad\qquad \text{Divide both sides by 4.}$$

$$x = 3 \qquad\qquad \text{Simplify.}$$

Checking Solutions to Equations

Checking can be done by substituting the solution found back into the original equation to see if the resulting statement is true. (See Examples 6 and 7.)

If there are fractions in an equation, multiply each term on both sides of the equation by the LCM (least common multiple) of all the denominators. Then solve the equation just as before. The following example illustrates how to proceed with fractions to give an equation with integer coefficients and constants.

EXAMPLE 9

$$\frac{1}{2}x + \frac{3}{4}x = \frac{1}{6}x - 26$$

$$12\left(\frac{1}{2}x\right) + 12\left(\frac{3}{4}x\right) = 12\left(\frac{1}{6}x\right) - 12\,(26) \quad \text{(12 is the LCM of 2, 4, 6)}$$

$$6x + 9x = 2x - 312 \qquad \text{This equation has integer coefficients.}$$

$$15x - 2x = 2x - 312 - 2x$$

$$13x = -312$$

$$\frac{13x}{13} = \frac{-312}{13}$$

$$x = -24$$

CLASSROOM PRACTICE

Solve the following equations.

1. $10x + 4 = 14$

2. $x + 5 - 2x = 7 + x$

3. $2(x - 7) = 5(x + 2)$

4. $\frac{2}{3}x = \frac{1}{2}x + 1$

ANSWERS: **1.** $x = 1$ **2.** $x = -1$ **3.** $x = -8$ **4.** $x = 6$

Exercises E.2

Solve the following equations.

1. $x + 4 = 10$ **2.** $x + 13 = 20$ **3.** $y - 5 = 17$

4. $y - 12 = 4$ **5.** $y + 8 = 3$ **6.** $x + 10 = 7$

7. $x - 5 = -7$ **8.** $x - 14 = -10$ **9.** $y - 8 = -6$

10. $x - 12 = -5$ **11.** $5x = 30$ **12.** $3y = 15$

13. $10y = -40$ **14.** $6x = -48$ **15.** $-2x = 12$

16. $-4x = 24$ **17.** $-8y = -40$ **18.** $-12y = -36$

19. $16 = x + 3$ **20.** $25 = x + 14$ **21.** $2x + 3 = 5$

22. $3x - 4 = 8$ **23.** $4y + 1 = 9$ **24.** $3x - 10 = 11$

25. $6x + 4 = -14$ **26.** $7y - 8 = -1$ **27.** $3 + 6y = 15$

28. $6 + 5y = 21$ **29.** $2x + 3 = -9$ **30.** $3x - 1 = -4$

31. $5y + 12 = -3$ **32.** $10y + 3 = -17$ **33.** $15 = 2x - 3$

34. $20 = 3x - 1$ **35.** $-17 = 5y - 2$ **36.** $30 = 4y + 6$

37. $4 = 5x + 9$ **38.** $28 = 10x - 2$ **39.** $-24 = 7x - 3$

40. $96 = 25y - 4$ **41.** $3x = x - 10$ **42.** $5y = 2y + 12$

43. $7y = 6y + 5$ **44.** $6x = 2x + 20$ **45.** $5x = 2x$

46. $4x = 3x$ **47.** $4x + 3 = 2x + 9$

48. $5y - 2 = 4y - 6$ **49.** $7x + 14 = 10x + 5$

50. $5x + 20 = 8x - 4$ **51.** $5(x - 2) = 3(x - 8)$

52. $2(y + 1) = 3y + 3$ **53.** $4(x - 1) = 2x + 6$

54. $6y - 3 = 3(y + 2)$ **55.** $7y - 6y + 12 = 4y$

56. $6x + 5 + 3x = 3x - 13$ **57.** $5x - 2x + 4 = 3x + x - 1$

58. $7x + x - 6 = 2(x + 9)$ **59.** $x - 5 + 4x = 4(x - 3)$

60. $3(-x + 6) = 3x + 2(x + 1)$ **61.** $\frac{1}{2}x + \frac{3}{4}x = -15$

62. $\frac{2}{3}y - 5 = \frac{1}{3}y + 20$ **63.** $\frac{3}{5}x - 4 = \frac{1}{5}x - 5$

64. $\frac{5}{8}y - \frac{1}{4} = \frac{2}{5}y + \frac{1}{3}$ **65.** $\frac{x}{8} + \frac{1}{6} = \frac{x}{10} + 2$

E.3   ## Solving Word Problems

OBJECTIVES

1. Know how to represent consecutive integers, consecutive even integers, and consecutive odd integers.
2. Be able to solve number problems by using equations.

Translating English Phrases

As we discussed in Section 9.5, many word problems can be solved by using the following three basic skills:

1. Know how to translate English phrases into algebraic expressions.

2. Be able to set up an equation using the translated expressions.

3. Solve the resulting equation.

The following list of key words to look for in word problems is repeated from Section 9.5. These words indicate what operations are to be performed with constants and variables.

Key Words (that indicate operations)

Addition	Subtraction	Multiplication	Division
add	subtract	multiply	divide
sum	difference	product	quotient
plus	minus	times	
more than	less than	twice, double	
increased by	decreased by	of (with fractions)	

The following examples illustrate how English phrases can be translated into algebraic expressions. Different letters are used as variables to represent the unknown numbers and the key words are in boldface print. (Also, see Section 9.5 for more examples.)

English Phrase	Algebraic Expression

1. A number **minus** 18
The **difference** between a number and 18 $\longrightarrow$ $n - 18$
18 **less than** a number

2. -6 **times** a number
The **product** of a number and -6 $\longrightarrow$ $-6x$
A number **multiplied by** -6

3. The **quotient** of a number and 10 $\longrightarrow$ $\dfrac{x}{10}$
A number **divided by** 10

4. 10 **divided by** a number $\longrightarrow$ $\dfrac{10}{x}$

5. **Twice** the **sum** of a number and 3 $\longrightarrow$ $2(y + 3)$
Two **times** the quantity $y + 3$

Solving Number Problems and Consecutive Integer Problems

The problem-solving approach for the number problems discussed in this section is outlined below.

Basic Steps for Solving Number Problems

1. Read the problem carefully at least twice. Look for key words and phrases that can be translated into algebraic expressions.

2. Assign a variable as the unknown quantity and form an equation using the expressions you translated.

3. Solve the equation.

4. Look back over the problem and check that the answer is reasonable.

EXAMPLE 1

Eight less than four times a number is equal to twice the number increased by 6. Find the number.

Solution

Let $n =$ the unknown number.

Translate "eight less than four times a number" to $\underline{4n - 8}$.

Translate "twice the number increased by 6" to $\underline{2n + 6}$.

Form the equation: $\underline{4n - 8 = 2n + 6}$

Solve the equation:

$$4n - 8 = 2n + 6$$
$$4n - 8 + 8 = 2n + 6 + 8$$
$$4n = 2n + 14$$
$$4n - 2n = 2n + 14 - 2n$$
$$2n = 14$$
$$\frac{2n}{2} = \frac{14}{2}$$
$$n = 7$$

CHECK:

$$4n - 8 = 2n + 6$$
$$4 \cdot 7 - 8 \overset{?}{=} 2 \cdot 7 + 6$$
$$28 - 8 \overset{?}{=} 14 + 6$$
$$20 = 20 \qquad \text{a true statement}$$

Remember that the set of integers consists of the whole numbers and their opposites:

$$\{\ldots, -4, -3, -2, -1, 0, 1, 2, 3, 4, \ldots\}$$

We use the following terms for integers that follow certain patterns; these appear frequently in applications.

Consecutive Integers: Integers are consecutive if each is one more than the previous integer. For example,

10, 11, and 12 are three consecutive integers.

They can be represented as

$n, n + 1,$ and $n + 2$

Consecutive Odd Integers: Odd integers are consecutive odd integers if each is two more than the previous odd integer. For example,

13, 15, and 17 are three consecutive odd integers.

They can be represented as

$n, n + 2,$ and $n + 4$

continued

Consecutive Even Integers: Even integers are consecutive even integers if each is two more than the previous even integer. For example,

20, 22, and 24 are three consecutive even integers.

They can be represented as

$n, n + 2,$ and $n + 4$

Note that consecutive even and consecutive odd integers are represented in the same way. The value of the first integer, n, determines whether the expressions n, $n + 2$, and $n + 4$ represent consecutive even or odd integers.

EXAMPLE 2

Find three consecutive integers whose sum is -39.

Solution

Let

$$n = \text{the first integer}$$
$$n + 1 = \text{the second integer}$$
$$n + 2 = \text{the third integer}$$

The equation to be solved is

$$n + (n + 1) + (n + 2) = -39$$
$$n + n + 1 + n + 2 = -39$$
$$3n + 3 = -39$$
$$3n + 3 - 3 = -39 - 3$$
$$3n = -42$$
$$\frac{3n}{3} = \frac{-42}{3}$$
$$n = -14 \qquad \text{the first integer}$$
$$n + 1 = -13 \qquad \text{the second integer}$$
$$n + 2 = -12 \qquad \text{the third integer}$$

CHECK:

The numbers -14, -13, and -12 are consecutive integers, and

$$(-14) + (-13) + (-12) = -39$$

EXAMPLE 3

Find three consecutive odd integers such that the sum of the first and third is 33 more than the second.

Solution

Let

$$n = \text{the first odd integer}$$
$$n + 2 = \text{the second odd integer}$$
$$n + 4 = \text{the third odd integer}$$

Tne equation to be solved is

$$n + (n + 4) = (n + 2) + 33$$
$$n + n + 4 = n + 2 + 33$$
$$2n + 4 = n + 35$$
$$2n + 4 - 4 = n + 35 - 4$$
$$2n = n + 31$$
$$2n - n = n + 31 - n$$
$$n = 31 \qquad \text{the first odd integer}$$
$$n + 2 = 33 \qquad \text{the second odd integer}$$
$$n + 4 = 35 \qquad \text{the third odd integer}$$

Solving Geometry Problems

Formulas for the area and perimeter for six geometric figures were presented in Table 9.1 in Section 9.6. Use that table for reference as you need.

EXAMPLE 4

A rectangular swimming pool has a perimeter of 160 meters and a length that is 20 meters more than its width. Find the width and length of the pool.

Solution

Use the formula for the perimeter of a rectangle: $P = 2l + 2w$. Sketch a picture and label the sides where

$$w = \text{the width}$$
$$w + 20 = \text{the length}$$

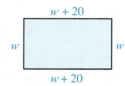

Substituting into the formula gives the equation to be solved:

$$160 = 2(w + 20) + 2w$$

$$160 = 2w + 40 + 2w$$

$$160 = 4w + 40$$

$$160 - 40 = 4w + 40 - 40$$

$$120 = 4w$$

$$\frac{120}{4} = \frac{4w}{4}$$

$$30 = w$$

The width of the pool is 30 meters and the length is $w + 20 = 50$ meters.

Exercises E.3

Write an algebraic expression described by each of the following English phrases.

1. the product of a number and -2

2. the sum of a number and -3

3. 4 more than twice a number

4. the quotient of a number and -5

5. 3 times the sum of a number and 12

6. -4 times the difference of a number and 3

7. twice the quantity of 9 plus a number

8. twice the quantity of 9 minus a number

9. 5 times a number decreased by the product of the number and -2

10. 3 times the difference between twice a number and the number

Follow the steps outlined in this section in solving Exercises 11 and 12.

11. The difference between a number and 16 is -48. What is the number?
                        (a)

Let x = the unknown number.

Translation of (a): _____

Equation: _____

Solve the equation.

12. If the product of a number and 4 is decreased by 10, the result is 50.
                        (a)

Find the number.

Let x = the unknown number.

Translation of (a): _____

Equation: _____

Solve the equation.

Before solving the following problems, make a guess as to what you think the answer is. [For example, do you think the number is positive (greater than 0) or negative (less than 0)? Do you think the number is greater than 50 or less than 50?]

13. If the product of a number and 5 is added to 12, the result is 7. What is the number?

14. The product of a number and 8 increased by 24 is equal to twice the number. Find the number.

15. The sum of a number and 2 is equal to three times the number. What is the number?

16. If a number is equal to 4 less than twice the number, what is the number?

17. Find a number such that three times the sum of the number and 4 is equal to -60.

18. Five more than twice a number is equal to 20 more than the number. What is the number?

19. Twenty plus a number is equal to the sum of twice the number and three times the same number. Find the number.

20. Four times the difference between a number and 7 is equal to 2 plus the number. Find the number.

21. Find two consecutive integers whose sum is 37.

22. Find three consecutive integers whose sums is -42.

23. The sum of three consecutive odd integers is 27. Find the three integers.

24. Find three consecutive odd integers such that three times the first is 19 more than the sum of the other two.

25. Find four consecutive even integers whose sum is 54 more than the smallest of the four integers.

26. Find three consecutive even integers such that 6 times the second integer plus 32 is equal to the sum of the first and third integers.

27. If the sum of two consecutive integers is multiplied by 3, the result is -15. What are the two integers?

28. If 7 times a number is decreased by 4 times the number, the difference is equal to the sum of the number and 10. What is the number?

29. What is the number whose product with 6 is equal to 12 less than twice the number?

30. The difference of a number and 3 is equal to the difference between 5 times the number and 15. What is the number?

31. The perimeter of a rectangular garden is 240 feet. If the length is 40 feet more than the width, what are the dimensions (width and length) of the garden?

32. The perimeter of a triangle is 52 centimeters. If two sides are equal and the third side is 4 centimeters less than the sum of the other two sides, what are the lengths of the three sides?

33. A wire 66 inches long is bent to form a triangle such that the lengths of the sides are three consecutive even integers. What are the lengths of the sides?

34. If the sum of two consecutive integers is multiplied by 3, the result is -15. What are the two integers?

35. The circumference of a circular swimming pool is measured to be 78.5 feet. What is the diameter of the pool? [HINT: Use $C = \pi d$ with $\pi = 3.14$.]

36. A mother is 5 inches taller than her daughter. If her daughter is 13 years old and the sum of their heights is 123 inches, how tall is her daughter?

37. The length of a rectangle is 6 feet more than twice the width. If the perimeter of the rectangle is 72 feet, what are the length and the width of the rectangle?

38. A classic car is now selling for $500 more than three times its original price. If the selling price is now $11,000, what was the car's original price?

39. A student is going to buy three textbooks: one for chemistry, one for English, and one for music. The chemistry book costs one dollar more than twice the cost of the English book, and the music book costs three dollars less than the cost of the English book. If the total cost for the three books is $78, what is the price of each book?

40. A real estate agent says that the current value of a home is $80,000 more than twice its value when it was new. If the current value is $260,000, what was the value of the home when it was new?

E.4   Working with Formulas

OBJECTIVES

1. Know how to substitute given values into a formula and find the value of the remaining unknown variable.
2. Be able to solve a formula for a designated variable in terms of the other variables in the formula.

What Is a Formula?

A **formula** is a general statement (usually an equation) that relates two or more variables. We have already discussed a variety of formulas throughout the text, particularly in terms of geometric figures, and evaluated formulas for given values of the variables. In this section, we show two ways to work with formulas: (1) by evaluating formulas and (2) by solving formulas for certain terms in the formulas.

Some formulas and their meanings are stated here for interest and clarification.

Formula	**Meaning**

1. $I = Prt$

The simple interest (I) earned investing money is equal to the product of the principal (P), the rate of interest (r), and the time (t) in years.

2. $d = rt$

The distance (d) traveled is equal to the product of the rate of speed (r) and the time (t).

3. $C = \pi d$

The circumference (C) of a circle is equal to the product of pi (π) and the diameter (d).

4. $C = \dfrac{5}{9}(F - 32)$

Temperature in degrees Celsius (C) is equal to $\dfrac{5}{9}$ of the difference between the Fahrenheit temperature (F) and 32.

5. $\alpha + \beta + \gamma = 180°$

The sum of the angles (α, β, and γ) of a triangle is 180°. (α, β, and γ are the Greek letters alpha, beta, and gamma, respectively.)

6. $P = 2\ell + 2w$

The perimeter (P) of a rectangle is equal to twice the length (ℓ) plus twice the width (w).

Evaluating Formulas

If we know values for all but one variable in a formula, then we can substitute these values and solve the equation for the unknown variable. If the equation is first-degree, then we simply use the methods for solving equations discussed in Section E.2.

EXAMPLE 1   Given the formula $C = \dfrac{5}{9}(F - 32)$, find the Fahrenheit temperature that corresponds to a Celsius temperature of 100°.

Solution

Substitute 100 for C in the formula and solve the resulting equation for F.

$$100 = \frac{5}{9}(F - 32)$$

$$9 \cdot 100 = 9 \cdot \frac{5}{9}(F - 32) \qquad \text{Multiply both sides by 9.}$$

$$900 = 5(F - 32) \qquad \text{Simplify.}$$

$$900 = 5F - 160 \qquad \text{Use the distributive property.}$$

$$900 + 160 = 5F \qquad \text{Add 160 to both sides.}$$

$$\frac{1060}{5} = \frac{\cancel{5}\,F}{\cancel{5}} \qquad \text{Divide both sides by 5.}$$

$$212 = F$$

Thus, $100°$ C is equal to $212°$ F, the boiling point of water at sea level.

EXAMPLE 2 Two angles of a triangle are measured as $62°$ and $83°$. What is the measure of the third angle?

Solution

Use the formula $\alpha + \beta + \gamma = 180°$. Substitute $\alpha = 62$ and $\beta = 83$ and solve for γ.

$$62 + 83 + \gamma = 180$$
$$145 + \gamma = 180$$
$$\gamma = 180 - 145$$
$$\gamma = 35$$

The measure of the third angle is $35°$.

Solving for Any Term in a Formula

Sometimes a particular formula might be expressed in one form when another form would be more convenient. To change the form of a formula, we need to be able to **solve the formula for any variable in that formula.** In such cases, we solve the equation just as we did in Section E.2, but we **temporarily treat the variables other than the chosen one as if they were constants.** The following examples illustrate the technique.

EXAMPLE 3 Given the formula $d = rt$, solve for t in terms of d and r.

Solution

$$d = rt \qquad \text{Treat } d \text{ and } r \text{ as if they were constants.}$$

$$\frac{d}{r} = \frac{\cancel{r}t}{\cancel{r}} \qquad \text{Divide both sides by } r.$$

$$\frac{d}{r} = t \quad \text{Simplify.}$$

The formula now indicates that time is equal to distance divided by rate. (Note that the relationship among distance, rate, and time is not changed. Only the form of the formula is changed.)

EXAMPLE 4

Given the equation $3x + 4y = 12$, (a) solve for x in terms of y, and (b) solve for y in terms of x.

Solution

(a) Solving for x:

$$3x + 4y = 12$$

$$3x + 4y - 4y = 12 - 4y \quad \text{Subtract } 4y \text{ from both sides.}$$

$$3x = 12 - 4y \quad \text{Simplify.}$$

$$\frac{3x}{3} = \frac{12 - 4y}{3} \quad \text{Divide both sides by 3.}$$

$$x = \frac{12 - 4y}{3}$$

(b) Solving for y:

$$3x + 4y = 12$$

$$3x + 4y - 3x = 12 - 3x$$

$$4y = 12 - 3x$$

$$\frac{4y}{4} = \frac{12 - 3x}{4}$$

$$y = \frac{12 - 3x}{4}$$

EXAMPLE 5

Given $C = \frac{5}{9}(F - 32)$, solve for F in terms of C.

Solution

$$C = \frac{5}{9}(F - 32) \quad \text{Treat } C \text{ as a constant.}$$

$$\frac{9}{5} \cdot C = \frac{9}{5} \cdot \frac{5}{9}(F - 32) \quad \text{Multiply both sides by } \frac{9}{5}.$$

$$\frac{9}{5}C = F - 32 \quad \text{Simplify.}$$

$$\frac{9}{5}C + 32 = F \quad \text{Add 32 to both sides.}$$

Thus, two forms of the same formula are

$$C = \frac{5}{9}(F - 32) \quad \text{(solved for } C)$$

$$F = \frac{9}{5}C + 32 \quad \text{(solved for } F)$$

Exercises E.4

In each formula, substitute the given values and find the value of the unknown variable.

1. $d = rt$; $r = 50$ mph, $t = 2.5$ hr

2. $A = lw$; $l = 10$ ft, $w = 40$ ft

3. $A = \pi r^2$; $r = 3$ cm; use $\pi = 3.14$

4. $C = \pi d$; $d = 5$ m; use $\pi = 3.14$

5. $2x + 3y = 10$; $x = -1$

6. $I = Prt$; $I = \$150$, $r = 10\%$, $t = \frac{1}{2}$ yr
   [HINT: Change 10% to a decimal.]

7. $P = 2l + 2w$; $P = 50$ cm, $l = 15$ cm

8. $3x - y = 15$; $y = 3$

9. $\alpha + \beta + \gamma = 180°$; $\beta = 10°$, $\gamma = 80°$

10. $V = lwh$; $V = 30$ mm^3, $l = 5$ mm, $w = 1.2$ mm

Solve each formula for the indicated variable in terms of the other variables.

11. $f = ma$; solve for m

12. $C = 2\pi r$; solve for r

13. $L = 2\pi rh$; solve for h

14. $I = Prt$; solve for P

15. $y = mx + b$; solve for x

16. $p = b + 2s$; solve for s

17. $p = a + b + c$; solve for a

18. $\alpha + \beta + \gamma = 180°$; solve for β

19. $P = 2l + 2w$; solve for l

20. $A = P + Prt$; solve for t

21. $V = lwh$; solve for w

22. $V = \pi r^2 h$; solve for h

23. $2x + y = 4$; solve for y

24. $2x + y = 4$; solve for x

25. $x - y = 7$; solve for y

26. $3x - y = 5$; solve for y

27. $3x + 2y = 6$; solve for y

28. $2x - 4y = 5$; solve for x

29. $6x - 2y = -3$; solve for x

30. $5x - 3y = -1$; solve for y

Writing and Thinking about Mathematics

31. The formula $z = \dfrac{x - \bar{x}}{s}$ is used extensively in statistics. In this formula x represents one of a set of numbers, $\bar{x}$ represents the average (or mean) of those numbers, and s represents a value called the standard deviation of the numbers. (The standard deviation is a positive number and is a measure of how "spread out" the numbers are.) The values for z are called **z-scores,** and they measure the number of standard deviation units a number x is from the mean $\bar{x}$.

(a) If $\bar{x} = 60$, what will be the z-score that corresponds to $x = 60$? Does this z-score depend on the value of s? Explain.

(b) For what values of x will the corresponding z-scores be negative?

(c) Calculate your z-score on the last two exams in this class. (Your instructor will give you the mean and standard deviation for each test.) What do these scores tell you about your performances on the two exams?

32. Suppose that, for a particular set of exam scores, $\bar{x} = 75$ and $s = 6$. Find the z-score that corresponds to a score of

(a) 81 (b) 69 (c) 84 (d) 57

Chapter Summary

Key Terms and Ideas

A **variable** is a symbol or letter that can represent more than one number. An expression that involves only multiplication and/or division with constants and/or variables is called a **term.**

Like terms (or **similar terms**) can be constants, or they can be terms that contain variables that are of the same power in each term.

For signed numbers a, b, and c, the **distributive property** of multiplication over addition states that

$$a(b + c) = ab + ac.$$

A **first-degree equation in x** is any equation that can be written in the form

$$ax + b = c \quad \text{where } a, b, \text{ and } c \text{ are constants and } a \neq 0.$$

Look for Key Words in Word Problems:

Key Words

Addition	Subtraction	Multiplication	Division
add	subtract	multiply	divide
sum	difference	product	quotient

continued

Addition	**Subtraction**	**Multiplication**	**Division**
plus	minus	times	
more than	less than	twice, double	
increased by	decreased by	of (with fractions)	

$n, n + 1, n + 2$	represent **consecutive integers.**
$n, n + 2, n + 4$	represent **consecutive odd integers** if n is odd.
$n, n + 2, n + 4$	represent **consecutive integers** if n is even.

Procedures

To evaluate an algebraic expression:

1. Combine like terms.

2. Substitute the given value(s) for the variable(s).

3. Follow the rules for order of operations.

**Principles Used in Solving
a First-Degree Equation**

In the four basic principles stated here, A and B represent algebraic expressions or constants.

1. **The Addition Principle:**
   If $A = B$ is true, then $A + C = B + C$ is also true for any number C.
   (The same number may be added to both sides of an equation.)

2. **The Subtraction Principle:**
   If $A = B$ is true, then $A - C = B - C$ is also true for any number C.
   (The same number may be subtracted from both sides of an equation.)

3. **The Multiplication Principle:**
   If $A = B$ is true, then $C \cdot A = C \cdot B$ is also true for any number C. (Both sides of an equation may be multiplied by the same number.)

4. **The Division Principle:**
   If $A = B$ is true, then $\dfrac{A}{C} = \dfrac{B}{C}$ is also true for any nonzero number C. (Both sides of an equation may be divided by the same nonzero number.)

Chapter Review Questions

Simplify the following expressions by combining like terms whenever possible.

1. $8x + 7x$

2. $9y + 3y$

3. $4x - x$

4. $-5x - x$

5. $13w + w$

6. $10y - 10y$

7. $8y - 11y + y$

8. $-30p + 2p - 6p$

9. $-7a - 2a + a + 6$

10. $5x + 13 - 7x + 1$

11. $18y - 10 + 3 - 4y$

12. $-2x - 8x - 2 - 8$

13. $3x^2 - x^2$

14. $y^2 - 4y^2$

15. $5x^2 - 3x + 1$

16. $4y^2 + 2y - 1$

17. $3x^2 - 4 - x^2 + 1$

18. $4y^2 - 5 + y^2 + 9$

19. $5(x + y) + 4(x - y)$
[HINT: Use the distributive property.]

20. $3(x - 2y) + 2(2x - 3)$
[HINT: Use the distributive property.]

Evaluate each expression for $x = 3$ and $y = -2$.

21. $20 - 2y + x$

22. $-10 + 6x + 2y$

23. $7x - 3y + x - 5$

24. $4y^2 - 2y + 1$

25. $x^2 + y^2 - 4x^2 + 4y$

Solve the following equations.

26. $x + 7 = 11$

27. $y - 8 = 8$

28. $x + 9 = 5$

29. $y - 3 = 0$

30. $x + 4 = -3$

31. $y - 2 = -4$

32. $7x = 28$

33. $4x = -20$

34. $-2x = 12$

35. $-3y = -15$

36. $3x + 4 = 13$

37. $4x - 1 = 15$

38. $2y + 5 = -5$

39. $3x - 7 = -7$

40. $5x = 4x + 9$

41. $5y = 2y - 9$

42. $7x = 4x$

43. $5y = -3y$

44. $4x + 3 = x + 6$

45. $4y - 3 = 11y - 17$

46. $2(x - 3) = 3(x + 2)$

47. $5(y - 2) + y = 4(y - 1)$

48. $\dfrac{3}{4}x - 1 = \dfrac{2}{3}x$

49. $\dfrac{2}{5}y - \dfrac{1}{4} = \dfrac{3}{10}y + \dfrac{1}{2}$

Write the following English phrases as algebraic expressions. Use x for the unknown number.

50. 8 more than a number

51. 3 less than 5 times a number

52. the quotient of a number and 9

53. −3 times the sum of a number and 2

54. the sum of 10 and 4 times a number

55. 18 decreased by twice a number

For each of the following, write an equation and use it to solve the problem.

56. If the product of a number and 10 is decreased by 15, the result is −35. Find the number.

57. If 14 is added to twice a number, the result is that number minus 13. Find the number.

58. The sum of three consecutive integers is 78. Find the three integers.

59. Find three consecutive odd integers with the property that $\frac{1}{3}$ of the first number is equal to the sum of the other two minus 41.

60. The product of 7 and a number is equal to the product of 5 with the sum of the number and 2. What is the number?

61. What is the number whose sum with −8 is equal to 6 less than 4 times the number?

62. The perimeter of a rectangle is 680 meters. Find the width if the length of the rectangle is 200 m.

In each formula, substitute the given values and find the value of the unknown variable.

63. $\alpha + \beta + \gamma = 180°$; $\alpha = 23°$, $\gamma = 57°$

64. $d = rt$; $d = 300$ mi, $t = 12$ hr

Solve each formula for the indicated variable in terms of the other variables.

65. $C = \pi d$; solve for d **66.** $3x - 4y = 2$; solve for y

Chapter Test

Simplify the following expressions by combining like terms.

1. $-5 - 6x - 2 + 3x$

2. $x^2 + 1 - 4x + 3x^2 - 5$

3. $a + 2b - a + b$

4. $6(x + 4y) + 3(2x - y)$

Evaluate the following expressions for $x = -1$, $y = 4$, $a = 2$, and $b = -2$.

5. $-3y - 2x + 18 + 5x - 4$

6. $5x^2 - 3xy - y^2$

7. $a^2 - b^2 + 2ab$

8. $4a^2 + 2b + b^2$

Solve the following equations.

9. $x - 20 = 20$

10. $y + 4 = 4$

11. $-7x = -42$

12. $4y = -2y$

13. $81 = 21 - 5y$

14. $5x + 6 = 4x - 4$

15. $2(2x + 3) = 3(-7 - x)$

16. $\frac{1}{2}y + 2 = \frac{3}{4}y - 3$

Write the following phrases as algebraic expressions. Use x for the unknown number.

17. 3 less than the quotient of a number and 4

18. 1 more than twice the sum of a number and 3

19. the difference of $\frac{3}{4}$ of a number and 5

20. the quantity $x + 3$ divided by -2

Solve the word problems in Exercises 21–23.

21. The sum of three consecutive even integers is equal to 252. Find the integers.

22. Twice the sum of a number and 3 is equal to the difference of the number and 10. What is the number?

23. Find the width of a rectangle if the perimeter is 30 centimeters and the length is $8\frac{1}{4}$ centimeters.

24. Substitute the given values and solve for t.

$I = Prt$; $I = \$480$, $P = \$2400$, $r = 10\%$

25. Solve for y: $4x - 3y = -2$.

1. Find the LCM for the numbers 24, 28, and 40.

2. (a) List the factors of 75. (b) List the multiples of 75.

3. Find two factors of 96 whose product is 96 and whose sum is 35.

Perform the indicated operations and simplify.

4. $1 - \dfrac{1}{16}$

5. $\begin{array}{r} 340\frac{1}{10} \\ -150\frac{3}{5} \\ \hline \end{array}$

6. $\begin{array}{r} 25\frac{1}{6} \\ \times 13\frac{3}{8} \\ \hline \end{array}$

7. $\dfrac{4 - \frac{1}{2}}{\frac{1}{24} + \frac{1}{6}}$

8. Divide (to the nearest hundredth): $176 \div 32.1$

9. Find $\dfrac{2}{3}$ of $\dfrac{7}{8}$.

10. Find $33\frac{1}{3}\%$ of 60.

11. 135% of _____ is 270.

12. 25% of 200 is _____.

Find the equivalent metric units as indicated.

13. 3000 m = _____ km

14. 8.6 cm = _____ mm

15. 9.2 m² = _____ cm²

16. 7.54 L = _____ mL

Simplify the following radical expressions.

17. $\sqrt{256}$

18. $\sqrt{600}$

19. $2\sqrt{10} + \sqrt{90}$

20. Evaluate the expression $(-5)^2 - 20 - 3(4 - 5)$.

Solve the following equations.

21. $20 = 3x + 5$

22. $7(x - 1) = 2x - 16$

23. $\dfrac{3}{4}x = x - 3$

24. Solve for y: $x + 2y = 5$.

25. Find (a) the mean, (b) the median, (c) the mode, and (d) the range for the following set of mathematics exam scores.

    75, 86, 93, 93, 87, 65, 68, 70, 72, 93, 95, 90, 85, 72, 60, 76

26. If a car averages 23.6 miles per gallon, how many miles will it go on 15 gallons of gas?

27. If one leg of a right triangle is 3 ft long and the other leg is 1 ft 3 in. long, how long is the hypotenuse?

28. The cost of a sofa to a furniture store owner was $750 and he sold the sofa for $1250. (a) What was his profit? (b) What was his percent of profit based on cost? (c) What was his percent of profit based on selling price?

Additional Topics
from Algebra

Mathematics at Work!

An Eskimo ice fishes along the Arctic coast, where mean winter temperatures are about −30° F (−34° C).

In many real-life applications, the relationship between two variables can be expressed in terms of a formula in which both variables are first-degree. For example, the simple interest on any principal, P, for one year at 8% can be calculated according to the formula $I = 0.08P$. The distance you travel on your bicycle at an average speed of 15 miles per hour can be represented by the formula $d = 15t$, where t is the time (in hours) that you ride.

As we will see in Sections T.2 and T.3, the relationships indicated by equations and formulas such as these can be "pictured" in two-dimensional graph form in which the graphs are straight lines. Once the correct graph has been drawn, this graph can serve in place of the formula in finding a value (or an approximate value) of one of the variables corresponding to a known value of the other variable.

The formula $C = \dfrac{5}{9}(F - 32)$ represents the relationship between degrees Fahrenheit (F) and degrees Celsius (C). A portion of the graph of this relationship is shown below. (See Section T.2, Exercise 33.)

(a) From the graph, estimate the Celsius temperature corresponding to 32° F. (That is, estimate the value of C given that $F = 32$.)

(b) From the graph, estimate the Celsius temperature corresponding to 59° F. (That is, estimate the value of C given that $F = 59$.)

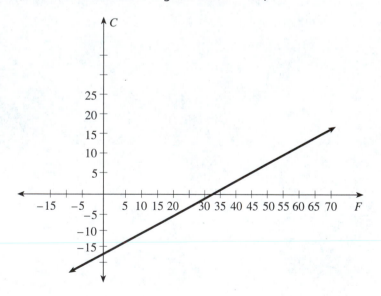

What to Expect in this Chapter

Some algebraic concepts such as integers, sums of terms, and solving equations were introduced in Chapter 9. This chapter is designed to provide a first look at a few slightly more advanced topics from a beginning algebra course. That is, this chapter provides a "jump start" into algebra for those students who plan to continue their studies in mathematics.

Section T.1 shows how to solve first-degree inequalities (this is similar to solving first-degree equations) and how to graph the solutions on number lines. Section T.2 and T.3 deal with concepts related to graphing in two dimensions: relating points in a plane to ordered pairs of real numbers (called coordinates of the points) and graphing straight lines.

Section T.4 provides an introduction to polynomials: classification of polynomials, addition with polynomials, and subtraction with polynomials. Multiplication with polynomials, by using the distributive property, is the final topic of the chapter in Section T.5.

T.1   First-Degree Inequalities
($ax + b < c$)

OBJECTIVES

1. Be able to read inequalities both from left to right and from right to left.
2. Be able to determine whether a given inequality is true or false.
3. Know the different types of intervals of real numbers.
4. Learn how to solve and graph the solutions to inequalities of the form $ax + b < c$.

A Review of Real Numbers

Real numbers include radicals such as $\sqrt{2}$, $\sqrt{3}$, $\sqrt[3]{15}$, and $\sqrt[4]{10}$, integers such as 5, 0, and -3, fractions such as $\frac{1}{5}$, $\frac{3}{100}$, and $-\frac{5}{8}$, and decimals such as 0.007, 4.56, and 0.3333. . . . In fact, real numbers include all those numbers that can be classified as either **rational numbers** or **irrational numbers.** All of the numbers that we have discussed in this text are real numbers.

   Rational numbers are numbers that can be represented either as **terminating decimals** or as **infinite repeating decimals.** The whole numbers, integers, and fractions that we have studied are **all** rational numbers. Examples of rational numbers are

$$0, \quad -6, \quad \frac{2}{3}, \quad 27.1, \quad -90, \quad \text{and} \quad 19.3$$

   Irrational numbers are numbers that can be represented as **infinite nonrepeating decimals.** Examples of irrational numbers are

$$\pi = 3.1415926535\ldots$$
$$\sqrt{2} = 1.414213562\ldots$$
$$-\sqrt{5} = -2.236067977\ldots$$

Most calculators give irrational numbers to eight- or nine-digit accuracy.

   The diagram in Figure T.1 illustrates the relationships among various types of real numbers.

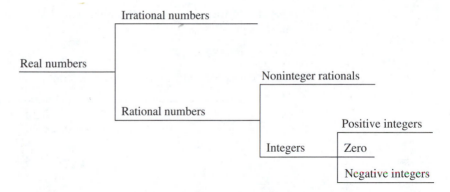

Figure T.1

EXAMPLE 1

Given the set of real numbers

$$A = \left\{ -5, -\pi, -1.3, 0, \sqrt{2}, \frac{4}{5}, 8 \right\}$$

determine which numbers in A are (a) integers, (b) rational numbers, (c) irrational numbers.

Solution

(a) Integers: $-5, 0, 8$

(b) Rational numbers: $-5, 0, 8, -1.3, \dfrac{4}{5}$

(Note that each integer is also a rational number.)

(c) Irrational numbers: $-\pi, \sqrt{2}$

Real Number Lines and Inequalities

Number lines are called **real number lines** because of the following important relationship between real numbers and points on a line.

> **There is a one-to-one correspondence between the real numbers and the points on a line.** That is, each point on a number line corresponds to one real number, and each real number corresponds to one point on a number line.

The locations of some real numbers, including integers, are shown in Figure T.2. Sometimes a calculator may be used to find the approximate value (and therefore the approximate location) of an unfamiliar real number. For example, a calculator will show that

$$\sqrt{6} = 2.449489743\ldots$$

Therefore, $\sqrt{6}$ is located between 2.4 and 2.5 on a real number line.

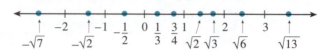

Figure T.2

To compare two real numbers, we use the following symbols of equality and inequality (reading from **left to right**):

$a = b$ a is equal to b

$a < b$ a is less than b

$a \leq b$ a is less than or equal to b

$a > b$ a is greater than b

$a \geq b$ a is greater than or equal to b

These symbols can also be read from **right to left.** For example, we can read

$a < b$ as   "b is greater than a"

and

$a > b$ as   "b is less than a"

A slash, /, through a symbol negates that symbol. Thus, for example, $\neq$ is read "is not equal to" and $\nless$ is read "is not less than."

EXAMPLE 2

Write the meaning of each of the following inequalities.

(a) $6 < 7.5$

(b) $-3 > -10$

(c) $-14 \neq |-14|$

Solution

(a) $6 < 7.5$ "6 is less than 7.5"   or   "7.5 is greater than 6"

(b) $-3 > -10$ "-3 is greater than -10"   or   "-10 is less than -3"

(c) $-14 \neq |-14|$ "-14 is not equal to the absolute value of -14"

On a number line, smaller numbers are to the left of larger numbers. (Or, larger numbers are to the right of smaller numbers.) Thus, as shown in Figure T.3 we have

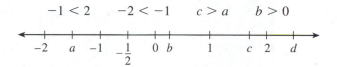

Figure T.3

Suppose that a and b are two real numbers and that $a < b$. The set of all real numbers between a and b is called an **interval of real numbers.** Intervals are classified and graphed as indicated in the box that follows. You should keep in mind the following facts as you consider the boxed information.

1. x is understood to represent real numbers.

2. Open dots at endpoints a and b indicate that these points are **not** included in the graph.

3. Solid dots at endpoints a and b indicate that these points **are** included in the graph.

Intervals of Real Numbers

Name	Symbolic Representation	Graph
Open Interval	$a < x < b$	○——○ at a, b
Closed Interval	$a \leq x \leq b$	●——● at a, b
Half-open Interval	$a \leq x < b$	●——○ at a, b
	$a < x \leq b$	○——● at a, b
Open Interval	$x > a$	○——→ at a
	$x < a$	←——○ at a
Half-open Interval	$x \geq a$	●——→ at a
	$x \leq a$	←——● at a

EXAMPLE 3

Graph the closed interval $4 \leq x \leq 6$.

Solution

Note that 4 and 6 are in the interval, as are all the real numbers between 4 and 6. For example, $4\frac{1}{2}$ and 5.99 are in the interval since

$$4 \leq 4\frac{1}{2} \leq 6 \quad \text{and} \quad 4 \leq 5.99 \leq 6$$

EXAMPLE 4

Represent the following graph using algebraic notation and tell what kind of interval it is.

Solution

$-2 < x < 0$ is an open interval.

Solving Linear Inequalities

Inequalities of the forms

$$ax + b < c \quad \text{and} \quad ax + b \leq c$$
$$ax + b > c \quad \text{and} \quad ax + b \geq c$$
$$c < ax + b < d \quad \text{and} \quad c \leq ax + b \leq d$$

are all called **linear inequalities** or **first-degree inequalities.**

The solutions to linear inequalities are intervals of real numbers, and the methods for solving linear inequalities are smilar to those used in solving linear (or first-degree) equations. The rules are the same with one important exception:

Multiplying both sides of an inequality by a negative number "reverses the sense" of the inequality.

Consider the following examples.
We know that $4 < 10$:

Add 3	Multiply by 2	Add -5
$4 < 10$	$4 < 10$	$4 < 10$
$4 + 3 \; ? \; 10 + 3$	$2 \cdot 4 \; ? \; 2 \cdot 10$	$4 + (-5) \; ? \; 10 + (-5)$
$7 < 13$	$8 < 20$	$-1 < 5$

In each instance, the sense of the inequality stayed the same, namely $<$.

Now we see that multiplying both sides by a negative number reverses the sense from $<$ to $>$:

Multiply by -6

$$4 < 10$$
$$-6 \cdot 4 \; ? \; -6 \cdot 10$$
$$-24 > -60 \qquad \text{The sense is reversed from } < \text{ to } >.$$

Rules for Solving Linear (or First-Degree) Inequalities

1. The same number (positive or negative) may be added to both sides, and the sense of the inequality will remain the same.

2. Both sides may be multiplied by (or divided by) the same **positive** number, and the sense of the inequality will remain the same.

3. Both sides may be multiplied by (or divided by) the same **negative** number, but the sense of the inequality must be **reversed.**

As with solving an equation, the object of solving an inequality is to find equivalent inequalities that are simpler than the original and to isolate the variables on one side of the inequality. The difference between the solutions of inequalities and equations is that the solution to an inequality consists of an interval of numbers (an infinite set of numbers) whereas the solution to a first-degree equation is a single number.

EXAMPLE 5

Solve the inequality $x - 3 < 2$ and graph the solution on a number line.

Solution

$$x - 3 < 2$$
$$x - 3 + 3 < 2 + 3 \qquad \text{Add 3 to both sides.}$$
$$x < 5$$

EXAMPLE 6

Solve the inequality $-2x + 7 \geq 4$ and graph the solution on a number line.

Solution

$$-2x + 7 \geq 4$$

$$-2x + 7 - 7 \geq 4 - 7 \quad \text{Add } -7 \text{ to both sides.}$$

$$-2x \geq -3$$

$$\frac{-2x}{-2} \leq \frac{-3}{-2} \quad \text{Divide both sides by } -2 \text{ and \textbf{reverse the sense.}}$$

$$x \leq \frac{3}{2}$$

$$\frac{3}{2}$$

EXAMPLE 7

Solve the inequality $7y - 8 > y + 10$ and graph the solution on a number line.

Solution

$$7y - 8 > y + 10$$

$$7y - 8 - y > y + 10 - y \quad \text{Add } -y \text{ to both sides.}$$

$$6y - 8 > 10 \qquad\qquad \text{Simplify.}$$

$$6y - 8 + 8 > 10 + 8 \qquad \text{Add \textbf{8} to both sides.}$$

$$6y > 18 \qquad\qquad \text{Simplify.}$$

$$\frac{6y}{6} > \frac{18}{6} \qquad\qquad \text{Divide both sides by \textbf{6.}}$$

$$y > 3 \qquad\qquad \text{Simplify.}$$

$$3$$

EXAMPLE 8

Find the values of x that satisfy **both** of the inequalities:

$$5 < 2x + 3 \quad \textbf{and} \quad 2x + 3 < 10$$

Graph the solution on a number line.

Solution

Since the variable expression $2x + 3$ is the same in both inequalities and $5 < 10$, we can write the two inequalities in one expression and solve both inequalities at the same time.

$$5 < 2x + 3 < 10$$

$$5 - 3 < 2x + 3 - 3 < 10 - 3 \qquad \text{Add } -3 \text{ to each part.}$$

$$2 < 2x < 7 \qquad \text{Simplify.}$$

$$\frac{2}{2} < \frac{2x}{2} < \frac{7}{2} \qquad \text{Divide each part by 2.}$$

$$1 < x < 3.5 \qquad \text{Thus, } x \text{ is greater than } 1 \text{ \textbf{and} less than 3.5.}$$

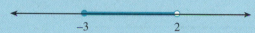

CLASSROOM PRACTICE

1. Graph the closed interval $-2 \leq x \leq 1$.

2. Graph the open interval $x > 3$.

3. Represent the graph shown here using algebraic notation and tell what type of interval it is:

Solve each inequality and graph the solution on a number line.

4. $2x - 1 > 7$ 　　　　　　　　　　　　5. $3x + 2 \leq 5x + 1$

6. $-4 < 5y + 1 \leq 11$

ANSWERS: **1.**

2.

3. $-3 \leq x < 2$; half-open interval

4. $x > 4$

5. $\frac{1}{2} \leq x$

6. $-1 < y \leq 2$

Exercises T.1

Write each of the following inequalities in words and state whether it is true or false. If an inequality is false, rewrite it in a correct form. (There may be more than one way to correct a false statement.)

1. $3 \neq -3$ **2.** $-5 < -2$ **3.** $-13 > -1$ **4.** $|-7| \neq |+7|$

5. $|-7| < +5$ **6.** $|-4| > |+3|$ **7.** $-\dfrac{1}{2} < -\dfrac{3}{4}$ **8.** $\sqrt{2} > \sqrt{3}$

9. $-4 < -6$ **10.** $|-6| > 0$

Represent each of the following graphs with algebraic notation and tell what kind of interval it is.

11.

12.

13.

14.

15.

In Exercises 16–27, graph the interval on a number line and then tell what kind of interval each is.

16. $5 < x < 8$ **17.** $-2 < x < 0$ **18.** $3 \leq y \leq 6$

19. $-3 \leq y \leq 5$ **20.** $-2 < y \leq 1$ **21.** $4 \leq x < 7$

22. $47 \leq x \leq 52$ **23.** $-12 < z < -7$ **24.** $x > -5$

25. $x \geq 0$ **26.** $x < -\sqrt{3}$ **27.** $x \leq \dfrac{2}{3}$

In Exercises 28–55, solve each inequality. Write the solution and then graph the solution on a number line.

28. $x + 5 < 6$ **29.** $x - 6 > -2$

30. $y - 4 \geq -1$ **31.** $y + 5 \leq 2$

32. $3y \leq 4$ **33.** $5y > -6$

34. $10y + 1 > 5$ **35.** $7x - 2 < 9$

36. $x + 3 < 4x + 3$ **37.** $x - 4 > 2x + 1$

38. $3x - 5 \geq x + 5$ **39.** $3x - 8 \leq x + 2$

40. $\frac{1}{2}x - 2 < 6$

41. $\frac{1}{3}x + 1 > -1$

42. $-x + 4 < -1$

43. $-x - 5 \geq -4$

44. $8y - 2 \geq 5y + 1$

45. $6x + 3 > x - 2$

46. $-5x - 7 \geq -7 + 2x$

47. $3x + 15 < x + 5$

48. $6 \leq x + 9 \leq 7$

49. $-2 \leq x - 3 \leq 1$

50. $-5 \leq 4y + 3 \leq 0$

51. $0 \leq 3x - 1 \leq 5$

52. $-2 < 5x + 3 \leq 8$

53. $-5 \leq y - 2 \leq -1$

54. $5 \leq -2x - 5 < 7$

55. $14 < -5x - 1 \leq 24$

T.2 Graphing Ordered Pairs of Real Numbers

OBJECTIVES

1. Learn the terminology related to graphs in two dimensions: such as **ordered pairs, first coordinate, second coordinate, axis,** and **quadrant.**
2. Be able to list the set of ordered pairs represented on a graph.
3. Know how to graph a set of ordered pairs of real numbers.

Equations in Two Variables

Equations such as

$$d = 40t, \qquad I = 0.18P, \qquad \text{and} \qquad y = 2x - 5$$

represent relationships between pairs of variables. These equations are said to be **equations in two variables.** The first equation, $d = 40t$, can be interpreted as follows: The distance d traveled in time t at a rate of 40 miles per hour is found by multiplying 40 by t (where t is measured in hours). Thus, if $t = 3$ hours, then $d = 40(3) = 120$ miles. The pair (3, 120) is called an **ordered pair** and is in the form (t, d).

We say that the ordered pair (3, 120) **is a solution of** (or **satisfies**) the equation $d = 40t$. Similarly, (5, 200) represents $t = 5$ and $d = 200$ and satisfies the equation $d = 40t$. In the same way, (100, 18) satisfies the equation $I = 0.18P$, where $P = 100$ and $I = 0.18(100) = 18$. In this equation, the interest I is equal to 18% (or 0.18) times the principal P and the ordered pair (100, 18) is in the form (P, I).

For the equation $y = 2x - 5$, ordered pairs are in the form (x, y) and $(3, 1)$ satisfies the equation: if $x = 3$, then $y = 2(3) - 5 = 1$. In the ordered pair (x, y), x is called the **first coordinate** (or **first component**) and y is called the **second coordinate** (or **second component**). To find ordered pairs that satisfy an equation in two variables, we can **choose any value** for one variable and find the corresponding value for the other variable by substituting into the equation. For example,

if $y = 2x - 5$, then

for $x = 2$ we have $y = 2(2) - 5 = 4 - 5 = -1$

for $x = 0$ we have $y = 2(0) - 5 = 0 - 5 = -5$

for $x = 6$ we have $y = 2(6) - 5 = 12 - 5 = 7$

All the ordered pairs $(2, -1)$, $(0, -5)$, and $(6, 7)$ satisfy the equation $y = 2x - 5$. There are an infinite number of such ordered pairs.

Since the equation $y = 2x - 5$ is solved for y, we say that the value of y "depends" on the choice of x. Thus, in an ordered pair of the form (x, y), the second coordinate y is called the **dependent variable** and the first coordinate x is called the **independent variable.**

Some examples of ordered pairs for each of the three equations discussed are shown below in table form. Remember that the choices for the value of the independent variable are arbitrary; other values could have been chosen.

$d = 40t$		$I = 0.18P$		$y = 2x - 5$	
t	d	P	I	x	y
1	40	100	18	-2	-9
2	80	200	36	0	-5
3	120	1000	180	1	-3
4	160	5000	900	5	5

Graphing Ordered Pairs

Ordered pairs of real numbers can be graphed as points in a plane by using the **Cartesian coordinate system** [named after the famous French mathematician Rene Descartes (1596–1650)]. In this system, the plane is separated into four **quadrants** by two number lines that are perpendicular to each other. The lines intersect at a point called the **origin,** represented by the ordered pair $(0, 0)$. The horizontal number line represents the independent variable and is called the **horizontal axis** (or the **x-axis**). The vertical number line represents the dependent variable and is called the **vertical axis** (or the **y-axis**). (See Figure T.4.)

On the x-axis, positive values of x are indicated to the right of the origin and negative values of x are to the left of the origin. On the y-axis, positive values of y are indicated above the origin and negative values of y are below the origin. The relationship between ordered pairs of real numbers and the points in a plane is similar to the correspondence between real numbers and points on a number line and is expressed in the following important statement.

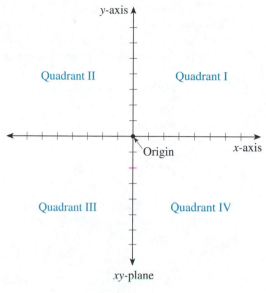

Figure T.4

> **There is a one-to-one correspondence between the points in a plane and ordered pairs of real numbers.** That is, each point in a plane corresponds to one ordered pair of real numbers, and each ordered pair of real numbers corresponds to one point in a plane. (See Figure T.5.)

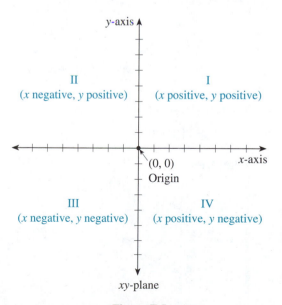

Figure T.5

The graphs of the points A (3, 1), B (−2, 3), C (−3, −1), D (1, −2), and E (2, 0) are shown in Figure T.6. The point E (2, 0) is on an axis and not in any quadrant. Each ordered pair gives the **coordinates** of the corresponding point. For example, the coordinates of point A are given by the ordered pair (3, 1).

POINT	QUADRANT
A(3, 1)	I
B(−2, 3)	II
C(−3, −1)	III
D(1, −2)	IV
E(2, 0)	x-axis

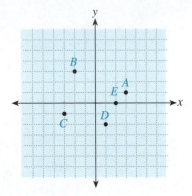

Figure T.6

Note: Unless a scale is labeled on the x-axis or on the y-axis, the grid lines are assumed to be one unit apart in both the horizontal and vertical directions.

EXAMPLE 1   Graph the set of ordered pairs:

$$\{(-2, 1), (0, 3), (1, 2), (2, -2)\}$$

Solution

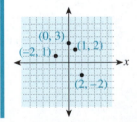

EXAMPLE 2   Graph the set of ordered pairs:

$$\{(-3, -5), (-2, -3), (-1, -1), (0, 1), (1, 3)\}$$

Solution

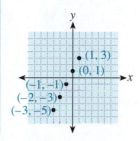

EXAMPLE 3   The graph of a set of points is given. List the set of ordered pairs that correspond to the points in the graph.

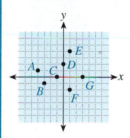

Solution

$$\{A\ (-4, 1), B\ (-3, -1), C\ (-1, 0), D\ (0, 2), E\ (1, 4), F\ (1, -2), G\ (3, 0)\}$$

Exercises T.2

In Exercises 1–10, list the set of ordered pairs that correspond to the points in the graph.

1.

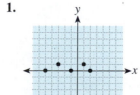

2.

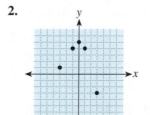

3.

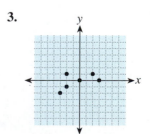

4.

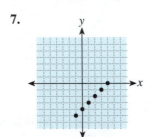

5.

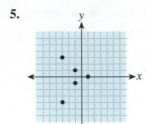

6.

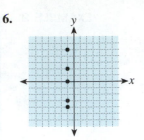

7.

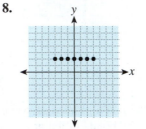

8.

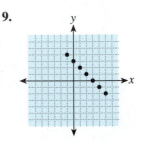

9.

10.

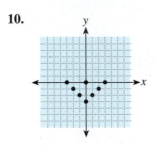

Graph each of the following sets of ordered pairs.

11. $\{(-2, 4), (-1, 3), (0, 1), (1, -2), (1, 3)\}$

12. $\{(-5, 1), (-3, 2), (-2, -1), (0, 2), (2, -1)\}$

13. $\{(-1, 2), (1, 3), (2, -2), (3, 4), (4, -2)\}$

14. $\{(-2, 3), (-1, 0), (0, -3), (2, 3), (4, -1)\}$

15. $\{(0, -3), (1, -1), (2, 1), (3, 3), (4, 5)\}$

16. $\{(-3, 3), (-2, 2), (-1, 1), (0, 0), (1, -1)\}$

17. $\{(-2, -1), (0, -1), (2, -1), (4, -1), (6, -1)\}$

18. $\{(-3, 1), (-2, 1), (-1, 1), (0, 1), (1, 1)\}$

19. $\{(-3, 3), (-1, 1), (0, 0), (1, 1), (3, 3)\}$

20. $\{(-2, -2), (-1, -1), (0, 0), (1, -1), (2, -2)\}$

21. $\{(-3, 9), (-2, 4), (-1, 1), (1, 1), (2, 4), (3, 9)\}$

22. $\{(-3, -9), (-2, -4), (-1, -1), (1, -1), (2, -4), (3, -9)\}$

23. $\{(-4, 0), (-2, 0), (0, 0), (2, 0), (4, 0)\}$

24. $\{(0, -3), (0, -1), (0, 0), (0, 1), (0, 3)\}$

25. $\{(-2, 1), (1, 4), (2, 5), (3, 6), (4, 7)\}$

26. $\{(-1, -5), (0, -2), (1, 1), (2, 4), (3, 7)\}$

27. $\{(-2, -7), (-1, -5), (2, 1), (3, 3), (4, 5)\}$

28. $\{(0, 1), (1, -1), (2, -3), (3, -5), (5, -9)\}$

29. $\{(-3, 11), (-2, 8), (0, 2), (2, -4), (3, -7)\}$

30. $\{(-2, -1), (-1, 1), (1, 5), (2, 7), (3, 9)\}$

31. Given the equation $I = 0.12P$, where I is the interest earned on a principal P at a rate of 12%:

(a) Make a table of ordered pairs for the values of P and I if P has the values $100, $200, $300, $400, and $500.

(b) Graph the points corresponding to the ordered pairs. Place P on the horizontal axis and I on the vertical axis.

32. Given the equation $d = 16t^2$, where d is the distance an object falls in feet and t is the time in seconds that the object falls:

(a) Make a table of ordered pairs for the values of t and d if t has the values 1, 2, 3, 4, and 5 seconds.

(b) Graph the points corresponding to the ordered pairs. Place t on the horizontal axis and d on the vertical axis.

33. Given the equation $F = \dfrac{9}{5}C + 32$, where C is temperature in degrees Celsius and F is the corresponding temperature in degrees Fahrenheit:

(a) Make a table of ordered pairs for the values of C and F if C has the values $-20°$, $-15°$, $-10°$, $-5°$, $0°$, $5°$, and $10°$.

(b) Graph the points corresponding to the ordered pairs. Place C on the horizontal axis and F on the vertical axis.

34. Given the equation $V = 25h$, where V is the volume (in cm^3) of a box with height h in centimeters and a fixed base of area 25 cm^2:

(a) Make a table of ordered pairs for the values of h and V if h has the values of 3 cm, 5 cm, 6 cm, 8 cm, and 10 cm.

(b) Graph the points corresponding to the ordered pairs. Place h on the horizontal axis and V on the vertical axis.

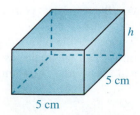

5 cm

5 cm

T.3   Graphing Linear Equations (Ax + By = C)

OBJECTIVES

1. Be familiar with the **standard form** for the equation of a line: Ax + By = C.
2. Know how to find the y-intercept and the x-intercept of a line.
3. Know how to plot points that satisfy a linear equation and draw the graph of the corresponding line.

Linear Equations in Standard Form

There are an infinite number of ordered pairs of real numbers that satisfy the equation $y = 3x + 1$. In Section T.2, we substituted only integer values for x in introducing the concepts of ordered pairs and graphing ordered pairs. However, the discussion was based on ordered pairs of real numbers, and this means that we can also substitute values for x that are rational numbers or irrational numbers. That is, fractions and radicals can be substituted just as well as integers. The corresponding y values may also be expressions with fractions and radicals. For example,

$$\text{if } x = \frac{1}{3}, \quad \text{then} \quad y = 3 \cdot \frac{1}{3} + 1 = 1 + 1 = 2$$

$$\text{if } x = -\frac{3}{4}, \quad \text{then} \quad y = 3\left(-\frac{3}{4}\right) + 1 = -\frac{9}{4} + \frac{4}{4} = -\frac{5}{4}$$

$$\text{if } x = \sqrt{2}, \quad \text{then} \quad y = 3\sqrt{2} + 1$$

The important idea here is that even though we cannot actually substitute all real numbers for x (we do not have enough time or paper), there is a corresponding real value for y for any real value of x we choose. In Figure T.7 a few points that satisfy the equation $y = 3x + 1$ have been graphed so that you can observe a pattern.

The points in Figure T.7 appear to lie on a straight line and, in fact, they do. We can draw a straight line through all the points, as shown in Figure T.8, and **any point that lies on the line will satisfy the equation.**

x	$y = 3x + 1$
-2	$y = 3(-2) + 1 = -5$
-1	$y = 3(-1) + 1 = -2$
0	$y = 3(0) + 1 = 1$
$\frac{2}{3}$	$y = 3\left(\frac{2}{3}\right) + 1 = 3$
2	$y = 3(2) + 1 = 7$

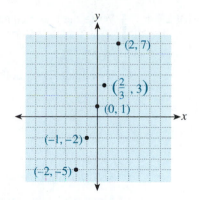

Figure T.7

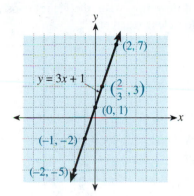

Figure T.8

The points (ordered pairs of real numbers) that satisfy any equation of the form

$Ax + By = C$ where A and B are not both 0

will lie on a straight line. The equation is called a **linear equation in two variables** and is in the **standard form** for the equation of a line.

The linear equation $y = 3x + 1$ is solved for y and is not in standard form. However, it can be written in the standard form as

$$-3x + y = 1 \quad \text{or} \quad 3x - y = -1$$

All three forms are acceptable and correct.

EXAMPLE 1   Write the linear equation $y = -2x + 5$ in standard form.

Solution

By adding $2x$ to both sides we get the standard form:

$$2x + y = 5$$

Graphing Linear Equations

Since we know that the graph of a linear equation is a straight line, we need only graph two points (because two points determine a line) and draw the line through these two points. A good check against possible error is to locate three points instead of only two. Also, the values chosen for x should be such that the points are not too close together.

EXAMPLE 2   Draw the graph of the linear equation $x + 2y = 6$.

Solution

We find and plot three ordered pairs that satisfy the equation and then sketch the graph.

$$
\begin{array}{ccc}
x = -2 & x = 0 & x = 2 \\
-2 + 2y = 6 & 0 + 2y = 6 & 2 + 2y = 6 \\
2y = 8 & 2y = 6 & 2y = 4 \\
y = 4 & y = 3 & y = 2
\end{array}
$$

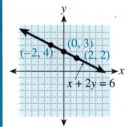

(Locating three points helps in avoiding errors. Avoid choosing points close together.)

The **y-intercept** of a line is the point where the line crosses the y-axis. This point can be located by letting $x = 0$. Similarly, the **x-intercept** is the point where the line crosses the x-axis and is found by letting $y = 0$. When the line is not vertical or horizontal, the y-intercept and the x-intercept are generally easy to locate and are frequently used for drawing the graph of a linear equation.

EXAMPLE 3

Draw the graph of the linear equation $x - 2y = 8$ by locating the y-intercept and the x-intercept.

Solution

Find the y-intercept:

$$x = 0$$
$$0 - 2y = 8$$
$$y = -4$$

Find the x-intercept:

$$y = 0$$
$$x - 2 \cdot 0 = 8$$
$$x = 8$$

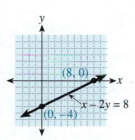

EXAMPLE 4

Locate the y-intercept and the x-intercept and draw the graph of the linear equation $3x - y = 3$.

Solution

Find the y-intercept:

$$x = 0$$
$$3 \cdot 0 - y = 3$$
$$y = -3$$

Find the x-intercept:

$$y = 0$$
$$3x - 0 = 3$$
$$x = 1$$

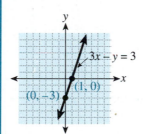

If $A = 0$ and $B \neq 0$ in the standard form $Ax + By = C$, the equation takes the form $By = C$ and can be solved for y as $y = \dfrac{C}{B}$. For example, we can write $0x + 3y = 6$ as $y = 2$. Thus, no matter what value x has, the value of y is 2. The graph of the equation $y = 2$ is a **horizontal line,** as shown in Figure T.9. For horizontal lines (other than the x-axis itself) there is no x-intercept.

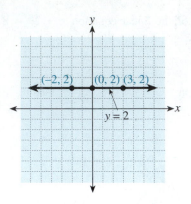

Figure T.9

The y-coordinate is 2 for every point on the line $y = 2$.

If $B = 0$ and $A \neq 0$ in the standard form $Ax + By = C$, the equation takes the form $Ax = C$ or $x = \dfrac{C}{A}$. For example, we can write $5x + 0y = -5$ as $x = -1$. Thus, no matter what value y has, the value of x is -1. The graph of the equation $x = -1$ is a **vertical line,** as shown in Figure T.10. For vertical lines (other than the y-axis itself) there is no y-intercept.

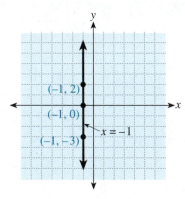

Figure T.10

The x-coordinate is -1 for every point on the line $x = -1$.

EXAMPLE 5   Graph the horizontal line $y = -1$ and the vertical line $x = 3$ on the same coordinate system.

Solution

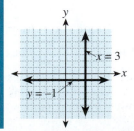

Exercises T.3

Graph the following linear equations. Label at least three points on each line.

1. $y = x + 1$ **2.** $y = x + 2$ **3.** $y = x - 4$

4. $y = x - 6$ **5.** $y = 2x$ **6.** $y = 3x$

7. $y = -x$ **8.** $y = -4x$ **9.** $y = x$

10. $y = 2 - x$ **11.** $y = 3 - x$ **12.** $y = 5 - x$

13. $y = 2x + 1$ **14.** $y = 2x - 1$ **15.** $y = 2x - 3$

16. $y = 2x + 5$ **17.** $y = -2x + 1$ **18.** $y = -2x - 2$

19. $y = -3x + 2$ **20.** $y = -3x - 4$ **21.** $x - 2y = 4$

22. $x - 3y = 6$ **23.** $-2x + 3y = 6$ **24.** $2x - 5y = 10$

25. $-2x + y = 4$ **26.** $2x + 3y = 6$ **27.** $3x + 5y = 15$

28. $4x + y = 8$ **29.** $x + 4y = 8$ **30.** $3x - 4y = 12$

31. $x = 4$ **32.** $y = 5$ **33.** $y = -5$

34. $x = -2$ **35.** $x - 5 = 0$ **36.** $y + 3 = 0$

37. $2y - 3 = 0$ **38.** $3x = -1$ **39.** $4x - 5 = 0$

40. $3y + 7 = 0$

T.4 Addition and Subtraction with Polynomials

OBJECTIVES

1. Understand how to classify polynomials.
2. Learn how to add polynomials.
3. Learn how to subtract polynomials.

Classification of Polynomials

An **algebraic term** is an expression that involves only multiplication and/or division with constants and/or variables. **Like terms** (or **similar terms**) are terms that contain the same variable(s) to the same power or are terms that are constants. The **distributive property** can be used to combine like terms. For example,

$$7x + 5x = (7 + 5)x = 12x$$

and

$$15x^2 - 20x^2 = (15 - 20)x^2 = -5x^2$$

The distributive property is restated here for your reference and convenience.

> ## Distributive Property of Multiplication over Addition
>
> For real numbers a, b, and c,
>
> $$a(b + c) = ab + ac$$

A **monomial** is a single term with no variable in a denominator and with variables that have only whole number exponents.

> The general form of a **monomial in x** is
>
> kx^n where n is a whole number and k is any real number
>
> n is called the **degree** of the monomial, and k is the **coefficient.**

Consider a constant monomial such as 7. We can write

$$7 = 7 \cdot 1 = 7x^0.$$

With this reasoning, we say that a nonzero constant is a **monomial of 0 degree.** The constant 0 is a special case. Since we can write

$$0 = 0x^2 = 0x^4 = 0x^{63}$$

(in effect, any exponent can be used on x), we say that **0 is a monomial of no degree.**

A **polynomial** is a monomial or the sum of monomials. The **degree of a polynomial** is the largest of the degrees of its terms after all like terms have been combined. Generally, for easy reading, a polynomial is written so that the degrees of its terms either decrease from left to right or increase from left to right. If the degrees decrease, we say that the terms are written in **descending order.** If the degrees increase, we say that the terms are written in **ascending order.** For example,

$3x^4 + 5x^2 - 8x + 34$ is a fourth-degree polynomial written in descending order.

$15 - 3x + 4x^2 + x^3$ is a third-degree polynomial written in ascending order.

For consistency and because of the style used in multiplying polynomials, the polynomials in this text will be written in descending order.

> **Note:** Although polynomials may have more than one variable, in this text we will limit our discussion to polynomials in only one variable.

Some forms of polynomials are used so frequently that they have been given special names, as indicated in the following box.

Classification of Polynomials	**Example**
Monomial: Polynomial with one term	$-2x^3$
Binomial: Polynomial with two terms	$7x + 23$
Trinomial: Polynomial with three terms	$a^2 + 5a + 6$

EXAMPLE 1

Simplify each of the following polynomials and tell its degree and what type of polynomial it is.

(a) $4x^2 - 7x^2$

(b) $x^3 + 8x - 9 - x^3$

Solution

(a) $4x^2 - 7x^2 = (4 - 7)x^2 = -3x^2$ second-degree monomial

(b) $x^3 + 8x - 9 - x^3 = (1 - 1)x^3 + 8x - 9$

$$= 0x^3 + 8x - 9$$

$$= 8x - 9 \qquad \text{first-degree binomial}$$

Addition with Polynomials

The **sum** of two or more polynomials is found by combining like terms. The polynomials may be written horizontally (as in Example 2) or vertically with like terms aligned (as in Example 3).

EXAMPLE 2

Add the polynomials as indicated.

$$(4x^3 - 5x^2 + 15x - 10) + (-7x^2 - 2x + 3) + (4x^2 + 9)$$

$$= 4x^3 + (-5x^2 - 7x^2 + 4x^2) + (15x - 2x) + (-10 + 3 + 9)$$

$$= 4x^3 - 8x^2 + 13x + 2$$

EXAMPLE 3

Find the sum.

$$x^3 - 4x^2 + 2x + 13$$
$$2x^3 + \ x^2 - 5x + \ 6$$
$$\overline{3x^3 - 3x^2 - 3x + 19}$$

EXAMPLE 4

Write the sum

$$(5x^3 - 8x^2 - 10x + 2) + (3x^3 + 4x^2 - x - 7)$$

in the vertical format and find the sum.

Solution

$$5x^3 - 8x^2 - 10x + 2$$
$$3x^3 + 4x^2 - \quad x - 7$$
$$\overline{8x^3 - 4x^2 - 11x - 5}$$

Subtraction with Polynomials

The opposite of a polynomial can be indicated by writing a negative sign in front of the polynomial. In this case the sign of every term in the polynomial is changed. For example,

$$-(5x^2 - 6x - 1) = -5x^2 + 6x + 1$$

We can also think of this type of expression as indicating multiplication by -1 and use the distributive property as follows:

$$-(5x^2 - 6x - 1) = -1(5x^2 - 6x - 1)$$
$$= -1(5x^2) - 1(-6x) - 1(-1)$$
$$= -5x^2 + 6x + 1$$

The answer is the same in either case. Therefore, the **difference** of two polynomials can be found by **adding the opposite** of the polynomial being subtracted. The polynomials can be written horizontally (as in Example 5) or vertically with like terms aligned (as in Example 6).

EXAMPLE 5

Subtract the polynomials as indicated.

$$(9x^3 + 4x^2 - 15) - (-2x^3 + x^2 - 5x - 6)$$
$$= 9x^3 + 4x^2 - 15 + 2x^3 - x^2 + 5x + 6$$
$$= 11x^3 + 3x^2 + 5x - 9$$

When using the vertical alignment format, write a 0 as a placeholder for any missing powers of the variable in order to maintain proper alignment and ensure that like terms will be subtracted (or added).

EXAMPLE 6

Find the difference:

$$8x^4 + 2x^3 - 5x^2 + 0x - 7$$
$$-(3x^4 + 5x^3 - x^2 + 6x - 11)$$

Be sure to change the sign of each term in the polynomial being subtracted and then combine like terms.

$$8x^4 + 2x^3 - 5x^2 + 0x - 7$$
$$\underline{-3x^4 - 5x^3 + x^2 - 6x + 11} \quad \leftarrow \text{signs changed}$$
$$5x^4 - 3x^3 - 4x^2 - 6x + 4 \quad \leftarrow \text{difference}$$

CLASSROOM PRACTICE

Find the indicated sums.

1. $(2x^2 - 5x + 3) + (x^2 - 7) + (2x + 10)$

2. $\quad x^4 - 5x^3 + 7x^2 \qquad + 12$
$\quad \underline{3x^4 - 6x^3 - 9x^2 - 2x - 14}$

Find the indicated differences.

3. $\quad (-2x^3 + 5x^2 + 8x - 1)$
$\quad \underline{-(\ 2x^3 - x^2 - 6x + 13)}$

4. $\quad 15x^3 + 10x^2 - 17x - 25$
$\quad \underline{-(12x^3 + 10x^2 + 3x - 16)}$

ANSWERS: **1.** $3x^2 - 3x + 6$ **2.** $4x^4 - 11x^3 - 2x^2 - 2x - 2$
3. $-4x^3 + 6x^2 + 14x - 14$ **4.** $3x^3 - 20x - 9$

Exercises T.4

Simplify each polynomial and tell what type of polynomial it is and its degree.

1. $5x + 6x - 10 + 3$

2. $8x - 9x + 14 - 5$

3. $3x^2 - x^2 + 7x - x + 2$

4. $4x^2 + 3x^2 - x + 2x + 18$

5. $a^3 + 4a^2 + a^2 - a^3$

6. $y^4 - 2y^3 + 3y^3 - y^4$

7. $-2y^2 - y^2 + 10y - 3y + 2 + 5$

8. $-5a^2 + 2a^2 - 4a - 2a + 4 + 1$

9. $5x^3 + 2x - 8x + 17 + 3x$

10. $-4x^3 + 5x^2 + 12x - x$

Add or subtract as indicated and simplify if possible.

11. $(3x - 5) + (2x - 5)$

12. $(7x + 8) + (-3x + 8)$

13. $(x^2 + 4x - 6) + (x^2 - 4x + 2)$

14. $(x^2 - 3x - 10) + (2x^2 - 3x - 10)$

15. $(4y^3 + 2y - 7) + (3y^2 - 2)$

16. $(-2y^3 + y^2 - 4) + (3y^2 - 6)$

17. $(8x + 3) - (7x + 5)$

18. $(4x - 9) - (5x + 2)$

19. $(2a^2 + 3a - 1) - (a^2 - 2a - 1)$

20. $(a^2 - 5a + 3) - (2a^2 + 5a - 3)$

21. $(9x^3 + x^2 - x) - (-3x^3 + 5x)$

22. $(4x^3 - 9x + 11) - (-x^3 + 2x + 1)$

Add in Exercises 23–26.

23. $\quad x^2 + 5x - 7$
$\underline{-3x^2 + 2x - 1}$

24. $2x^2 + 4x - 6$
$\underline{3x^2 - 9x + 2}$

25. $\quad x^3 + 2x^2 + \ x$
$\underline{2x^3 - 2x^2 - 2x + 6}$

26. $x^3 + 6x^2 + 7x - 8$
$\underline{\qquad 7x^2 + 2x + 1}$

Subtract in Exercises 27–30.

27. $\quad 9x^2 + 3x - 2$
$\underline{-(4x^2 + 5x + 3)}$

28. $\quad -3x^2 + 6x - 7$
$\underline{-(\ 2x^2 - \ x + 7)}$

29. $\quad 4x^3 \qquad + 10x - 15$
$\underline{-(\ x^3 - 5x^2 - \ 3x - \ 9)}$

30. $\quad x^3 - 8x^2 + 11x + 6$
$\underline{-(-3x^3 + 8x^2 - \ 2x + 6)}$

Writing and Thinking about Mathematics

31. (a) Find the sum of these two polynomials.

$2x^3 - 4x^2 + 3x + 20$
$\underline{3x^3 + \ x^2 - 5x + 10}$

(b) Now, substitute 1 for x in each of the polynomials in part (a), including the sum. Does the sum of the first two values equal the value of the sum?

(c) Repeat the process in part (b) using $x = 3$.

(d) Substituting a value for x in each polynomial seems to provide a method for checking answers. Do you think that this method will catch all errors? Would substituting $x = 0$ be a good idea? Briefly discuss your reasoning.

32. (a) Find the difference of these two polynomials.

$2x^3 - 4x^2 + 3x + 20$
$\underline{-(3x^3 + \ x^2 - 5x + 10)}$

(b) Now, substitute 2 for x in each of the polynomials in part (a), including the difference. Does the difference of the first two values equal the value of the difference?

(c) Repeat the process in part (b) using $x = 3$.

(d) Substituting a value for x in each polynomial seems to provide a method for checking answers. Do you think that this method will catch all errors? Would substituting $x = 0$ be a good idea? Briefly discuss your reasoning.

T.5   Multiplication with Polynomials

OBJECTIVES

1. Learn how to use the distributive property to multiply a monomial and a polynomial with two or more terms.
2. Learn how to multiply two binomials.
3. Know how to use a vertical format to multiply two polynomials.

Review of Exponents and the Product Rule for Exponents

Exponents have been discussed in several sections and used in conjunction with a variety of topics throughout the text. Whole number exponents were introduced in Chapter 2, where we defined

$$\underbrace{a \cdot a \cdot a \cdots a}_{n \text{ factors}} = a^n \quad \text{and} \quad a = a^1$$

We also stated that for $a \neq 0$, $a^0 = 1$. (0^0 is undefined.)

Expressions with exponents can be evaluated with a calculator that has a key marked $\boxed{x^y}$ (or a similar symbol). In Chapter 5, we noted that calculators use scientific notation for very large and very small decimal numbers, and scientific notation involves both positive and negative integer exponents. For example,

$$3{,}200{,}000 = 3.2 \times 10^6 \quad \text{and} \quad 0.000047 = 4.7 \times 10^{-5}.$$

In Section 8.2, we again used exponents to find compound interest with the formula $A = P\left(1 + \dfrac{r}{n}\right)^{nt}$.

One of the basic rules (or properties) of exponents deals with multiplying powers that have the same base. For example,

$$a^2 \cdot a^4 = \underbrace{(a \cdot a)}_{\text{2 factors}} \cdot \underbrace{(a \cdot a \cdot a \cdot a)}_{\text{4 factors}} = \underbrace{(a \cdot a \cdot a \cdot a \cdot a \cdot a)}_{\text{6 factors}} = a^6$$

or

$$a^2 \cdot a^4 = a^{2+4} = a^6$$

The property of exponents illustrated by this example is called the **Product Rule for Exponents.**

Product Rule for Exponents

For any real number a and whole numbers m and n,

$$a^m \cdot a^n = a^{m+n}$$

(To multiply two powers with the same base, keep the base and add the exponents.)

Note: The Product Rule for Exponents is stated here only for whole number exponents because those are the only exponents we use with polynomials. However, this rule, and many other rules for exponents, are valid for negative integer exponents and fractional exponents as well. You will study these ideas in detail in later courses in mathematics.

EXAMPLE 1

Use the Product Rule for Exponents to simplify each of the following expressions.

(a) $7^2 \cdot 7^3$ (b) $x^3 \cdot x^5$ (c) $3y \cdot 5y^9$

Solution

Use the Product Rule in each case.

(a) $7^2 \cdot 7^3 = 7^{2+3} = 7^5$ (or 16,807)

(b) $x^3 \cdot x^5 = x^{3+5} = x^8$

(c) $3y \cdot 5y^9 = (3 \cdot 5)(y^1 \cdot y^9) = 15 \cdot y^{1+9} = 15y^{10}$

Multiplication with Polynomials

We can find **the product of a monomial and a polynomial of two or more terms** by using the distributive property and the Product Rule for Exponents as follows:

$$6x(3x + 5) = 6x \cdot 3x + 6x \cdot 5 = 18x^2 + 30x$$

$$4x^2(3x^2 - 5x + 2) = 4x^2(3x^2) + 4x^2(-5x) + 4x^2(+2)$$
$$= 12x^4 - 20x^3 + 8x^2$$

$$-2a^3(5a^2 + 3a - 7) = (-2a^3)(5a^2) + (-2a^3)(+3a) + (-2a^3)(-7)$$
$$= -10a^5 - 6a^4 + 14a^3$$

Now, **to multiply two binomials,** such as $(x + 5)(x + 8)$, we can apply the distributive property three times. For the first step we treat the binomial $(x + 8)$ as a single term and multiply on the right as follows:

$$(b + c)a \quad = \quad ba \quad + \quad ca$$

$$(x + 5)(x + 8) = x(x + 8) + 5(x + 8)$$

The product is completed by applying the distributive property twice more and combining like terms:

$$(x + 5)(x + 8) = x(x + 8) + 5(x + 8)$$
$$= x \cdot x + x \cdot 8 + 5 \cdot x + 5 \cdot 8$$
$$= x^2 + 8x + 5x + 40$$
$$= x^2 + 13x + 40$$

EXAMPLE 2

Find each product.

(a) $3a(2a^2 + 3a - 4)$

(b) $-2x^3(7x^2 + 4x - 9)$

Solution

(a) $3a(2a^2 + 3a - 4) = 3a(2a^2) + 3a(3a) + 3a(-4)$
$$= 6a^3 + 9a^2 - 12a$$

(b) $-2x^3(7x^2 + 4x - 9) = -2x^3(7x^2) + (-2x^3)(+4x) + (-2x^3)(-9)$
$$= -14x^5 - 8x^4 + 18x^3$$

EXAMPLE 3

Find each product.

(a) $(x + 5)(x - 10)$

(b) $(2x - 3)(2x - 7)$

Solution

(a) $(x + 5)(x - 10) = x(x - 10) + 5(x - 10)$
$$= x^2 - 10x + 5x - 50$$
$$= x^2 - 5x - 50$$

(b) $(2x - 3)(2x - 7) = 2x(2x - 7) - 3(2x - 7)$
$$= 4x^2 - 14x - 6x + 21$$
$$= 4x^2 - 20x + 21$$

Another technique for finding the product of two polynomials is outlined as follows:

> **Multiplying Polynomials Vertically**
>
> **1.** Write the polynomials in a vertical format with one polynomial directly below the other.
>
> **2.** Multiply each term of the top polynomial by each term in the bottom polynomial and align like terms.
>
> **3.** Combine like terms.

This technique is illustrated in Examples 4–6.

EXAMPLE 4

Find the product $(2x + 3)(3x^2 + 4x - 5)$.

Solution

Arrange the polynomials in a vertical format and multiply each term in the top polynomial by $2x$.

$$
\begin{array}{r}
3x^2 + 4x - 5 \\
2x + 3 \\
\hline
6x^3 + 8x^2 - 10x
\end{array}
$$

Next multiply each term in the top polynomial by 3 and align like terms.

$$
\begin{array}{r}
3x^2 + 4x - 5 \\
2x + 3 \\
\hline
6x^3 + 8x^2 - 10x \\
9x^2 + 12x - 15
\end{array}
$$

Now combine like terms to find the product.

$$
\begin{array}{r}
3x^2 + 4x - 5 \\
2x + 3 \\
\hline
6x^3 + 8x^2 - 10x \\
9x^2 + 12x - 15 \\
\hline
6x^3 + 17x^2 + 2x - 15 \quad \text{product}
\end{array}
$$

EXAMPLE 5   Multiply: $(3x - 1)(x^3 - 5x^2 + 2x + 6)$.

Solution

$$
\begin{array}{r}
x^3 - 5x^2 + 2x + 6 \\
3x - 1 \\
\hline
3x^4 - 15x^3 + 6x^2 + 18x \\
-\ x^3 + 5x^2 - 2x - 6 \\
\hline
3x^4 - 16x^3 + 11x^2 + 16x - 6
\end{array}
$$

EXAMPLE 6   Find the product: $(x^2 + x + 2)(x^2 + 3x - 4)$.

Solution

$$
\begin{array}{ll}
x^2 + 3x - 4 & \\
x^2 + x + 2 & \\
\hline
x^4 + 3x^3 - 4x^2 & \text{Multiply by } x^2. \\
\quad\ x^3 + 3x^2 - 4x & \text{Multiply by } x. \\
\quad\quad\quad 2x^2 + 6x - 8 & \text{Multiply by 2.} \\
\hline
x^4 + 4x^3 + x^2 + 2x - 8 & \text{Combine like terms.}
\end{array}
$$

CLASSROOM PRACTICE

Use the Product Rule for Exponents to simplify each of the following expressions.

1. $8^2 \cdot 8$ **2.** $(-2)^3 \cdot (-2)^4$ **3.** $-5x^3 \cdot 2x^3$

Find each indicated product and simplify.

4. $5a(3a^2 + 9a - 2)$ **5.** $7x^2(-2x^3 + 8x^2 + 3x - 5)$

6. $(x - 9)(2x + 1)$ **7.** $(2x + 5)(x^2 - 8x + 1)$

8. $2x^3 + 4x^2 - 7$ (**Note:** In Exercise 8, be sure to align like terms
    $5x + 6$ as you multiply.)

ANSWERS: **1.** 8^3 (or 512)   **2.** $(-2)^7$ (or -128)   **3.** $-10x^6$
4. $15a^3 + 45a^2 - 10a$ **5.** $-14x^5 + 56x^4 + 21x^3 - 35x^2$
6. $2x^2 - 17x - 9$ **7.** $2x^3 - 11x^2 - 38x + 5$
8. $10x^4 + 32x^3 + 24x^2 - 35x - 42$

Exercises T.5

Find each indicated product and simplify if possible.

1. $-4x^2(-3x^2)$ **2.** $(-5x^3)(-2x^2)$ **3.** $9a^2(2a)$

4. $-7a^2(2a^4)$ **5.** $4y(2y^2 + y + 2)$ **6.** $5y(3y^2 - 2y + 1)$

7. $-1(4x^3 - 2x^2 + 3x - 5)$ **8.** $-1(7x^3 + 3x^2 - 4x - 2)$

9. $7x^2(-2x^2 + 3x - 12)$ **10.** $-3x^2(x^2 - x + 13)$

11. $(x + 2)(x + 5)$ **12.** $(x + 3)(x + 10)$ **13.** $(x - 4)(x - 3)$

14. $(x - 7)(x - 2)$ **15.** $(a + 6)(a - 2)$ **16.** $(a - 7)(a + 5)$

17. $(7x + 1)(x - 3)$ **18.** $(5x - 6)(3x + 2)$ **19.** $(2x + 3)(2x - 3)$

20. $(6x + 5)(6x - 5)$ **21.** $(x + 3)(x - 3)$ **22.** $(x + 5)(x - 5)$

23. $(x + 1.5)(x - 1.5)$ **24.** $(x + 2.5)(x - 2.5)$ **25.** $(3x + 7)(3x - 7)$

26. $(5x + 1)(5x - 1)$

Find the indicated products. Be sure to align like terms, particularly if some degrees are missing.

27. $x^3 + 5x + 3$
   $\underline{x + 4}$

28. $x^3 - 2x + 1$
   $\underline{x + 5}$

29. $2x^3 + x - 7$
   $\underline{x - 2}$

30. $4x^3 - x - 8$
   $\underline{x - 6}$

31. $x^3 - 6x - 10$
   $\underline{3x + 1}$

32. $x^3 - 5x - 11$
   $\underline{2x + 9}$

33. $x^4 + 2x^3 - 5x^2 + 3$
   $\underline{x^2 + x - 1}$

34. $x^4 - 3x^2 + 4x - 1$
   $\underline{x^2 - 2x + 1}$

35. $x^3 - 7x + 2$
   $\underline{x^2 + 2x + 3}$

36. $x^3 + 3x - 4$
   $\underline{2x^2 - x + 3}$

37. $x^4 + 3x^2 - 5$
   $\underline{x^4 - x^2 + 1}$

38. $x^4 - 4x^2 + 7$
   $\underline{x^4 + x^2 - 1}$

39. $x^3 + x^2 + x + 1$
   $\underline{x^3 + x^2 + x + 1}$

40. $x^3 - x^2 + x - 1$
   $\underline{x^3 - x^2 + x - 1}$

41. The sides of the square shown in the figure are of length $(a + b)$. The area of the square is $(a + b)^2$. Show that $(a + b)^2$ is equal to the sum of the indicated areas in the figure.

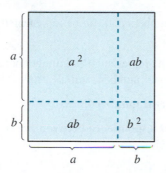

42. (a) Find the product of these two polynomials:

$$2x^3 - 4x^2 - 5x + 6$$
$$5x - 2$$

(b) Now, substitute 2 for x in each of the polynomials in part (a), including the product. Does the product of the first two values equal the value of the product?

(c) Repeat the process in part (b) using $x = -2$.

(d) Substituting a value for x in each polynomial seems to provide a method for checking answers. Do you think that this method will catch all errors? Would substituting $x = 0$ be a good idea? Briefly discuss your reasoning.

Chapter Summary

Key Terms and Ideas

Rational numbers are numbers that can be represented as terminating decimals or infinite repeating decimals.

Irrational numbers can be represented as infinite nonrepeating decimals. The **real numbers** include all rational numbers and all irrational numbers.

Number lines are called **real number lines.**

There is a one-to-one correspondence between the real numbers and the points on a line.

Intervals are classified as

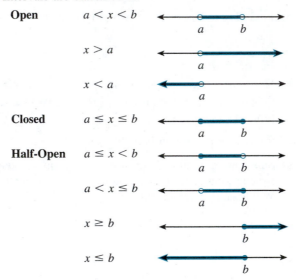

Open	$a < x < b$
	$x > a$
	$x < a$
Closed	$a \leq x \leq b$
Half-Open	$a \leq x < b$
	$a < x \leq b$
	$x \geq b$
	$x \leq b$

Linear inequalities are of the form $ax + b < c.$ ($>$, $\leq$, or $\geq$ might appear instead of $<$.)

In an **ordered pair** such as (x, y), x is called the **first component** (or **first coordinate**), and y is called the **second component** (or **second coordinate**). x is also called the **independent variable** and y is called the **dependent variable.**

In the **Cartesian coordinate system,** two perpendicular number lines (called **axes**) intersect at a point (called the **origin**) and separate a plane into four **quadrants.** The horizontal number line is called the **horizontal axis** or **x-axis.** The vertical number line is called the **vertical axis** or **y-axis.**

Each point in a plane corresponds to one ordered pair of real numbers, and each ordered pair of real numbers corresponds to one point in a plane.

An equation in the form $Ax + By = C$ is said to be in the **standard form** on a **linear equation in two variables.** The points that satisfy any such equation will lie on a straight line. The **y-intercept** of a line is the point where the line crosses the y-axis. The **x-intercept** is the point where the line crosses the x-axis.

The general form of a **monomial in x** is

  kx^n where n is a whole number and k is any real number.

n is called the **degree** of the monomial, and k is the **coefficient.**

A **polynomial** is a monomial or the sum of monomials. The **degree of a polynomial** is the largest of the degrees of its terms after all like terms have been combined. Some classifications of polynomials are

Monomial: Polynomial with one term
Binomial: Polynomial with two terms
Trinomial: Polynomial with three terms

Polynomials can be added by combining like terms. Polynomials can be multiplied by using Product Rule for Exponents and the distributive property.

Rules and Properties

Rules for Solving Linear (or First-Degree) Inequalities

1. The same number (positive or negative) may be added to both sides, and the sense of the inequality will remain the same.

2. Both sides may be multiplied by (or divided by) the same **positive** number, and the sense of the inequality will remain the same.

3. Both sides may be multiplied by (or divided by) the same **negative** number, but the sense of the inequality must be **reversed.**

Product Rule for Exponents

For any nonzero real number a and whole numbers m and n,

$$a^m \cdot a^n = a^{m+n}$$

Procedures

> **Multiplying Polynomials Vertically**
>
> **1.** Write the polynomials in a vertical format with one polynomial directly below the other.
>
> **2.** Multiply each term of the top polynomial by each term in the bottom polynomial and align like terms.
>
> **3.** Combine like terms.

Chapter Review Questions

1. Answer the following questions using the set $B = \left\{ -\sqrt{5},\ -\frac{1}{2},\ 0,\ 6.13 \right\}$.

   (a) Which of the numbers in set B is an integer?

   (b) Which of the numbers in set B is a rational number?

   (c) Which of the numbers in set B is a real number?

Graph each of the following intervals and tell what kind of interval it is.

2. $-\frac{1}{2} < x < \frac{3}{4}$ 3. $0 \le x < 5$ 4. $x \ge \sqrt{2}$

5. $-3 \le x \le 3.1$ 6. $y < \frac{1}{3}$ 7. $14 < y \le 15$

Represent each of the following graphs using interval notation and tell what kind of interval it is.

8.
9.

Solve and graph the solution for each of the following inequalities.

10. $x + 3 < -1$ 11. $y - 5 \le 2$ 12. $3y \ge 8$

13. $x - 1 > 3x + 5$ 14. $-4x + 6 < 16 + x$ 15. $9 \le 5x - 1 \le 14$

List the set of ordered pairs that correspond to the points in the graph.

16.
17.
18.

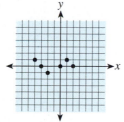 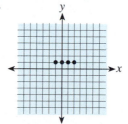 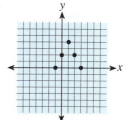

Graph each of the following sets of ordered pairs.

19. $\{(-3, 1), (-2, 1), (-1, 2), (0, 2), (1, 3)\}$

20. $\{(-4, 5), (-3, 2), (0, -4), (1, 1), (3, 1)\}$

21. $\{(1, 4), (1, 3), (1, 1), (1, 0)\}$

Graph the following linear equations.

22. $y = 2x - 1$ **23.** $y = x + 5$ **24.** $y = -x - 2$

25. $3x + y = 6$ **26.** $2x - 3y = 12$ **27.** $y = -3$

Use the Product Rule for Exponents to simplify each of the following expressions.

28. $2^2 \cdot 2^4$ **29.** $3x^3 \cdot 4x^4$

30. (a) Simplify the following polynomial and state its degree. (b) Evaluate the simplified polynomial for $x = -3$.

$$2x^3 - x^3 + 5x^2 + 6x - 5x^2 + 8 + 2x$$

Add or subtract as indicated and simplify if possible.

31. $(5x + 3) + (2x - 7) + (-3x + 11)$

32. $(6y^3 + 9y^2 - 10y - 16) + (-4y^3 - y^2 + 5y - 3)$

33. $(12x + 14) - (7x - 8)$

34. $(3a^2 + 2a - 5) - (5a^2 + 3a - 13)$

Find each of the following products.

35. $7x(4x^2 + 3x - 6.1)$ **36.** $(6y + 5)(2y - 3)$

37. $\begin{array}{r} x^3 - 3x^2 + 2x + 5 \\ \underline{4x - 3} \end{array}$ **38.** $\begin{array}{r} 2x^2 + 5x + 10 \\ \underline{x^2 - 3x + 2} \end{array}$

Chapter Test

1. Answer the following questions by using the set

$$A = \left\{ -5, -\sqrt{9}, -1.2, \sqrt{5}, 1\frac{3}{4} \right\}$$

   (a) Which of the numbers in set A is an integer?

   (b) Which of the numbers in set A is a rational number?

   (c) Which of the numbers in set A is an irrational number?

   (d) Which of the numbers in set A is a real number?

Represent each of the following graphs by using algebraic notation and tell what type of interval it is.

2.

3.

4. Graph the interval $1.5 \le x \le 3$ on a real number line and tell what type of interval it is.

5. Graph the interval $x \ge 0$ on a real number line and tell what type of interval it is.

Solve and graph the solution for each of the following inequalities.

6. $x - 2 > 3x + 6$ 7. $2x - 5 \le 5x - 2$

8. $-16 \le 10x - 1 \le 19$

List the set of ordered pairs that correspond to the points on each graph.

9.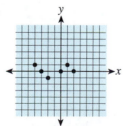

10.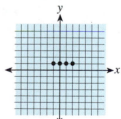

11. Graph the following set of ordered pairs:
    $\{(-2, 5), (-1, 3), (0, 1), (1, -1), (2, 0)\}$

Graph the following linear equations.

12. $y = -2x + 1$ 13. $2x - 3y = 6$ 14. $x = 2.5$

Use the Product Rule for Exponents to simplify each of the following expressions.

15. $3^2 \cdot 3^2$ 16. $2x^2 \cdot 15x^3$

17. (a) Simplify the following polynomial and state its degree. (b) Evaluate the simplified polynomial for $x = -2$.

$$5x^3 - x^3 + 4x^2 + 5x - 3x^2 + 7 - 12x$$

Add or subtract as indicated and simplify if possible.

18. $(4y^3 + 7y^2 - 8y - 14) + (-5y^3 - 2y^2 + 6y - 4)$

19. $(8x + 17) - (4x - 10)$

20. $(2a^2 + 5a - 3) - (4a^2 + 9a - 3)$

Find each of the following products.

21. $5x(3x^2 + 2x - 4.1)$ **22.** $(4y - 11)(2y + 3)$

23. $x^3 + 2x^2 - 3x + 2$ **24.** $2x^2 - 6x + 7$
     $\underline{5x - 4}$ $\underline{x^2 + 3x + 1}$

All answers should be in simplest form. Reduce all fractions to lowest terms and express all improper fractions as mixed numbers.

Find each sum.

1. $\dfrac{2}{3} + \dfrac{5}{6} + \dfrac{2}{9}$

2. $17\dfrac{5}{8} + 12\dfrac{7}{10}$

3. $6.09 + 10.6 + 7$

4. (4 yd 2 ft 8 in.) + (1 yd 6 in.)

5. $(-2.03) + (16.7) + (-5.602)$

Find each difference.

6. $\dfrac{3}{4} - \dfrac{5}{9}$

7. $4\dfrac{1}{6} - 2\dfrac{2}{3}$

8. $142.01 - 67.135$

9. (3 days 9 hr 15 min) − (9 hr 30 min)

10. Subtract -4.5 from -0.03.

Find each product.

11. $2400 \cdot 30{,}000$

12. $\dfrac{5}{12} \cdot \dfrac{3}{5} \cdot \dfrac{4}{7}$

13. $\left(3\dfrac{2}{3}\right)\left(4\dfrac{4}{11}\right)$

14. $(7.03)(0.28)$

15. $(-9)(1.6)$

Find each quotient.

16. $\dfrac{4}{9} \div \dfrac{8}{15}$

17. $5\dfrac{1}{4} \div \dfrac{7}{10}$

18. $1.0336 \div 1.7$

19. $-3.7 \div 0$

20. Find the prime factorization of 792.

21. Which of the numbers 2, 3, 4, 5, 9, 10 will exactly divide 31,752?

22. True or false: $\dfrac{2}{3} : 72 = \dfrac{3}{4} : 56$.

Round off each of the following as indicated.

23. 5987.042 to the nearest ten

24. 723.042 to the nearest tenth

25. Find the mean of the following decimal numbers (correct to the nearest hundredth): 6.07, 9.72, 8.51, and 7.23.

Evaluate each of the following expressions.

26. $(2^2 \cdot 3 \div 4 + 2) \cdot 5 - 3$

27. $4(x - 2) + 2(3y + 4)$, if $x = -3$ and $y = 2$

28. Simplify: $3\sqrt{20} + 4\sqrt{75} - \sqrt{45}$.

29. (a) Find the length of the hypotenuse for the right triangle shown.

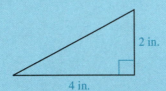

2 in.

4 in.

    (b) Use a calculator to find this value to the nearest hundredth.

30. Fill in the blank with the appropriate symbol: $<$, $>$, or $=$.

$$-|-7| \text{_____} -(-7)$$

31. Write an algebraic expression for the following English phrase: 5 less than the sum of twice a number and 6.

Solve each of the following equations.

32. $-2(3 - x) = 3(x + 2)$

33. $\frac{5}{6}x - 1 = \frac{2}{3}x + 2$

34. Solve and graph the solution for $2 \le 3 - 4x < 5$.

35. Solve $y = mx + b$ for x in terms of the other variables.

Graph each of the following equations.

36. $3x + 4y = 0$

37. $y = 2x + 5$

38. $16\frac{1}{2}\%$ of what number is $6\frac{3}{5}$?

39. What percent of 2.1 is 4.41?

40. Given that 1 pound = 0.454 kilograms, 1 lb 4 oz = _567_ g?   1 oz = 28.35 g

41. Given that 1 inch = 2.54 centimeters, 1 yd 2 ft 10 in. = _____ m?   70 in

42. Use either $C = \dfrac{5(F - 32)}{9}$ or $F = \dfrac{9}{5}C + 32$ to convert $24.5°$ C to degrees Fahrenheit.

43. Find the volume of a rectangular box 80 cm long, 40 cm wide, and 20 cm high.

44. Find the area in square centimeters of a rectangle 0.12 m long and 0.07 m wide.

45. Find the circumference of circle that has a radius of 20 centimeters. (Use $\pi = 3.14$.)

46. Find the area in square meters of a circle that has a diameter of 50 cm. (Use $\pi = 3.14$.)

47. Two towns are shown as 13.5 centimeters apart on a map that has a scale of 3 centimeters to 50 miles. What is the actual distance between the towns (in miles)?

48. During the last week, Cara jogged the following distances: 2 miles on both Thursday and Saturday; 0 miles on Sunday; 3 miles per day on Monday, Tuesday and Wednesday; and 1 mile on Friday. What was the average number of miles she jogged per day for last week?

49. If $1550 is deposited in an account paying 8% simple interest, what is the total amount in the account at the end of the first year (to the nearest dollar)?

50. Find two consecutive integers such that 3 more than twice the smaller equals 13 less than the larger integer.

51. The circle graph shows a family budget for one year. What amount will be spent in each category if the family income is $45,000?

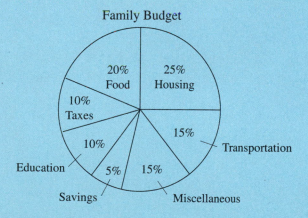

52. Simplify each expression.

(a) $(-5)^2 \cdot (-5)$

(b) $-7x^3 \cdot 4x^2$

53. Perform the indicated operations and simplify.

(a) $(5x^2 + 7x - 10) + (4x^2 - 7x + 3)$

(b) $(-2x^3 + x^2 - 8x + 2) - (-5x^3 - x^2 - 4x + 3)$

54. Find each product and simplify.

(a) $6x^2(2x - 20)$

(b) $(3x + 5)(2x^2 - 9x - 4)$

APPENDIX I

Ancient Numeration Systems

 I.1 Egyptian, Mayan, Attic Greek, and Roman Systems

The number systems used by ancient peoples are interesting from a historical point of view, but from a mathematical point of view they are difficult to work with. One of the many things that determine the progress of any civilization is its system of numeration. Humankind has made its most rapid progress since the invention of the zero and the place value system (which we will discuss in the next section) by the Hindu-Arabic peoples about A.D. 800.

Egyptian Numerals (Hieroglyphics)

The ancient Egyptians used a set of symbols called hieroglyphics as early as 3500 B.C. (See Table I.1 on page A2.) To write the numeral for a number, the Egyptians wrote the symbols next to each other from left to right, and the number represented was the sum of the values of the symbols. The most times any symbol was used was nine. Instead of using a symbol ten times, they used the symbol for the next higher number. They also grouped the symbols in threes or fours.

EXAMPLE 1

represents the number one thousand six hundred twenty-seven, or

$$1000 + 600 + 20 + 7 = 1627$$

Table I.1 Egyptian Hieroglyphic Numerals			
SYMBOL	NAME		VALUE
|	Staff (vertical stroke)	1	one
∩	Heel bone (arch)	10	ten
?	Coil of rope (scroll)	100	one hundred
𐦀	Lotus flower	1000	one thousand
𓏭	Pointing finger	10,000	ten thousand
﹏	Bourbot (tadpole)	100,000	one hundred thousand
𓁧	Astonished man	1,000,000	one million

Mayan System

The Mayans used a system of dots and bars (for numbers from 1 to 19) combined with a place value system. A dot represented one and a bar represented five. They had a symbol, ⬭, for zero and based their system, with one exception, on twenty. (See Table I.2.) The symbols were arranged vertically, smaller values starting at the bottom. The value of the third place up was 360 (18 times the value of the second place), but all other places were 20 times the value of the previous place.

Table I.2 Mayan Numerals	
SYMBOL	VALUE
•	1 one
——	5 five
⬭	0 zero

EXAMPLE 2

(a) $\cdots$ $(3 + 5 = 8)$

(b) $\stackrel{\cdots\cdots}{=}$ $(3 \cdot 5 + 4 = 19)$

(c) $\cdots$ 3 20's
  ⬭   0 units
  $(3 \cdot 20 + 0 = 60)$

(d) $\cdot\cdot$ 2 7200's
  ⬭   0 360's
  —   6 20's
  $\cdot\cdot$ 7 units

$(2 \cdot 7200 + 0 \cdot 360 + 6 \cdot 20 + 7 = 14{,}527)$

[**NOTE:** The symbol for zero, ⬭, is used as a place holder.]

Attic Greek System

The Greeks used two numeration systems, the Attic (see Table I.3) and the Alexandrian (see Section I.2 for information on the Alexandrian system). In the Attic system, no numeral was used more than four times. When a symbol was needed five or more times, the symbol for five was used, as shown in the examples.

Table I.3 Attic Greek Numerals		
SYMBOL	**VALUE**	
I	1	one
Γ	5	five
Δ	10	ten
H	100	one hundred
X	1000	one thousand
M	10,000	ten thousand

EXAMPLE 3

(a) X X ꟾꓧ H H ꓕ I I I I $(2 \cdot 1000 + 7 \cdot 100 + 5 \cdot 10 + 4 = 2754)$

(b) ꓨ H H H H Δ Δ Γ $(5 \cdot 1000 + 4 \cdot 100 + 2 \cdot 10 + 5 = 5425)$

Roman System

The Romans used a system (Table I.4) that we still see in evidence as hours on clocks and dates on buildings.

Table I.4 Roman Numerals		
SYMBOL	**VALUE**	
I	1	one
V	5	five
X	10	ten
L	50	fifty
C	100	one hundred
D	500	five hundred
M	1000	one thousand

The symbols were written largest to smallest, from left to right. The value of the numeral was the sum of the values of the individual symbols. Each symbol was used as many times as necessary, with the following exceptions: When the Romans got to 4, 9, 40, 90, 400, or 900, they used a system of subtraction.

$$IV = 5 - 1 = 4 \qquad XC = 100 - 10 = 90$$
$$IX = 10 - 1 = 9 \qquad CD = 500 - 100 = 400$$
$$XL = 50 - 10 = 40 \qquad CM = 1000 - 100 = 900$$

EXAMPLE 4

(a) VII   represents 7

(b) DXLIV   represents 544

(c) MCCCXXVIII   represents 1328

Exercises I.1

Find the values of the following ancient numbers.

1. 𝟿 𝟿 ∩∩∩ ∩∩ | | |

2. ⌒ 𝑓𝑓𝑓/𝑓𝑓𝑓 ⚱⚱⚱ ∩∩∩∩|

3. 𝑓𝑓 ⚱ 𝟿𝟿𝟿 ∩∩ | | | | / | | |

4. ⫶̄

5. ⊹ / ⬯

6. ⁚̄

7. Γˉ I I I

8. Γᴴ H H Δ Δ Δ Γˉ I

9. X X H H H Γˉ Δ I I

10. XCVII

11. DCCXLIV

12. MMMCDLXV

13. CMLXXVIII

14. Write 64
    (a) as an Egyptian numeral   (b) as a Mayan numeral
    (c) as an Attic Greek numeral   (d) as a Roman numeral

15. Follow the instructions for Exercise 14, using 532 in place of 64.

16. Follow the same instructions, using 1969.

17. Follow the same instructions, using 846.

I.2 Babylonian, Alexandrian Greek, and Chinese-Japanese Systems

Babylonian System (Cuneiform Numerals)

The Babylonians (about 3500 B.C.) used a place value system based on the number sixty, called a sexagesimal system. They had only two symbols, $\vee$ and $<$. (See Table I.5.) These wedge shapes are called cuneiform numerals, since **cuneus** means **wedge** in Latin.

Table I.5	Cuneiform Numerals
SYMBOL	**VALUE**
$\vee$	1 one
$<$	10 ten

The symbol for one was used as many as nine times, and the symbol for ten as many as five times; however, since there was no symbol for zero, many Babylonian numbers could be read several ways. For our purposes, we will group the symbols to avoid some of the ambiguities inherent in the system.

EXAMPLE 1

$$\vee\vee\vee \quad << \quad {\vee\vee\vee \atop \vee\vee} \quad <<< \quad \vee\vee$$

$$
\begin{aligned}
(3 \cdot 60^2) + (25 \cdot 60^1) + (32 \cdot 1) &= (3 \cdot 3600) + (25 \cdot 60) + 32 \\
&= 10{,}800 + 1500 + 32 \\
&= 12{,}332
\end{aligned}
$$

EXAMPLE 2

$$\vee \quad <<<< \quad {\vee\vee\vee \atop \vee\vee\vee} \quad < \quad {\vee\vee\vee\vee \atop \vee\vee\vee}$$

$$(1 \cdot 60^2) + (46 \cdot 60^1) + (17 \cdot 1) = 3600 + 2760 + 17 = 6377$$

Alexandrian Greek System

The Greeks used two numeration systems, the Attic and the Alexandrian. We discussed the Attic Greek system in Section I.1.

In the Alexandrian system (Table I.6), the letters were written next to each other, largest to smallest, from left to right. Since the numerals were also part of the Greek alphabet, an accent mark or bar was sometimes used above a letter to indicate that it represented a number. Multiples of 1000 were indicated by strikes in front of the unit symbols, and multiples of 10,000 were indicated by placing the unit symbols above the symbol M.

Table I.6   Alexandrian Greek Symbols

SYMBOL	NAME	VALUE		SYMBOL	NAME	VALUE	
A	Alpha	1	one	Ξ	Xi	60	sixty
B	Beta	2	two	O	Omicron	70	seventy
Γ	Gamma	3	three	Π	Pi	80	eighty
Δ	Delta	4	four	C	Koppa	90	ninety
E	Epsilon	5	five	P	Rho	100	one hundred
F	Digamma (or Vau)	6	six	Σ	Sigma	200	two hundred
Z	Zeta	7	seven	T	Tau	300	three hundred
H	Eta	8	eight	Y	Upsilon	400	four hundred
Θ	Theta	9	nine	Φ	Phi	500	five hundred
I	Iota	10	ten	X	Chi	600	six hundred
K	Kappa	20	twenty	Ψ	Psi	700	seven hundred
Λ	Lambda	30	thirty	Ω	Omega	800	eight hundred
M	Mu	40	forty	Π	Sampi	900	nine hundred
N	Nu	50	fifty				

EXAMPLE 3

(a) $\overline{\Phi \Xi Z}$ $(500 + 60 + 7 = 567)$

(b) $\overline{B}$
    $M T N \Delta$ $(20,000 + 300 + 50 + 4 = 20,354)$

Chinese-Japanese System

The Chinese-Japanese system (Table I.7) uses a different numeral for each of the digits up to ten, then a symbol for each power of ten. A digit written above a power of ten is to be multiplied by that power, and all such results are to be added to find the value of the numeral.

Table I.7	Chinese-Japanese Numerals		
SYMBOL	**VALUE**	**SYMBOL**	**VALUE**
一	1 one	七	7 seven
二	2 two	八	8 eight
三	3 three	九	9 nine
四	4 four	十	10 ten
五	5 five	百	100 one hundred
六	6 six	千	1000 one thousand

EXAMPLE 4

三
十 } 30

九   9

(30 + 9 = 39)

EXAMPLE 5

五
千 } 5000

四
百 } 400

八
十 } 80

二 } 2

(5000 + 400 + 80 + 2 = 5482)

Exercises 1.2

Find the value of each of the following ancient numerals.

1. $\lor$ $<< \lor\lor\lor$ $\lor\lor\lor$

2. $<<< \lor\lor$

3. $<<<$ $\lor\lor$

4. $\overline{Y\,N\,E}$

5. $\overline{\Delta}$

   M /Z π

6. $\overline{\Sigma\,K\,B}$

7. 四

   十

   大

8. 五

   千

   一

   6

   八

9. ん

   6

   ん

   十

   ん

Write the following numbers as (a) Babylonian numerals, (b) Alexandrian Greek numerals, and (c) Chinese-Japanese numerals.

10. 472   **11.** 596   **12.** 5047   **13.** 3665   **14.** 7293   **15.** 10,852

APPENDIX II

Base Two and Base Five

II.1 The Binary System (Base Two)

In the decimal system, ten is the base. You might ask if another number could be chosen as the base in a place value system. And, if so, would the system be any better or more useful than the decimal system? The fact is that computers do operate under a place value system with base two. In the **binary system** (or base two system), only two digits are needed, 0 and 1. These two digits correspond to the two possible conditions of an electric current, either **on** or **off.**

Any number can be represented in base two or in base ten. However, base ten has a definite advantage when large numbers are involved, as you will see. The advantage of base two over base ten is that for base two only two digits are needed, while ten digits are needed for base ten.

If the base of a place value system were not ten but two, then the beginning point would be not a decimal point but a **binary point.** The value of each place would be a power of two, as shown in Figure II.1.

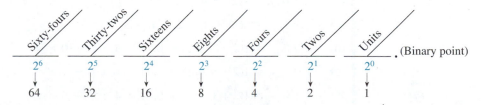

Figure II.1

> **To write numbers in the base two system, remember three things:**
>
> **1.** $\{0, 1\}$ is the set of digits that can be used.
>
> **2.** The value of each place from the binary point is in powers of two.
>
> **3.** The symbol 2 does not exist in the binary system, just as there is no digit for ten in the decimal system.

To avoid confusion with the base ten numerals, we will write $_{(2)}$ to the lower right of each base two numeral. We could write $_{(10)}$ to the lower right of each base ten numeral, but this would not be practical since most of the numerals we work with are in base ten. Therefore, **if no base is indicated, the numeral will be understood to be in base ten.**

EXAMPLE 1

Find the value of the numeral $1101._{(2)}$.

Solution

Writing the value of each place under the digit gives

$$\begin{array}{cccc} 1 & 1 & 0 & 1 \\ \overline{2^3} & \overline{2^2} & \overline{2^1} & \overline{2^0} \end{array} \cdot_{(2)}$$

In expanded notation,

$$\begin{aligned} 1101._{(2)} &= 1(2^3) + 1(2^2) + 0(2^1) + 1(2^0) \\ &= 1(8) + 1(4) + 0(2) + 1(1) \\ &= 8 + 4 + 0 + 1 \\ &= 13 \end{aligned}$$

Thus, to a computer, the symbol $1101._{(2)}$ means "thirteen."

EXAMPLE 2

(a) $\quad\underline{1}._{(2)} = 1$

(b) $\quad\underline{1}\ \underline{0}._{(2)} = 1(2) + 0 = 2$

(c) $\quad\underline{1}\ \underline{1}._{(2)} = 1(2) + 1 = 2 + 1 = 3$

(d) $\quad\underline{1}\ \underline{0}\ \underline{0}._{(2)} = 1(2^2) + 0(2) + 0(1) = 4 + 0 + 0 = 4$

(e) $\quad\underline{1}\ \underline{0}\ \underline{1}._{(2)} = 1(2^2) = 0(2) + 1(1) = 4 + 0 + 1 = 5$

(f) $\quad\underline{1}\ \underline{1}\ \underline{0}._{(2)} = 1(2^2) + 1(2) + 0(1) = 4 + 2 + 0 = 6$

(g) $\quad\underline{1}\ \underline{1}\ \underline{1}._{(2)} = 1(2^2) + 1(2) + 1(1) = 4 + 2 + 1 = 7$

(h) $\underline{1}\ \underline{0}\ \underline{0}\ \underline{0}._{(2)} = 1(2^3) + 0(2^2) + 0(2) + 0(1) = 8 + 0 + 0 + 0 = 8$

(i) $\underline{1}\ \underline{0}\ \underline{0}\ \underline{1}._{(2)} = 1(2^3) + 0(2^2) + 0(2) + 1(1) = 8 + 0 + 0 + 1 = 9$

(j) $\underline{1}\ \underline{0}\ \underline{1}\ \underline{0}._{(2)} = 1(2^3) + 0(2^2) + 1(2) + 0(1) = 8 + 0 + 2 + 0 = 10$

Do **not** read $100_{(2)}$ as "one hundred" because the 1 is not in the hundreds place. The 1 is in the fours place. So, $100_{(2)}$ is read "four" or "one, zero, zero–base two." Similarly, $111_{(2)}$ is read "seven" or "one, one, one–base two."

Exercises II.1

Write the following base ten numerals in expanded form using exponents.

Example: $273 = 2(10^2) + 7(10^1) + 3(10^0)$

1. 35   **2.** 761   **3.** 8469   **4.** 500   **5.** 62,322

Write the following base two numerals in expanded form and find the value of each numeral.

Example: $110_{(2)} = 1(2^2) + 1(2^1) + 0(2^0)$
$\qquad\qquad = 1(4) + 1(2) + 0(1)$
$\qquad\qquad = 4 + 2 + 0$
$\qquad\qquad = 6$

6. $11_{(2)}$ **7.** $101_{(2)}$ **8.** $111_{(2)}$ **9.** $1011_{(2)}$

10. $1101_{(2)}$ **11.** $110111_{(2)}$ **12.** $11110_{(2)}$ **13.** $101011_{(2)}$

14. $11010_{(2)}$ **15.** $1000_{(2)}$ **16.** $1000010_{(2)}$ **17.** $11101_{(2)}$

18. $10110_{(2)}$ **19.** $111111_{(2)}$ **20.** $1111_{(2)}$

21. A computer is directed to place some information in memory space number $1101111_{(2)}$. What is the number of this memory space in base ten?

II.2   The Quinary System (Base Five)

Many numbers may be used as bases for place value systems. To illustrate this point and to emphasize the concept of place value, we will discuss one more base system, base five. Interested students may want to try writing numerals in base three or base eight or base eleven.

Again, the system relies on powers of the base and a set of digits. In the **quinary system** (base five system), the powers of 5 are 5^0, 5^1, 5^2, 5^3, 5^4, and so on, and the digits to be used are $\{0, 1, 2, 3, 4\}$. The **quinary point** is the beginning point, as shown in Figure II.2.

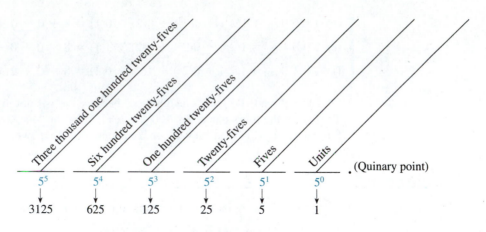

Figure II.2

EXAMPLE 1

(a) $\underline{1}_{.(5)} = 1$

(b) $\underline{2}_{.(5)} = 2$

(c) $\underline{3}_{.(5)} = 3$

(d) $\underline{4}_{.(5)} = 4$

EXAMPLE 2

(a) $\underline{1}\ 0_{.(5)} = 1(5) + 0(1) = 5 + 0 = 5$

(b) $\underline{1}\ 1_{.(5)} = 1(5) + 1(1) = 5 + 1 = 6$

(c) $\underline{1}\ 2_{.(5)} = 1(5) + 2(1) = 5 + 2 = 7$

(d) $\underline{1}\ 3_{.(5)} = 1(5) + 3(1) = 5 + 3 = 8$

(e) $\underline{1}\ 4_{.(5)} = 1(5) + 4(1) = 5 + 4 = 9$

(f) $\underline{2}\ 0_{.(5)} = 2(5) + 0(1) = 10 + 0 = 10$

(g) $\underline{2}\ 1_{.(5)} = 2(5) + 1(1) = 10 + 1 = 11$

(h) $\underline{2}\ 2_{.(5)} = 2(5) + 2(1) = 10 + 2 = 12$

EXAMPLE 3

$$\underline{3}\ \underline{2}\ 4_{.(5)} = 3(5^2) + 2(5^1) + 4(5^0)$$
$$= 3(25) + 2(5) + 4(1)$$
$$= 75 + 10 + 4$$
$$= 89$$

Exercises II.2

Write the following base five numerals in expanded form and find the value of each.

1. $24_{(5)}$ **2.** $13_{(5)}$ **3.** $10_{(5)}$ **4.** $43_{(5)}$ **5.** $104_{(5)}$

6. $312_{(5)}$ **7.** $32_{(5)}$ **8.** $230_{(5)}$ **9.** $423_{(5)}$ **10.** $444_{(5)}$

11. $1034_{(5)}$ **12.** $4124_{(5)}$ **13.** $244_{(5)}$ **14.** $3204_{(5)}$ **15.** $13042_{(5)}$

16. Do the numerals $101_{(2)}$ and $10_{(5)}$ represent the same number? If so, what is the number?

17. Answer the questions in Exercise 16 about the numerals $1101_{(2)}$ and $23_{(5)}$.

18. Answer the questions in Exercise 16 about the numerals $11100_{(2)}$ and $103_{(5)}$.

19. What set of digits do you think would be used in a base eight system?

20. What set of digits do you think would be in a base twelve system?
[HINT: New symbols for some new digits must be introduced.]

II.3   Addition and Multiplication in Base Two and Base Five

Now that we have two new numeration systems, base two and base five, a natural question to ask is, How are addition and multiplication* performed in these systems? The basic techniques are the same as for base ten, since place value is involved. However, because different bases are involved, the numerals will be different. For example, to add five plus seven in base ten, we write $5 + 7 = 12$. In base two, this same sum is written $101_{(2)} + 111_{(2)} = 1100_{(2)}$.

Writing the numerals vertically (one under the other) gives

$$
\begin{array}{rl}
5 & \quad 101_{(2)} \\
\underline{7} & \quad \underline{111_{(2)}} \\
12 & \quad 1100_{(2)}
\end{array}
$$

Now a step-by-step analysis of the sum in base two will be provided.

---

* Subtraction and division may also be performed in base two and base five, but will not be discussed here for reasons of time. Some students may want to investigate these operations on their own.

EXAMPLE 1

(a) $101_{(2)}$
$111_{(2)}$

The numerals are written so that the digits of the same place value line up.

(b) $\overset{1}{}$
$101_{(2)}$
$111_{(2)}$
―――――
$0_{(2)}$

Adding $1 + 1$ in the units column gives "two," which is written $10_{(2)}$. 0 is written in the units column, and 1 is "carried" to the "twos" column.

(c) $\overset{1\,1}{}$
$101_{(2)}$
$111_{(2)}$
―――――
$00_{(2)}$

Now, in the twos column, $1 + 0 + 1$ is again "two," or $10_{(2)}$. Again 0 is written, and 1 is "carried" to the next column, the "fours" column.

(d) $\overset{1\,1}{}$
$101_{(2)}$
$111_{(2)}$
―――――
$1100_{(2)}$

In the fours column (or third column), $1 + 1 + 1$ is "three," or $11_{(2)}$. Since there are no digits in the "eights" column (or fourth column), 11 is written, and the sum is $1100_{(2)}$.

Checking in Base Ten

$$101_{(2)} = \qquad 1(2^2) + 0(2^1) + 1(2^0) = \quad 4 + 0 + 1 = \ 5 \qquad 5$$
$$111_{(2)} = \qquad 1(2^2) + 1(2^1) + 1(2^0) = \quad 4 + 2 + 1 = \ 7 \qquad \underline{7}$$
$$1100_{(2)} = 1(2^3) + 1(2^2) + 0(2^1) + 0(2^0) = 8 + 4 + 0 + 0 = 12 \quad 12$$

Addition in base five is similar. Although the thinking is done in base ten, which is familiar to us, the numerals written must be in base five.

EXAMPLE 2

(a) $143_{(5)}$
$34_{(5)}$

The numerals are written so that the digits of the same place value line up.

(b) $\overset{1}{}$
$143_{(5)}$
$34_{(5)}$
―――――
$2_{(5)}$

Adding $3 + 4$ in the units column gives "seven," which is written $12_{(5)}$. 2 is written in the units column, and 1 is carried to the fives column.

(c) $\overset{1\,1}{}$
$143_{(5)}$
$34_{(5)}$
―――――
$32_{(5)}$

In the fives column, $1 + 4 + 3$ gives "eight," which is $13_{(5)}$. The 3 is written, and 1 is carried to the next column (the twenty-fives column).

(d) $\overset{1\,1}{}$
$143_{(5)}$
$34_{(5)}$
―――――
$232_{(5)}$

In the third column, $1 + 1$ gives 2. The sum is $232_{(5)}$.

Checking in Base Ten

$$143_{(5)} = 1(5^2) + 4(5^1) + 3(5^0) = 25 + 20 + 3 = 48 \quad 48$$
$$\underline{34_{(5)} = \qquad\quad 3(5^1) + 4(5^0) = \qquad 15 + 4 = 19 \quad 19}$$
$$232_{(5)} = 2(5^2) + 3(5^1) + 2(5^0) = 50 + 15 + 2 = 67 \quad 67$$

Multiplication in each base is performed and checked in the same manner as addition. Of course, the difference is that you multiply instead of add. When you multiply, be sure to write the correct symbol for the number in the base being used. Also remember to add in the correct base.

EXAMPLE 3

$$101_{(2)}$$
$$\underline{111_{(2)}}$$
$$101$$
^{1}
$$101$$
^{1}
$$\underline{101}$$

$$100011_{(2)}$$

Multiplication in base two is easy, since we are multiplying by only 1's or 0's. The adding must be done in base two.

Checking gives

$$\left.\begin{array}{r}101_{(2)} = 5 \\ \underline{111_{(2)} = 7}\end{array}\right\}$$
$$\begin{array}{cc}101 & 35\end{array}$$
$$101$$
$$\underline{101}$$
$$100011_{(2)}$$

Remember to **multiply** the checking numbers.

$$100,011_{(2)} = 1(2^5) + 0(2^4) + 0(2^3) + 0(2^2) + 1(2) + 1(1)$$
$$= 32 + 2 + 1 = 35$$

EXAMPLE 4

^{2}
$$^{1}34_{(5)}$$
$$\underline{23_{(5)}}$$
$$212$$
$$\underline{123}$$
$$1442_{(5)}$$

Multiplying 3×4 gives "twelve," which is $22_{(5)}$. Write 2 and carry 2 just as in regular multiplication. Then, 3×3 is "nine," and "nine" plus 2 is "eleven"; but in base five, "eleven" is $21_{(5)}$. Similarly, 2×4 is "eight," or $13_{(5)}$. Write the 3, carry the 1. 2×3 is "six," and "six" plus 1 is "seven," or $12_{(5)}$.

Checking gives

$$\left.\begin{array}{r}34_{(5)} = 19 \\ \underline{23_{(5)} = 13}\end{array}\right\}$$
$$\begin{array}{cc}212 & 57\end{array}$$
$$\underline{123} \quad \underline{19}$$
$$1442_{(5)} \quad 247$$

Remember to **multiply** the checking numbers.

$$1442_{(5)} = 1(5^3) + 4(5^2) + 4(5) + 2(1)$$
$$= 125 + 100 + 20 + 2 = 247$$

Exercises II.3

Add in the base indicated and check your work in base ten.

1. $101_{(2)}$
$\underline{11_{(2)}}$

2. $43_{(5)}$
$\underline{213_{(5)}}$

3. $1101_{(2)}$
$\underline{1011_{(2)}}$

4. $111_{(2)}$
$\underline{1010_{(2)}}$

5. $134_{(5)}$
$\underline{243_{(5)}}$

6. $11_{(2)}$
$10_{(2)}$
$\underline{11_{(2)}}$

7. $11_{(2)}$
$11_{(2)}$
$\underline{101_{(2)}}$

8. $214_{(5)}$
$\underline{343_{(5)}}$

9. $14_{(5)}$
$321_{(5)}$
$\underline{43_{(5)}}$

10. $431_{(5)}$
$214_{(5)}$
$\underline{102_{(5)}}$

11. $11_{(2)}$
$101_{(2)}$
$111_{(2)}$
$\underline{101_{(2)}}$

12. $111_{(2)}$
$11_{(2)}$
$110_{(2)}$
$\underline{111_{(2)}}$

13. $101_{(2)}$
$101_{(2)}$
$101_{(2)}$
$\underline{101_{(2)}}$

14. $23_{(5)}$
$103_{(5)}$
$214_{(5)}$
$\underline{322_{(5)}}$

15. $414_{(5)}$
$211_{(5)}$
$334_{(5)}$
$\underline{222_{(5)}}$

Multiply in the base indicated and check your work in base ten.

16. $1101_{(2)}$
$\underline{111_{(2)}}$

17. $1011_{(2)}$
$\underline{101_{(2)}}$

18. $423_{(5)}$
$\underline{30_{(5)}}$

19. $104_{(5)}$
$\underline{23_{(5)}}$

20. $223_{(5)}$
$\underline{44_{(5)}}$

21. $423_{(5)}$
$\underline{32_{(5)}}$

22. $1111_{(2)}$
$\underline{111_{(2)}}$

23. $111_{(2)}$
$\underline{111_{(2)}}$

24. $2212_{(5)}$
$\underline{43_{(5)}}$

25. $10111_{(2)}$
$\underline{110_{(2)}}$

APPENDIX III

Greatest Common Divisor (GCD)

Consider the two numbers 12 and 18. Is there a number (or numbers) that will divide into **both** 12 and 18? To help answer this question, the divisors for 12 and 18 are listed below.

Set of divisors for 12: {1, 2, 3, 4, 6, 12}
Set of divisors for 18: {1, 2, 3, 6, 9, 18}

The **common divisors** for 12 and 18 are 1, 2, 3, and 6. The **greatest common divisor (GCD)** for 12 and 18 is 6: that is, of all the common divisors of 12 and 18, 6 is the largest divisor.

EXAMPLE 1

List the divisors of each number in the set {36, 24, 48} and find the greatest common divisor (GCD).

Set of divisors for 36: {**1, 2, 3, 4, 6,** 9, **12,** 18, 36}
Set of divisors for 24: {**1, 2, 3, 4, 6,** 8, **12,** 24}
Set of divisors for 48: {**1, 2, 3, 4, 6,** 8, **12,** 16, 24, 48}

The common divisors are **1, 2, 3, 4, 6,** and **12. GCD = 12.**

Definition

The Greatest Common Divisor (GCD)* of a set of natural numbers is the largest natural number that will divide into all the numbers in the set.

---

* The largest common divisor is, of course, the largest common factor, and the GCD could be called the **greatest common factor,** and be abbreviated GCF.

As Example 1 illustrates, listing all the divisors of each number before finding the GCD can be tedious and difficult. **The use of prime factorizations leads to a simple technique for finding the GCD.**

Technique for Finding the GCD of a Set of Counting Numbers

1. Find the prime factorization of each number.

2. Find the prime factors common to all factorizations.

3. Form the product of these primes, using each prime the number of times it is common to **all** factorizations.

4. This product is the GCD. If there are no primes common to all factorizations, the GCD is 1.

EXAMPLE 2   Find the GCD for {36, 24, 48}.

$$36 = 2 \cdot 2 \cdot 3 \cdot 3$$
$$24 = 2 \cdot 2 \cdot 2 \cdot 3 \quad \Big\} \; GCD = 2 \cdot 2 \cdot 3 = 12$$
$$48 = 2 \cdot 2 \cdot 2 \cdot 2 \cdot 3$$

In **all** the prime factorizations, factor 2 appears twice, and factor 3 appears once.

EXAMPLE 3   Find the GCD for {360, 75, 30}.

$$360 = 36 \cdot 10 = 4 \cdot 9 \cdot 2 \cdot 5 = 2 \cdot 2 \cdot 2 \cdot 3 \cdot 3 \cdot 5$$
$$75 = 3 \cdot 25 = 3 \cdot 5 \cdot 5 \quad \Big\} \quad GCD = 3 \cdot 5$$
$$30 = 6 \cdot 5 = 2 \cdot 3 \cdot 5 \qquad\qquad\qquad = 15$$

Each of the factors 3 and 5 appears only once in **all** the prime factorizations.

EXAMPLE 4   Find the GCD for {168, 420, 504}.

$$168 = 8 \cdot 21 = 2 \cdot 2 \cdot 2 \cdot 3 \cdot 7$$
$$420 = 10 \cdot 42 = 2 \cdot 5 \cdot 6 \cdot 7$$
$$= 2 \cdot 2 \cdot 3 \cdot 5 \cdot 7 \quad \Big\} \quad GCD = 2 \cdot 2 \cdot 3 \cdot 7 = 84$$
$$504 = 4 \cdot 126 = 2 \cdot 2 \cdot 6 \cdot 21$$
$$= 2 \cdot 2 \cdot 2 \cdot 3 \cdot 3 \cdot 7$$

In **all** the prime factorizations, 2 appears twice, 3 once, and 7 once.

If the GCD of two numbers is 1 (that is, they have no common prime factors), then the two numbers are said to be **relatively prime.** The numbers themselves may be prime or they may be composite.

EXAMPLE 5 | Find the GCD for {15, 8}.

$$\left.\begin{array}{l} 15 = 3 \cdot 5 \\ 8 = 2 \cdot 2 \cdot 2 \end{array}\right\} \quad \text{GCD} = 1 \qquad \text{8 and 15 are relatively prime.}$$

EXAMPLE 6 | Find the GCD for {20, 21}.

$$\left.\begin{array}{l} 20 = 2 \cdot 2 \cdot 5 \\ 21 = 3 \cdot 7 \end{array}\right\} \quad \text{GCD} = 1 \qquad \text{20 and 21 are relatively prime.}$$

Exercises III

Find the GCD for each of the following sets of numbers.

1. {12, 8}   2. {16, 28}   3. {85, 51}

4. {20, 75}   5. {20, 30}   6. {42, 48}

7. {15, 21}   8. {27, 18}   9. {18, 24}

10. {77, 66}   11. {182, 184}   12. {110, 66}

13. {8, 16, 64}   14. {121, 44}   15. {28, 52, 56}

16. {98, 147}   17. {60, 24, 96}   18. {33, 55, 77}

19. {25, 50, 75}   20. {30, 78, 60}   21. {17, 15, 21}

22. {520, 220}   23. {14, 55}   24. {210, 231, 84}

25. {140, 245, 420}

Which of the following pairs of numbers are relatively prime?

26. {35, 24}   27. {11, 23}   28. {14, 36}   29. {72, 35}

30. {42, 77}   31. {16, 51}   32. {20, 21}   33. {8, 15}

34. {66, 22}   35. {10, 27}

Answer Key

Numbers in brackets following each answer in the Chapter Tests refer to the section that covers the corresponding test question.

CHAPTER 1

Exercises 1.1, p. 6

1. 30 + 7; thirty-seven **2.** 80 + 4; eighty-four **3.** 90 + 8; ninety-eight
5. 100 + 20 + 2; one hundred twenty-two **6.** 400 + 90 + 3; four hundred ninety-three
7. 800 + 20 + 1; eight hundred twenty-one
9. 1000 + 800 + 90 + 2; one thousand eight hundred ninety-two
10. 5000 + 400 + 90 + 6; five thousand four hundred ninety-six
11. 10,000 + 2000 + 500 + 10 + 7; twelve thousand, five hundred seventeen
13. 200,000 + 40,000 + 3000 + 400; two hundred forty-three thousand, four hundred
14. 800,000 + 90,000 + 1000 + 500 + 40; eight hundred ninety-one thousand, five hundred forty
15. 40,000 + 3000 + 600 + 50 + 5; forty-three thousand, six hundred fifty-five
17. 8,000,000 + 400,000 + 800 + 10; eight million, four hundred thousand, eight hundred ten
18. 5,000,000 + 600,000 + 60,000 + 3000 + 700 + 1; five million, six hundred sixty-three thousand, seven hundred one
19. 10,000,000 + 6,000,000 + 300,000 + 2000 + 500 + 90;
sixteen million, three hundred two thousand, five hundred ninety
21. 80,000,000 + 3,000,000 + 600 + 5; eighty-three million, six hundred five
22. 100,000,000 + 50,000,000 + 2,000,000 + 400,000 + 3000 + 600 + 70 + 2; one hundred fifty-two million, four hundred three thousand, six hundred seventy-two
23. 600,000,000 + 70,000,000 + 9,000,000 + 70,000 + 8000 + 100; six hundred seventy-nine million, seventy-eight thousand, one hundred
25. 8,000,000,000 + 500,000,000 + 70,000,000 + 2,000,000 + 3000 + 400 + 20 + 5; eight billion, five hundred seventy-two million, three thousand, four hundred twenty-five
26. 76 **27.** 132 **29.** 3842 **30.** 2005 **31.** 192,151 **33.** 21,400 **34.** 33,333 **35.** 5,045,000
37. 10,639,582 **38.** 281,300,501 **39.** 530,000,700 **41.** 90,090,090 **42.** 82,700,000
43. 175,000,002 **45.** 757
46. 1 hundred thousands; 9 thousands; 7 hundreds; 5 tens
47. three million, four hundred eighty-five thousand, three hundred ninety-eight
49. one hundred; two thousand seven hundred
50. three hundred fifty-two thousand, one hundred forty-three; nine hundred twelve thousand, fifty

Exercises 1.2, p. 13

1. 16 **2.** 15 **3.** 13 **5.** 21 **6.** 20 **7.** 19 **9.** 12 **10.** 17 **11.** 18 **13.** 12 **14.** 21
15. 17 **17.** 27 **18.** 18 **19.** 21 **21.** commutative **22.** commutative **23.** associative
25. associative **26.** associative **27.** identity **29.** identity **30.** associative **31.** 162 **33.** 239
34. 835 **35.** 1298 **37.** 1236 **38.** 4168 **39.** 6869 **41.** 1,603,426 **42.** 1,463,930
43. 2,610,667 **45.** 2762 miles **46.** $6,313,323 **47.** $18,463 **49.** 1518 students
50. 33,830 appliances

51.

+	5	8	7	9
3	8	11	10	12
6	11	14	13	15
5	10	13	12	14
2	7	10	9	11

Exercises 1.3, p. 19

1. 3   **2.** 13   **3.** 0   **5.** 9   **6.** 17   **7.** 9   **9.** 5   **10.** 6   **11.** 0   **13.** 13   **14.** 20
15. 20   **17.** 5   **18.** 45   **19.** 13   **21.** 94   **22.** 126   **23.** 218   **25.** 475   **26.** 376   **27.** 593
29. 188   **30.** 478   **31.** 1569   **33.** 1568   **34.** 1531   **35.** 0   **37.** 694   **38.** 5871   **39.** 2517
41. 2,806,644   **42.** 3,800,559   **43.** 1,006,958   **45.** 5,671,011   **46.** 222   **47.** 140   **49.** 32 years
50. 44 points   **51.** $250,404   **53.** $934   **54.** $235,456   **55.** $39,100   **57.** $3700   **58.** $868

Exercises 1.4, p. 26

1. 760   **2.** 30   **3.** 80   **5.** 300   **6.** 720   **7.** 990   **9.** 4200   **10.** 4500   **11.** 500   **13.** 600
14. 3800   **15.** 76,500   **17.** 7000   **18.** 6000   **19.** 8000   **21.** 13,000   **22.** 14,000   **23.** 62,000
25. 80,000   **26.** 130,000   **27.** 260,000   **29.** 120,000   **30.** 310,000   **31.** 180,000   **33.** 100
34. 0   **35.** 1000   **37.** 180 (estimate); 167 (sum)   **38.** 800 (estimate); 789 (sum)
39. 1200 (estimate); 1173 (sum)   **41.** 1900 (estimate); 1881 (sum)   **42.** 21,000 (estimate); 21,141 (sum)
43. 9500 (estimate); 9224 (sum)   **45.** 6000 (estimate); 5467 (difference)
46. 4000 (estimate); 4769 (difference)   **47.** 5000 (estimate); 5931 (difference)
49. 20,000 (estimate); 20,804 (difference)   **50.** 40,000 (estimate); 41,223 (difference)   **51.** 910,200,000
53. 60 centimeters   **54.** 40 meters   **55.** $200

Exercises 1.5, p. 39

1. 72   **2.** 42   **3.** 24   **5.** 7   **6.** 45   **7.** 0   **9.** 28; commutative property of multiplication
10. 12; associative property of multiplication   **11.** 5; multiplicative identity   **13.** 250   **14.** 47,000
15. 4000   **17.** 16,000   **18.** 80,000   **19.** 900   **21.** 16,000,000   **22.** 2,700,000   **23.** 5000
25. 36,000   **26.** 8100   **27.** 150,000   **29.** 240 (estimate); 224 (product)
30. 180 (estimate); 162 (product)   **31.** 450 (estimate); 432 (product)   **33.** 240 (estimate); 252 (product)
34. 800 (estimate); 760 (product)   **35.** 2400 (estimate); 2352 (product)   **37.** 1000 (estimate); 960 (product)
38. 2700 (estimate); 2790 (product)   **39.** 7200 (estimate); 7055 (product)   **41.** 600 (estimate); 544 (product)
42. 2700 (estimate); 2548 (product)   **43.** 800 (estimate); 880 (product)   **45.** 600 (estimate); 375 (product)
46. 4500 (estimate); 4371 (product)   **47.** 1800 (estimate); 2064 (product)   **49.** 100 (estimate); 156 (product)
50. 3200 (estimate); 2916 (product)   **51.** 4000 (estimate); 5166 (product)
53. 3000 (estimate); 2850 (product)   **54.** 7000 (estimate); 7632 (product)
55. 20,000 (estimate); 29,601 (product)   **57.** 10,000 (estimate); 9800 (product)
58. 160,000 (estimate); 174,045 (product)   **59.** 140,000 (estimate); 125,178 (product)   **61.** 231; 58; 13,398
62. 89; 337; 29,993   **63.** $2760; $165,600   **65.** $230,400   **66.** 36,400 square feet
67. 17,630 square meters   **69.** $15,200,000   **70.** 1,492,259,238,000 miles   **71.** $200
73. (a) C.   (b) B.   (c) A.   (d) D.

Exercises 1.6, p. 50

1. 40 R0   **2.** 21 R0   **3.** 30 R0   **5.** 21 R0   **6.** 14 R0   **7.** 12 R0   **9.** 5 R0   **10.** 17 R0
11. 6 R4   **13.** 24 R0   **14.** 20 R13   **15.** 10 R11   **17.** 11 R7   **18.** 35 R10   **19.** 41 R0
21. 5 R2   **22.** 2 R3   **23.** 6 R1   **25.** 6   **26.** 12   **27.** 9   **29.** 32 R2   **30.** 3 R10   **31.** 9
33. 8   **34.** 15 R5   **35.** 11 R8   **37.** 42 R3   **38.** 50   **39.** 20   **41.** 30   **42.** 400 R3
43. 300 R13   **45.** 301 R4   **46.** 2 R2   **47.** 3 R3   **49.** 61 R15   **50.** 54 R3   **51.** 2   **53.** 22 R74
54. 4 R192   **55.** 7 R358   **57.** 196 R370   **58.** 221 R308   **59.** 107 R215   **61.** (a) $3000 (b) $2548
62. (a) $33 (b) $38   **63.** $1008 \div 28 = 36$; $1008 \div 36 = 28$ **65.** about 50
66. (a) C. (b) A. (c) E. (d) D. (e) B.   **67.** (a) B. (b) D. (c) A. (d) C.

Exercises 1.7, p. 56

1. $160   **2.** $262   **3.** $786   **5.** $874   **6.** $85   **7.** $485   **9.** $316   **10.** $4865
11. (a) 104 m (b) 555 sq. m   **13.** 120 in.   **14.** 168 sq. in.   **15.** 103   **17.** 6   **18.** 6   **19.** 485
21. 85   **22.** $932   **23.** $18 per share; $1200   **25.** (a) $665 (b) $2495   **26.** 7 hours   **27.** 52,050
29. (a) 12,370 (b) $10,624 (c) $12,626 (d) $4872

Review Questions: Chapter 1, p. 63

1. $400 + 90 + 5$; four hundred ninety-five
2. $1000 + 900 + 70 + 5$; one thousand nine hundred seventy-five
3. $60,000 + 300 + 8$; sixty thousand, three hundred eight   **4.** 4856   **5.** 15,032,197
6. 672,340,083   **7.** 630   **8.** 15,000   **9.** 700   **10.** 2600   **11.** commutative property of addition
12. associative property of multiplication   **13.** associative property of addition
14. commutative property of multiplication   **15.** 9800 (estimate); 10,541 (sum)
16. 1740 (estimate); 1674 (sum)   **17.** 508   **18.** 2384   **19.** 2102   **20.** 0   **21.** 5600   **22.** 360,000
23. 5000 (estimate); 5096 (product)   **24.** 3,600,000 (estimate); 3,913,100 (product)
25. 25,000,000 (estimate); 24,185,000 (product)   **26.** 285 (estimate); 292 (quotient)
27. 500 (estimate); 606 (quotient)   **28.** 140 (estimate); 135 R81   **29.** 1059   **30.** 35   **31.** 9
32. $1485; $99   **33.** 83   **34.** 70   **35.** $7700
36.

Given Number	Add 100	Double	Subtract 200
3	103	206	6
20	120	240	40
15	115	230	30
8	108	216	16

Test: Chapter 1, p. 65

1. $8000 + 900 + 50 + 2$; eight thousand nine hundred fifty-two [1.1]
2. identity [1.5]   **3.** $7 \cdot 9 = 9 \cdot 7$ [1.5]   **4.** 1000 [1.4]   **5.** 140,000 [1.4]
6. 12,300 (estimate); 12,009 (sum) [1.4]   **7.** 1840 (estimate); 1735 (sum) [1.4]
8. 14,000,000 (estimate); 13,781,661 (sum) [1.4]   **9.** 488 [1.4]   **10.** 1229 [1.4]   **11.** 5707 [1.4]
12. 2400 (estimate); 2584 (product) [1.6]   **13.** 270,000 (estimate); 220,405 (product) [1.6]
14. 240,000 (estimate); 210,938 (product) [1.6]   **15.** 403 [1.7]   **16.** 172 R388 [1.7]   **17.** 2005 [1.7]
18. 74 [1.8]   **19.** 54 [1.8]   **20.** (a) $306 (b) $51 (c) $224 [1.8]   **21.** 1428 sq. in. [1.8]
22. $2025 [1.8]

CHAPTER 2

Exercises 2.1, p. 72

1. (a) 3 (b) 2; 8 **2.** (a) 5 (b) 2; 32 **3.** (a) 2 (b) 5; 25 **5.** (a) 0 (b) 7; 1
6. (a) 2 (b) 11; 121 **7.** (a) 4 (b) 1; 1 **9.** (a) 0 (b) 4; 1 **10.** (a) 6 (b) 3; 729
11. (a) 2 (b) 3; 9 **13.** (a) 0 (b) 5; 1 **14.** (a) 50 (b) 1; 1 **15.** (a) 1 (b) 62; 62
17. (a) 2 (b) 10; 100 **18.** (a) 3 (b) 10; 1000 **19.** (a) 2 (b) 4; 16 **21.** (a) 4 (b) 10; 10,000
22. (a) 3 (b) 5; 125 **23.** (a) 3 (b) 6; 216 **25.** (a) 0 (b) 19; 1 **26.** 2^2 **27.** 5^2 **29.** 3^3
30. 2^5 **31.** 11^2 **33.** 2^3 **34.** 3^2 **35.** 6^2 **37.** 9^2 or 3^4 **38.** 8^2 or 4^3 or 2^6 **39.** 10^2 **41.** 10^4
42. 6^3 **43.** 12^2 **45.** 3^5 **46.** 25^2 or 5^4 **47.** 15^2 **49.** 7^3 **50.** 10^5 **51.** 6^5 **53.** $2^2 \cdot 7^2$
54. $5^2 \cdot 9^3$ **55.** $2^2 \cdot 3^3$ **57.** $7^2 \cdot 13$ **58.** 11^3 **59.** $2 \cdot 3^2 \cdot 11^2$ **61.** 64 **62.** 9 **63.** 49
65. 225 **66.** 196 **67.** 324 **69.** 144 **70.** 400 **71.** 100 **73.** 900 **74.** 1600 **75.** 2500

Exercises 2.2, p. 76

1. (a) multiplication (b) division (c) addition
2. (a) subtraction, addition (within parentheses) (b) division (c) subtraction
3. (a) division (within parentheses) (b) addition (within parentheses) (c) division
5. 3 **6.** 15 **7.** 7 **9.** 22 **10.** 31 **11.** 3 **13.** 5 **14.** 10 **15.** 5 **17.** 5 **18.** 0
19. 3 **21.** 7 **22.** 0 **23.** 0 **25.** 6 **26.** 3 **27.** 3 **29.** 27 **30.** 0 **31.** 26 **33.** 0
34. 5 **35.** 69 **37.** 68 **38.** 80 **39.** 140 **41.** 5 **42.** 9 **43.** 0 **45.** 9 **46.** 93
47. 12 **49.** 24 **50.** 118 **51.** 70 **53.** 230 **54.** 34 **55.** 110 **57.** 110 **58.** 0 **59.** 2980

Exercises 2.3, p. 84

1. 2, 3, 4, 9 **2.** 3, 9 **3.** 3, 5 **5.** 2, 3, 5, 10 **6.** 3 **7.** 2, 4 **9.** 2, 3, 4 **10.** 3, 5
11. none **13.** 3, 9 **14.** 2, 3, 4, 5, 10 **15.** none **17.** none **18.** 2 **19.** 3 **21.** 3, 5 **22.** 3
23. 2, 3, 4, 5, 9, 10 **25.** 2, 3 **26.** 2 **27.** none **29.** 2 **30.** 2, 3 **31.** 3, 5 **33.** 3, 9
34. 3, 5 **35.** 2, 3, 9 **37.** 2, 4, 5, 10 **38.** 2, 3, 4, 9 **39.** 3, 9 **41.** 2, 3, 4, 5, 10 **42.** none
43. 2, 4 **45.** 2, 3, 4, 5, 10 **46.** 2, 3, 5, 10 **47.** 3, 5 **49.** none **50.** 2 **51.** 2, 3, 9
53. 2, 3, 4 **54.** 5 **55.** 2, 3, 4, 9 **57.** 3, 9 **58.** 2 **59.** 2, 3, 4
61. Yes; $2 \cdot 3 \cdot 3 \cdot 5 = 6 \cdot 15$. The product is 90 and 6 divides the product 15 times.
62. Yes; $2 \cdot 3 \cdot 3 \cdot 5 = 10 \cdot 9$. The product is 90 and 10 divides the product 9 times.
63. Yes; $2 \cdot 3 \cdot 5 \cdot 7 = 14 \cdot 15$. The product is 210 and 14 divides the product 15 times.
65. No. $3 \cdot 3 \cdot 5 \cdot 7 = 315$ and 10 is not a factor of 315.
66. No. $2 \cdot 3 \cdot 5 \cdot 7 \cdot 11 = 2310$ and 25 is not a factor of 2310.
67. Yes; $2 \cdot 2 \cdot 3 \cdot 5 \cdot 5 = 25 \cdot 12$. The product is 300 and 25 divides the product 12 times.
69. Yes; $3 \cdot 3 \cdot 5 \cdot 7 \cdot 11 = 21 \cdot 165$. The product is 3465 and 21 divides the product 165 times.
70. Yes; $2 \cdot 3 \cdot 4 \cdot 5 \cdot 13 = 30 \cdot 52$. The product is 1560 and 30 divides the product 52 times.

The Recycle Bin, p. 85

1. 850 **2.** 1930 **3.** 400 **4.** 2600 **5.** 14,000 **6.** 21,000 **7.** 690 (estimate); 693 (sum)
8. 4700 (estimate); 4450 (sum) **9.** 5000 (estimate); 5225 (difference)
10. 20,000 (estimate); 28,522 (difference)

Exercises 2.4, p. 91

1. 5, 10, 15, 20, . . . **2.** 7, 14, 21, 28, . . . **3.** 11, 22, 33, 44, . . . **5.** 12, 24, 36, 48, . . .
6. 9, 18, 27, 36, . . . **7.** 20, 40, 60, 80, . . . **9.** 16, 32, 48, 64, . . . **10.** 25, 50, 75, 100, . . .

11.

1	②	③	4	⑤	6	⑦	8	9	10
⑪	12	⑬	14	15	16	⑰	18	⑲	20
21	22	㉓	24	25	26	27	28	㉙	30
㉛	32	33	34	35	36	㊲	38	39	40
㊶	42	㊸	44	45	46	㊼	48	49	50
51	52	㊾	54	55	56	57	58	㊾	60
㊿	62	63	64	65	66	㊌	68	69	70
㋁	72	㋂	74	75	76	77	78	㋅	80
81	82	㋓	84	85	86	87	88	㋙	90
91	92	93	94	95	96	㋝	98	99	100

13. prime   **14.** composite; 1 · 28, 2 · 14, 4 · 7   **15.** composite; 1 · 32, 2 · 16, 4 · 8   **17.** prime
18. composite; 1 · 16, 2 · 8, 4 · 4   **19.** composite; 1 · 63, 3 · 21, 7 · 9   **21.** composite; 1 · 51, 3 · 17
22. prime   **23.** prime   **25.** prime   **26.** composite; 1 · 52, 2 · 26, 4 · 13   **27.** composite; 1 · 57, 3 · 19
29. composite; 1 · 86, 2 · 43   **30.** prime   **31.** prime   **33.** 3, 4   **34.** 8, 2   **35.** 12, 1   **37.** 25, 2
38. 5, 4   **39.** 8, 3   **41.** 12, 3   **42.** 7, 1   **43.** 21, 3   **45.** 5, 5   **46.** 4, 4   **47.** 12, 5   **49.** 9, 3
50. 18, 4

The Recycle Bin, p. 92

1. commutative property of multiplication   **2.** associative property of multiplication   **3.** 16,000
4. 150,000   **5.** 840,000   **6.** 22,922   **7.** 84,812   **8.** 178,500

Exercises 2.5, p. 97

1. $2^3 \cdot 3$ **2.** $2^2 \cdot 7$ **3.** 3^3 **5.** $2^2 \cdot 3^2$ **6.** $2^2 \cdot 3 \cdot 5$ **7.** $2^3 \cdot 3^2$ **9.** 3^4 **10.** $3 \cdot 5 \cdot 7$
11. 5^3 **13.** $3 \cdot 5^2$ **14.** $2 \cdot 3 \cdot 5^2$ **15.** $2 \cdot 3 \cdot 5 \cdot 7$ **17.** $2 \cdot 5^3$ **18.** $3 \cdot 31$ **19.** $2^3 \cdot 3 \cdot 7$
21. $2 \cdot 3^2 \cdot 7$ **22.** $2^4 \cdot 3$ **23.** 17   **25.** $3 \cdot 17$ **26.** $2^4 \cdot 3^2$ **27.** 11^2 **29.** $3^2 \cdot 5^2$ **30.** $2^2 \cdot 13$
31. 2^5 **33.** $2^2 \cdot 3^3$ **34.** 103   **35.** 101   **37.** $2 \cdot 3 \cdot 13$ **38.** $2^2 \cdot 5^3$ **39.** $2^4 \cdot 5^4$
41. 1, 2, 3, 4, 6, 12   **42.** 1, 2, 3, 6, 9, 18   **43.** 1, 2, 4, 7, 14, 28   **45.** 1, 11, 121   **46.** 1, 3, 5, 9, 15, 45
47. 1, 3, 5, 7, 15, 21, 35, 105   **49.** 1, 97   **50.** 1, 2, 3, 4, 6, 8, 9, 12, 16, 18, 24, 36, 48, 72, 144

The Recycle Bin, p. 98

1. 1575   **2.** $12,942   **3.** 104   **4.** (a) 30 meters   (b) 30 square meters

Exercises 2.6, p. 104

1. 24   **2.** 105   **3.** 36   **5.** 110   **6.** 252   **7.** 60   **9.** 200   **10.** 600   **11.** 196   **13.** 240
14. 112   **15.** 100   **17.** 700   **18.** 432   **19.** 252   **21.** 8   **22.** 210   **23.** 1560   **25.** 60
26. 120   **27.** 240   **29.** 2250   **30.** 918   **31.** 726   **33.** 2610   **34.** 324   **35.** 675   **37.** 120
38. 600   **39.** 1680   **41.** 120: 8 · 15; 10 · 12; 15 · 8   **42.** 30: 6 · 5; 15 · 2; 30 · 1
43. 120: 10 · 12; 15 · 8; 24 · 5   **45.** 270: 6 · 45; 18 · 15; 27 · 10; 45 · 6
46. 1140: 12 · 95; 95 · 12; 228 · 5   **47.** 4410: 45 · 98; 63 · 70; 98 · 45
49. 14,157: 99 · 143; 143 · 99; 363 · 39   **50.** 3375: 125 · 27; 135 · 25; 225 · 15
51. (a) 70 min   (b) 7 times, 5 times
53. (a) 48 hr   (b) 4 orbits and 3 orbits, respectively
54. (a) once every 30 days   (b) once every 30 days
55. (a) 180 days   (b) 18 trips, 15 trips, 12 trips, 10 trips

Review Questions: Chapter 2, p. 108

1. base, exponent, power   **2.** different factors   **3.** prime   **4.** composite   **5.** 16   **6.** 27   **7.** 169
8. 441   **9.** 15   **10.** 46   **11.** 35   **12.** 13   **13.** 2   **14.** 2   **15.** 3, 5, 9   **16.** 2, 3, 4, 9
17. none   **18.** 2, 3, 4, 5, 9, 10   **19.** 2, 4   **20.** 3   **21.** 3, 6, 9, 12, . . . ; yes, 3 is prime   **22.** yes
23. 2, 3, 5, 7, 11, 13, 17, 19, 23, 29, 31, 37, 41, 43, 47, 53, 59   **24.** 6 and 4   **25.** 5 and 12   **26.** $2 \cdot 3 \cdot 5^2$
27. $5 \cdot 13$ **28.** $2^2 \cdot 3 \cdot 7$ **29.** $2^2 \cdot 23$ **30.** 168   **31.** 1800   **32.** 270
33. 1638; $18 \cdot 91$, $39 \cdot 42$, $63 \cdot 26$ **34.** 210 sec; 14 laps, 12 laps

Test: Chapter 2, p. 109

1. (a) 3   (b) 7   (c) 343 [2.1]
2. $7^2 = 49$; $11^2 = 121$; $13^2 = 169$; $17^2 = 289$; $19^2 = 361$ [2.1]
3. 53 [2.2]   **4.** 66 [2.2]   **5.** 14 [2.2]   **6.** 5 [2.2]   **7.** 2, 3, 4, and 9 [2.3]   **8.** none [2.3]
9. 2, 3, 4, 5, 9, and 10 [2.3]   **10.** 19, 38, 57, 76, 95 [2.4]   **11.** 6, 9, 15, 27, 39, 51 [2.4]
12. $2 \cdot 5 \cdot 7$ [2.5]   **13.** $2 \cdot 3^2 \cdot 13$ [2.5]   **14.** $2 \cdot 3 \cdot 5^3$ [2.5]   **15.** 1, 3, 5, 9, 15, 45 [2.5]
16. 120 [2.6]   **17.** 108 [2.6]   **18.** 75 [2.6]   **19.** 1008 [2.6]
20. (a) 378 seconds   (b) 7 laps and 6 laps [2.6]

Cumulative Review, p. 110

1. $40,000 + 500 + 80 + 6$; forty thousand, five hundred eighty-six
2. commutative property of multiplication   **3.** associative property of addition   **4.** multiplicative identity
5. 1277   **6.** 4107   **7.** 2275   **8.** 40,000   **9.** 4000 (estimate); 4316 (product)
10. 100 (estimate); 130 (quotient)   **11.** 309 R25   **12.** 1,521,000   **13.** 130   **14.** 2, 3, 5, 9, 10
15. true   **16.** $3 \cdot 5^3$ **17.** $2^2 \cdot 3^2 \cdot 5^2 = 900$ **18.** 79   **19.** \$133   **20.** 864 sq. in.

CHAPTER 3

Exercises 3.1, p. 119

1. 0   **2.** These expressions are undefined.   **3.** $\dfrac{1}{3}$ **5.** $\dfrac{1}{4}$ **6.** $\dfrac{3}{16}$ **7.** $\dfrac{1}{4}$

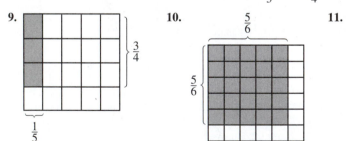

9.   **10.**   **11.**

13. $\dfrac{3}{25}$ **14.** $\dfrac{1}{16}$ **15.** $\dfrac{2}{27}$ **17.** $\dfrac{1}{4}$ **18.** $\dfrac{1}{9}$ **19.** $\dfrac{9}{16}$ **21.** $\dfrac{4}{81}$ **22.** $\dfrac{12}{35}$ **23.** $\dfrac{5}{32}$ **25.** 0

26. 0   **27.** $\dfrac{9}{25}$ **29.** $\dfrac{9}{100}$ **30.** $\dfrac{12}{1}$ or 12   **31.** $\dfrac{10}{1}$ or 10   **33.** $\dfrac{7}{50}$ **34.** $\dfrac{3}{50}$ **35.** $\dfrac{4}{21}$ **37.** $\dfrac{99}{20}$

38. $\dfrac{6}{385}$ **39.** $\dfrac{48}{455}$ **41.** $\dfrac{1}{360}$ **42.** $\dfrac{27}{100}$ **43.** $\dfrac{840}{1} = 840$ **45.** 0   **46.** 0   **47.** 0   **49.** $\dfrac{728}{45}$

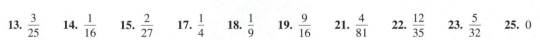

50. $\dfrac{9}{500}$ **51.** commutative property of multiplication   **53.** associative property of multiplication

54. commutative property of multiplication   **55.** See page 116 in the text.   **57.** $\dfrac{29}{35}$ **58.** $\dfrac{20}{8}$ in. $= \dfrac{5}{2}$ in.

59. $\dfrac{3}{8}$ cup

Exercises 3.2, p. 126

1. $\dfrac{3}{4} = \dfrac{3}{4} \cdot \dfrac{3}{3} = \dfrac{9}{12}$ **2.** $\dfrac{3}{5} = \dfrac{3}{5} \cdot \dfrac{2}{2} = \dfrac{6}{10}$ **3.** $\dfrac{2}{3} = \dfrac{2}{3} \cdot \dfrac{4}{4} = \dfrac{8}{12}$ **5.** $\dfrac{6}{7} = \dfrac{6}{7} \cdot \dfrac{2}{2} = \dfrac{12}{14}$ **6.** $\dfrac{5}{8} = \dfrac{5}{8} \cdot \dfrac{2}{2} = \dfrac{10}{16}$

7. $\dfrac{5}{9} = \dfrac{5}{9} \cdot \dfrac{3}{3} = \dfrac{15}{27}$ **9.** $\dfrac{1}{2} = \dfrac{1}{2} \cdot \dfrac{10}{10} = \dfrac{10}{20}$ **10.** $\dfrac{1}{4} = \dfrac{1}{4} \cdot \dfrac{10}{10} = \dfrac{10}{40}$ **11.** $\dfrac{3}{8} = \dfrac{3}{8} \cdot \dfrac{5}{5} = \dfrac{15}{40}$

13. $\dfrac{2}{3} = \dfrac{2}{3} \cdot \dfrac{5}{5} = \dfrac{10}{15}$ **14.** $\dfrac{1}{6} = \dfrac{1}{6} \cdot \dfrac{2}{2} = \dfrac{2}{12}$ **15.** $\dfrac{3}{5} = \dfrac{3}{5} \cdot \dfrac{9}{9} = \dfrac{27}{45}$ **17.** $\dfrac{10}{11} = \dfrac{10}{11} \cdot \dfrac{4}{4} = \dfrac{40}{44}$

18. $\dfrac{5}{11} = \dfrac{5}{11} \cdot \dfrac{3}{3} = \dfrac{15}{33}$ **19.** $\dfrac{5}{21} = \dfrac{5}{21} \cdot \dfrac{2}{2} = \dfrac{10}{42}$ **21.** $\dfrac{5}{7} = \dfrac{5}{7} \cdot \dfrac{6}{6} = \dfrac{30}{42}$ **22.** $\dfrac{3}{8} = \dfrac{3}{8} \cdot \dfrac{3}{3} = \dfrac{9}{24}$

23. $\dfrac{3}{16} = \dfrac{3}{16} \cdot \dfrac{5}{5} = \dfrac{15}{80}$ **25.** $\dfrac{7}{2} = \dfrac{7}{2} \cdot \dfrac{10}{10} = \dfrac{70}{20}$ **26.** $\dfrac{14}{3} = \dfrac{14}{3} \cdot \dfrac{3}{3} = \dfrac{42}{9}$ **27.** $\dfrac{5}{2} = \dfrac{5}{2} \cdot \dfrac{10}{10} = \dfrac{50}{20}$

29. $\dfrac{3}{1} = \dfrac{3}{1} \cdot \dfrac{5}{5} = \dfrac{15}{5}$ **30.** $\dfrac{4}{1} = \dfrac{4}{1} \cdot \dfrac{8}{8} = \dfrac{32}{8}$ **31.** $\dfrac{8}{10} = \dfrac{8}{10} \cdot \dfrac{10}{10} = \dfrac{80}{100}$ **33.** $\dfrac{6}{10} = \dfrac{6}{10} \cdot \dfrac{10}{10} = \dfrac{60}{100}$

34. $\dfrac{7}{10} = \dfrac{7}{10} \cdot \dfrac{10}{10} = \dfrac{70}{100}$ **35.** $\dfrac{11}{10} = \dfrac{11}{10} \cdot \dfrac{6}{6} = \dfrac{66}{60}$ **37.** $\dfrac{9}{100} = \dfrac{9}{100} \cdot \dfrac{10}{10} = \dfrac{90}{1000}$

38. $\dfrac{9}{10} = \dfrac{9}{10} \cdot \dfrac{100}{100} = \dfrac{900}{1000}$ **39.** $\dfrac{3}{10} = \dfrac{3}{10} \cdot \dfrac{100}{100} = \dfrac{300}{1000}$ **41.** $\dfrac{1}{3}$ **42.** $\dfrac{2}{3}$ **43.** $\dfrac{3}{4}$ **45.** $\dfrac{2}{5}$ **46.** $\dfrac{4}{5}$

47. $\dfrac{7}{18}$ **49.** $\dfrac{0}{25} = 0$ **50.** $\dfrac{3}{4}$ **51.** $\dfrac{2}{5}$ **53.** $\dfrac{5}{6}$ **54.** $\dfrac{1}{4}$ **55.** $\dfrac{2}{3}$ **57.** $\dfrac{2}{3}$ **58.** $\dfrac{3}{4}$ **59.** $\dfrac{6}{25}$

61. $\dfrac{2}{3}$ **62.** $\dfrac{2}{3}$ **63.** $\dfrac{12}{35}$ **65.** $\dfrac{2}{9}$ **66.** $\dfrac{3}{7}$ **67.** $\dfrac{25}{76}$ **69.** $\dfrac{1}{2}$ **70.** $\dfrac{4}{1} = 4$ **71.** $\dfrac{6}{1} = 6$ **73.** $\dfrac{2}{17}$

74. $\dfrac{3}{8}$ **75.** $\dfrac{9}{20}$ **77.** $\dfrac{10}{9}$ **78.** $\dfrac{11}{15}$ **79.** $\dfrac{5}{4}$ **81.** $\dfrac{5}{18}$ **82.** $\dfrac{1}{6}$ **83.** $\dfrac{1}{4}$ **85.** $\dfrac{21}{4}$ **86.** $\dfrac{189}{52}$

87. $\dfrac{77}{4}$ **89.** $\dfrac{8}{5}$ **90.** $\dfrac{3}{10}$ **91.** $\dfrac{2}{7}$ **93.** $\dfrac{2}{3}$ **94.** $\dfrac{1}{80}$ **95.** $\dfrac{21}{16}$ **97.** 110   **98.** 10 ft, 5 ft, $\dfrac{5}{2}$ ft, $\dfrac{5}{4}$ ft

99. 75 miles   **101.** 72   **102.** 225 "earth-days"
Check Your Number Sense   **103.** The statement is true. The answer is "yes" to all questions.

The Recycle Bin, p. 129

1. $3^2 \cdot 5$ **2.** $2 \cdot 3 \cdot 13$ **3.** $2 \cdot 5 \cdot 13$ **4.** $2^2 \cdot 5 \cdot 23$ **5.** $2^6 \cdot 5^6$ **6.** 1, 2, 3, 5, 6, 10, 15, 30
7. 1, 2, 3, 4, 6, 8, 12, 16, 24, 48   **8.** 1, 3, 7, 9, 21, 63

Exercises 3.3, p. 134

1. $\dfrac{7}{5}$ **2.** reciprocal   **3.** undefined   **5.** $\dfrac{9}{8}$ **6.** $\dfrac{25}{24}$ **7.** $\dfrac{5}{3}$ **9.** $\dfrac{4}{7}$ **10.** $\dfrac{7}{12}$ **11.** $\dfrac{3}{4}$ **13.** $\dfrac{5}{9}$

14. $\dfrac{21}{16}$ **15.** $\dfrac{7}{12}$ **17.** 5   **18.** 3   **19.** 7   **21.** 1   **22.** 1   **23.** $\dfrac{9}{16}$ **25.** $\dfrac{3}{20}$ **26.** $\dfrac{1}{10}$ **27.** $\dfrac{1}{10}$

29. undefined   **30.** 0   **31.** 0   **33.** $\dfrac{16}{5}$ **34.** 5   **35.** $\dfrac{5}{9}$ **37.** $\dfrac{1}{5}$ **38.** $\dfrac{2}{5}$ **39.** $\dfrac{4}{9}$ **41.** $\dfrac{65}{24}$

42. $\frac{8}{13}$ **43.** $\frac{176}{105}$ **45.** $\frac{22}{7}$ **46.** 98 **47.** 125 **49.** 100 **50.** $\frac{50}{27}$ **51.** $\frac{62}{43}$ **53.** 225 **54.** $\frac{5}{12}$

55. 1250 **57.** more; 60 **58.** less; 2

59. More; more, because $\frac{1}{10}$ of a number larger than 1200 will be more than 120.

The Recycle Bin, p. 136

1. 390 **2.** 252 **3.** 100 **4.** 663 **5.** 300 **6.** 968

7. (a) 250 (b) 250 = 50 · 5; 250 = 125 · 2 **8.** (a) 120 (b) 120 = 20 · 6; 120 = 24 · 5; 120 = 30 · 4

9. (a) 420 (b) 420 = 12 · 35; 420 = 70 · 6

10. (a) 3465 (b) 3465 = 45 · 77; 3465 = 63 · 55; 3465 = 99 · 35 **11.** 90 days

Exercises 3.4, p. 142

1. 16 **2.** 39 **3.** 54 **5.** 1000 **6.** $\frac{10}{10} = 1$ **7.** $\frac{5}{14}$ **9.** $\frac{3}{2}$ **10.** $\frac{3}{2}$ **11.** $\frac{10}{5} = 2$ **13.** $\frac{5}{3}$

14. $\frac{3}{5}$ **15.** $\frac{13}{18}$ **17.** $\frac{11}{16}$ **18.** $\frac{23}{20}$ **19.** $\frac{3}{5}$ **21.** $\frac{12}{12} = 1$ **22.** $\frac{11}{16}$ **23.** $\frac{17}{20}$ **25.** $\frac{17}{21}$ **26.** $\frac{3}{4}$

27. $\frac{9}{13}$ **29.** $\frac{23}{54}$ **30.** $\frac{151}{140}$ **31.** $\frac{23}{72}$ **33.** $\frac{9}{4}$ **34.** $\frac{3}{5}$ **35.** $\frac{1}{2}$ **37.** $\frac{5}{8}$ **38.** $\frac{49}{60}$ **39.** $\frac{5}{6}$

41. $\frac{343}{432}$ **42.** $\frac{31}{96}$ **43.** $\frac{13}{9}$ **45.** $\frac{1}{5}$ **46.** $\frac{7}{6}$ **47.** $\frac{317}{1000}$ **49.** $\frac{271}{10,000}$ **50.** $\frac{631}{100}$ **51.** $\frac{8191}{1000}$

53. $\frac{753}{1000}$ **54.** $\frac{63}{50}$ **55.** $\frac{89}{200}$ **57.** $\frac{613}{1000}$ **58.** $\frac{27,683}{10,000}$ **59.** $\frac{5134}{1000} = \frac{2567}{500}$ **61.** $\frac{5}{3}$ **62.** $\frac{7}{6}$

63. $\frac{119}{72}$ **65.** $\frac{19}{100}$ **66.** $\frac{531}{1000}$ **67.** $\frac{49}{36}$ **69.** $\frac{73}{96}$ **70.** $\frac{39}{100}$ **71.** 1 oz **73.** $\frac{23}{20}$ in.

74. $\frac{37}{40}$ in. **75.** $\frac{5}{8}$ in.

Exercises 3.5, p. 149

1. $\frac{3}{7}$ **2.** $\frac{2}{7}$ **3.** $\frac{3}{5}$ **5.** $\frac{1}{2}$ **6.** $\frac{1}{4}$ **7.** $\frac{1}{3}$ **9.** $\frac{3}{5}$ **10.** $\frac{2}{3}$ **11.** $\frac{1}{2}$ **13.** $\frac{13}{30}$ **14.** $\frac{3}{5}$ **15.** $\frac{1}{12}$

17. $\frac{9}{32}$ **18.** $\frac{5}{16}$ **19.** $\frac{13}{20}$ **21.** $\frac{7}{54}$ **22.** $\frac{11}{18}$ **23.** $\frac{1}{40}$ **25.** $\frac{7}{60}$ **26.** $\frac{17}{4}$ **27.** $\frac{27}{8}$ **29.** $\frac{3}{16}$

30. $\frac{16}{3}$ **31.** $\frac{87}{100}$ **33.** $\frac{3}{50}$ **34.** $\frac{9}{1000}$ **35.** $\frac{1}{25}$ **37.** $\frac{3}{20}$ **38.** 0 **39.** 0 **41.** $\frac{1}{10}$ **42.** $\frac{1}{8}$

43. $\frac{1}{3}$ **45.** $\frac{17}{20}$ **46.** $\frac{5}{9}$ **47.** $\frac{5}{24}$ **49.** $\frac{1}{50}$ **50.** $\frac{29}{1000}$ **51.** $\frac{5}{16}$ **53.** $\frac{11}{12}$ **54.** $\frac{9}{25}$ **55.** $\frac{11}{10}$

The Recycle Bin, p. 150

1. 81 **2.** 169 **3.** 225 **4.** 324 **5.** 0 **6.** 29 **7.** 74 **8.** 0 **9.** 70 **10.** 4

Exercises 3.6, p. 156

1. $\frac{3}{4}$ by $\frac{1}{12}$ **2.** $\frac{7}{8}$ by $\frac{1}{24}$ **3.** $\frac{17}{20}$ by $\frac{1}{20}$ **5.** $\frac{13}{20}$ by $\frac{1}{40}$ **6.** $\frac{21}{25}$ by $\frac{11}{400}$ **7.** equal **9.** $\frac{11}{48}$ by $\frac{1}{60}$

10. $\frac{37}{100}$ by $\frac{1}{20}$ **11.** $\frac{3}{5}, \frac{2}{3}, \frac{7}{10}$ **13.** $\frac{11}{12}, \frac{19}{20}, \frac{7}{6}$ **14.** $\frac{5}{42}, \frac{1}{3}, \frac{3}{7}$ **15.** $\frac{1}{4}, \frac{1}{3}, \frac{1}{2}$ **17.** $\frac{13}{18}, \frac{7}{9}, \frac{31}{36}$

18. $\frac{40}{36}, \frac{31}{24}, \frac{17}{12}$ **19.** $\frac{20}{10,000}, \frac{3}{1000}, \frac{1}{100}$ **21.** $\frac{2}{3}$ **22.** $\frac{1}{5}$ **23.** $\frac{13}{9}$ **25.** $\frac{187}{32}$ **26.** $\frac{43}{450}$ **27.** $\frac{1}{4}$

29. $\frac{8}{39}$ **30.** $\frac{14}{45}$ **31.** $\frac{15}{64}$ **33.** $\frac{29}{36}$ **34.** $\frac{31}{36}$ **35.** $\frac{3}{16}$ **37.** $\frac{11}{70}$ **38.** $\frac{5}{7}$ **39.** $\frac{25}{21}$ **41.** $\frac{33}{4}$

42. $\frac{328}{133}$

Check Your Number Sense   **43.** (a) yes (b) no
45. Yes. For 0, the quotient will remain 0.   **46.** Smaller. This will always be so.

Review Questions: Chapter 3, p. 160

1. 0 **2.** undefined **3.** $\frac{3}{2}, \frac{2}{3}$ **4.** associative **5.** $\frac{4}{15}$ **6.** $\frac{1}{30}$ **7.** $\frac{3}{49}$ **8.** $\frac{1}{12}$ **9.** 2 **10.** 54

11. 75 **12.** $\frac{1}{2}$ **13.** $\frac{9}{8}$ **14.** 0 **15.** $\frac{5}{4}$ **16.** $\frac{5}{7}$ **17.** $\frac{2}{3}$ **18.** $\frac{1}{4}$ **19.** $\frac{49}{72}$ **20.** $\frac{7}{22}$ **21.** $\frac{25}{54}$

22. $\frac{7}{20}$ **23.** $\frac{1}{3}$ **24.** $\frac{7}{8}$ **25.** $\frac{1}{9}$ **26.** $\frac{5}{3}$ **27.** 1 **28.** $\frac{5}{4}$ **29.** $\frac{4}{5}$ **30.** $\frac{4}{5}$ by $\frac{2}{15}$ **31.** $\frac{11}{20}, \frac{5}{9}, \frac{7}{12}$

32. $\frac{25}{112}$ **33.** $\frac{32}{125}$ **34.** $\frac{19}{30}$ **35.** $\frac{7}{18}$ **36.** $\frac{35}{16}$

Test: Chapter 3, p. 162

1. $\frac{8}{5}$ [3.3]   **2.** (a) commutative property of addition [3.4] (b) associative property of multiplication [3.1]

3. $\frac{3}{5}$ [3.2] **4.** $\frac{9}{11}$ [3.2] **5.** $\frac{5}{6}$ [3.2] **6.** $\frac{2}{5}$ [3.4] **7.** $\frac{4}{5}$ [3.4] **8.** $\frac{29}{30}$ [3.4] **9.** 2 [3.5]

10. $\frac{4}{15}$ [3.5] **11.** $\frac{7}{36}$ [3.5] **12.** $\frac{151}{120}$ [3.4] **13.** 1 [3.3] **14.** $\frac{1}{2}$ [3.2] **15.** $\frac{9}{20}$ [3.3] **16.** $\frac{19}{20}$ [3.6]

17. $\frac{11}{30}$ [3.6] **18.** $\frac{1}{16}$ [3.6] **19.** $\frac{239}{70}$ [3.6] **20.** $\frac{2}{45}$ [3.6] **21.** $\frac{85}{54}$ [3.6] **22.** $\frac{5}{7}$ [3.1]

23. $\frac{7}{12}, \frac{3}{5}, \frac{5}{8}$ [3.6] **24.** $\frac{7}{16}$ [3.5] **25.** $\frac{6}{5}$ [3.5] **26.** $\frac{97}{100}$ [3.5]

Cumulative Review, p. 163

1. 1,300,000 **2.** 14,000 (estimate); 15,795 (product) **3.** 25 (estimate); 24 (quotient)
4. distributive property of multiplication over addition **5.** 304 R10 **6.** 30 **7.** $3 \cdot 5^3$ **8.** 252
9. 120,000 **10.** 2, 3, 5, 9, 10 **11.** 331 is prime. **12.** 355 **13.** 1020 **14.** 1268 **15.** 407 R141

16. 1 **17.** $\frac{5}{6}$ **18.** $\frac{1}{24}$ **19.** $\frac{5}{6}$ **20.** $\frac{5}{34}$ **21.** 94 **22.** $3824 **23.** 794 **24.** $\frac{11}{8}$

25. 10,000 sq. ft

CHAPTER 4

Exercises 4.1, p. 172

1. $\frac{4}{3}$ **2.** $\frac{5}{2}$ **3.** $\frac{4}{3}$ **5.** $\frac{3}{2}$ **6.** $\frac{3}{2}$ **7.** $\frac{7}{5}$ **9.** $\frac{5}{4}$ **10.** $\frac{5}{4}$ **11.** $4\frac{1}{6}$ **13.** $1\frac{1}{3}$ **14.** $1\frac{1}{4}$

15. $1\frac{1}{2}$ **17.** $6\frac{1}{7}$ **18.** $2\frac{1}{8}$ **19.** $7\frac{1}{2}$ **21.** $3\frac{1}{9}$ **22.** $2\frac{1}{15}$ **23.** 3 **25.** $4\frac{1}{2}$ **26.** $2\frac{1}{17}$

27. $\frac{3}{1} = 3$ **29.** $1\frac{3}{4}$ **30.** $1\frac{17}{20}$ **31.** $\frac{37}{8}$ **33.** $\frac{76}{15}$ **34.** $\frac{8}{5}$ **35.** $\frac{46}{11}$ **37.** $\frac{7}{3}$ **38.** $\frac{34}{7}$ **39.** $\frac{32}{3}$

41. $\frac{34}{5}$ **42.** $\frac{71}{5}$ **43.** $\frac{50}{3}$ **45.** $\frac{101}{5}$ **46.** $\frac{47}{5}$ **47.** $\frac{92}{7}$ **49.** $\frac{17}{1} = 17$ **50.** $\frac{151}{50}$ **51.** $\frac{603}{100}$

53. $\frac{8921}{1000}$ **54.** $\frac{3477}{1000}$ **55.** $\frac{2631}{1000}$ **57.** $1\frac{15}{16}$ in. **58.** $\$1\frac{3}{4}$ **59.** $2\frac{3}{4}$ in.

The Recycle Bin, p. 174

1. $\frac{7}{9}$ **2.** $\frac{2}{5}$ **3.** $\frac{3}{1}$ (or 3) **4.** $\frac{6}{13}$ **5.** $\frac{1}{5}$ **6.** $\frac{3}{16}$ **7.** $\frac{5}{9}$ **8.** $\frac{15}{14}$

Exercises 4.2, p. 180

1. $7\frac{7}{12}$ **2.** $1\frac{13}{35}$ **3.** $10\frac{1}{2}$ **5.** $22\frac{1}{2}$ **6.** 12 **7.** $31\frac{1}{6}$ **9.** 8 **10.** 12 **11.** 30 **13.** $21\frac{3}{4}$

14. 1 **15.** 1 **17.** $27\frac{1}{2}$ **18.** $10\frac{1}{5}$ **19.** $19\frac{1}{4}$ **21.** $\frac{1}{7}$ **22.** $\frac{1}{5}$ **23.** $\frac{1}{10}$ **25.** $\frac{36}{385}$ **26.** $39\frac{3}{8}$

27. 11 **29.** $3016\frac{26}{75}$ **30.** $640\frac{1}{16}$ **31.** 40 **33.** 20 **34.** 60 **35.** $1\frac{5}{16}$ **37.** $3\frac{12}{35}$ **38.** $5\frac{19}{20}$

39. $\frac{1}{3}$ **41.** $\frac{5}{9}$ **42.** 1 **43.** $\frac{10}{39}$ **45.** $\frac{29}{155}$ **46.** $\frac{63}{50} = 1\frac{13}{50}$ **47.** $\frac{28}{17} = 1\frac{11}{17}$ **49.** $\frac{200}{301}$

50. $\frac{41}{12} = 3\frac{5}{12}$ **51.** $\frac{7}{5} = 1\frac{2}{5}$ **53.** $\frac{41}{3} = 13\frac{2}{3}$ **54.** $\frac{37}{2} = 18\frac{1}{2}$ **55.** $\frac{63}{5} = 12\frac{3}{5}$ **57.** $\frac{20}{11} = 1\frac{9}{11}$

58. $\frac{22}{7} = 3\frac{1}{7}$ **59.** 10 ft; 22 ft **61.** 177 mi **62.** $1012\frac{7}{8}$ ft **63.** 90 pages; 18 hr **65.** $1\frac{19}{43}$

66. 175 passengers **67.** $180 **69.** $5\frac{19}{20}$ **70.** $8\frac{2}{5}$ **71.** $3200 **73.** (a) $460\frac{1}{4}$ mi (b) $2325\frac{3}{4}¢$

74. $47\frac{1}{10}$ centimeters **75.** $162\frac{1}{2}$ mi **77.** $53\frac{9}{10}$ meters per second

Check Your Number Sense **78.** (a) more than 1 (b) less than 2 (c) $1\frac{16}{19}$

The Recycle Bin, p. 184

1. $\frac{17}{21}$ **2.** $\frac{61}{40}$ **3.** $\frac{31}{30}$ **4.** $\frac{59}{45}$ **5.** $\frac{17}{40}$ **6.** $\frac{67}{84}$ **7.** $\frac{17}{64}$ **8.** $\frac{1}{6}$

Exercises 4.3, p. 189

1. $14\frac{5}{7}$ **2.** 10 **3.** 8 **5.** $12\frac{3}{4}$ **6.** $15\frac{9}{10}$ **7.** $8\frac{2}{3}$ **9.** $34\frac{7}{18}$ **10.** $10\frac{1}{2}$ **11.** $7\frac{2}{3}$ **13.** $42\frac{7}{20}$

14. $8\frac{10}{21}$ **15.** $10\frac{3}{4}$ **17.** $12\frac{7}{27}$ **18.** $14\frac{1}{16}$ **19.** $18\frac{23}{45}$ **21.** $28\frac{1}{2}$ **22.** $7\frac{1}{4}$ **23.** $9\frac{29}{40}$ **25.** $12\frac{47}{60}$

26. $4\frac{2}{3}$ **27.** $63\frac{11}{24}$ **29.** $53\frac{83}{192}$ **30.** $12\frac{5}{12}$ **31.** $18\frac{7}{8}$ **33.** $36\frac{23}{24}$ **34.** $9\frac{7}{20}$ **35.** $90\frac{1}{8}$

37. $35\frac{7}{10}$ km **38.** $89\frac{1}{8}$ ft **39.** $12\frac{1}{8}$ in. **41.** $45\frac{3}{4}$ miles **42.** $152\frac{1}{2}$ miles **43.** $26\frac{1}{10}$ cm

45. $\$142\frac{4}{5}$ million **46.** $3\frac{1}{2}$ hr

Check Your Number Sense **47.** (a) 6 (b) 600 (c) 60,000

Exercises 4.4, p. 196

1. $4\frac{1}{2}$ **2.** $5\frac{3}{4}$ **3.** $1\frac{5}{12}$ **5.** 4   **6.** 4   **7.** $3\frac{1}{2}$ **9.** $3\frac{1}{4}$ **10.** $5\frac{1}{8}$ **11.** $3\frac{5}{8}$ **13.** $4\frac{2}{3}$ **14.** $6\frac{1}{6}$

15. $6\frac{7}{12}$ **17.** $10\frac{4}{5}$ **18.** $3\frac{29}{32}$ **19.** $5\frac{7}{12}$ **21.** $1\frac{7}{16}$ **22.** $3\frac{9}{10}$ **23.** $\frac{1}{3}$ **25.** $\frac{5}{8}$ **26.** $57\frac{1}{6}$

27. $1\frac{11}{16}$ **29.** $1\frac{13}{16}$ **30.** 17   **31.** 5   **33.** $7\frac{3}{40}$ **34.** $16\frac{5}{24}$ **35.** $7\frac{4}{5}$ **37.** $\frac{3}{8}$ **38.** $\frac{1}{4}$ **39.** $\frac{1}{10}$

41. $\frac{1}{5}$ **42.** $\frac{3}{5}$ **43.** $\frac{5}{6}$ **45.** $1\frac{1}{4}$ **46.** $56\frac{13}{24}$ **47.** $136\frac{23}{30}$ **49.** $165\frac{13}{24}$ **50.** $476\frac{13}{15}$ gal

51. $1\frac{3}{4}$ pounds   **53.** $1\frac{3}{20}$ hours (or 1 hr 9 min)   **54.** $\frac{5}{6}$ hour (or 50 min)   **55.** $6\frac{3}{5}$ min   **57.** $3\frac{1}{4}$ pounds

58. $219\frac{13}{16}$ pounds   **59.** $3\frac{1}{4}$ ft (or 3 ft 3 in.)   **61.** $4\frac{13}{20}$ parts water   **62.** $9\frac{1}{5}$ calories

The Recycle Bin, p. 199

1. 18   **2.** 101   **3.** $\frac{47}{18}$ **4.** 0   **5.** $\frac{13}{12}$

Exercises 4.5, p. 203

1. $\frac{44}{25} = 1\frac{19}{25}$ **2.** $\frac{26}{135}$ **3.** $\frac{52}{75}$ **5.** $\frac{25}{21} = 1\frac{4}{21}$ **6.** $\frac{8}{27}$ **7.** 4   **9.** $\frac{7}{45}$ **10.** $\frac{13}{100}$

11. $\frac{429}{365} = 1\frac{64}{365}$ **13.** $\frac{69}{35} = 1\frac{34}{35}$ **14.** $\frac{60}{17} = 3\frac{9}{17}$ **15.** $\frac{1}{5}$ **17.** $\frac{172}{9} = 19\frac{1}{9}$ **18.** $\frac{111}{20} = 5\frac{11}{20}$

19. $\frac{791}{120} = 6\frac{71}{120}$ **21.** $\frac{3}{5}$ **22.** $\frac{12}{7} = 1\frac{5}{7}$ **23.** $\frac{5}{2} = 2\frac{1}{2}$ **25.** $\frac{543}{40} = 13\frac{23}{40}$ **26.** $\frac{119}{20} = 5\frac{19}{20}$

27. $\frac{42}{5} = 8\frac{2}{5}$ **29.** $\frac{47}{40} = 1\frac{7}{40}$ **30.** $\frac{22}{45}$ **31.** $\frac{223}{32} = 6\frac{31}{32}$

Review Questions: Chapter 4, p. 206

1. $6\frac{5}{8}$ **2.** 7   **3.** $3\frac{21}{50}$ **4.** $\frac{51}{10}$ **5.** $\frac{35}{12}$ **6.** $\frac{67}{5}$ **7.** $\frac{123}{4} = 30\frac{3}{4}$ **8.** $\frac{11}{15}$ **9.** 12   **10.** $\frac{2}{9}$

11. $\frac{59}{8} = 7\frac{3}{8}$ **12.** $\frac{56}{3} = 18\frac{2}{3}$ **13.** 104   **14.** $\frac{7}{10}$ **15.** $\frac{79}{12} = 6\frac{7}{12}$ **16.** $\frac{13}{8} = 1\frac{5}{8}$ **17.** $10\frac{2}{3}$

18. $120\frac{17}{18}$ **19.** $\frac{90}{7} = 12\frac{6}{7}$ **20.** $\frac{49}{15} = 3\frac{4}{15}$ **21.** $\frac{11}{70}$ **22.** $\frac{33}{4} = 8\frac{1}{4}$ **23.** $\frac{91}{120}$ **24.** $\frac{7}{4} = 1\frac{3}{4}$

25. 64   **26.** $\frac{5}{2} = 2\frac{1}{2}$ **27.** $\frac{39}{124}$ **28.** $\frac{11}{4} = 2\frac{3}{4}$ lb   **29.** \$625   **30.** $\frac{61}{8} = 7\frac{5}{8}$ **31.** $7\frac{5}{8}$ yd

32. \$6000   **33.** 6 ft   **34.** 6   **35.** $189\frac{1}{2}$ miles

Test: Chapter 4, p. 208

1. $4\frac{6}{7}$ [4.1]   **2.** $\frac{59}{8}$ [4.1]   **3.** $43\frac{1}{14}$ [4.3]   **4.** $10\frac{7}{10}$ [4.4]   **5.** $\frac{21}{2}$ or $10\frac{1}{2}$ [4.2]   **6.** $\frac{5}{7}$ [4.2]

7. $7\frac{5}{42}$ [4.3]   **8.** $1\frac{2}{11}$ [4.4]   **9.** $\frac{101}{12} = 8\frac{5}{12}$ [4.3]   **10.** $\frac{37}{24} = 1\frac{13}{24}$ [4.4]   **11.** $\frac{113}{8} = 14\frac{1}{8}$ [4.3]

12. $\frac{17}{36}$ [4.4]   **13.** 9 [4.2]   **14.** $\frac{16}{27}$ [4.2]   **15.** $\frac{255}{16} = 15\frac{15}{16}$ [4.2]   **16.** $\frac{131}{10} = 13\frac{1}{10}$ [4.5]

17. $\frac{25}{24} = 1\frac{1}{24}$ [4.5]   **18.** $\frac{20}{7} = 2\frac{6}{7}$ [4.2]   **19.** $\frac{63}{20} = 3\frac{3}{20}$ [4.5]   **20.** $\frac{63}{8} = 7\frac{7}{8}$ [4.5]

21. (a) $56\frac{3}{4}$ miles   (b) $18\frac{11}{12}$ miles [4.5]   **22.** $76\frac{1}{4}$ miles [4.3]   **23.** (a) 39 in.   (b) 92 sq. in. [4.3, 4.2]

Cumulative Review; p. 210

1. 40,000 + 3000 + 500 + 70; forty-three thousand, five hundred seventy   **2.** 6475   **3.** 180,000
4. commutative property of addition   **5.** (a) B.   (b) A.   (c) D.   (d) C.   **6.** 72,000   **7.** 91

8. 2, 3, 5, 7, 11, 13, 17, 19   **9.** $45 = 3^2 \cdot 5, 105 = 3 \cdot 5 \cdot 7$ **10.** $\frac{1}{10}$ **11.** 0, undefined   **12.** 44

13. 11,338   **14.** 317   **15.** 25,935   **16.** 204   **17.** $\frac{1}{16}$ **18.** $\frac{1}{18}$ **19.** $\frac{5}{24}$ **20.** $2\frac{2}{15}$ **21.** $48\frac{23}{60}$

22. $1\frac{3}{35}$ **23.** $9\frac{1}{3}$ **24.** $1\frac{3}{4}$ **25.** 90   **26.** 3   **27.** $2\frac{5}{6}$ **28.** $\frac{7}{10}, \frac{3}{4}, \frac{5}{6}$ **29.** $1\frac{3}{8}$ **30.** $221

31. $21   **32.** $11\frac{1}{4}$ gal   **33.** $3\frac{5}{8}$ pounds   **34.** $12\frac{7}{8}$ in.   **35.** $950

CHAPTER 5

Exercises 5.1, p. 219

1. 37.498   **2.** 18.76   **3.** 4.11   **5.** 87.003   **6.** 95.2   **7.** 62.7   **9.** 100.38   **10.** 250.623

11. $82\frac{56}{100}$ **13.** $10\frac{576}{1000}$ **14.** $100\frac{6}{10}$ **15.** $65\frac{3}{1000}$ **17.** 0.014   **18.** 0.17   **19.** 6.28   **21.** 72.392

22. 850.0036   **23.** 700.77   **25.** 600,500.402   **26.** five tenths   **27.** ninety-three hundredths
29. thirty-two and fifty-eight hundredths   **30.** seventy-one and six hundredths
31. thirty-five and seventy-eight thousandths   **33.** eighteen and one hundred two thousandths
34. fifty and eight thousandths   **35.** six hundred seven and six hundred seven thousandths
37. five hundred ninety-three and eighty-six hundredths
38. four thousand, seven hundred and six hundred seventeen thousandths
39. five thousand and five thousandths
41. nine hundred and four thousand, six hundred thirty-eight ten-thousandths
42. $356\frac{45}{100}$; three hundred fifty-six and $\frac{45}{100}$ **43.** $651\frac{50}{100}$; six hundred fifty-one and $\frac{50}{100}$
45. two and fifty-four hundredths   **46.** eight and thirty-three hundredths
47. one hundred seventy-six and four hundred fifty-seven thousandths
49. three and fourteen thousand, one hundred fifty-nine hundred-thousandths
50. two and seventy-one thousand, eight hundred twenty-eight hundred-thousandths
51. four hundred five thousandths

The Recycle Bin, p. 222

1. (a) 9000 (estimate)   (b) 9107 (sum)   **2.** (a) 1600 (estimate)   (b) 1555 (sum)
3. (a) 550,000 (estimate)   (b) 562,712 (sum)   **4.** (a) 200 (estimate)   (b) 188 (difference)
5. (a) 4000 (estimate)   (b) 4001 (difference)   **6.** (a) 85,000 (estimate)   (b) 84,891 (difference)   **7.** 350

Exercises 5.2, p. 225

1. (a) 7 (b) 6 (c) 6; 7; 8 (d) 34.8
2. (a) 3 (b) 2 (c) 2; 3; 2 (d) 6.83
3. (a) 4 (b) 3 (c) 3; 4; 3 (d) 1.0064; ten-thousandth
5. 89.0 **6.** 7.6 **7.** 18.0 **9.** 14.3 **10.** 0.0 **11.** 0.39 **13.** 5.72 **14.** 8.99 **15.** 7.00
17. 0.08 **18.** 6.00 **19.** 5.71 **21.** 0.067 **22.** 0.056 **23.** 0.634 **25.** 32.479 **26.** 9.430
27. 17.364 **29.** 0.002 **30.** 20.770 **31.** 479. **33.** 18. **34.** 20. **35.** 382. **37.** 440.
38. 701. **39.** 6333. **41.** 5160. **42.** 6480. **43.** 500. **45.** 1000. **46.** 380. **47.** 5480.
49. 92,540. **50.** 7010. **51.** 7000. **53.** 48,000. **54.** 103,000. **55.** 217,000. **57.** 380,000.
58. 4,501,000. **59.** 7,305,000. **61.** 0.00058 **62.** 0.54 **63.** 470. **65.** 500. **66.** 6000.
67. 3.230 **69.** 80,000. **70.** 78,420.

Exercises 5.3, p. 232

1. 2.3 **2.** 11.5 **3.** 7.55 **5.** 72.31 **6.** 4.6926 **7.** 276.096 **9.** 44.6516 **10.** 481.25
11. 118.333 **13.** 7.148 **14.** 7.4914 **15.** 93.877 **17.** 103.429 **18.** 46.943 **19.** 137.150
21. 1.44 **22.** 8.93 **23.** 15.89 **25.** 64.947 **26.** 4.895 **27.** 4.7974 **29.** 2.9434 **30.** 34.186
31. $4.50 **33.** (a) $94.85 (b) $5.15 **34.** 12.28 in. **35.** (a) $95.50 (b) $4.50
37. $1100 estimated (In this case $1800 was not rounded off.)
38. $6200 estimated (In this case $15,000 was not rounded off.) **39.** 2687.725 million bushels
41. 0.188919 **42.** 9.73 in.; 18.4 in.

The Recycle Bin, p. 235

1. 350,000 **2.** 525 **3.** 7020 **4.** 2,000,000 (estimate); 1,714,125 (product) **5.** 13,272

Exercises 5.4, p. 242

1. (a) B. (b) C. (c) E. (d) A. (e) D. **2.** (a) B. (b) A. (c) C. (d) D. **3.** 0.42 **5.** 0.04
6. 0.09 **7.** 21.6 **9.** 0.42 **10.** 0.9 **11.** 0.004 **13.** 0.108 **14.** 0.073 **15.** 0.0276
17. 0.0486 **18.** 0.0568 **19.** 0.0006 **21.** 0.375 **22.** 1.5 **23.** 1.4 **25.** 1.725 **26.** 1.2
27. 5.063 **29.** 0.08 **30.** 0.14 **31.** 346 **33.** 782 **34.** 693 **35.** 1610 **37.** 4.35 **38.** 7.19
39. 18.6 **41.** 380 **42.** 470 **43.** 50 **45.** 74,000 **46.** 50 mm **47.** 130 mm **49.** 1500 cm
50. 3200 mm **51.** 6170 mm **53.** 16 000 m **54.** 500 m **55.** 600 m = 60 000 cm = 600 000 mm
57. 20 m = 2000 cm = 20 000 mm **58.** 10 700 m = 1 070 000 cm = 10 700 000 mm
59. 0.009 (estimate); 0.00954 (product) **61.** 2 (estimate); 2.032(product)
62. 0.0009 (estimate); 0.0009222 (product) **63.** 0.0014 (estimate); 0.0013845 (product)
65. 2.0 (estimate); 1.717202 (product) **66.** 5.4 (estimate); 4.9077 (product)
67. 1.6 (estimate); 1.4094 (product) **69.** 3.6 (estimate); 3.4314 (product)
70. 2.4 (estimate); 2.32986 (product) **71.** $240.90 **73.** (a) $636 (b) $617.98 **74.** $418.75
75. 93.33 cubic inches **77.** 12 feet perimeter; 9 square feet area **78.** $714,870 **79.** $30,855

The Recycle Bin, p. 245

1. 8 R0 **2.** 15 R15 **3.** 81 R7 **4.** 203 R100
5. about 1 square mile per person; about 2.5 square kilometers per person

Exercises 5.5, p. 254

1. (a) B. (b) E. (c) D. (d) C. (e) A. **2.** (a) A. (b) C. (c) D. (d) B. **3.** 2.34 **5.** 0.99
6. 0.18 **7.** 0.08 **9.** 2056 **10.** 534 **11.** 20 **13.** 56.9 **14.** 65.8 **15.** 0.7 **17.** 0.1

18. 12.7   **19.** 21.0   **21.** 0.01   **22.** 12.12   **23.** 5.70   **25.** 2.74   **26.** 0.22   **27.** 5.04
29. 0.784   **30.** 0.164963   **31.** 0.5036   **33.** 0.7385   **34.** 0.0186   **35.** 16.7   **37.** 0.785
38. 0.000154   **39.** 0.01699   **41.** 0.5 cm   **42.** 1.1 cm   **43.** 0.83 m   **45.** 0.344 m   **46.** 0.255 m
47. 1.5 km   **49.** 9.72 cm   **50.** 1.85 cm   **51.** 3.2 cm = 0.32 dm = 0.032 m
53. 3.5 (estimate); 3.087 (quotient   **54.** 4.3 (estimate); 4.246 (quotient)   **55.** 0.2 (estimate); 0.285 (quotient)
57. 0.1 (estimate); 0.079 (quotient   **58.** 0.4 (estimate); 0.291 (quotient)
59. 0.005 (estimate); 0.007 (quotient)   **61.** (a) 400 mi   (b) 442.8 mi   **62.** 226.8 mi   **63.** 5620
65. 285   **66.** 99.1 m   **67.** 42.2 km   **69.** (a) $5000   (b) $6500

Exercises 5.6, p. 259

1. 5.0×10^6 **2.** 4.3×10^6 **3.** 7.5×10^6 **5.** 6.7×10^7 **6.** 4.51×10^7 **7.** 1.75×10^8
9. 2.137×10^{11} **10.** 8.245×10^{11} **11.** 6.2×10^{-4} **13.** 2.5×10^{-5} **14.** 3.4×10^{-5}
15. 8.0×10^{-7} **17.** 6.71×10^{-9} **18.** 2.55×10^{-9} **19.** 3.21×10^{-12} **21.** 5700   **22.** 63,000
23. 75,400   **25.** 4,720,000   **26.** 899,000,000   **27.** 0.0057   **29.** 0.0000184   **30.** 0.000000000000317
31. 0.0000000524   **33.** no; 5.671×10^4 **34.** no; 1.9823×10^5 **35.** yes   **37.** no; 4.743×10^{-3}
38. no; 3.21×10^{-2} **39.** no; 1.786×10^{-5} **41.** (a) 4. 04   (b) 40,000   **42.** (a) 3.1 05   (b) 310,000
43. (a) 5.358 09   (b) 5,358,000,000   **45.** (a) 3.08 08   (b) 308,000,000
46. (a) 1.125 08   (b) 112,500,000   **47.** (a) 2. −07   (b) 0.0000002   **49.** (a) 3.39 08   (b) 339,000,000
50. (a) 1.32 08   (b) 132,000,000
51. (a) 3×10^{10} (b) 300,000,000 m per sec and 30,000,000,000 cm per sec
53. 0.00000000000000000000000019926   **54.** 4.0×10^{25} **55.** 250,000,000
57. 81,170,000,000,000,000; 67,470,000,000,000,000
58. 97,600,000,000,000; 169,300,000,000,000; 72,700,000,000,000

Exercises 5.7, p. 266

1. $\frac{9}{10}$ **2.** $\frac{3}{10}$ **3.** $\frac{5}{10}$ **5.** $\frac{62}{100}$ **6.** $\frac{38}{100}$ **7.** $\frac{57}{100}$ **9.** $\frac{526}{1000}$ **10.** $\frac{625}{1000}$ **11.** $\frac{16}{1000}$ **13.** $\frac{51}{10}$
14. $\frac{72}{10}$ **15.** $\frac{815}{100}$ **17.** $\frac{1}{8}$ **18.** $\frac{9}{25}$ **19.** $\frac{9}{50}$ **21.** $\frac{9}{40}$ **22.** $\frac{91}{200}$ **23.** $\frac{17}{100}$ **25.** $\frac{16}{5} = 3\frac{1}{5}$
26. $\frac{5}{4} = 1\frac{1}{4}$ **27.** $\frac{25}{4} = 6\frac{1}{4}$ **29.** $0.\overline{6}$ **30.** 0.3125   **31.** $0.\overline{63}$ **33.** 0.6875   **34.** 0.5625
35. $0.\overline{428571}$ **37.** $0.1\overline{6}$ **38.** $0.2\overline{7}$ **39.** $0.\overline{5}$ **41.** 0.292   **42.** 0.485   **43.** 0.417   **45.** 0.031
46. 0.071   **47.** 1.231   **49.** 1.429   **50.** 13.333   **51.** 0.7   **53.** 1.635   **54.** 11.59   **55.** 14.98
57. 1.125   **58.** 0.75   **59.** 13.51   **61.** 0.089425   **62.** 58.523   **63.** 11.083   **65.** 2.638
66. 180.4   **67.** 27.3   **69.** 2.3 is larger; 0.05 difference   **70.** 0.878 is larger; 0.003 difference
71. 0.28 is larger; $0.00\overline{72}$ difference   **73.** $70\frac{3}{10}$ **74.** $9\frac{1}{10}$ **75.** $26\frac{3}{10}; 24\frac{1}{10}$ **77.** $22\frac{8}{10}$
78. $\frac{38}{100}; \frac{1}{100}$ **79.** 0.06   **81.** 5.72   **82.** 0.4324

Review Questions: Chapter 5, p. 272

1. decimal   **2.** four tenths   **3.** seven and eight hundredths
4. ninety-two and one hundred thirty-seven thousandths
5. eighteen and five thousand, five hundred twenty-six ten-thousandths   **6.** $81\frac{47}{100}$ **7.** $100\frac{3}{100}$ **8.** $9\frac{592}{1000}$
9. $200\frac{5}{10}$ **10.** 2.17   **11.** 84.075   **12.** 3003.003   **13.** 5900   **14.** 7.6   **15.** 0.039   **16.** 2.06988

17. 26.82   **18.** 93.418   **19.** 9.02   **20.** 3.9623   **21.** 104.272   **22.** 22.708   **23.** 0.72   **24.** 0.0064
25. 235   **26.** 1.7632   **27.** 5964.1   **28.** 1.728   **29.** 171.55   **30.** 0.71   **31.** 880.53   **32.** 2.961
33. 0.00567   **34.** 23.5   **35.** 1820 mm   **36.** 135 cm   **37.** 35 cm   **38.** 1.8 km   **39.** 7.975×10^6
40. 4.38×10^{-5} **41.** $\frac{7}{100}$ **42.** $2\frac{1}{40}$ **43.** $\frac{3}{200}$ **44.** $0.\overline{3}$ **45.** 0.625   **46.** $2.\overline{4}$ **47.** 0.882
48. 0.980   **49.** (a) $200,000 (b) $228,650   **50.** (a) $700 (b) $660

Test: Chapter 5, p. 274

1. thirty and six hundred fifty-seven thousandths [5.1]   **2.** $\frac{3}{8}$ [5.7]   **3.** 0.3125 [5.7]   **4.** 203.02 [5.2]
5. 2000 [5.2]   **6.** 0.1 [5.2]   **7.** 103.758 [5.3]   **8.** 1.888 [5.3]   **9.** 122.11 [5.3]   **10.** 18.145 [5.3]
11. 37.313 [5.7]   **12.** 0.90395 [5.7]   **13.** 0.294 [5.4]   **14.** 12.80335 [5.4]   **15.** 1920 [5.4]
16. 0.03614 [5.5]   **17.** 17.83 [5.5]   **18.** 66.28 [5.5]   **19.** 681 m = 68 100 cm = 681 000 mm [5.4]
20. 35.5 cm = 0.355 m [5.5]   **21.** 8.65×10^7 [5.6]   **22.** 3.72×10^{-4} [5.6]   **23.** $847.60 [5.4]
24. (a) 400 mi (b) 392 mi [5.4]   **25.** 16 [5.5]

Cumulative Review, p. 276

1. (a) commutative property of addition (b) associative property of multiplication (c) additive identity
2. 346   **3.** 210,000   **4.** 13   **5.** 28   **6.** 52   **7.** 45   **8.** $2^2 \cdot 3 \cdot 5 \cdot 7$ **9.** (a) 140 (b) 64
10. $\frac{1}{30}$ **11.** $8\frac{21}{40}$ **12.** $55\frac{29}{40}$ **13.** $32\frac{7}{20}$ **14.** $\frac{14}{5} = 2\frac{4}{5}$ **15.** $\frac{11}{30}$ **16.** $26\frac{1}{4}$ **17.** $\frac{32}{9} = 3\frac{5}{9}$
18. 72   **19.** 18.9   **20.** 2.185   **21.** (a) 1.87×10^7 (b) 6.32×10^{-7} **22.** 117,800
23. (a) $100,000 (b) $107,700   **24.** $300

CHAPTER 6

Exercises 6.1, p. 286

1. $\frac{1}{2}$ **2.** $\frac{1}{3}$ **3.** $\frac{4}{1}$ (or 4)   **5.** $\frac{50 \text{ miles}}{1 \text{ hour}}$ **6.** $\frac{60 \text{ miles}}{1 \text{ hour}}$ **7.** $\frac{25 \text{ miles}}{1 \text{ gallon}}$ **9.** $\frac{6 \text{ chairs}}{5 \text{ people}}$ **10.** $\frac{5 \text{ people}}{6 \text{ chairs}}$
11. $\frac{1}{2}$ **13.** $\frac{8}{7}$ **14.** $\frac{3}{4}$ **15.** $\frac{\$2 \text{ profit}}{\$5 \text{ invested}}$ **17.** $\frac{1}{1}$ (or 1)   **18.** $\frac{100}{1}$ (or 100)   **19.** $\frac{1 \text{ hit}}{4 \text{ times at bat}}$
21. $\frac{7}{25}$ **22.** $\frac{10}{3}$ **23.** $\frac{9}{41}$ **25.** 5.6¢/oz; 5.3¢/oz; 32 oz box   **26.** 7.2¢/oz; 10.0¢/oz; 42 oz box
27. 8.7¢/oz; 11.8¢/oz; 8 oz container   **29.** 43.7¢/oz; 64.5¢/oz; 8 oz jar   **30.** 37.5¢/oz; 46.2¢/oz; 12 oz jar
31. 10.6¢/fl oz; 8.3¢/fl oz; 11.5¢/fl oz; 12 fl oz can   **33.** 15.6¢/oz; 14.1¢/oz; 25.6 oz box
34. 12.5¢/oz; 13.0¢/oz; 16 oz pkg.   **35.** 12.4¢/oz; 7.5¢/oz; 16 oz box
37. 2.1¢/sq ft; 2.2¢/sq ft; 2.8¢/sq ft; 200 sq ft box   **38.** 7.2¢/oz; 7.7¢/oz; 9.1¢/oz; 32 oz bottle
39. 13.7¢/oz; 13.3¢/oz; 12 oz pkg.   **41.** 4.1¢/oz; 5.6¢/oz; 5.7¢/oz; 7.3¢/oz; 9 lb 3 oz box
42. 10.8¢/oz; 10.5¢/oz; 10.0¢/oz; 9.8¢/oz; 40 oz jar   **43.** 0.9¢/fl oz; 1.4¢/fl oz; 1.9¢/fl oz; 1 gal bottle
45. 10.7¢/oz; 12.2¢/oz; 13.8¢/oz; 14 oz box   **46.** 8.2¢/oz; 9.1¢/oz; 5.0¢/oz; 24 oz jar
47. 7.7¢/oz; 5.3¢/oz; 32 oz jar   **49.** 8.7¢/oz; 6.9¢/oz; 16 oz jar   **50.** 6.0¢/oz; 3.1¢/oz; 32 oz bottle
51. 5.8¢/oz; 6.0¢/oz; 40 oz can   **53.** 3.0¢/oz; 3.9¢/oz; 30 oz container   **54.** 2.6¢/oz; 3.4¢/oz; 31 oz can

The Recycle Bin, p. 289

1. 11 **2.** $\dfrac{185}{304}$ **3.** $\dfrac{1}{20}$ **4.** 80 **5.** 149.76 **6.** 0.02118 **7.** 5.5 **8.** 400

Exercises 6.2, p. 292

1. (a) extremes: 7 and 544 (b) means: 8 and 476
2. (a) extremes: x and z (b) means: y and w (c) y and z cannot be 0. **3.** true **5.** true **6.** true
7. false **9.** true **10.** true **11.** true **13.** true **14.** true **15.** true **17.** true **18.** false
19. false **21.** false **22.** false **23.** true **25.** false **26.** true **27.** true **29.** true **30.** false
31. true **33.** true **34.** true **35.** true **37.** true **38.** true **39.** false **41.** true **42.** false
43. false

Exercises 6.3, p. 299

1. $x = 12$ **2.** $y = 2$ **3.** $z = 20$ **5.** $B = 40$ **6.** $B = 21$ **7.** $x = 50$ **9.** $A = \dfrac{21}{2}$ **10.** $x = 5$

11. $D = 100$ **13.** $x = 1$ **14.** $y = 28\dfrac{2}{9}$ **15.** $x = \dfrac{3}{5}$ **17.** $w = 24$ **18.** $w = 150$ **19.** $y = 6$

21. $x = \dfrac{3}{2}$ **22.** $R = 40$ **23.** $R = 60$ **25.** $A = 2$ **26.** $B = 80$ **27.** $B = 120$ **29.** $R = \dfrac{100}{3}$

30. $R = \dfrac{200}{3}$ **31.** $x = 22$ **33.** $x = 1$ **34.** $x = 60$ **35.** $x = 15$ **37.** $B = 7.8$ **38.** $x = 8.2$

39. $x = 1.56$ **41.** $R = 50$ **42.** $R = 20$ **43.** $B = 48$ **45.** $A = 27.3$ or $27\dfrac{3}{10}$ **46.** $A = 180$

47. $A = 1.8$ **49.** $x = 3.3$ **50.** $x = 6.6$
Check Your Number Sense **51.** (b) **53.** (c) **54.** (d) **55.** (c)

The Recycle Bin, p. 301

1. 8.51 **2.** 42.6 **3.** 13.7 **4.** 18.2925 **5.** 21.95 **6.** 2.98

Exercises 6.4, p. 305

1. 45 mi **2.** $11.80 **3.** 7 gal **5.** $180 **6.** $3.27 **7.** 20 yd **9.** 437.5 g **10.** $120

11. $12 **13.** $1275 **14.** $160,000 **15.** $34 **17.** 42.86 ft or $42\dfrac{6}{7}$ ft **18.** $7\dfrac{1}{2}$

19. 7.5 mph; 52.5 mph **21.** 3360 mi **22.** $11\dfrac{1}{4}$ hr **23.** 9 hr **25.** 22.3 gal **26.** 259,200 revolutions

27. $648 **29.** $1\dfrac{7}{8}$ in. **30.** 30.48 cm **31.** 90 lb **33.** $1\dfrac{1}{2}$ cups **34.** 18 biscuits **35.** 4700 g

37. 2.7 kg **38.** 0.004 mg **39.** 500 mg **41.** 4 drams **42.** 3 drams **43.** 1920 minims

45. $8\dfrac{1}{3}$ grains **46.** 60 g **47.** 75 minims **49.** 0.5 ounce **50.** 750 mL

51. (a) 56 lb (b) 4 bags (c) $56 **53.** (a) 4 gal (b) yes (c) 0.88 gal **54.** (a) 96 lb (b) 5 bags
Check Your Number Sense **55.** (c) **57.** (c) **58.** (a)

The Recycle Bin, p. 310

1. (a) close to 6; $6\frac{3}{10}$ **2.** (b) close to 20; $19\frac{21}{30}$ **3.** (b) less than 6; 5   **4.** $1\frac{1}{3}$

Review Questions: Chapter 6, p. 312

1. $\frac{4 \text{ nickels}}{5 \text{ nickels}} = \frac{4}{5}$ **2.** $\frac{5 \text{ in.}}{9 \text{ in.}} = \frac{5}{9}$ **3.** $\frac{2.54 \text{ cm}}{1 \text{ in.}}$ **4.** $\frac{17 \text{ mi}}{1 \text{ gal}}$ **5.** $\frac{3 \text{ hr}}{8 \text{ hr}} = \frac{3}{8}$ **6.** $\frac{2 \text{ girls}}{3 \text{ boys}}$

7. 13.2¢/oz; 12.5¢/oz; "deli" cheese   **8.** 1.7¢/fl oz; 1.6¢/fl oz; 1 gal of milk

9. 4.7¢/oz; 4.8¢/oz; 1 lb 2 oz loaf   **10.** 21.5¢/oz; 18.7¢/oz; 8 oz bologna

11. means: 5, 16; extremes: 4, 20   **12.** means: $3, \frac{1}{9}$; extremes: $\frac{1}{6}, 2$ **13.** means: 7, 0.75; extremes: 3, 1.75

14. true   **15.** true   **16.** false   **17.** $x = 5$ **18.** $y = 300$ **19.** $w = 5\frac{1}{16}$ **20.** $a = \frac{7}{15}$

21. 149.8 mi   **22.** 4.6855   **23.** 166.85 mi   **24.** \$700   **25.** 40,000 safety pins   **26.** 45 mph

27. $26\frac{2}{3}$ ft   **28.** $266\frac{2}{3}$ mi   **29.** 64 km per hr   **30.** 468 girls

Test: Chapter 6, p. 313

1. $\frac{7}{6}$ [6.1]   **2.** extremes [6.2]   **3.** false; $4 \cdot 14 = 56$, $6 \cdot 9 = 54$, and $56 \neq 54$ **4.** $\frac{3}{5}$ [6.1]   **5.** $\frac{2}{5}$ [6.1]

6. $\frac{55 \text{ miles}}{1 \text{ hour}}$ [6.1]   **7.** 6.8¢/oz; 6.5¢/oz; 32 oz carton [6.1]   **8.** $x = 27$ [6.3]   **9.** $y = 16\frac{2}{3}$ [6.3]

10. $x = 75$ [6.3]   **11.** $y = 19.5$ [6.3]   **12.** $x = \frac{5}{6}$ [6.3]   **13.** $x = \frac{1}{90}$ [6.3]   **14.** $y = 4.5$ [6.3]

15. $x = 36$ [6.3]   **16.** 171.4 miles [6.4]   **17.** 24 miles [6.4]   **18.** \$295 [6.4]   **19.** 45 mph [6.4]

20. 27 boys [6.4]   **21.** 4700 g [6.4]   **22.** 25 weeks [6.4]   **23.** 312 pounds [6.4]   **24.** \$7.50 [6.4]

Cumulative Review, p. 315

1. 5740   **2.** three and seventy-five thousandths   **3.** $2 \cdot 5 \cdot 11 \cdot 41$ **4.** 79   **5.** 26

6. $2 \cdot 3 \cdot 5 \cdot 7 = 210$ **7.** 60   **8.** $\frac{1}{16}$ **9.** $2\frac{5}{16}$ **10.** $14\frac{2}{3}$ **11.** $3\frac{3}{5}$ **12.** $3\frac{1}{4}$ **13.** $38\frac{1}{4}$

14. 57   **15.** 0.036   **16.** (a) six hundred and six thousandths   (b) six hundred six thousandths

17. 1639.438   **18.** 16.172   **19.** 4077.36   **20.** 0.79   **21.** (a) 9.35×10^7 (b) 4.82×10^{-7}

22. $x = 2\frac{1}{2}$ **23.** $A = 0.32$ **24.** (a) 45.5 mph   (b) 318.5 miles   **25.** \$2600

CHAPTER 7

Exercises 7.1, p. 323

1. 60%   **2.** 20%   **3.** 65%   **5.** 100%   **6.** 0.5%   **7.** 20%   **9.** 15%   **10.** 62%   **11.** 53%

13. 125%   **14.** 200%   **15.** 336%   **17.** 0.48%   **18.** 0.5%   **19.** 2.14%

21. (a) $\frac{18}{100} = 18\%$ (b) $\frac{17}{100} = 17\%$; investment (a) is better.

22. (a) $\frac{10}{100} = 10\%$ (b) $\frac{10}{100} = 10\%$; both investments yield the same percent of profit.

23. (a) $\dfrac{10}{100} = 10\%$ (b) $\dfrac{10}{100} = 10\%$; both investments yield the same percent of profit.   **25.** 2%   **26.** 9%

27. 10%   **29.** 36%   **30.** 52%   **31.** 40%   **33.** 2.5%   **34.** 3.5%   **35.** 5.5%   **37.** 110%

38. 175%   **39.** 200%   **41.** 0.02   **42.** 0.07   **43.** 0.18   **45.** 0.3   **46.** 0.8   **47.** 0.0026

49. 1.25   **50.** 1.2   **51.** 2.32   **53.** 0.173   **54.** 0.101   **55.** 0.132   **57.** 0.0725   **58.** 0.064

59. 0.085   **61.** 28%   **62.** 2%   **63.** 0.45   **65.** 6.5%   **66.** 0.222   **67.** 1.33

The Recycle Bin, p. 326

1. $R = 80$ **2.** $R = 66\dfrac{2}{3}$ **3.** $A = 115$ **4.** $A = 45$ **5.** $B = 40$ **6.** $B = 74$

Exercises 7.2, p. 332

1. 3%   **2.** 16%   **3.** 7%   **5.** 50%   **6.** 75%   **7.** 25%   **9.** 55%   **10.** 70%   **11.** 30%

13. 20%   **14.** 40%   **15.** 80%   **17.** 26%   **18.** 4%   **19.** 48%   **21.** 12.5%   **22.** 62.5%

23. 87.5%   **25.** $55\dfrac{5}{9}\%$ **26.** $28\dfrac{4}{7}\%$ **27.** $42\dfrac{6}{7}\%$ **29.** $63\dfrac{7}{11}\%$ **30.** $45\dfrac{5}{11}\%$ **31.** $107\dfrac{1}{7}\%$

33. 105%   **34.** 125%   **35.** 175%   **37.** 137.5%   **38.** 250%   **39.** 210%   **41.** $\dfrac{1}{10}$ **42.** $\dfrac{1}{20}$

43. $\dfrac{3}{20}$ **45.** $\dfrac{1}{4}$ **46.** $\dfrac{3}{10}$ **47.** $\dfrac{1}{2}$ **49.** $\dfrac{3}{8}$ **50.** $\dfrac{1}{6}$ **51.** $\dfrac{1}{3}$ **53.** $\dfrac{33}{100}$ **54.** $\dfrac{1}{200}$ **55.** $\dfrac{1}{400}$

57. 1   **58.** $1\dfrac{1}{4}$ **59.** $1\dfrac{1}{5}$ **61.** $\dfrac{3}{1000}$ **62.** $\dfrac{1}{40}$ **63.** $\dfrac{5}{8}$ **65.** $\dfrac{3}{400}$ **66.** (a) 0.625 (b) 62.5%

67. (a) 0.55 (b) 55%   **69.** (a) $\dfrac{7}{4}$ or $1\dfrac{3}{4}$ (b) 175%   **70.** (a) $\dfrac{9}{25}$ (b) 0.36   **71.** (a) $\dfrac{21}{200}$ (b) 0.105

73. 4%   **74.** 1%   **75.** (a) $8\dfrac{1}{3}\%$ (or 8.3%) (b) 25% (c) 6.25%   **77.** $\dfrac{1}{7}\%$ (or 0.14%)

Check Your Number Sense   **78.** (a) $\dfrac{7}{20}$ (b) $\dfrac{3}{10}$ (c) $\dfrac{1}{25}$ (Each denominator is a factor of 100.)

The Recycle Bin, p. 334

1. $7200   **2.** 216 feet   **3.** 15 gal   **4.** (a) 24 lb (b) 3 bags

Exercises 7.3, p. 338

1. 9   **2.** 3.6   **3.** 7.5   **5.** 24   **6.** 24   **7.** 47   **9.** 250   **10.** 70   **11.** 900   **13.** 62

14. 70   **15.** 46   **17.** 150%   **18.** 40%   **19.** 30%   **21.** 150%   **22.** 20%   **23.** $33\dfrac{1}{3}\%$

25. 17.5   **26.** 26.35   **27.** 160   **29.** 230%   **30.** 180%   **31.** 120   **33.** 620   **34.** 81

35. 200%   **37.** 33.6   **38.** 43.68   **39.** 76.5   **41.** 62.1   **42.** 12.825   **43.** 105   **45.** 160

46. 300   **47.** 66.5%   **49.** $16\dfrac{2}{3}\%$ **50.** 200   **51.** 152,429   **53.** 23,220   **54.** 34,530

Exercises 7.4, p. 346

1. 7   **2.** 3.1   **3.** 9   **5.** 9   **6.** 18   **7.** 36   **9.** 150   **10.** 85   **11.** 700   **13.** 75   **14.** 84

15. 42   **17.** 150%   **18.** 40%   **19.** 20%   **21.** 50%   **22.** 20%   **23.** $33\dfrac{1}{3}\%$ **25.** 12.5

26. 23.56   **27.** 110   **29.** 130%   **30.** 150%   **31.** 72   **33.** 520   **34.** 38   **35.** 200%
37. 16.32   **38.** 26.88   **39.** 58.5   **41.** 43.92   **42.** 11.4   **43.** 72   **45.** 80   **46.** 16   **47.** 40
49. 10   **50.** 25   **51.** 37.5   **53.** 70   **54.** 163.2   **55.** 76.3   **57.** 6800   **58.** 4.8%   **59.** 44%

Exercises 7.5, p. 348

1. (c)   **2.** (c)   **3.** (a)   **5.** (a)   **6.** (b)   **7.** (b)   **9.** (b)   **10.** (b)   **11.** (b)   **13.** (d)
14. (c)   **15.** (a)   **17.** (b)   **18.** (d)   **19.** (d)   **21.** (c)   **22.** (c)   **23.** (b)   **25.** (b)   **26.** (a)
27. (d)   **29.** (c)   **30.** (c)   **31.** (a)   **33.** (b)   **34.** (b)   **35.** (b)   **37.** (c)   **38.** (c)   **39.** (c)
41. (c)   **42.** (d)   **43.** (c)   **45.** (d)   **46.** (c)   **47.** (c)   **49.** (d)   **50.** (b)

Exercises 7.6, p. 357

1. (a) $465   (b) $15,035   **2.** (a) $1.82   (b) $32.02   **3.** (a) $6.93   (b) $122.38
5. (a) $750   (b) $600   (c) $636   **6.** (a) $250   (b) 20%   **7.** (a) $10.53   (b) 60%   **9.** $4.30
10. $6.55   **11.** $1.95   **13.** $5.70; $44.10   **14.** $1.05; $1.35   **15.** $29.90
17. (a) the paperback   (b) $4.12   **18.** 20   **19.** $32,800   **21.** (a) $161   (b) $5474

Exercises 7.7, p. 362

1. $5700   **2.** $35,000   **3.** $1310   **5.** $28,000   **6.** (a) 40%   (b) 60%   **7.** 153
9. (a) $50   (b) 20%   (c) $16\frac{2}{3}$% (or 16.7%)   **10.** (a) $140   (b) 40%   (c) $28\frac{4}{7}$% (or 28.6%)
11. (a) $1875   (b) $2025   **13.** $120,000   **14.** $22,018   **15.** (a) 88%   (b) 250   (c) 30   **17.** 770,000
18. 52,000   **19.** 423%

Review Questions: Chapter 7, p. 367

1. hundredths   **2.** 85%   **3.** 18%   **4.** 37%   **5.** $16\frac{1}{2}$%   **6.** 15.2%   **7.** 115%   **8.** 6%
9. 30%   **10.** 67%   **11.** 2.7%   **12.** 300%   **13.** 120%   **14.** 0.35   **15.** 0.04   **16.** 0.0025
17. 0.0025   **18.** 0.071   **19.** 1.32   **20.** 60%   **21.** 15%   **22.** 16%   **23.** 37.5%   **24.** $41\frac{2}{3}$%
25. $126\frac{2}{3}$%   **26.** $\frac{7}{50}$ **27.** $\frac{2}{5}$ **28.** $\frac{33}{50}$ **29.** $\frac{1}{8}$ **30.** 4   **31.** $\frac{67}{200}$ **32.** 15.6   **33.** 2.55
34. $233\frac{1}{3}$ **35.** $42\frac{6}{7}$ **36.** 20%   **37.** $33\frac{1}{3}$%   **38.** 25%   **39.** 1.095   **40.** 50   **41.** $254\frac{6}{11}$
42. 0.975   **43.** 200%   **44.** (c)   **45.** (b)   **46.** (a)   **47.** (b)   **48.** (c)   **49.** $11.93
50. (a) 50% on cost   (b) $33\frac{1}{3}$% on selling price   **51.** 8 problems   **52.** $1950   **53.** 1.2%
54. $5\frac{5}{9}$%—movie; $38\frac{8}{9}$%—clothes; $22\frac{2}{9}$%—anniversary   **55.** (a) $10,000   (b) $9150

Test: Chapter 7, p. 369

1. 101% [7.1]   **2.** 0.3% [7.1]   **3.** 17.3% [7.1]   **4.** 32% [7.2]   **5.** 237.5% [7.2]   **6.** 0.7 [7.1]

7. 1.8 [7.1]   **8.** 0.093 [7.1]   **9.** $1\frac{3}{10}$ [7.2]   **10.** $\frac{71}{200}$ [7.2]   **11.** $\frac{43}{500}$ [7.2]   **12.** 5.6 [7.3, 7.4]

13. $33\frac{1}{3}$% [7.3, 7.4]   **14.** 30 [7.3, 7.4]   **15.** 294.5 [7.3, 7.4]   **16.** 20.5 [7.3, 7.4]   **17.** 33.1% [7.3, 7.4]

18. (c) [7.2]   **19.** (d) [7.2]   **20.** (b) [7.3, 7.4]   **21.** (c) [7.3, 7.4]

22. (a) $3.75   (b) $3.90   (c) $2.70 [7.6]   **23.** (a) $62.50   (b) 25%   (c) 20% [7.7]

24. (a) $1200   (b) $24 [7.6]   **25.** $3300 [7.7]

Cumulative Review, p. 371

1. identity   **2.** commutative property of addition   **3.** (a) 13,000   (b) 3.097

4. 10,500 (estimate); 10,358 (sum)   **5.** 4000 (estimate); 3825 (difference)   **6.** 100 (estimate); 93 (quotient)

7. 8000 (estimate); 6552 (product)   **8.** 3   **9.** 4, 9, 25, 49, 121, 169, 289, 361, 529, 841

10. (a) 4200   (b) $4200 = 56 \cdot 75 = 60 \cdot 70$ **11.** $\frac{3}{20}$ **12.** $\frac{5}{2} = 2\frac{1}{2}$ **13.** 9.475   **14.** $\frac{11}{18}$

15. $x = 22$ **16.** 78   **17.** 400   **18.** (a) 5.63×10^8 (b) 9.42×10^{-5}

19. (a) 3450 mm   (b) 1600 cm   (c) 83 cm   (d) 5.2 km   **20.** (a) $100   (b) $33\frac{1}{3}$%   (c) 25%

21. (a) $4.20   (b) $31.75   **22.** (a) $25,000   (b) $21,300   (c) 25%   (d) 20%

CHAPTER 8

Exercises 8.1, p. 378

1. $30   **2.** $160   **3.** $90   **5.** $26.67   **6.** $100   **7.** $2500   **9.** 1 year

10. 0.5 year (or 180 days or 6 months)   **11.** $32   **13.** $20.25   **14.** $200   **15.** $11.25   **17.** $32.67

18. (a) $66.25   (b) $5366.25   **19.** $730   **21.** (a) $463.50   (b) $36.50   **22.** (a) $20   (b) $6.67

23. $37,500   **25.** 72 days (0.2 year)   **26.** (a) $16   (b) $100   (c) 30 days (1 month)   (d) 8.5%

27. (a) $7.50   (b) 60 days (2 months)   (c) 18%   (d) $100   **29.** 10%   **30.** 180 days   **31.** $460,000

33. 1 year   **34.** $337,500   **35.** (a) $17.50   (b) 60 days (2 months)   (c) 11.5%   (d) $1133.34

Exercises 8.2, p. 387

1. (a) $13,261.30   (b) $13,527.86   **2.** $9277.38   **3.** (a) $306.83   (b) $5306.83   (c) $6.83

5. (a) $339.08   (b) $343.87   **6.** $16,153.37   **7.** $406.75   **9.** $600   **10.** (a) 240 days   (b) $11.00

11. (a) $634.13   (b) no, only $618.00   (c) There are fewer periods and the principal is different for each period.

13. (a) $6863.93   (b) about $20   (c) $21.47   **14.** (a) $28,051.04   (b) $329.35

15. (a) $3153.49   (b) no, $2234.08   **17.** (a) $1638.62   (b) $638.62   **18.** (a) $1645.31   (b) $645.31

19. (a) $3297.33   (b) $1297.33

Exercises 8.3, p. 393

1.

YOUR CHECKBOOK REGISTER

Check No.	Date	Transaction Description	Payment (−)	(√)	Deposit (+)	Balance
					Balance brought forward	⊖
	7–15	Deposit ( )		√	700.00	+700.00 / 700.00
1	7–15	Quiet Town Apt. (rent/deposit)	520.00	√		−520.00 / 180.00
2	7–15	Pa Bell Telephone (phone installation)	32.16	√		−32.16 / 147.84
3	2–15	XYZ Power Co. (gas/elect. hook up)	46.49	√		−46.49 / 101.35
4	7–16	Foodway Stores (groceries)	51.90	√		−51.90 / 49.45
	7–20	Deposit ( )		√	350.00	+350.00 / 399.45
5	7–23	Comfy Furniture (sofa, chair)	300.50			−300.50 / 98.95
	8–1	Deposit ( )			350.00	+350.00 / 448.95
	7–31	Interest ( )		√	2.50	+2.50 / 451.45
	7–31	Service ( )	2.00	√		−2.00 / 449.45
		True Balance				449.45

BANK STATEMENT
Checking Account Activity

Transaction Description	Amount	(√)	Running Balance	Date
Beginning Balance	0.00	√	0.00	7-1
Deposit	700.00	√	700.00	7-15
Check #1	520.00	√	180.00	7-16
Check #4	51.90	√	128.10	7-17
Check #2	32.16	√	95.94	7-18
Check #3	46.49	√	49.45	7-18
Deposit	350.00	√	399.45	7-20
Interest	2.50	√	401.95	7-31
Service Charge	2.00	√	399.95	7-31
Ending Balance			399.95	

RECONCILIATION SHEET

A. First mark √ beside each check and deposit listed in both your checkbook register and on the bank statement.
B. Second, in your checkbook register, add any interest paid and subtract any service charge listed on the bank statement.
C. Third, find the total of all outstanding checks.

Outstanding Checks		
No.	Amount	
5	300.50	
Total	300.50	

Statement Balance	399.95
Add deposits not credited	+ 350.00
Total	749.95
Subtract total amount of checks outstanding	− 300.50
True Balance	449.45

2.

YOUR CHECKBOOK REGISTER

Check No.	Date	Transaction Description	Payment (−)	(√)	Deposit (+)	Balance
					Balance brought forward	1610.39
1234	12-7	Pearl City (pearl ring)	524.00	√		−524.00 / 1086.39
1235	12-7	Comp-U-Tate (home computer)	801.60	√		−801.60 / 284.79
1236	12-8	Sportz Hutz (skis)	206.25	√		−206.25 / 78.54
1237	12-8	Guild Card Shop (Christmas cards)	25.50	√		−25.50 / 53.04
	12-10	Deposit ( )		√	1000.00	+1000 / 1053.04
1238	12-14	Toys-R-We (stuffed panda)	80.41	√		−80.41 / 972.63
1239	12-24	Meat Markette (turkey)	18.39	√		−18.39 / 954.24
1240	12-24	Poodle Shoppe (pedigreed puppy)	300.00	√		−300.00 / 654.24
1241	12-31	Homey Sav.&Loan (mortgage payment)	600.00			−600.00 / 54.24
	12-31	Service Charge ( )	⊖			——— / 54.24
		True Balance				54.24

BANK STATEMENT
Checking Account Activity

Transaction Description	Amount	(√)	Running Balance	Date
Beginning Balance		√	1610.39	12-01
Check #1234	524.00	√	1086.39	12-08
Check #1236	206.25	√	880.14	12-09
Check #1237	25.50	√	854.64	12-09
Deposit	1000.00	√	1854.64	12-10
Check #1235	801.60	√	1053.04	12-11
Check #1238	80.41	√	972.63	12-15
Check #1239	18.39	√	954.24	12-27
Check #1240	300.00	√	654.24	12-28
Ending Balance			654.24	

RECONCILIATION SHEET

A. First mark √ beside each check and deposit listed in both your checkbook register and on the bank statement.
B. Second, in your checkbook register, add any interest paid and subtract any service charge listed on the bank statement.
C. Third, find the total of all outstanding checks.

Outstanding Checks		
No.	Amount	
1241	600.00	
Total		

Statement Balance	654.24
Add deposits not credited	+ ———
Total	654.24
Subtract total amount of checks outstanding	− 600.00
True Balance	54.24

3.

YOUR CHECKBOOK REGISTER

Check No.	Date	Transaction Description	Payment (−)	(√)	Deposit (+)	Balance
		Balance brought forward				756.14
271	6-15	*Parts, Parts, Parts* (spark plugs)	12.72	√		− 12.72 / 743.42
272	6-24	*Firerock Tire Co.* (2 tires)	121.40	√		− 121.40 / 622.02
273	6-30	*Gus' Gas Station* (tune-up)	75.68			− 75.68 / 546.34
	7-1	*Deposit* ( )			250.00	+ 250.00 / 796.34
274	7-1	*Prudent Ins Co.* (car insurance)	300.00			− 300.00 / 496.34
	6-30	*Service Charge* ( )	1.00	√		− 1.00 / 495.34
						—
						—
						—
						—
		True Balance				495.34

BANK STATEMENT
Checking Account Activity

Transaction Description	Amount	(√)	Running Balance	Date
Beginning Balance		√	756.14	6-01
Check #271	12.72	√	743.42	6-16
Check #272	121.40	√	622.02	6-26
Service Charge	1.00	√	621.02	6-30
Ending Balance			621.02	

RECONCILIATION SHEET

A. First mark √ beside each check and deposit listed in both your checkbook register and on the bank statement.
B. Second, in your checkbook register, add any interest paid and subtract any service charge listed on the bank statement.
C. Third, find the total of all outstanding checks.

Outstanding Checks

No.	Amount
273	75.68
274	300.00
	375.68
Total _____	

Statement Balance	621.02
Add deposits not credited	+ 250.00
Total	871.02
Subtract total amount of checks outstanding	− 375.68
True Balance	495.34

5.

YOUR CHECKBOOK REGISTER

Check No.	Date	Transaction Description	Payment (−)	(√)	Deposit (+)	Balance
		Balance brought forward				967.22
772	4-13	*C.P. Hay* (accountant)	85.00	√		− 85.00 / 882.22
	4-14	*Deposit* ( )		√	1200.00	+1200.00 / 2082.22
773	4-14	*E.Z. Pharmacy* (aspirin)	4.71	√		− 4.71 / 2077.51
774	4-15	*I.R.S* (income tax)	2000.00			−2000.00 / 77.51
775	4-30	*Heavy Finance Co.* (loan payment)	52.50			− 52.50 / 25.01
	5-1	*Deposit* ( )			600.00	+600.00 / 625.01
	4-30	*Interest* ( )		√	2.82	+ 2.82 / 627.83
	4-30	*Service Charge* ( )	4.00	√		4.00 / 623.83
		( )				—
		( )				—
		True Balance				623.83

BANK STATEMENT
Checking Account Activity

Transaction Description	Amount	(√)	Running Balance	Date
Beginning Balance		√	967.22	4–01
Deposit	1200.00	√	2167.22	4–14
Check #772	85.00	√	2082.22	4–15
Check #773	4.71	√	2077.51	4–15
Interest	2.82	√	2080.33	4–30
Service Charge	4.00	√	2076.33	4–30
Ending Balance			2076.33	

RECONCILIATION SHEET

A. First mark √ beside each check and deposit listed in both your checkbook register and on the bank statement.
B. Second, in your checkbook register, add any interest paid and subtract any service charge listed on the bank statement.
C. Third, find the total of all outstanding checks.

Outstanding Checks

No.	Amount
774	2000.00
775	52.50
Total	2052.50

Statement Balance	2076.33
Add deposits not credited	+ 600.00
Total	2676.33
Subtract total amount of checks outstanding	− 2052.50
True Balance	623.83

6.

YOUR CHECKBOOK REGISTER

Check No.	Date	Transaction Description	Payment (−)	(√)	Deposit (+)	Balance
			Balance brought forward			1403.49
86	9-1	Now Stationers (school supplies)	17.12	√		−17.12 / 1386.37
87	9-2	Young-At-Heart (clothes)	192.50	√		−192.50 / 1193.87
88	9-4	H.S.U. Bookstore (books)	56.28	√		−56.28 / 1137.59
89	9-7	Regent's Office (tuition)	380.00			−380.00 / 757.59
90	9-7	Off-Campus Apts. (rent)	240.00	√		−240.00 / 517.59
91	9-27	State Telephone Co. (phone bill)	24.62			−24.62 / 492.97
92	9-30	Up-N-Up Foods (groceries)	47.80			−47.80 / 445.17
	9-30	Service Charge ( )	4.00	√		−4.00 / 441.17
		( )				
		( )				
		True Balance				441.17

BANK STATEMENT
Checking Account Activity

Transaction Description	Amount	(√)	Running Balance	Date
Beginning Balance		√	1403.49	9-01
Check #86	17.12	√	1386.37	9-02
Check #87	192.50	√	1193.87	9-05
Check #88	56.28	√	1137.59	9-05
Check #90	240.00	√	897.59	9-10
Service Charge	4.00	√	893.59	9-30
Ending Balance			893.59	

RECONCILIATION SHEET

A. First mark √ beside each check and deposit listed in both your checkbook register and on the bank statement.
B. Second, in your checkbook register, add any interest paid and subtract any service charge listed on the bank statement.
C. Third, find the total of all outstanding checks.

Outstanding Checks

No.	Amount
89	380.00
91	24.62
92	47.80
Total	452.42

Statement Balance	893.59
Add deposits not credited	+
Total	893.59
Subtract total amount of checks outstanding	−452.42
True Balance	441.17

7.

YOUR CHECKBOOK REGISTER

Check No.	Date	Transaction Description	Payment (−)	(√)	Deposit (+)	Balance
			Balance brought forward			602.82
14	6-20	Aisle Bridal (flowers)	402.40	√		−402.40 / 200.42
	6-22	Deposit ( )		√	1000.00	+1000.00 / 1200.42
15	6-24	Tuxedo Junction (tux)	155.65	√		−155.65 / 1044.77
16	6-28	D. Lohengrin (organist)	55.00			−55.00 / 989.77
17	6-28	D-Lux Limo (limo rental)	125.00			−125.00 / 864.77
18	6-30	C.C. Catering (food caterer)	700.00			−700.00 / 164.77
19	7-1	Luv-Lee Stationers (thank-you cards)	35.20			−35.20 / 129.57
	6-30	Service Charge ( )	1.00	√		−1.00 / 128.57
		( )				
		( )				
		True Balance				128.57

BANK STATEMENT
Checking Account Activity

Transaction Description	Amount	(√)	Running Balance	Date
Beginning Balance		√	602.82	6-01
Deposit	1000.00	√	1602.82	6-22
Check #14	402.40	√	1200.42	6-22
Check #15	155.65	√	1014.77	6-26
Service Charge	1.00	√	1013.77	6-30
Ending Balance			1013.77	

RECONCILIATION SHEET

A. First mark √ beside each check and deposit listed in both your checkbook register and on the bank statement.
B. Second, in your checkbook register, add any interest paid and subtract any service charge listed on the bank statement.
C. Third, find the total of all outstanding checks.

Outstanding Checks

No.	Amount
16	55.00
17	125.00
18	700.00
19	35.20
Total	915.20

Statement Balance	1043.77
Add deposits not credited	+
Total	1043.77
Subtract total amount of checks outstanding	−915.20
True Balance	128.57

9.

YOUR CHECKBOOK REGISTER

Check No.	Date	Transaction Description	Payment (−)	(√)	Deposit (+)	Balance
			Balance brought forward			147.02
203	2-3	Food Stoppe (groceries)	26.90	√		−26.90 / 120.12
204	2-8	Ekkon Oil (gasoline bill)	71.45	√		−71.45 / 48.67
205	2-14	Rose's Roses (flowers)	25.00	√		−25.00 / 23.67
206	2-14	I.M.R.U. (alumni dues)	20.00			−20.00 / 3.67
	2-15	Deposit ( )		√	600.00	+ 600.00 / 603.67
207	2-26	SRO (theater tickets)	52.50			− 52.50 / 551.17
208	2-28	MPG Mtg. (house payment)	500.00			− 500.00 / 51.17
	2-28	Service Charge ( )	3.00	√		− 3.00 / 48.17
		( )				
		( )				
		True Balance				48.17

BANK STATEMENT
Checking Account Activity

Transaction Description	Amount	(√)	Running Balance	Date
Beginning Balance		√	147.02	2–01
Check #203	26.90	√	120.12	2–04
Check #204	71.45	√	48.67	2–14
Deposit	600.00	√	648.67	2–15
Check #205	25.00	√	623.67	2–15
Service Charge	3.00	√	620.67	2–28
Ending Balance			620.67	

RECONCILIATION SHEET

A. First mark √ beside each check and deposit listed in both your checkbook register and on the bank statement.

B. Second, in your checkbook register, add any interest paid and subtract any service charge listed on the bank statement.

C. Third, find the total of all outstanding checks.

Outstanding Checks	
No.	Amount
206	20.00
207	52.50
208	500.00
Total	572.50

Statement Balance	620.67
Add deposits not credited	+ _____
Total	620.67
Subtract total amount of checks outstanding	−572.50
True Balance	48.17

10.

YOUR CHECKBOOK REGISTER

Check No.	Date	Transaction Description	Payment (−)	(√)	Deposit (+)	Balance
			Balance brought forward			4071.82
996	10-1	Red-E Credit (loan payment)	200.75	√		−200.75 / 3871.07
997	10-10	United Ways (charity donation)	25.00			−25.00 / 3846.07
998	10-21	MacIntosh Farms (barrel of apples)	42.20	√		−42.20 / 3803.87
999	10-26	Fun Haus (costume rental)	35.00	√		−35.00 / 3768.87
1000	10-28	Yum Yum Shoppe (Halloween candy)	12.14			−12.14 / 3756.73
1001	10-29	B-Sharp, Inc. (piano tuners)	20.00			−20.00 / 3736.73
1002	10-30	Food-2-Go! (party platter)	78.50			−78.50 / 3658.23
1003	10-31	Cash ( )	300.00	√		−300.00 / 3358.23
1004	10-31	Principal S&L (house payment)	1250.60			−1250.60 / 2107.63
	10-31	Interest ( )		√	16.29	16.29 / 2123.92
		True Balance				2123.92

BANK STATEMENT
Checking Account Activity

Transaction Description	Amount	(√)	Running Balance	Date
Beginning Balance		√	4071.82	10-01
Check #996	200.75	√	3871.07	10-05
Check #998	42.40	√	3828.87	10-23
Check #999	35.00	√	3793.87	10-30
Check #1003	300.00	√	3493.87	10-31
Interest	16.29	√	3510.16	10-31
Ending Balance			3510.16	

RECONCILIATION SHEET

A. First mark √ beside each check and deposit listed in both your checkbook register and on the bank statement.

B. Second, in your checkbook register, add any interest paid and subtract any service charge listed on the bank statement.

C. Third, find the total of all outstanding checks.

Outstanding Checks	
No.	Amount
997	25.00
1000	12.14
1001	20.00
1002	78.50
1004	1250.60
Total	1386.24

Statement Balance	3510.16
Add deposits not credited	+ _____
Total	3510.16
Subtract total amount of checks outstanding	− 1386.24
True Balance	2123.92

Exercises 8.4, p. 400

1. $1099.20 **2.** $5475.00 **3.** $4820.00 **5.** $919.40 **6.** (a) $625 (b) 31.25% **7.** $1938
9. No. Her projected expenses are 23.5% of her income. **10.** (a) $105 (b) 84 gal (c) 1596 miles

Exercises 8.5, p. 407

1. (a) social science (b) chemistry & phys. and humanities (c) about 3280 (d) about 21%
2. (a) second Monday; 800 (b) 150 (c) 6400 (d) about 12.5%
3. (a) Sue (b) Bob and Sue (c) about 86% (d) Bob and Sue; Bob and Sue; Yes, in most cases.
(e) No. The vertical scales represent two different types of quantities.
5. news—300 min, commercials—180 min, soaps—180 min, sitcoms—156 min, drama—144 min,
movies—120 min, children's shows—120 min
6. (a) rent—$7200, food—$4800, ins. & health—$3600, clothing—$2400, utilities—$1200, fun—$2400,
savings—$2400 (b) utilities (c) $14,400
7. (a) Bank A, $5 million (b) Bank B, $2 million (c) about 15.7%
9. (a) 1989 (b) 18 inches (c) 1991 and 1992 (d) 16 in. average rainfall
10. (a) July (b) March and December (c) about 16.6%
11. (a) August (b) 3.5 million (c) 0.5 million (d) 5.5 million (e) $33\frac{1}{3}$% (f) $33\frac{1}{3}$%
13. (a) 8% Information; 18% Service; 40% Industry; 42% Agriculture
(b) 45% Information; 30% Service; 20% Industry; 5% Agriculture (c) Service (d) 12%
(e) Information (f) 8%, 1860 (g) 45%, 1980 (h) Agriculture
14. (a) 5 classes (b) 5000 (c) 3rd class (d) 40 (e) 15,000.5 and 20,000.5 (f) 100 (g) 5% (h) 35%
15. (a) 8 classes (b) 3 (c) 8th class (d) 2 (e) 27 and 29 (f) 50 (g) 10 (h) 16%

Exercises 8.6, p. 419

1. (a) 79 (b) 84 (c) 85 (d) 38 **2.** (a) 45 (b) 48 (c) 52 (d) 33 **3.** (a) 6.2 (b) 6 (c) 6 (d) 6
5. (a) $400 (b) $375 (c) $325 (d) $225 **6.** (a) $81\frac{5}{6}$ (b) $83\frac{1}{2}$ (c) 82 (d) 12
7. (a) 81 (b) 79 (c) 88 (d) 26 **9.** (a) 22 (b) 19 (c) 18 (d) 21
10. (a) 32 (b) 30 (c) 30 (d) 40 **11.** (a) 18.5 in. (b) 14.9 in. (c) none (d) 22.0 in.

Review Questions: Chapter 8, p. 423

1. $97.50 **2.** $807 **3.** $7.50 **4.** $2000 **5.** (a) $12 (b) 1.5 years (c) 8.5% (d) $2000
6. (a) $2198.47 (b) $2254.56 (c) $2267.53 (d) $2273.87

7.

	Principal	Amount in Account	Interest Earned
1st year	$10,000	$10,800	$800
2nd year	$10,800	$11,664	$864
3rd year	$11,664	$12,597.12	$933.12

8. $3868.25
9. housing—$10,500; food—$7000; taxes—$3500; transportation—$3500; clothing—$3500;
entertainment—$2800; education—$2450; savings—$1750
10. $4000 **11.** $\frac{3}{5}$ **12.** $18,000 **13.** 87 **14.** 88 **15.** 88.5 **16.** 26

Test: Chapter 8, p. 424

1. $67.50 [8.1] **2.** $800 [8.1] **3.** $\frac{1}{2}$ yr (or 6 months) [8.1] **4.** 14% [8.1]
5. (a) $1061.37 (b) $1346.86 [8.2] **6.** $2.97 [8.2] **7.** (a) $13,863 (b) $518.90 [8.4] **8.** $\frac{13}{25}$ [8.5]

9. $750,000 [8.5]   **10.** $468,000 [8.5]   **11.** $15.5 million [8.5]   **12.** 10% [8.5]   **13.** 100% [8.5]
14. (a) 2 (b) 2 (c) 2 (d) 4 [8.6]   **15.** (a) 10.16 (b) 8.89 (c) 7.62 (d) 17.78 [8.6]
16. [8.5]

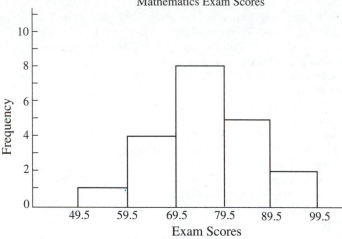

Mathematics Exam Scores

Cumulative Review, p. 427

1. 200,016   **2.** 300.004   **3.** 17.00   **4.** 0.4   **5.** 0.525   **6.** 180%   **7.** 0.015   **8.** $\frac{4}{3} = 1\frac{1}{3}$

9. $\frac{41}{11} = 3\frac{8}{11}$ **10.** $\frac{8}{105}$ **11.** $\frac{152}{15} = 10\frac{2}{15}$ **12.** $46\frac{5}{12}$ **13.** 12   **14.** $\frac{9}{5} = 1\frac{4}{5}$ **15.** 5,600,000

16. 398.988   **17.** 75.744   **18.** 0.01161   **19.** 80.6   **20.** 7   **21.** 2, 3, and 4   **22.** $2^2 \cdot 3^2 \cdot 11$

23. 210   **24.** zero   **25.** 7 and 160   **26.** 50   **27.** 18.5   **28.** 250%   **29.** $x = \frac{3}{8}$ **30.** 196

31. 325   **32.** 79   **33.** $208   **34.** $4536   **35.** 9%   **36.** (a) $10,199.44 (b) $10,271.35
37. $22,225.82

CHAPTER 9

Exercises 9.1, p. 436

1. -14 **2.** -12 **3.** $+10$ or 10   **5.** $-1\frac{1}{3}$ **6.** $-2\frac{1}{4}$ **7.** $+5.3$ or 5.3   **9.** -30 **10.** -40

11.

13.

14.

15.

17.

18.

19.

$-2\frac{1}{2}$ -1.3 $-\frac{1}{8}$ 0 0.75

21. $>$ **22.** $<$ **23.** $<$ **25.** $>$ **26.** $>$ **27.** $<$ **29.** $=$ **30.** $>$ **31.** true   **33.** false
34. true   **35.** true   **37.** false   **38.** true   **39.** true   **41.** false   **42.** false   **43.** $x = 6$ or $x = -6$
45. $x = 4.3$ or $x = -4.3$ **46.** $x = \frac{3}{2}$ or $x = -\frac{3}{2}$ **47.** no values for x **49.** $x = \frac{2}{3}$ or $x = -\frac{2}{3}$
50. $x = \frac{1}{4}$ or $x = -\frac{1}{4}$

Exercises 9.2, p. 441

1. 2   **2.** 1   **3.** 10   **5.** 19   **6.** -10 **7.** -9 **9.** 0   **10.** 25   **11.** -4 **13.** 2   **14.** -3
15. -16 **17.** -4 **18.** -4 **19.** 0   **21.** 4   **22.** -6 **23.** 14   **25.** -1 **26.** 4   **27.** -15
29. 41   **30.** 6   **31.** 13   **33.** 10   **34.** 8   **35.** -12 **37.** -10 **38.** 0   **39.** 0   **41.** -235
42. -165 **43.** 120   **45.** -121 **46.** $-5\frac{1}{2}$ **47.** $-3\frac{4}{5}$ **49.** $2\frac{1}{10}$ **50.** $-5\frac{1}{4}$ **51.** -18.81
53. 1.6   **54.** -6.1 **55.** -69.28 **57.** 5.9   **58.** -8 **59.** -1.1

Exercises 9.3, p. 446

1. 3   **2.** 13   **3.** 11   **5.** -7 **6.** -13 **7.** -9 **9.** 4   **10.** 10   **11.** -10 **13.** 1   **14.** 3
15. 18   **17.** 17   **18.** 23   **19.** -31 **21.** -5 **22.** -4 **23.** 0   **25.** -5 **26.** -2.1
27. -12.51 **29.** 0   **30.** $7\frac{1}{4}$ **31.** 30   **33.** 4   **34.** 5   **35.** -9 **37.** 23.9   **38.** -11.9
39. -28.79 **41.** 8   **42.** 12   **43.** 6   **45.** 10   **46.** 17   **47.** -4 **49.** 6   **50.** 6   **51.** -9
53. -1 **54.** -15 **55.** -8 **57.** -18 **58.** -10 **59.** 3   **61.** 1   **62.** 2   **63.** 0   **65.** 15
66. 4   **67.** 4   **69.** -3 **70.** 22.3   **71.** -28.6 **73.** $24\frac{1}{4}$ **74.** $\frac{1}{2}$ **75.** 0   **77.** 5   **78.** 0
79. -5

Exercises 9.4, p. 453

1. -15 **2.** -24 **3.** 24   **5.** -20 **6.** -24 **7.** 28   **9.** -50 **10.** -33 **11.** -21
13. -48 **14.** -36 **15.** 63   **17.** 0   **18.** 0   **19.** 90   **21.** 24   **22.** 30   **23.** -42 **25.** 0
26. 0   **27.** -1 **29.** 16   **30.** -64 **31.** -3 **33.** -2 **34.** -4 **35.** 4   **37.** 5   **38.** -5
39. -5 **41.** 1   **42.** 2   **43.** 4   **45.** -2 **46.** undefined   **47.** undefined   **49.** 0   **50.** 0
51. -23 **53.** 33   **54.** 31   **55.** -8 **57.** -12 **58.** -100 **59.** -37 **61.** -48 **62.** -20
63. -100 **65.** -99 **66.** -6 **67.** -13.44 **69.** -7 **70.** -4 **71.** 16   **73.** 108   **74.** 33
75. -0.27 **77.** 12.6   **78.** 13.5   **79.** $-\frac{8}{21}$ **81.** positive   **82.** positive   **83.** positive
85. negative   **86.** negative   **87.** zero   **89.** undefined   **90.** positive

Exercises 9.5, p. 460

1. $x + 5$ **2.** $x + 8$ **3.** $n + 9$ **5.** $\frac{n}{9}$ **6.** $x - 9$ **7.** $y - 0.13$ **9.** $0.13 - y$ **10.** $16 + n$
11. $2x + 4$ **13.** $8x + 6$ **14.** $5x + 6$ **15.** $\frac{x}{2} - 1.8$ **17.** $3 - 2n$ **18.** $3n + 1$ **19.** $6x + 4x$
21. 5 times a number   **22.** the sum of a number and 15   **23.** the difference between a number and 6

25. a number plus 4.91   **26.** the product of 7 and a number   **27.** a number divided by 17
29. 1 more than twice a number   **30.** 1 less than three times a number
31. the quotient of a number and 3 increased by 20   **33.** the product of a number and 15 decreased by 15
34. the quotient of 6 and a number   **35.** 13 minus a number   **37.** 5   **38.** 3   **39.** 6
41. Write the equation.
　　Subtract 10 from both sides.
　　Simplify both sides.
　　Divide both sides by 3.
　　Simplify both sides.
42. Write the equation.
　　Add 1.5 to both sides.
　　Simplify both sides.
　　Divide both sides by 2.
　　Simplify both sides.
43. $x = 18$ **45.** $n = 10$ **46.** $n = 8$ **47.** $y = 6$ **49.** $n = 22.4$ **50.** $x = 30$ **51.** $x = 6.5$
53. $6 = y$ **54.** $5.6 = n$ **55.** $n = 24$ **57.** $x = 2.4$ **58.** $x = 20$ **59.** $38 = x$

Exercises 9.6, p. 467

1. The sum of a number and 5 is equal to 16.   **2.** The difference between a number and 7 is equal to 8.
3. Four times a number decreased by 8 is equal to 20.
5. One more than the quotient of a number and 3 is equal to 13.
6. Fifteen is equal to 25 minus twice a number.   **7.** $x - 12$; $x - 12 = 16$; $x = 28$
9. $3y - 4$; $3y - 4 = 2.6$; $y = 2.2$ **10.** $2x + 1$; $20 = 2x + 1$; $9.5 = x$ **11.** 11   **13.** 17   **14.** 144
15. 5   **17.** 18   **18.** 48   **19.** 100   **21.** 5 meters   **22.** 11 inches   **23.** 6 inches   **25.** 40 yards
26. 100 square millimeters   **27.** 3 inches   **29.** 90°   **30.** 100°   **31.** 5 inches   **33.** 14 centimeters
34. 7 inches   **35.** \$60,000   **37.** 5650 feet   **38.** \$210,069   **39.** \$77,500   **41.** 2.25 inches
42. rectangle: 6 ft by 8 ft; triangle: 6 ft sides   **43.** 312.5 miles; this is not a better deal.

Review Questions: Chapter 9, p. 474

1. +4 or 4   **2.** $-\dfrac{3}{8}$ **3.** 0   **4.**

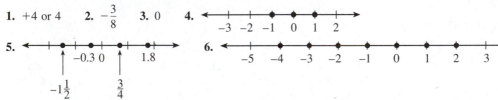

5.

6.

7. $<$ **8.** $>$ **9.** $=$ **10.** true   **11.** true   **12.** false   **13.** $x = 0.4$ or $x = -0.4$ **14.** no values
15. $x = \dfrac{9}{10}$ or $x = -\dfrac{9}{10}$ **16.** 4   **17.** -5 **18.** -20 **19.** 0   **20.** $-4\dfrac{3}{10}$ **21.** -3.3 **22.** -124
23. 5.62   **24.** 5   **25.** 16   **26.** -21 **27.** 0   **28.** 26   **29.** 6   **30.** -50 **31.** -74
32. -10.29 **33.** $\dfrac{35}{24} = 1\dfrac{11}{24}$ **34.** 0   **35.** -72 **36.** 64   **37.** $\dfrac{1}{2}$ **38.** 0   **39.** -0.024
40. -125 **41.** -17 **42.** $\dfrac{1}{3}$ **43.** 0   **44.** undefined   **45.** 16   **46.** -290 **47.** -6 **48.** 8
49. $\dfrac{7}{4} = 1\dfrac{3}{4}$ **50.** $-\dfrac{5}{4} = -1\dfrac{1}{4}$ **51.** undefined   **52.** 0   **53.** $x = -7.4$ **54.** $x = 25$ **55.** $n = -30$
56. $n = 20$ **57.** $n = 10$ **58.** $-10 = x$ **59.** 0.6   **60.** -7

Test: Chapter 9, p. 476

1. -0.34 [9.1] **2.**

3. $>$ [9.1] **4.** $<$ [9.1] **5.** $x = 17$ or $x = -17$ [9.1] **6.** true [9.1] **7.** false [9.1] **8.** true [9.1]

9. -4.6 [9.2] **10.** -9 [9.2] **11.** -23 [9.2] **12.** -2.6 [9.2] **13.** -4 [9.3] **14.** $-2\frac{1}{4}$ [9.3]

15. 0.028 [9.4] **16.** -90 [9.4] **17.** $\frac{1}{4}$ [9.4] **18.** 4.41 [9.4] **19.** 0.8 [9.4] **20.** undefined [9.4]

21. -5 [9.4] **22.** $-2\frac{1}{6}$ [9.4] **23.** $-8\frac{9}{14}$ [9.5] **24.** -32 [9.5] **25.** $x = 6$ [9.5] **26.** $n = 2.17$ [9.5]

27. $x = 0.3$ [9.5] **28.** $x = 2$ [9.5] **29.** 16 [9.6] **30.** 50 feet [9.6]

Cumulative Review, p. 478

1. c **2.** a **3.** d **4.** b **5.** four hundred twenty-three and eighty-five hundredths **6.** 200

7. 7349 **8.** 304.06 **9.** 1523.1 **10.** 118.26 **11.** $-\frac{7}{30}$ **12.** $-3,500,000$ **13.** undefined

14. 0 **15.** 0 **16.** 14.64 **17.** 219.8 **18.** 22 **19.** 12.5% **20.** $13,498.26 **21.** 354 miles

22. (a) $500 (b) $66\frac{2}{3}\%$ (c) 40% **23.** $57\frac{1}{7}$ miles **24.** $x = 3.1$ **25.** $n = 21.875$ **26.** 9

MEASUREMENT

Exercises M.1, p. M13

1. kilo-, hecto-, deka-, deci-, centi-, milli- **2.** 100 cm **3.** 500 cm **5.** 600 cm **6.** 2000 mm
7. 300 mm **9.** 1400 mm **10.** 16 mm **11.** 18 mm **13.** 350 mm **14.** 40 dm
15. 160 dm **17.** 210 cm **18.** 3000 m **19.** 5000 m **21.** 6400 m **22.** 1.1 cm **23.** 2.6 cm
25. 4.8 cm **26.** 0.06 dm **27.** 0.12 dm **29.** 0.03 m **30.** 0.145 m **31.** 0.256 m **33.** 0.32 m
34. 1.5 m **35.** 1.7 m **37.** 2.4 km **38.** 0.5 km **39.** 0.4 km **41.** 462 cm **42.** 0.063 m
43. 0.052 m **45.** 6410 mm **46.** 0.3 cm **47.** 0.5 cm **49.** 0.057 m **50.** 20 km **51.** 35 km
53. 2300 m **54.** 0.0005 km **55.** 0.0015 km **57.** 560 mm^2 **58.** 870 mm^2 **59.** 361 mm^2
61. 0.28 cm^2 **62.** 14 cm^2 **63.** 200 cm^2 **65.** 730 cm^2 = 73 000 mm^2 **66.** 5700 cm^2 = 570 000 mm^2
67. 60 cm^2 = 6000 mm^2 **69.** 290 dm^2 = 29 000 cm^2 = 2 900 000 mm^2
70. 3 dm^2 = 300 cm^2 = 30 000 mm^2 **71.** 780 m^2 **73.** 4 m^2 **74.** 53 m^2 **75.** 869 a = 86 900 m^2
77. 16 a = 1600 m^2 **78.** 2 a = 200 m^2 **79.** 0.01 ha **81.** 500 ha **82.** 476 ha **83.** 30 ha

Exercises M.2, p. M22

1. 2000 mg **2.** 7000 g **3.** 3.7 t **5.** 5.6 kg **6.** 4 t **7.** 0.091 t **9.** 700 mg **10.** 540 mg
11. 5000 kg **13.** 2000 kg **14.** 0.896 g **15.** 896 000 mg **17.** 75 kg **18.** 3 g **19.** 7 000 000 g
21. 0.000 34 g **22.** 780 mg **23.** 0.016 g **25.** 0.0923 kg **26.** 3940 mg **27.** 7580 kg
29. 2.963 t **30.** 3.547 t

31.

$1 \text{ cm}^3 = 1000 \text{ mm}^3$
$1 \text{ dm}^3 = 1000 \text{ cm}^3$
$1 \text{ m}^3 = 1000 \text{ dm}^3$
$1 \text{ km}^3 = 1\,000\,000\,000 \text{ m}^3$

33.

$1 \text{ m} = 10 \text{ dm}$
$1 \text{ m} = 100 \text{ cm}$
$1 \text{ m}^2 = 100 \text{ dm}^2$
$1 \text{ m}^2 = 10\,000 \text{ cm}^2$
$1 \text{ m}^3 = 1000 \text{ dm}^3$
$1 \text{ m}^3 = 1\,000\,000 \text{ cm}^3$

34.

$1 \text{ km} = 1000 \text{ m}$
$1 \text{ km}^2 = 1\,000\,000 \text{ m}^2$
$1 \text{ km}^3 = 1\,000\,000\,000 \text{ m}^3$
$1 \text{ km} = 10\,000 \text{ dm}$
$1 \text{ km}^2 = 100 \text{ ha}$
$1 \text{ km}^3 = 1\,000\,000\,000 \text{ kL}$

35. $73\,000 \text{ dm}^3$ **37.** $0.000\,525 \text{ m}^3$ **38.** $400\,000\,000 \text{ cm}^3$ **39.** $8\,700\,000 \text{ cm}^3$ **41.** 0.045 cm^3
42. 3100 mm^3 **43.** $0.000\,000\,19 \text{ dm}^3$ **45.** 2000 cm^3 **46.** 0.0764 L   **47.** 5300 mL   **49.** 0.03 L
50. 0.0053 L   **51.** 48 000 L   **53.** 0.29 kL   **54.** 0.569 L   **55.** $80\,000 \text{ mL} = 80\,000 \text{ cm}^3$

Exercises M.3, p. M33

1. 60 in.   **2.** 36 in.   **3.** 18 in.   **5.** 4 ft   **6.** 10 ft   **7.** $2\frac{1}{2}$ ft   **9.** 9 ft   **10.** 12 ft   **11.** 7 ft

13. 15,840 ft   **14.** 21,120 ft   **15.** 2 yd   **17.** 2 mi   **18.** 3 mi   **19.** 32 fl oz   **21.** 6 pt   **22.** 3 pt

23. 20 qt   **25.** 3 gal   **26.** 5 gal   **27.** $3\frac{3}{4}$ gal   **29.** 80 oz   **30.** 48 oz   **31.** 56 oz   **33.** 5000 lb

34. 6000 lb   **35.** 300 min   **37.** 90 min   **38.** 210 min   **39.** $\frac{1}{2}$ hr   **41.** 72 hr   **42.** 96 hr

43. 1800 sec   **45.** 300 sec   **46.** 180 sec   **47.** 4 min   **49.** 2 days   **50.** 3 days   **51.** 77° F
53. 50° F   **54.** 95° F   **55.** 10° C   **57.** 0° C   **58.** 5° C   **59.** 2.742 m   **61.** 96.6 km
62. 161 km   **63.** 124 mi   **65.** 19.7 in.   **66.** 39.4 in.   **67.** 19.35 cm^2 **69.** 55.8 m^2 **70.** 27.9 m^2
71. 83.6 m^2 **73.** 405 ha   **74.** 101.25 ha   **75.** 741 acres   **77.** 53.82 ft^2 **78.** 11.96 yd^2
79. 4.65 in^2 **81.** 9.46 L   **82.** 18.92 L   **83.** 10.6 qt   **85.** 11.088 gal   **86.** 13.2 gal   **87.** 4.54 kg

89. 453.6 g   **90.** 3.5 oz   **91.** $\frac{3}{4}$ acre (or 0.75 acre)   **93.** (a) 1728 in.^2 (b) $1\frac{1}{3} \text{ yd}^2$

94. (a) 4356 ft^2 (b) 484 yd^2

Exercises M.4, p. M38

1. 4 ft 8 in.   **2.** 5 ft 6 in.   **3.** 7 lb 4 oz   **5.** 6 min 20 sec   **6.** 15 min 30 sec   **7.** 3 days 6 hr
9. 9 gal 1 qt   **10.** 3 gal 2 qt   **11.** 5 pt 4 fl oz   **13.** 9 ft 7 in.   **14.** 12 ft 1 in.   **15.** 18 lb 2 oz
17. 18 min 10 sec   **18.** 19 min 45 sec   **19.** 4 hr 11 min 15 sec   **21.** 6 days 2 hr 35 min
22. 4 days 2 hr   **23.** 13 gal 3 qt   **25.** 7 gal 2 qt 1 pt 4 fl oz   **26.** 7 gal 2 qt 7 fl oz
27. 12 yd 2 ft 6 in.   **29.** 2 yd 2 ft 2 in.   **30.** 1 yd 7 in.   **31.** 3 gal 2 qt 1 pt 14 fl oz   **33.** 2 hr 45 min
34. 1 hr 40 min   **35.** 4 min 50 sec   **37.** 4 lb 8 oz   **38.** 9 lb 12 oz   **39.** 3 ft 8 in.   **41.** 5 ft 2 in.
42. 16 hr 30 min   **43.** 2 hr 40 min   **45.** 5 gal   **46.** 10 gal 1 qt

Chapter Review Questions, p. M42

1. 1500 cm   **2.** 0.35 dm   **3.** 3700 mm^2 **4.** 0.17 cm^2 **5.** 300 a   **6.** 30 000 m^2 **7.** 5000 cm^3
8. 36 000 mL   **9.** 13 000 cm^3 **10.** 68 000 mm^3 **11.** 5000 g   **12.** 3400 mg   **13.** 6710 kg

14. 0.019 g   **15.** 8000 g   **16.** 4.29 kg   **17.** 33   **18.** $\frac{39}{2} = 19\frac{1}{2}$ **19.** $\frac{37}{6} = 6\frac{1}{6}$ **20.** $\frac{2}{3}$ **21.** 76

22. 2600   **23.** $\frac{27}{8} = 3\frac{3}{8}$ **24.** $0.4 = \frac{2}{5}$ **25.** 192   **26.** 58   **27.** $2.8 = 2\frac{4}{5}$ **28.** $3.5 = 3\frac{1}{2}$

29. $25.6 = 25\frac{3}{5}$ **30.** $\frac{19}{2} = 9\frac{1}{2}$ **31.** $\frac{13}{16}$ **32.** $\frac{11}{8} = 1\frac{3}{8}$ **33.** 2 yd 7 in.   **34.** 4 yd 1 ft 4 in.

35. 4 lb 6 oz   **36.** 3 t 1120 lb   **37.** 6 days 4 hr   **38.** 1 hr 3 min 13 sec   **39.** 1 qt 1 pt 8 fl oz
40. 5 gal 1 pt   **41.** 5 lb 11 oz   **42.** 18 hr 44 min 16 sec   **43.** 1 yd 2 ft 5 in.   **44.** 4 gal 2 qt 1 pt

45. 2 lb 7 oz   **46.** 2 gal 1 qt   **47.** 49 min   **48.** 2 yd 2 ft 8 in.   **49.** $\frac{200}{9} = 22\frac{2}{9}$ **50.** $37.4 = 37\frac{2}{5}$

51. 12.7   **52.** 68.58   **53.** 10.0076   **54.** 7.88   **55.** 1.371   **56.** 0.68625   **57.** 22.96   **58.** 3.488
59. 322   **60.** 20.5275   **61.** 54.56   **62.** 0.248   **63.** 2.43   **64.** 4.18   **65.** 22.6044   **66.** 1.482
67. 18.16   **68.** 510.3   **69.** 15.89   **70.** 0.1764

Chapter Test, p. M45

1. 20 cm by 180 mm [M.1]   **2.** 10 kg by 9990 g [M.2]   **3.** volume is the same [M.2]   **4.** 0.37 m [M.1]
5. 2300 cm [M.1]   **6.** 2000 cm^3 [M.2]   **7.** 1.2 kg [M.2]   **8.** 5600 kg [M.2]   **9.** 7500 m^2 [M.1]
10. 11 m [M.1]   **11.** 4000 mm^3 [M.2]   **12.** 9.6 cm^2 [M.1]   **13.** 0.0835 g [M.2]

14. 92 [M.3]   **15.** $\frac{11}{9} = 1\frac{2}{9}$ [M.3]   **16.** 1680 [M.3]   **17.** 19.7 [M.3]   **18.** 17.199 [M.3]

19. 6.6 [M.3]   **20.** 6.723 [M.3]   **21.** 124 [M.3]   **22.** 28.38 L [M.3]   **23.** 1.75 oz [M.3]
24. 20° C [M.3]   **25.** 86°F [M.3]   **26.** 2 gal 2 qt 2 fl oz [M.4]   **27.** 7 days 12 hr 48 min [M.4]
28. 1 yd 1 ft 1 in. [M.4]   **29.** 2 qt 7 fl oz [M.4]

Cumulative Review, p. M46

1. 1   **2.** 630.055   **3.** 760.7   **4.** 5.1   **5.** 735   **6.** 7   **7.** 1.15   **8.** 0.81
9. (a) 45   (b) 46   (c) 41   (d) 35   **10.** $2000   **11.** $10,798.61   **12.** $2907.04   **13.** 20%
14. 130   **15.** 38.775   **16.** 12,040   **17.** (a) $630   (b) 25%   **18.** 405 miles   **19.** 27,000 revolutions
20. (b) at 9.08¢/oz   **21.** (a) $18   (b) $45   **22.** 19 hr 3 min   **23.** 2 yd 2 ft 8 in.
24. (a) 356 mm   (b) 0.1872 m   (c) 19 500 mm^2 (d) 540 mL   (e) 0.078 m^3
25. Housing—$11,250; Transportation—$6750; Misc.—$6750; Savings—$2250; Education—$4500;
Taxes—$4500; Food—$9000   **26.** (a) 8.9×10^8 (b) 6.32×10^{-5} **27.** (a) -21 (b) -16
28. (a) $x = -7$ (b) $y = 9.7$ **29.** 11   **30.** -2.7

GEOMETRY

Exercises G.1, p. G8

1. (a) B.   (b) C.   (c) D.   (d) A.   (e) F   (f) E.   **2.** 18.4 cm   **3.** 104 mm   **5.** 31.4 ft   **6.** 900 mm

7. 19.468 yd   **9.** 13.5 cm   **10.** 4.396 m   **11.** 16 000 m   **13.** 20.80 in.   **14.** $9\frac{1}{3}$ ft or 112 in.

15. 11.9634 cm   **17.** 35.98 in.   **18.** 27.42 m   **19.** 27.42 m   **21.** 60 cm   **22.** 12.71 ft
23. 3.57 km

Exercises G.2, p. G16

1. (a) C.   (b) B.   (c) D.   (d) A.   (e) F.   (f) E.   **2.** 875 in.^{2} **3.** 6 cm^2 **5.** 78.5 yd^2 **6.** 7.065 ft^2
7. 227.5 in.^{2} **9.** 500 mm^2 (or 5 cm^2)   **10.** 6 cm^2 **11.** 28.26 ft^2 **13.** 21.195 yd^2 **14.** 32.28 dm^2
15. 106 cm^2 **17.** 76.93 in.^{2} **18.** 50.13 m^2 **19.** 21.87 m^2 **21.** 15.8 km^2 = 158 000 a   **22.** 3.14 ft^2
23. 0.785 ft^2 **25.** 2500 cm^2 and 0.25 m^2 **26.** 210 in.^{2} and 1.46 ft^2 (rounded off)

Exercises G.3, p. G21

1. (a) C.   (b) E.   (c) D.   (d) B.   (e) A.   **2.** 70 in.^{3} **3.** 1695.6 in.^{3} **5.** 904.32 ft^3 **6.** 12.56 dm^3
7. 80 cm^3 **9.** 9106 dm^3 **10.** 282.6 cm^3 **11.** 113.04 in.^{3} **13.** 3456 in.^{3}

Exercises G.4, p. G28

1. (a) m$\angle 2$ = 75°   (b) m$\angle 2$ = 87°   (c) m$\angle 2$ = 45°   (d) m$\angle 2$ = 15°
2. (a) m$\angle 4$ = 135°   (b) m$\angle 4$ = 90°   (c) m$\angle 4$ = 70°   (d) m$\angle 4$ = 45°
3. (a) a right angle   (b) an acute angle   (c) an acute angle
5. 25°   **6.** 15°   **7.** 40°   **9.** 55°   **10.** 70°   **11.** 110°
13. m$\angle A$ = 55°; m$\angle B$ = 45°; m$\angle C$ = 80°
14. m$\angle P$ = 65°; m$\angle Q$ = 115°; m$\angle R$ = 105°; m$\angle S$ = 75°
15. m$\angle L$ = 115°; m$\angle M$ = 90°; m$\angle N$ = 105°; m$\angle O$ = 110°; m$\angle P$ = 120°
17. (a) straight angle   (b) right angle   (c) acute angle   (d) obtuse angle
18. (a) 180°   (b) 90°   (c) 30°   (d) 150°
19. (a) 150°   (b) yes; $\angle 3$ is supplementary to $\angle 2$ (c) $\angle 1$ and $\angle 2$; $\angle 2$ and $\angle 3$; $\angle 3$ and $\angle 4$; $\angle 4$ and $\angle 1$
21. m$\angle 2$ = 150°; m$\angle 3$ = 30°; m$\angle 4$ = 150°
22. (a) m$\angle 2$ = 70°; m$\angle 3$ = 90°; m$\angle 4$ = 20°; m$\angle 5$ = 70°   (b) $\angle 3$ (c) $\angle 2$ and $\angle 5$
23. m$\angle 3$ = m$\angle 5$; m$\angle 2$ = m$\angle 5$; m$\angle 1$ = m$\angle 4$; m$\angle 1$ = m$\angle 6$; m$\angle 4$ = m$\angle 6$

Exercises G.5, p. G38

1. scalene (and obtuse)   **2.** equilateral (and acute)   **3.** scalene (and right)   **5.** isosceles (and acute)
6. isosceles (and obtuse)   **7.** isosceles (and right)   **9.** scalene (and acute)   **10.** x= 50°; y = 60°
11. x = 7.5; y = 9   **13.** x = 3; y = 3
14. The triangles are not similar. The corresponding sides are not proportional.
15. The triangles are not similar. The corresponding sides are not proportional.
17. The triangles are similar. All pairs of corresponding sides are proportional in a ratio of 1 : 2. Therefore, $\Delta PQR \sim \Delta SUT$.
18. Yes, $\Delta PQR \sim \Delta PST$. m$\angle QPR$ = m$\angle SPT$ since they are vertical angles. Then the remaining pair of angles must also be equal since the sum of the measures of the angles of each triangle must be 180°. Since all pairs of corresponding angles have the same measures, the triangles are similar.
19. (a) Each of the other four angles measures 75°.   (b) The triangles are similar since the three pairs of corresponding angles are equal.

Exercises G.6, p. G47

1. yes (144 = 12^2)   **2.** yes (169 = 13^2)   **3.** yes (81 = 9^2)   **5.** yes (400 = 20^2)   **6.** yes (225 = 15^2)
7. no   **9.** no   **10.** no   **11.** (1.732)^2 = 2.999824 and (1.733)^2 = 3.003289   **13.** 2$\sqrt{3}$ **14.** 2$\sqrt{7}$
15. 2$\sqrt{6}$ **17.** 4$\sqrt{3}$ **18.** 12$\sqrt{2}$ **19.** 11$\sqrt{3}$ **21.** 10$\sqrt{5}$ **22.** 10$\sqrt{3}$ **23.** 8$\sqrt{2}$ **25.** 6$\sqrt{2}$
26. 7$\sqrt{2}$ **27.** 11$\sqrt{5}$ **29.** 13   **30.** 14   **31.** 20$\sqrt{2}$ **33.** 3$\sqrt{10}$ **34.** 2$\sqrt{10}$ **35.** 16
37. yes (6^2 + 8^2 = 10^2)   **38.** yes (20^2 + 21^2 = 29^2)   **39.** c = $\sqrt{5}$ **41.** x = 5$\sqrt{2}$ **42.** x = 10

43. $c = 2\sqrt{29}$ **45.** 2.8284   **46.** 5.8310   **47.** 6.7082   **49.** 6.3246   **50.** 4.4721   **51.** 8.6603
53. 17.3205   **54.** 8.9443   **55.** 22.3607   **57.** 63.2 ft   **58.** 22.4 m   **59.** 522.0 ft
61. (a) 127.3 ft   (b) closer to home plate
62. (a) $C = 22.8$ mm; $A = 314$ mm^2 (b) $P = 40\sqrt{2}$ mm (or 56.6 mm); $A = 200$ mm^2 **63.** 44.9 in.
65. 7.1 cm   **66.** 17.0 in.

Chapter Review Questions, p. G54

1. (a) 26 in.   (b) 24 in.^{2} **2.** (a) 16.28 m   (b) 13.14 m^2 **3.** (a) 8 ft   (b) 3 ft^2
4. (a) 41.12 m   (b) 100.48 m^2 **5.** 12.25 mm^2 **6.** 615.44 cm^2 **7.** 132 in.^{2} **8.** 11 ft
9. 4186.67 cm^3 **10.** 261.67 cm^3 **11.** 96 in.^{3} **12.** 75.988 m^3
13. (a) $\angle COD$ (or $\angle DOE$)   (b) $\angle AOE$ and $\angle BOC$ (c) $\angle DOE$ (d) $\angle AOE$ (or $\angle DOC$ or $\angle BOD$)
14. (a) 130°   (b) obtuse   (c) $\angle MOL$ (d) none   (e) $\angle KOL$ and $\angle NOL$
15. (a) acute   (b) right   (c) obtuse   (d) straight   **16.** m$\angle 2 = 40°$; m$\angle 4 = 70°$; m$\angle 5 = 110°$ **17.** obtuse
18. isosceles (or acute)   **19.** right (or isosceles)
20. (a) Yes. Since m$\angle ADE =$ m$\angle ABC$, m$\angle A =$ m$\angle A$, and m$\angle AED =$ m$\angle ACB$
    (The sum of the measures of the angles of a triangle is 180°.),
    the corresponding angles are equal and the triangles are similar.   (b) 6 in.
21. ΔPQR is a right triangle because $33^2 + 56^2 = 65^2$.   **22.** $10\sqrt{2}$ **23.** $5\sqrt{3}$ **24.** 11   **25.** $10\sqrt{6}$
26. $50\sqrt{2}$

Chapter Test, p. G57

1. (a) right angle   (b) obtuse angle   (c) $\angle AOB$ and $\angle DOE$; $\angle AOE$ and $\angle BOD$
    (d) $\angle AOB$ and $\angle AOE$; $\angle BOD$ and $\angle EOD$; $\angle AOB$ and BOD; $\angle AOE$ and $\angle EOD$ [G.4]
2. (a) 55°   (b) 145° [G.4]   **3.** (a) 18.84 in [G.1]   (b) 113.04 in.^{2} [G.2]
4. (a) 16.8 m [G.1]   (b) 12 m^2 [G.2]   **5.** (a) $40\frac{1}{3}$ in. [G.1]   (b) $99\frac{1}{6}$ in.^{2} [G.2]   **6.** 153.86 in.^{2} [G.2]
7. 10 cm^2 [G.2]   **8.** 88.5 cm^3 [G.3]   **9.** 16 dm^3 [G.3]   **10.** (a) isosceles   (b) obtuse   (c) scalene [G.5]
11. (a) 30°   (b) obtuse   (c) $\overline{RT}$ [G.5]   **12.** 2.5 [G.5]   **13.** $2\sqrt{5}$ [G.6]   **14.** $3\sqrt{5}$ [G.6]
15. $2\sqrt{17}$ [G.6]   **16.** $10\sqrt{7}$ [G.6]   **17.** 30 [G.6]   **18.** $13\sqrt{2}$ [G.6]
19. It is a right triangle; $12^2 + 16^2 = 20^2$. [G.6]   **20.** $2\sqrt{5}$ (or 4.4721) [G.6]
21. Yes, $\Delta AOB \sim \Delta COD$. m$\angle AOB =$ m$\angle COD$ since they are vertical angles. Then the remaining pair of angles
    must also be equal since the sum of the measures of the angles of each triangle must be 180°.
    Since all pairs of corresponding angles have the same measures, the triangles are similar. [G.5]
22. $x = 3$; $y = 4$ [G.5]

Cumulative Review, p. G61

1. (a) $2 \cdot 3 \cdot 5^2$ (b) 263 is prime.   (c) $2^3 \cdot 13$ **2.** $2 \cdot 3 \cdot 5 \cdot 5 \cdot 7 = 1050$ **3.** 2   **4.** 0.055   **5.** 690%
6. $1\frac{19}{40}$ **7.** $4\frac{3}{10}$ **8.** 25   **9.** $1\frac{1}{6}$ **10.** 6,400,000   **11.** 296.987   **12.** 5.21   **13.** 2.013   **14.** 72
15. 720   **16.** 7200   **17.** 72,000   **18.** 65   **19.** 15%   **20.** $x = 1$ **21.** $x = 5$ **22.** \$14,190.20
23. (a) \$494   (b) 30.3%   **24.** 23 000 mg   **25.** 6000 mm   **26.** 5600 m   **27.** 3800 mL
28. 0.0911 cm^2 **29.** 0.0045 m^3 **30.** 440 cm$^2 = 44\,000$ mm^2 **31.** 7349.25   **32.** undefined   **33.** 0

34. $-3,500,000$ **35.** 14.64   **36.** -42 **37.** 22   **38.** 219.8   **39.** $-\dfrac{7}{30}$

40. (a) 43.96 cm   (b) 153.86 cm^2 **41.** (a) 470 cm   (b) 7000 cm^2 **42.** 2034.72 in.^{3}

43. m$\angle D = 65°$; m$\angle E = 45°$; m$\angle DCE = $ m$\angle ACB = 70°$ **44.** $3\sqrt{2}$ m (or 4.2426 m)

SOLVING EQUATIONS

Exercises E.1, p. E8

1. $8x$ **2.** x **3.** $6x$ **5.** $-7a$ **6.** $-7y$ **7.** $-12y$ **9.** $-9x$ **10.** $-3x$ **11.** $-8x$ **13.** 0
14. $-p$ **15.** $-c$ **17.** $-9a$ **18.** $-2c$ **19.** 0   **21.** $5x - 7$ **22.** $-x + 2$ **23.** $-x + 5$
25. $-11a - 2$ **26.** $-3x - 2$ **27.** $3x + 1$ **29.** $4x + 7$ **30.** $4y - 1$ **31.** $5y^2$
33. $2x^2 + 3x + 1$ **34.** $5x^2 - 2x + 3$ **35.** $-y$ **37.** $3a^2 + 2ab$ **38.** $2y^2 + 5xy$ **39.** $5x + 4y$
41. -5 **42.** 0   **43.** 0   **45.** 9   **46.** -15 **47.** 22   **49.** -11 **50.** -22 **51.** -3 **53.** 12
54. 4   **55.** -8 **57.** 26   **58.** -9 **59.** -12

Exercises E2, p. E14

1. $x = 6$ **2.** $x = 7$ **3.** $y = 22$ **5.** $y = -5$ **6.** $x = -3$ **7.** $x = -2$ **9.** $y = 2$ **10.** $x = 7$
11. $x = 6$ **13.** $y = -4$ **14.** $x = -6$ **15.** $x = -6$ **17.** $y = 5$ **18.** $y = 3$ **19.** $x = 13$
21. $x = 1$ **22.** $x = 4$ **23.** $y = 2$ **25.** $x = -3$ **26.** $y = 1$ **27.** $y = 2$ **29.** $x = -6$
30. $x = -1$ **31.** $y = -3$ **33.** $x = 9$ **34.** $x = 7$ **35.** $y = -3$ **37.** $x = -1$ **38.** $x = 3$
39. $x = -3$ **41.** $x = -5$ **42.** $y = 4$ **43.** $y = 5$ **45.** $x = 0$ **46.** $x = 0$ **47.** $x = 3$
49. $x = 3$ **50.** $x = 8$ **51.** $x = -7$ **53.** $x = 5$ **54.** $y = 3$ **55.** $y = 4$ **57.** $x = 5$

58. $x = 4$ **59.** $x = -7$ **61.** $x = -12$ **62.** $y = 75$ **63.** $x = -\dfrac{5}{2}$ **65.** $x = \dfrac{220}{3}$

Exercises E.3, p. E21

1. $-2x$ **2.** $x + (-3)$ (or $x - 3$)   **3.** $2x + 4$ **5.** $3(n + 12)$ **6.** $-4(n - 3)$ **7.** $2(9 + y)$
9. $5y - (-2y)$ **10.** $3(2x - x)$ **11.** $x - 16$; $x - 16 = -48$; $x = -32$ **13.** -1 **14.** -4 **15.** 1
17. -24 **18.** 15   **19.** 5   **21.** 18, 19   **22.** $-15, -14, -13$ **23.** 7, 9, 11   **25.** 14, 16, 18, 20
26. $-10, -8, -6$ **27.** $-3, -2$ **29.** -3 **30.** 3   **31.** 40 ft, 80 ft   **33.** 20 in., 22 in., 24 in.
34. $-6, -5$ **35.** 25 ft   **37.** 10 ft, 26 ft   **38.** $3500
39. English book—$20; chemistry book—$41; music book—$17

Exercises E.4, p. E26

1. $d = 125$ mi   **2.** $A = 400$ ft^2 **3.** $A = 28.26$ cm^2 **5.** $y = 4$ **6.** $P = \$3000$ **7.** $w = 10$ cm

9. $\alpha = 90°$ **10.** $h = 5$ mm   **11.** $m = \dfrac{f}{a}$ **13.** $h = \dfrac{L}{2\pi r}$ **14.** $P = \dfrac{I}{rt}$ **15.** $x = \dfrac{y - b}{m}$

17. $a = P - b - c$ **18.** $\beta = 180 - \alpha - \gamma$ **19.** $l = \dfrac{P - 2w}{2}$ or $l = \dfrac{P}{2} - w$ **21.** $w = \dfrac{V}{lh}$

22. $h = \dfrac{V}{\pi r^2}$ **23.** $y = -2x + 4$ **25.** $y = x - 7$ **26.** $y = 3x - 5$

27. $y = \dfrac{-3x + 6}{2}$ or $y = -\dfrac{3}{2}x + 3$ **29.** $x = \dfrac{2y - 3}{6}$ or $x = \dfrac{1}{3}y - \dfrac{1}{2}$ **30.** $y = \dfrac{5x + 1}{3}$ or $y = \dfrac{5}{3}x + \dfrac{1}{3}$

Chapter Review Questions, p. E29

1. $15x$ **2.** $12y$ **3.** $3x$ **4.** $-6x$ **5.** $14w$ **6.** 0 **7.** $-2y$ **8.** $-34p$ **9.** $-8a + 6$
10. $-2x + 14$ **11.** $14y - 7$ **12.** $-10x - 10$ **13.** $2x^2$ **14.** $-3y^2$ **15.** $5x^2 - 3x + 1$
16. $4y^2 + 2y - 1$ **17.** $2x^2 - 3$ **18.** $5y^2 + 4$ **19.** $9x + y$ **20.** $7x - 6y - 6$ **21.** 27 **22.** 4
23. 25 **24.** 21 **25.** -31 **26.** $x = 4$ **27.** $y = 16$ **28.** $x = -4$
29. $y = 3$ **30.** $x = -7$ **31.** $y = -2$ **32.** $x = 4$ **33.** $x = -5$ **34.** $x = -6$ **35.** $y = 5$
36. $x = 3$ **37.** $x = 4$ **38.** $y = -5$ **39.** $x = 0$ **40.** $x = 9$ **41.** $y = -3$ **42.** $x = 0$
43. $y = 0$ **44.** $x = 1$ **45.** $y = 2$ **46.** $x = -12$ **47.** $y = 3$ **48.** $x = 12$ **49.** $y = \dfrac{15}{2}$
50. $x + 8$ **51.** $5x - 3$ **52.** $\dfrac{x}{9}$ **53.** $-3(x + 2)$ **54.** $10 + 4x$ **55.** $18 - 2x$
56. $10x - 15 = -35; -2$ **57.** $2x + 14 = x - 13; -27$ **58.** $x + (x + 1) + (x + 2) = 78; 25, 26, 27$
59. $\dfrac{1}{3}x = (x + 2) + (x + 4) - 41; 21, 23, 25$ **60.** $7x = 5(x + 2); 5$ **61.** $[x + (-8)] = 4x - 6; -\dfrac{2}{3}$
62. $680 = 2(200) + 2w; 140$ m   **63.** $\beta = 100°$ **64.** $r = 25$ mph   **65.** $d = \dfrac{C}{\pi}$
66. $y = \dfrac{3x - 2}{4}$ or $y = \dfrac{3}{4}x - \dfrac{1}{2}$

Chapter Test, p. E31

1. $-3x - 7$ [12.1]   **2.** $4x^2 - 4x - 4$ [12.1]   **3.** $3b$ [12.1]   **4.** $12x + 21y$ [12.1]   **5.** -1 [12.1]
6. 1 [12.1]   **7.** -8 [12.1]   **8.** 16 [12.1]   **9.** $x = 40$ [12.1]   **10.** $y = 0$ [12.1]   **11.** $x = 6$ [12.1]
12. $y = 0$ [12.1]   **13.** $y = -12$ [12.1]   **14.** $x = -10$ [12.1]   **15.** $x = -\dfrac{27}{7}$ [12.1]   **16.** $y = 20$ [12.2]
17. $\dfrac{x}{4} - 3$ [12.3]   **18.** $2(x + 3) + 1$ [12.3]   **19.** $\dfrac{3}{4}x - 5$ [12.3]   **20.** $\dfrac{x + 3}{-2}$ [12.3]
21. $82, 84, 86$ [12.3]   **22.** -16 [12.3]   **23.** $6\dfrac{3}{4}$ cm [12.3]   **24.** 2 yr [12.3]   **25.** $y = \dfrac{4x + 2}{3}$ [12.4]

Cumulative Review, p. E32

1. $2^3 \cdot 3 \cdot 5 \cdot 7 = 840$ **2.** (a) $1, 3, 5, 15, 25, 75$ (b) $75, 150, 225, 300, \ldots$ **3.** 3.32
4. $\dfrac{15}{16}$ **5.** $189\dfrac{1}{2}$ **6.** $336\dfrac{29}{48}$ **7.** $16\dfrac{4}{5}$ **8.** 5.48 **9.** $\dfrac{7}{12}$ **10.** 20 **11.** 200
12. 50 **13.** 3 km   **14.** 86 mm   **15.** $92\,000$ cm^2 **16.** 7540 mL   **17.** 16 **18.** $10\sqrt{6}$
19. $5\sqrt{10}$ **20.** 8 **21.** $x = 5$ **22.** $x = -\dfrac{9}{5}$ **23.** $x = 12$ **24.** $y = \dfrac{5 - x}{2}$
25. (a) 80 (b) 80.5 (c) 93 (d) 35 **26.** 354 miles   **27.** 39 in. (or 3 ft 3 in.)
28. (a) $\$500$ (b) $66\dfrac{2}{3}\%$ (c) 40%

ADDITIONAL TOPICS FROM ALGEBRA

Exercises T.1, p. T12

1. Three does not equal negative three. True.
2. Negative five is less than negative two. True.

3. Negative thirteen is greater than negative one. False. Correction: $-13 < -1$.

5. The absolute value of negative seven is less than positive five. False. Correction: $|-7| > +5$.

6. The absolute value of negative four is greater than the absolute value of positive three. True.

7. Negative one half is less than negative three fourths. False. Correction: $-\dfrac{1}{2} > -\dfrac{3}{4}$.

9. Negative four is less than negative six. False. Correction: $-4 > -6$.

10. The absolute value of negative six is greater than zero. True.

11. $-1 < x < 2$; open interval   **13.** $x < 0$; open interval   **14.** $-1 \le x \le 2$; closed interval

15. $x \ge 1$; half-open interval

17. open

18. closed

19. closed

21. half-open

22. closed

23. open

25. half-open

26. open

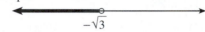

27. half-open

29. $x > 4$

30. $y \ge 3$

31. $y \le -3$

33. $y > -\dfrac{6}{5}$

34. $y > \dfrac{2}{5}$

35. $x < \dfrac{11}{7}$

37. $x < -5$

38. $x \ge 5$

39. $x \le 5$

41. $x > -6$

42. $x > 5$

43. $x \le -1$

45. $x > -1$

46. $x \leq 0$

47. $x < -5$

49. $1 \leq x \leq 4$

50. $-2 \leq y \leq -\dfrac{3}{4}$

51. $\dfrac{1}{3} \leq x \leq 2$

53. $-3 \leq y \leq 1$

54. $-6 < x \leq -5$

55. $-5 \leq x < -3$

Exercises T.2, p. T17

1. $\{(-5, 0), (-3, 1), (-1, 0), (1, 1), (2, 0)\}$ **2.** $\{(-3, 1), (-1, 4), (0, 5), (1, 4), (3, -3)\}$

3. $\{(-3, -2), (-2, -1), (-2, 1), (0, 0), (2, 1), (3, 0)\}$ **5.** $\{(-3, -4), (-3, 3), (-1, -1), (-1, 1), (1, 0)\}$

6. $\{(-1, -4), (-1, -3), (-1, 0), (-1, 2), (-1, 5)\}$ **7.** $\{(-1, -5), (0, -4), (1, -3), (2, -2), (3, -1), (4, 0)\}$

9. $\{(-1, 4), (0, 3), (1, 2), (2, 1), (3, 0), (4, -1), (5, -2)\}$

10. $\{(-3, 0), (-2, -1), (-1, -2), (0, -3), (0, 0), (1, -2), (2, -1), (3, 0)\}$

11.

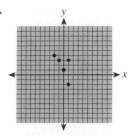

13.

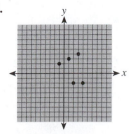

14.

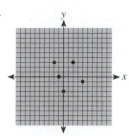

15.

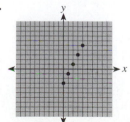

17.

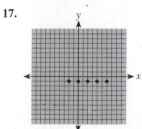

18.

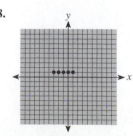

19.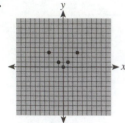

21.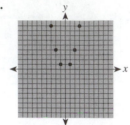

22.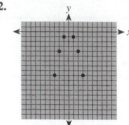

23.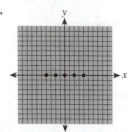

25.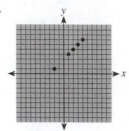

26.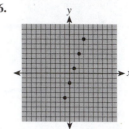

27.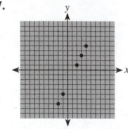

29.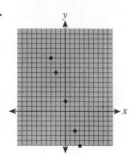

30.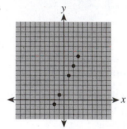

31.

P	I
100	12
200	24
300	36
400	48
500	60

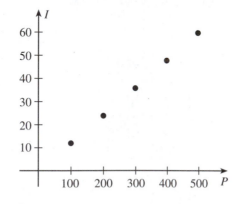

33.

C	F
−20	−4
−15	5
−10	14
−5	23
0	32
5	41
10	50

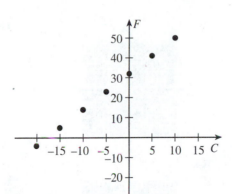

34.

h	V
3	75
5	125
6	150
8	200
10	250

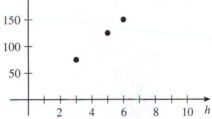

Exercises T.3, p. T25

1.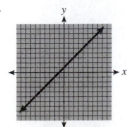

2.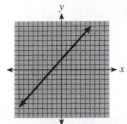

3.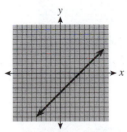

5.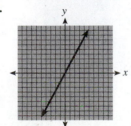

6.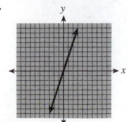

7.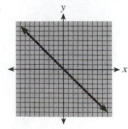

9.

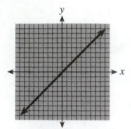

10.

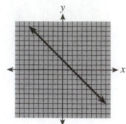

11.

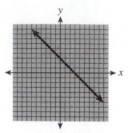

13.

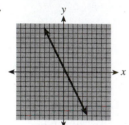

14.

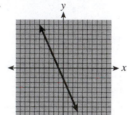

15.

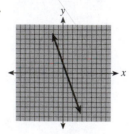

17.

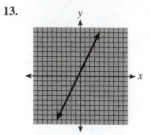

18.

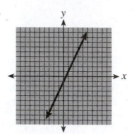

19.

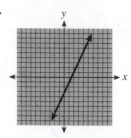

21.

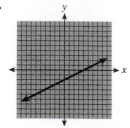

22.

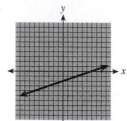

23.

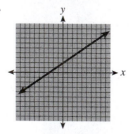

25.

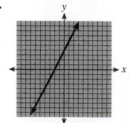

26.

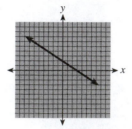

27.

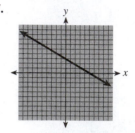

29.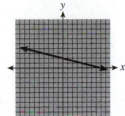

30.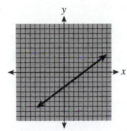

31.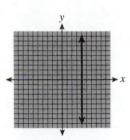

33.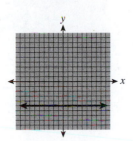

34.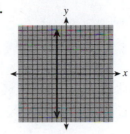

35.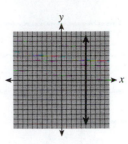

37.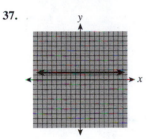

38.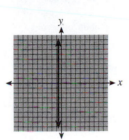

39.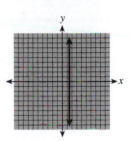

Exercises T.4, p. T29

1. $11x - 7$; binomial; first-degree   **2.** $-x + 9$; binomial; first-degree

3. $2x^2 + 6x + 2$; trinomial; second-degree   **5.** $5a^2$; monomial; second-degree

6. y^3; monomial; third-degree   **7.** $-3y^2 + 7y + 7$; trinomial; second-degree

9. $5x^3 - 3x + 17$; trinomial; third-degree   **10.** $-4x^3 + 2x^2 + 11x$; trinomial; third-degree

11. $5x - 10$ **13.** $2x^2 + 4$ **14.** $3x^2 - 6x - 20$ **15.** $4y^3 + 3y^2 + 2y - 9$ **17.** $x - 2$

18. $-x - 11$ **19.** $a^2 + 5a$ **21.** $12x^3 + x^2 - 6x$ **22.** $5x^3 - 11x + 10$ **23.** $-2x^2 + 7x - 8$

25. $3x^3 - x + 6$ **26.** $x^3 + 13x^2 + 9x - 7$ **27.** $5x^2 - 2x - 5$ **29.** $3x^3 + 5x^2 + 13x - 6$

30. $4x^3 - 16x^2 + 13x$

Exercises T.5, p. T35

1. $12x^4$ **2.** $10x^5$ **3.** $18a^3$ **5.** $8y^3 + 4y^2 + 8y$ **6.** $15y^3 - 10y^2 + 5y$ **7.** $-4x^3 + 2x^2 - 3x + 5$

9. $-14x^4 + 21x^3 - 84x^2$ **10.** $-3x^4 + 3x^3 - 39x^2$ **11.** $x^2 + 7x + 10$ **13.** $x^2 - 7x + 12$

14. $x^2 - 9x + 14$ **15.** $a^2 + 4a - 12$ **17.** $7x^2 - 20x - 3$ **18.** $15x^2 - 8x - 12$ **19.** $4x^2 - 9$

21. $x^2 - 9$ **22.** $x^2 - 25$ **23.** $x^2 - 2.25$ **25.** $9x^2 - 49$ **26.** $25x^2 - 1$

27. $x^4 + 4x^3 + 5x^2 + 23x + 12$ **29.** $2x^4 - 4x^3 + x^2 - 9x + 14$ **30.** $4x^4 - 24x^3 - x^2 - 2x + 48$

31. $3x^4 + x^3 - 18x^2 - 36x - 10$ **33.** $x^6 + 3x^5 - 4x^4 - 7x^3 + 8x^2 + 3x - 3$

34. $x^6 - 2x^5 - 2x^4 + 10x^3 - 12x^2 + 6x - 1$ **35.** $x^5 + 2x^4 - 4x^3 - 12x^2 - 17x + 6$
37. $x^8 + 2x^6 - 7x^4 + 8x^2 - 5$ **38.** $x^8 - 3x^6 + 2x^4 + 11x^2 - 7$
39. $x^6 + 2x^5 + 3x^4 + 4x^3 + 3x^2 + 2x + 1$

Chapter Review Questions, p. T41

1. (a) 0 (b) $-\dfrac{1}{2}, 0, 6.13$ (c) $-\sqrt{5}, -\dfrac{1}{2}, 0, 6.13$

2. open interval

3. half-open interval

4. half-open interval

5. closed interval

6. open interval

7. half-open interval

8. $-2 < x \le 3$, half-open interval   **9.** $x < \dfrac{1}{2}$, open interval

10. $x < -4$ **11.** $y \le 7$

12. $y \ge \dfrac{8}{3}$ **13.** $-3 > x$

14. $-2 < x$

15. $2 \le x \le 3$

16. $\{(-4, 1), (-3, 0), (-2, -1), (0, 0), (1, 1), (2, 0)\}$
17. $\{(-1, 1), (0, 1), (1, 1), (2, 1)\}$
18. $\{(-1, 0), (0, 2), (1, 4), (2, 2), (3, 0)\}$

19.   **20.**   **21.**

22.

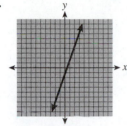

23.

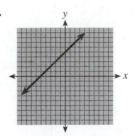

24.

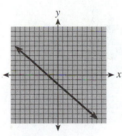

25.

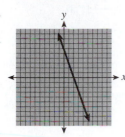

26.

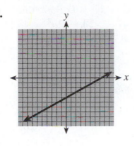

27.

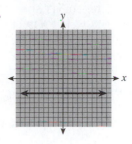

28. 2^6 (or 64)   **29.** $12x^7$ **30.** (a) $x^3 + 8x + 8$; third degree   (b) -43 **31.** $4x + 7$
32. $2y^3 + 8y^2 - 5y - 19$ **33.** $5x + 22$ **34.** $-2a^2 - a + 8$ **35.** $28x^3 + 21x^2 - 42.7x$
36. $12y^2 - 8y - 15$ **37.** $4x^4 - 15x^3 + 17x^2 + 14x - 15$ **38.** $2x^4 - x^3 - x^2 - 20x + 20$

Chapter Test, p. T43

1. (a) $-5, -\sqrt{9}$ (b) $-5, -\sqrt{9}, -1.2, 1\frac{3}{4}$ (c) $\sqrt{5}$ (d) $-5, -\sqrt{9}, -1.2, \sqrt{5}, 1\frac{3}{4}$ [T.1]

2. $4 < x \leq 6$; half-open interval [T.1]   **3.** $x < -1$; open interval [T.1]

4.   closed interval [T.1]

5. half-open interval [T.1]

6. $-4 > x$ [T.1]

7. $-1 \leq x$ [T.1]

8. $-1.5 \leq x \leq 2$ [T.1]

9. $\{(-4, 1), (-3, 0), (-2, -1), (0, 0), (1, 1), (2, 0)\}$ [T.2]   **10.** $\{(-1, 1), (0, 1), (1, 1), (2, 1)\}$ [T.2]

11. [T.2]

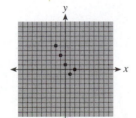

12. [T.3]

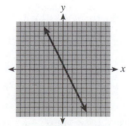

13. [T.3]

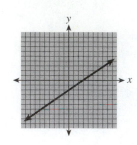

14. [T.3]

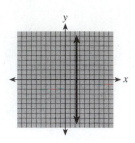

15. 3^4 (or 81) [T.5]   **16.** $30x^5$ [T.5]   **17.** (a) $4x^3 + x^2 - 7x + 7$; third degree   (b) -7 [T.4]
18. $-y^3 + 5y^2 - 2y - 18$ [T.4]   **19.** $4x + 27$ [T.4]   **20.** $-2a^2 - 4a$ [T.4]
21. $15x^3 + 10x^2 - 20.5x$ [T.5]   **22.** $8y^2 - 10y - 33$ [T.5]   **23.** $5x^4 + 6x^3 - 23x^2 + 22x - 8$ [T.5]
24. $2x^4 - 9x^2 + 15x + 7$ [T.5]

Cumulative Review, p. T45

1. $1\frac{13}{18}$ **2.** $30\frac{13}{40}$ **3.** 23.69   **4.** 6 yd 2 in.   **5.** 9.068   **6.** $\frac{7}{36}$ **7.** $1\frac{1}{2}$ **8.** 74.875

9. 2 days 23 hr 45 min   **10.** 4.47   **11.** 72,000,000   **12.** $\frac{1}{7}$ **13.** 16   **14.** 1.9684   **15.** -14.4

16. $\frac{5}{6}$ **17.** $7\frac{1}{2}$ **18.** 0.608   **19.** undefined   **20.** $2^3 \cdot 3^2 \cdot 11$ **21.** 2, 3, 4, and 9   **22.** false

23. 5990   **24.** 723.0   **25.** 7.88   **26.** 22   **27.** 0   **28.** $3\sqrt{5} + 20\sqrt{3}$

29. (a) $2\sqrt{5}$ in.   (b) 4.47 in.   **30.** $<$ **31.** $(2x + 6) - 5$ **32.** $x = -12$ **33.** $x = 18$

34. $-\frac{1}{2} < x \le \frac{1}{4}$

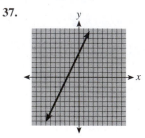

35. $x = \dfrac{y - b}{m}$

36.

37.

38. 40   **39.** 210%   **40.** 567.5   **41.** 1.778   **42.** 76.1°F   **43.** 64 L   **44.** 84 cm²   **45.** 125.6 cm
46. 0.19625 m²   **47.** 225 mi   **48.** 2 mi per day   **49.** \$1674   **50.** $-15, -14$
51. Housing—\$11,250; Transportation—\$6750; Misc.—\$6750; Savings—\$2250; Education—\$4500;
Taxes—\$4500; Food—\$9000
52. (a) -125 (b) $-28x^5$ **53.** (a) $9x^2 - 7$ (b) $3x^3 + 2x^2 - 4x - 1$
54. (a) $12x^3 - 120x^2$ (b) $6x^3 - 17x^2 - 57x - 20$

APPENDIX I

Exercises I.1, p. A4

1. 253   **2.** 163,041   **3.** 21,327   **5.** 140   **6.** 256   **7.** 53   **9.** 2362   **10.** 97   **11.** 744

13. 978   **14.** (64)   (a) ∩ ∩ ∩ | | | | (b) ∴∴ (c) ГˤΔIIII (d) LXIV
                              ∩ ∩ ∩

15. (532)   (a) ⟨ ⟨ ⟨ ∩ ∩ ∩ I I (b) ⩵̇ (c) ГˤΔΔΔII (d) DXXXII
                    ⟨ ⟨

17. (846)   (a) ⟨ ⟨ ⟨ ⟨ ∩ ∩ ∩ ∩ I I I (b) ⋮̇ (c) ГˤHHHΔΔΔΔГI (d) DCCCXLVI
                    ⟨ ⟨ ⟨ ⟨   I I I

Exercises I.2, p. A8

1. 4983   **2.** 32   **3.** 1802   **5.** 47,900   **6.** 222   **7.** 46   **9.** 999

10. (a) VVVV <<< VV (b) $\overline{\text{YOB}}$ (c) 罒
           VVV   <<   七
                                              七
                                              十
                                              二

11. (a) VVVVV <<< VVV (b) $\overline{\phi\text{CF}}$ (c) 五
            VVVV   <<   VVV   百
                                                  九
                                                  十
                                                  六

13. (a) V V VVV (b) $\overline{\text{ГХΞCE}}$ (c) 三
                 VV   千
                                              六
                                              百
                                              六
                                              十
                                              五

14. (a) ∨∨ ∨ <<< ∨∨∨   (b) $\overline{/Z\Sigma C\Gamma}$ (c) 七
七
二
丐
ん
十
三
十

15. (a) ∨∨∨ <<<< ∨∨∨   (b) $\overline{\text{A}}$ (c) 十
<<   MΩNB   七
八
丐
五
十
二

APPENDIX II

Exercises II.1, p. A11

1. $35 = 3(10^1) + 5(10^0)$ **2.** $761 = 7(10^2) + 6(10^1) + 1(10^0)$
3. $8469 = 8(10^3) + 4(10^2) + 6(10^1) + 9(10^0)$ **5.** $62{,}322 = 6(10^4) + 2(10^3) + 3(10^2) + 2(10^1) + 2(10^0)$
6. $11_{(2)} = 1(2^1) + 1(2^0) = 1(2) + 1(1) = 2 + 1 = 3$
7. $101_{(2)} = 1(2^2) + 0(2^1) + 1(2^0) = 1(4) + 0(2) + 1(1) = 4 + 0 + 1 = 5$
9. $1011_{(2)} = 1(2^3) + 0(2^2) + 1(2^1) + 1(2^0) = 1(8) + 0(4) + 1(2) + 1(1)$
$\qquad = 8 + 0 + 2 + 1 = 11$
10. $1101_{(2)} = 1(2^3) + 1(2^2) + 0(2^1) + 1(2^0) = 1(8) + 1(4) + 0(2) + 1(1)$
$\qquad = 8 + 4 + 0 + 1 = 13$
11. $110{,}111_{(2)} = 1(2^5) + 1(2^4) + 0(2^3) + 1(2^2) + 1(2^1) + 1(2^0)$
$\qquad = 1(32) + 1(16) + 0(8) + 1(4) + 1(2) + 1(1)$
$\qquad = 32 + 16 + 0 + 4 + 2 + 1 = 55$
13. $101{,}011_{(2)} = 1(2^5) + 0(2^4) + 1(2^3) + 0(2^2) + 1(2^1) + 1(2^0)$
$\qquad = 1(32) + 0(16) + 1(8) + 0(4) + 1(2) + 1(1)$
$\qquad = 32 + 0 + 8 + 0 + 2 + 1 = 43$
14. $11{,}010_{(2)} = 1(2^4) + 1(2^3) + 0(2^2) + 1(2^1) + 0(2^0)$
$\qquad = 1(16) + 1(8) + 0(4) + 1(2) + 0(1) = 16 + 8 + 0 + 2 + 0$
$\qquad = 26$

15. $1000_{(2)} = 1(2^3) + 0(2^2) + 0(2^1) + 0(2^0) = 1(8) + 0(4) + 0(2) + 0(1)$
$$= 8 + 0 + 0 + 0 = 8$$

17. $11,101_{(2)} = 1(2^4) + 1(2^3) + 1(2^2) + 0(2^1) + 1(2^0)$
$$= 1(16) + 1(8) + 1(4) + 0(2) + 1(1) = 16 + 8 + 4 + 0 + 1$$
$$= 29$$

18. $10,110_{(2)} = 1(2^4) + 0(2^3) + 1(2^2) + 1(2^1) + 0(2^0)$
$$= 1(16) + 0(8) + 1(4) + 1(2) + 0(1) = 16 + 0 + 4 + 2 + 0$$
$$= 22$$

19. $111,111_{(2)} = 1(2^5) + 1(2^4) + 1(2^3) + 1(2^2) + 1(2^1) + 1(2^0)$
$$= 1(32) + 1(16) + 1(8) + 1(4) + 1(2) + 1(1)$$
$$= 32 + 16 + 8 + 4 + 2 + 1 = 63$$

21. 111

Exercises II.2, p. A13

1. $24_{(5)} = 2(5^1) + 4(5^0) = 2(5) + 4(1) = 10 + 4 = 14$
2. $13_{(5)} = 1(5^1) + 3(5^0) = 1(5) + 3(1) = 5 + 3 = 8$
3. $10_{(5)} = 1(5^1) + 0(5^0) = 1(5) + 0(1) = 5 + 0 = 5$
5. $104_{(5)} = 1(5^2) + 0(5^1) + 4(5^0) = 1(25) + 0(5) + 4(1) = 25 + 0 + 4 = 29$
6. $312_{(5)} = 3(5^2) + 1(5^1) + 2(5^0) = 3(25) + 1(5) + 2(1) = 75 + 5 + 2 = 82$
7. $32_{(5)} = 3(5^1) + 2(5^0) = 3(5) + 2(1) = 15 + 2 = 17$
9. $423_{(5)} = 4(5^2) + 2(5^1) + 3(5^0) = 4(25) + 2(5) + 3(1) = 100 + 10 + 3 = 113$
10. $444_{(5)} = 4(5^2) + 4(5^1) + 4(5^0) = 4(25) + 4(5) + 4(1) = 100 + 20 + 4 = 124$
11. $1034_{(5)} = 1(5^3) + 0(5^2) + 3(5^1) + 4(5^0) = 1(125) + 0(25) + 3(5) + 4(1)$
$$= 125 + 0 + 15 + 4 = 144$$
13. $244_{(5)} = 2(5^2) + 4(5^1) + 4(5^0) = 2(25) + 4(5) + 4(1) = 50 + 20 + 4 = 74$
14. $3204_{(5)} = 3(5^3) + 2(5^2) + 0(5^1) + 4(5^0) = 3(125) + 2(25) + 0(5) + 4(1)$
$$= 375 + 50 + 0 + 4 = 429$$
15. $13,042_{(5)} = 1(5^4) + 3(5^3) + 0(5^2) + 4(5^1) + 2(5^0)$
$$= 1(625) + 3(125) + 0(25) + 4(5) + 2(1)$$
$$= 625 + 375 + 0 + 20 + 2 = 1022$$
17. yes; 13   **18.** yes; 28   **19.** {0, 1, 2, 3, 4, 5, 6, 7}

Exercises II.3, p. A16

1. $1000_{(2)} = 8$ **2.** $311_{(5)} = 81$ **3.** $11,000_{(2)} = 24$ **5.** $432_{(5)} = 117$ **6.** $1000_{(2)} = 8$
7. $1011_{(2)} = 11$ **9.** $433_{(5)} = 118$ **10.** $1302_{(5)} = 202$ **11.** $10,100_{(2)} = 20$ **13.** $10,100_{(2)} = 20$
14. $1222_{(5)} = 187$ **15.** $2241_{(5)} = 321$ **17.** $110,111_{(2)} = 55$ **18.** $23,240_{(5)} = 1695$ **19.** $3002_{(5)} = 377$
21. $30,141_{(5)} = 1921$ **22.** $1,101,001_{(2)} = 105$ **23.** $110,001_{(2)} = 49$ **25.** $10,001,010_{(2)} = 138$

APPENDIX III

Exercises III, p. A19

1. 4   **2.** 4   **3.** 17   **5.** 10   **6.** 6   **7.** 3   **9.** 6   **10.** 11   **11.** 2   **13.** 8   **14.** 11   **15.** 4
17. 12   **18.** 11   **19.** 25   **21.** 1   **22.** 20   **23.** 1   **25.** 35   **26.** relatively prime
27. relatively prime   **29.** relatively prime   **30.** not relatively prime; GCD is 7   **31.** relatively prime
33. relatively prime   **34.** not relatively prime; GCD is 22   **35.** relatively prime

Index

POWERS, ROOTS, AND PRIME FACTORIZATIONS

NO.	SQUARE	SQUARE ROOT	CUBE	CUBE ROOT	PRIME FACTORIZATION
1	1	1.0000	1	1.0000	—
2	4	1.4142	8	1.2599	prime
3	9	1.7321	27	1.4423	prime
4	16	2.0000	64	1.5874	2 · 2
5	25	2.2361	125	1.7100	prime
6	36	2.4495	216	1.8171	2 · 3
7	49	2.6458	343	1.9129	prime
8	64	2.8284	512	2.0000	2 · 2 · 2
9	81	3.0000	729	2.0801	3 · 3
10	100	3.1623	1000	2.1544	2 · 5
11	121	3.3166	1331	2.2240	prime
12	144	3.4641	1728	2.2894	2 · 2 · 3
13	169	3.6056	2197	2.3513	prime
14	196	3.7417	2744	2.4101	2 · 7
15	225	3.8730	3375	2.4662	3 · 5
16	256	4.0000	4096	2.5198	2 · 2 · 2 · 2
17	289	4.1231	4913	2.5713	prime
18	324	4.2426	5832	2.6207	2 · 3 · 3
19	361	4.3589	6859	2.6684	prime
20	400	4.4721	8000	2.7144	2 · 2 · 5
21	441	4.5826	9261	2.7589	3 · 7
22	484	4.6904	10,648	2.8020	2 · 11
23	529	4.7958	12,167	2.8439	prime
24	576	4.8990	13,824	2.8845	2 · 2 · 2 · 3
25	625	5.0000	15,625	2.9240	5 · 5
26	676	5.0990	17,576	2.9625	2 · 13
27	729	5.1962	19,683	3.0000	3 · 3 · 3
28	784	5.2915	21,952	3.0366	2 · 2 · 7
29	841	5.3852	24,389	3.0723	prime
30	900	5.4772	27,000	3.1072	2 · 3 · 5
31	961	5.5678	29,791	3.1414	prime
32	1024	5.6569	32,768	3.1748	2 · 2 · 2 · 2 · 2
33	1089	5.7446	35,937	3.2075	3 · 11
34	1156	5.8310	39,304	3.2396	2 · 17
35	1225	5.9161	42,875	3.2711	5 · 7
36	1296	6.0000	46,656	3.3019	2 · 2 · 3 · 3
37	1369	6.0828	50,653	3.3322	prime
38	1444	6.1644	54,872	3.3620	2 · 19
39	1521	6.2450	59,319	3.3912	3 · 13
40	1600	6.3246	64,000	3.4200	2 · 2 · 2 · 5
41	1681	6.4031	68,921	3.4482	prime
42	1764	6.4807	74,088	3.4760	2 · 3 · 7
43	1849	6.5574	79,507	3.5034	prime
44	1936	6.6333	85,184	3.5303	2 · 2 · 11
45	2025	6.7082	91,125	3.5569	3 · 3 · 5
46	2116	6.7823	97,336	3.5830	2 · 23
47	2209	6.8557	103,823	3.6088	prime
48	2304	6.9282	110,592	3.6342	2 · 2 · 2 · 2 · 3
49	2401	7.0000	117,649	3.6593	7 · 7
50	2500	7.0711	125,000	3.6840	2 · 5 · 5